Handbook of
Psychobiology

HANDBOOK OF PSYCHOBIOLOGY

EDITED BY

Michael S. Gazzaniga

Department of Psychology
State University of New York
Stony Brook, New York

Colin Blakemore

Physiological Laboratory
University of Cambridge
Cambridge, England

ACADEMIC PRESS   New York   San Francisco   London   1975

A Subsidiary of Harcourt Brace Jovanovich, Publishers

ACADEMIC PRESS, INC.
111 Fifth Avenue, New York, New York 10003

United Kingdom Edition published by
ACADEMIC PRESS, INC. (LONDON) LTD.
24/28 Oval Road, London NW1

Library of Congress Cataloging in Publication Data

Gazzaniga, Michael S
    Handbook of psychobiology.

    Includes bibliographies and index.
    1.   Psychobiology.   I.   Blakemore, Colin, joint
author.   II.   Title.
QP360.G38   596'.01'88   74-10193
ISBN 0–12–278656–4

Contents

PART III. VERTEBRATE SENSORY AND MOTOR SYSTEMS

Chapter 11 Auditory Localization
G. Bruce Henning

Chapter 12 The Somatosensory System
Patrick D. Wall

Chapter 13 The Chemical Senses: A Systematic Approach
Robert P. Erickson and Susan S. Schiffman

Chapter 14 Motor Coordination: Central and Peripheral Control during Eye–Head Movement
Emilio Bizzi

PART IV. INTEGRATION AND REGULATION IN THE BRAIN

List of Contributors

Numbers in parentheses indicate the pages on which the authors' contributions begin.

L. M. Aitkin (*325*), Neuropsychology Laboratory, Department of Physiology, Monash University, Clayton, Victoria, Australia

Stuart M. Anstis (*269*), Department of Psychology, University of Bristol, Bristol, England

G. Berlucchi (*481*), Istituto di Fisiologia dell' Università di Pisa, and Laboratorio di Neurofisiologia del CNR, Pisa, Italy

Emilio Bizzi (*427*), Department of Psychology, Massachusetts Institute of Technology, Cambridge, Massachusetts

Colin Blakemore (*241*), Physiological Laboratory, University of Cambridge, Cambridge, England

H. A. Buchtel (*481*), Department of Psychology, The National Hospital for Nervous Diseases, London, England

Terri Damstra (*201*), Division of Chemical Neurobiology, Department of Biochemistry and Nutrition, School of Medicine, University of North Carolina, Chapel Hill, North Carolina

Adrian Dunn (*201*), Division of Chemical Neurobiology, Department of Biochemistry and Nutrition, School of Medicine, University of North Carolina, Chapel Hill, North Carolina*

Dan Entingh (*201*), Division of Chemical Neurobiology, Department of Biochemistry and Nutrition, School of Medicine, University of North Carolina, Chapel Hill, North Carolina

Robert P. Erickson (*393*), Departments of Psychology and Physiology, Duke University, Durham, North Carolina

Michael S. Gazzaniga (*565*), Department of Psychology, State University of New York, Stony Brook, New York

Edward Glassman (*201*), Division of Chemical Neurobiology, Department of Biochemistry and Nutrition, School of Medicine, University of North Carolina, Chapel Hill, North Carolina

Richard L. Gregory (*607*), Brain and Perception Laboratory, Department of Anatomy, Medical School, University of Bristol, Bristol, England

Lennart Heimer (*73*), Department of Anatomy, School of Medicine, University of Virginia, Charlottesville, Virginia

G. Bruce Henning (*365*), Department of Experimental Psychology, University of Oxford, Oxford, England

Helmut V. B. Hirsch (*107*), Center for Neurobiology, Department of Biological Sciences, State University of New York, Albany, New York

Edward Hogan (*201*), Department of Neurology, Medical University of South Carolina, Charleston, South Carolina

Graham Hoyle (*3*), Department of Biology, University of Oregon, Eugene, Oregon

Leslie L. Iversen (*141, 153*), Medical Research Council, Neurochemical Pharmacology Unit, Department of Pharmacology, University of Cambridge, Cambridge, England

Susan D. Iversen (*141, 153*), Department of Experimental Psychology, University of Cambridge, Cambridge, England

Marcus Jacobson (*107*), Department of Physiology and Biophysics, University of Miami Medical School, Miami, Florida

* Present address: Department of Neuroscience, University of Florida College of Medicine, Gainesville, Florida.

Michel Jouvet (*499*), Department of Experimental Medicine, Claude Bernard University, Lyon, France

Michael F. Land (*49*), School of Biological Sciences, University of Sussex, Falmer, Brighton, Sussex, England

Anthony H. M. Lohman (*73*), Department of Anatomy, University of Nijmegen, Nijmegen, The Netherlands

Michael I. Posner (*441*), Department of Psychology, University of Oregon, Eugene, Oregon

David Premack (*591*), Department of Psychology, University of California, Santa Barbara, California

Stanley Schachter (*529*), Department of Psychology, Columbia University, New York, New York

Susan S. Schiffman (*393*), Department of Psychology, Duke University, Durham, North Carolina

Patrick D. Wall (*373*), Department of Anatomy, University College London, London, England

William R. Webster (*325*), Neuropsychology Laboratory, Department of Psychology, Monash University, Clayton, Victoria, Australia

John Eric Wilson (*201*), Division of Chemical Neurobiology, Department of Biochemistry and Nutrition, School of Medicine, University of North Carolina, Chapel Hill, North Carolina

Preface

Psychobiology, though a harsh word, sums up an extraordinarily active field of research that embraces an army of disciplines, from biochemistry to linguistics, and from invertebrate neurophysiology to perceptual phenomenology. And, like Topsy, psychobiology just keeps on growing.

This handbook is an attempt to distill the energies of the field and condense something of the vital spirit of brain research. The book is broad but could not be comprehensive; it is current but is no more than a frame from a moving picture. We have not tried, then, to produce a definitive and exhaustive account of this subject (if one could, it would be a dead subject). We have merely attempted to present an integrative overview of psychobiology, which may guide the newcomer and encourage the active researcher. Most of all, we hope this book may show the breadth of the subject, present a balanced account of some of its activities, and define its present boundaries and future goals.

The contributors to this book have been asked not to review their special fields of interest in detail but to discuss some central issues in general terms, to consider the limitations of ideas and techniques, and to point out ripe subjects for interdisciplinary attack.

The book begins with small, if not simple, nervous systems. The study of sensory mechanisms and the regulation of behavior in invertebrates has provided remarkable insight into comparable processes in grosser creatures, even in man. An account of the gamut of techniques for studying the structure of the brain, and a consideration of the development and plasticity of nervous systems, completes Section I.

Section II is devoted to neuropharmacology and neurochemistry, two areas that are currently very active. These chapters consider the organization of putative neurotransmitters within the brain, the role of specific transmitter systems in the regulation of behavior, and the biochemical changes that accompany activity in the nervous system.

Section III deals with sensory and motor systems in higher vertebrates: Vision, hearing, somatic sensation, and the chemical senses are discussed not only in physiological but in perceptual terms, and here perhaps most clearly can correlations be drawn across the boundaries of psychology and physiology. An account of motor coordination demonstrates the subtlety of present methods for studying the motor system, and their scope.

Finally, Section IV considers the mass of functions that one might call "integrative"—attention, memory and learning, sleep and dreaming, motivation and emotion, the neural control of behavior in vertebrates, and the origins of language.

In gathering the contributions to this book, we have been astounded by the wealth and energy of this field, and elated by the prospect of future research. If the book conveys this impression to the reader it will have fulfilled our aims.

Part I

FOUNDATIONS
OF PSYCHOBIOLOGY

Chapter 1

Neural Mechanisms Underlying
Behavior of Invertebrates

GRAHAM HOYLE

University of Oregon

INTRODUCTION

Invertebrate animals appropriately form the first chapter in this book, for among their nervous systems lie the only possible living parallels to the historical origins of the nervous systems of vertebrate animals including humans. Clues to fundamental principles of neural organization are sought among these "lower" forms, because the nervous systems of vertebrates are too complex and too highly evolved to permit comprehensive analysis at the cellular level.

Too often studies on invertebrates are "rationalized," or "excused," on the basis that they serve as "model" systems on account of alleged simplicity. Or they are used because of experimental convenience, in particular because of the large size of some individual invertebrate nerve cells. For example, certain molluscan ganglia contain a nerve cell body or two close to 1 mm in diameter, about ten times larger than the largest neuron somata of vertebrates. To the dedicated biologist no such excuses are needed. Every type of invertebrate organism has its own intrinsic interests and problems which command his attention, and the unbiased spirit of this type of approach is the one which seems the most likely to yield successful research results. The evolutionary biologist will view the information obtained from diverse phyla in two different ways. He will try, on the one hand, to fit it into a framework derived primarily from other, often extensive, anatomical features, to confirm or question his basic premises. On the other hand he will impartially survey the data and try to formulate the principles which have guided the evolution of nervous systems.

The psychologist will be interested in broad features of what is learned from invertebrates about the functioning of nerve cells individually, as exampled by the use of the squid giant axon in the elucidation of the mechanism of the nerve impulse and of the *Limulus* eye in quantifying sensory transduction. He will be especially interested in the neuron circuitry utilized by invertebrates, in so far as this has been ascertained with precision, especially when neural operations can be related to behavior. He will also wish to be apprised of most of the following:

1. Instinctive behaviors which are under endogenous control, requiring little or no sensory feedback for their execution.
2. The cellular apparatus in which such neural motor programs, "motor tapes," are stored and played back.
3. The selection procedure by which sensory clues select the appropriate program.
4. Neural details of the decision-making process: to behave or not to behave?
5. Behaviors which are controlled by reference to stored sequences of sensory information, "sensory tapes."
6. Behaviors which utilize combinations of sensory and "motor tape" information.
7. Genetic aspects of the neural machinery and its developmental biology.
8. He will be particularly concerned with the cellular basis of learning processes, since these are of supreme importance in mammals. The behavior of invertebrates is, in general, much less plastic than that of higher vertebrates, but various kinds of learning do occur in invertebrates, even in some which are quite low on the evolutionary ladder.

The supreme technical advantage offered by several invertebrate phyla is that of permitting direct electrical recording from specific, identifiable neurons. This advantage is further enhanced, in a few instances, as we shall see later, by freedom for the animal—or part of it, to behave relatively normally while recordings are made from the identified neurons.

It is argued by some psychologists that vertebrate nervous systems have evolved along quite different lines from those of any invertebrates, not only at the gross anatomical level, which is obvious enough, but at the level of circuitry. This, they believe, renders studies on invertebrates irrelevant from their point of view. It is undeniable that a high degree of redundancy occurs in vertebrate brains and that there is a notable lack of it in many invertebrate ganglia. At the present time such a negative view is not, however, justifiable on the basis of any positive knowledge regarding the vertebrate neuron organization, compared with that known for any invertebrate. Also, at the cellular level no exclusivity can be claimed, for so far no more than superficial differences have been disclosed. But at the subcellular level differences are rare, and an abundance of common particles and processes is already known.

There is no doubt that in the immediate future, rapid advances in knowledge of the previously listed matters will be made, for some invertebrate species from diverse phyla. In the same period forseeable progress seems less likely for vertebrates *at the same levels of inquiry*. Knowledge on these fundamental issues will, therefore, be almost entirely invertebrate-based knowledge.

The broadly based experimental biologist, sometimes termed comparative physiologist, has two options open in reviewing that data. He may consider it sequentially, phylum by phylum, or he can deal with it process by process, ranging across phyletic lines. For the psychologist, however, the former is totally unfamiliar and often seems pointless. Accordingly, we shall attempt to pick out some well-defined topics and show how current invertebrate knowledge is answering the questions raised, as well as to try to show what kind of knowledge we may expect to be attained in the future. But in certain respects this approach is not especially fruitful. Nevertheless, each different invertebrate group has a useful message to deliver. At the very least, the naive investigator will wish to be informed as to what each phylum has to offer. This we shall consider briefly first.

Coelenterates

The coelenterates, hydras, jellyfish, and sea anemones, which are bilayered, radially rather than bilaterally, symmetrical, lacking both a front end with which to meet the world and concentrations of neurons into ganglia or brains, nevertheless can offer insights into the capabilities of arrays of neurons. Their nervous system is a diffuse, nonpolarized net in which the component neurons are largely multipolar, displaying an apparent total absence of orderly array. Bipolar elements occur in special tracts, but the familiar monopolar neurons of higher anatomical levels of vertebrate, and even most invertebrate, nervous systems are totally lacking.

In spite of this apparent lack of organization the animals are beautifully coordinated. Furthermore, they are capable of complex, programmed behavior sequences. One of the subtlest of these is the escape swimming behavior shown by the sea anemone, *Stomphia coccinea,* in response to certain species of starfish; and the larger anemone, *Actinostola,* in response, surprisingly, to *Stomphia* (Figure 1). Once contact has been made briefly by the latter, the *Actinostola* reacts as follows: first, the anemone elongates markedly. While it is doing this, it expands its oral disk and at about the same time detaches its otherwise very firmly attached base from its shell substrate. This requires complex furrowing of the surface of the foot. Now the animal suddenly squeezes body cavity fluid toward the center of its pedal disk, which at the same time, by coordinated inhibition, relaxes, so that it balloons out into a conical projection. The movement is so sudden that the anemone is propelled upwards into the water through several inches. Now well on the way toward escape, the anemone violently contracts longitudinal muscles on one side of its body, and after a short delay contracts those on the opposite side. There follows a regular series of these quick movements, each successive one rotated by a few degrees around the radial axis so that the muscles of each group of longitudinal mesenteries of the body get their turn to contract. Purchase against the water is obtained by the umbrella-like expansion of the oral disk and the net result is that *Stomphia* swims away in an ever-increasing spiral. The whole act takes more than a minute to complete. No part of this behavior has ever been observed to occur out of sequence and each is invariably followed by its natural successor unless the anemone is highly fatigued. The sequence cannot be started somewhere in the middle, but it still occurs if the anemone is prevented from leaving the substrate by pins

Figure 1. Complex escape swimming behavior in an animal having a diffuse nerve net and lacking a brain: the sea anemone *Actinostola*. The animal undergoes an orderly sequence of complex, coordinated movements when touched by another sea anemone, *Stomphia coccinea*. A behavior built up of movements that are always executed in the same sequence, it is termed a *fixed action pattern* and is the mainstay of complex behavior of invertebrates and lower vertebrates. [From Ross and Sutton, 1967. Copyright 1967 by the American Association for the Advancement of Science.]

stuck through it onto a wax base. There is, therefore, no need for the natural proprioceptive afferent information to be received before the next movement occurs. The action does not comprise a chain of reflexes.

Very rarely a *Stomphia* takes off spontaneously; it also shows the swimming behavior in response to a prescribed train of electric shocks. The importance of such behavior to the animal is obvious. Its significance for the neuroethologist is that it is achieved by a nerve net which does not show signs of being arranged in an orderly manner. These observations show that in the evolution *of nervous systems* one of the first features which nerve cells achieved was the programming of behavioral sequences of considerable complexity and functional subtlety. There is no instinctive behavior of greater complexity than the *Stomphia* swim in a great many much more highly evolved organisms with well-structured brains!

The lack of influence of distorting sensory input on the sequence and its timing indicate that this behavior is of the type we term *programmed motor pattern sequence,* or *motor tape,* control. The nervous system, in some way which is at present totally unknown, excites the right motor neurons successively, as if driven from a set of instructions stored on a time recording device. The whole complex, storage selection, trigger, play-back, and read-out mechanism are genetically prescribed and built into the structures and physiological properties of neurons in the net.

The major initial tasks of neurobiological studies on invertebrates will be concerned with the elucidation of comparable behaviors in terms of underlying neuronal circuitry and physiological properties. We shall deal with examples from several phyla, but before departing upon this survey, let us consider the unique features of other invertebrate phyla, in order of presumed evolutionary advance.

Platyhelminthes

Platyhelminthes are bilaterally symmetrical and triploblastic, and they have complex head ganglia. They lack discrete muscles though, except for a retractor penis, and quantitative behavioral analysis is bound to be difficult. No programmed behaviors are known. Many are degenerate, specialized parasites, but others, the free-living Turbellaria, or Planarians, offer the special interest of extraordinary powers of cellular reorganization in regeneration, whole animals being formed from small pieces. Furthermore, they have the proven ability to learn simple avoidance and discrimination tasks. But so far, these organisms have been found to be completely intractable to physiological analysis.

Nematoda

The phylum *Nematoda* has been singled out by a group of geneticists under the leadership of S. Brenner, for intensive behavioral and potential neurobiological study on account of special feature of their development. There is a precise, small number of cell divisions resulting in a fixed number of nerve cells—162 in *Ascaris*—the configuration of each of which was described in 1909 by Goldschmidt in precise detail. Certain species of *Ascaris* can be utilized easily in large numbers and mutants having divergent "behavior" selected. These mutants wiggle differently from normal nematodes. The intention, then, is to match nervous system details to behavioral observations. Unfortunately, "behavior" is confined to a primitive form of locomotion, though the mutants, with their aberrant rhythms of twitching, etc., are impressive to watch. Studies of the nerve cells at the physiological level have not yet been achieved, however, so it is difficult to predict the extent to which the studies might yield information on the relation of circuitry to movement pattern. An additional problem is that nematodes suitable for genetic analysis are too small for physiological examination, so the proposed neurophysiological work will not directly match the genetic and anatomical data.

Annelids

Next in the presumed evolutionary line are annelids, and once again the target is understanding locomotory control mechanisms. There are also, however, some subtle behaviors in annelids, such as the feeding cycles of tubicolous polychaetes like *Arenicola,* which might provide additional, higher-level, behavior interest as well as preparations for studying the physiological basis of rhythms. Of special interest among annelids is the medicinal leech. This is because the segmental ganglia contain relatively few neurons (about 350) of relatively large size. Several of these neurons may be recognized individually and they occupy similar locations in different specimens. They can be penetrated with intracellular recording microelectrodes, permitting direct examination of their physiological properties by electrical stimulation and recording. Preparations have been developed in which the body wall of the leech remains functionally connected to the ganglia, permitting operational identification as well.

Surprise was generated by studies on the leech because six of the neuron somata of the ganglia turned out to be sensory from the body wall, responding to a light touch. It had previously been thought that all invertebrate sense organs have their cell bodies located peripherally. Fourteen pairs of excitatory motor neurons were located, and three pairs of peripheral inhibitors. These neurons together, supply five different muscle layers in each segment, controlling the body wall. Each was found to innervate a territory of muscle fibers of consistent location and size, and to cause a discrete movement when excited. The territories have a pattern closely resembling that found for the receptive fields of the touch-sensitive cells.

So far, no details of integration and motor control have been published, but the preparation is extremely promising, and neuronal programming of leech swimming should be one of the first mechanisms to be worked out in cellular detail.

The use of polychaetes seemed likely following a finely detailed neuro-anatomical study of the nervous system of body segments of nereid polychaete worms, by J. E. Smith in 1957, but the somata are unfortunately somewhat smaller than those in the leech, and no intracellular studies on them have been published. However, several are about 25 μm in diameter, which is adequate for intracellular techniques. So it may be possible to work with these nervous systems experimentally.

The same author has also given us remarkably fine studies of the anatomy of starfish nervous systems. The echinoderms would also be very interesting objects for neuroethology if they could be studied electrophysiologically, but prospects for this seem remote owing to the small size of their nervous systems.

Molluscs

For molluscs, exploration and analysis are already far advanced. Molluscs are an unique phylum in regard to the vast range of levels of organization they represent, from simple organisms, such as the nudibranches, with a small total number of central nerve cells, at one extreme, to the most complex invertebrates of all, the cephalopods, with highly complex brains comprising many millions of neurons, at the other. The latter possess subtle learning capabilities and it seemed two decades ago as though they might provide ideal experimental material for studying neuronal circuitry. But while the hopes they engendered promoted the collection of a great deal of distinguished anatomical, and some experimental, work on *Octopus* brains, they have not contributed to the extent that was envisaged, to our knowledge of fundamental neural machinery. Furthermore, the information the research did yield renders it unlikely that they ever will; for the neurons are numerous, often small, and not readily accessible. The lower molluscs, by contrast, have proved

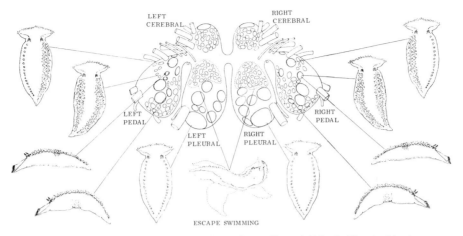

Figure 2. Identified neurons and behavior of the nudibranch *Tritonia*. The giant brain neurons shown are always of about the same size, and they occupy similar positions on the surfaces of the ganglia. The groups of neurons surrounded by broken lines are the trigger-group neurons (TGN's) that initiate swimming. Other neurons labeled cause unilateral withdrawal of branchial tufts, bilateral withdrawal of branchial tufts, turning of the tail to the left, turning of the tail to the right, and ventral curling.   [From A.O.D. Willows, "Giant brain cells in Mollusks, 1971. Copyright © 1971 by Scientific American, Inc. All rights reserved.]

to be extremely rewarding, as we shall see, especially large nudibranches such as *Tritonia* and *Hermissenda,* the large tectibranch *Aplysia,* and the gastropods *Helix, Helisoma, Pleurobranchaea,* and *Navanax.* This is because they all have large, pigmented neurons in their ganglia which are easy to identify and to record from electrically, some of which cause discrete movements when stimulated electrically (Figure 2).

A majority of work so far has been concerned simply to describe the properties of individual neurons and work out the physiological properties of the connections between them. But already there is also much that relates to behavior and there is a large recent increase in the amount of research aimed in this direction.

Arthropods

Finally, we come to the arthropods, which collectively form the most advanced invertebrate phylum. There has been a great deal of work at the level of identified cells in decapod *crustaceans,* especially the crayfish, lobster, and spiny lobster. Crabs have started to prove amenable to central nervous studies and a few behaviors, including running, swimming, sexual display, aggression, and shell selection (by hermit crabs) have been studied in ways which permit some guesses at least, as to the nature of the underlying neural mechanisms.

The behavior of *arachnids,* especially spiders and scorpions, is very interesting and it is a great pity that their nervous systems have not yet been subjected to experimental studies.

The special feature of crustaceans which has rendered them experimentally useful is the ease with which their nerve trunks can be split up into small bundles of axons. The larger of the axons, including all motorneurons, and many interneurons, can be isolated individually, recorded from in nearly intact preparations, and stimulated electrically. Also, some crustaceans offer unique anatomical features. The crab *Podophthalmus,* for example, has very long (3–5 cm) eyestalks, permitting easy recording from single integrating interneurons. Professor C. A. G. Wiersma, who studied them, has made extensive studies of individual interneurons in crayfish, crab, and lobster eyes also. In the crab brain there is an interneuron mediating the eye-withdrawal reflex which permits intracellular recording not only from its soma, but also from various regions of its integrating segment as well.

Individual motor neurons of lobsters were ones on which a dye injection technique was developed, permitting both neuronal morphology and synaptic contacts to be studied. The two most important dyes in current use are Procion yellow, introduced in 1968; and cobalt chloride, introduced in 1972. The former gives off a bright yellow fluorescence in ultraviolet light, so its detection requires the use of special microscopes. But it does diffuse into the axon and dendrites of the cell. It is especially useful in connection with serial sections following injection; from these the neuron can be reconstructed three-dimensionally, and the spatial relations to other neurons visualized. Cobalt salt is first injected and then given a few hours to diffuse. After precipitation as cobalt sulphide, the whole injected neuron, together with its branches, is examined in whole-mount preparations. Cobalt reveals finer branches than are detected using Procion dyes.

These methods are enabling the construction of neuron soma location and neurite pathway maps. In combination with physiological studies they are, in selected instances, for invertebrates from other phyla which we shall mention, permitting direct analysis of neuronal circuits.

This leaves us with the largest of all invertebrate groups, the *insecta.* Insect behavior has been intensively studied for two centuries and the modern generalizers among students of ethology, while almost completely ignoring most invertebrates, have all included insects in their surveys and drawn important conclusions from studies made on them. The small size of insects, unfortunately, prevented investigators from getting to the cellular level earlier. But since 1969, investigators have been able to overcome the technical difficulties, e.g., Figure 5. In the years ahead, insects seem destined once more to play the major role as they did in genetics, in the elucidation of basic principles, in this

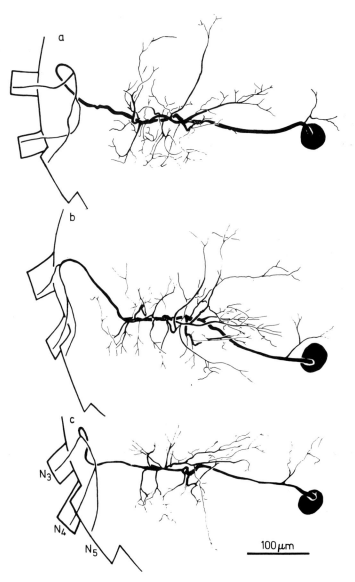

Figure 3. Profiles of the same neuron, the metathoracic common inhibitor of the desert locust, in three different animals after filling with cobalt salt. The location, pathway, and overall branching patterns are recognizably similar, though different in details. [From Burrows, 1973b.]

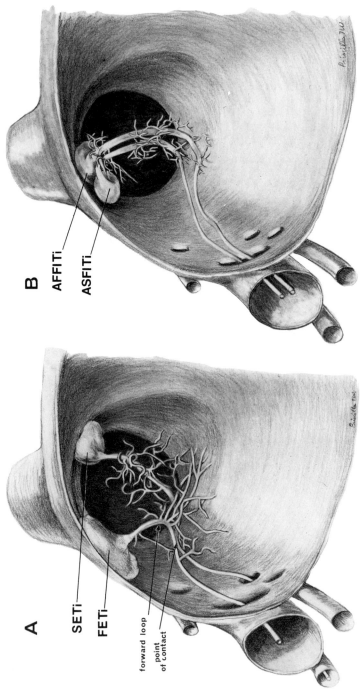

Figure 4. Spatial relationships of motorneurons determined after filling with Procion yellow dye, serial sectioning and reconstruction. A. The slow and fast extensor tibiae neurons of *Schistocerca gregaria* which, together with the homologue of the inhibitor shown in Figure 1, determine all postural and extension movements of the locust jumping leg. The extensor first travels inward and then makes a loop forward to make contact with its slow counterpart. The interaction between the neurons is inhibitory, for these neurons have totally independent functions. Their axons emerge through different nerve trunks. B. Slow and fast antagonistic (flexor) motorneurons of A. Note the close similarity of the pathways through the neuropile and of the branching patterns. These two neurons are closely synergistic. [From Burrows and Hoyle, 1973.]

A

SETi

FETi

forward loop
point
of contact

B

AFFITi

ASFITi

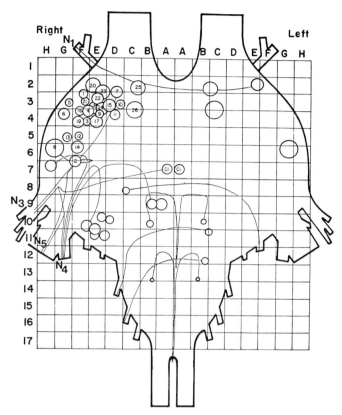

Figure 5. Identified motorneurons of the ventral surface of the metathoracic ganglion of the desert locust, *Schistocerca gregaria*. These neurons have been recorded by microelectrodes during tethered behavior. The numbered neurons are described in detail in papers by Hoyle and Burrows (1973). The common inhibitor (CI) is described by Burrows (1973a). The pathways of axons out of the ganglion are shown for a few key motor neurons.

case those of neural integration, behavior control mechanisms, and their genetic determination. Insects offer a vast range of complete, varied, subtle instinctive behaviors and also some learned behaviors. Their generation times are in many instances short, so their genetics is amenable to analysis in a way which will never be the case for the larger marine organisms.

Insect motorneurons and many interneurons are sufficiently large to be penetrated by intracellular electrodes in pairs before filling with dye for morphological and synaptological examination. Neurons occupy fixed locations in the ganglia, and research groups are preparing maps of these locations following identification in large insects such as the desert locust and the cockroach.

The motorneuron activity of insects is particularly easy to study in

the whole animal (chronic preparation). Most insect muscles are innervated by a single excitatory axon, while others have two, rarely more.
The myogram is, therefore, a digital pulse train which is analyzable
quantitatively, that can be recorded simply by placing a very fine (20
μm) insulated wire, with a cut end, into the muscle. In a small grasshopper, N. Elsner was able to study the neuronal basis of normal courtship behavior with no fewer than 30 independent electrodes implanted
in as many muscles (Figure 6).

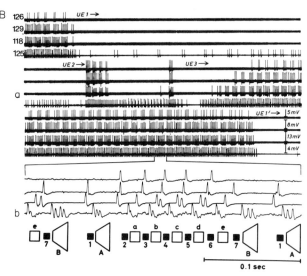

Figure 6. A. Male grasshopper, *Stenobrothus rubicundus*, with 30 independent recording
electrodes implanted in muscles involved in its courtship behavior. [Courtesy of Dr. N.
Elsner.] B. Electromyograms recorded from four of the leads of a similar preparation to
that shown in A, but *Gomphocerippus rufus* illustrating the digital nature of the motor output
underlying insect behavior, the precision of the pattern generation, and its sterotypy as the
movements are repeated. [From N. Elsner, 1973.]

STRATEGIES

It is not obvious, in this vast, complex field encompassing many experimental subdisciplines, how best to proceed toward realizing the ultimate goals. Let us first define the latter. They are: to obtain an understanding of the general principles of neuronal cellular physiology and connectivity which determine all aspects of behavior. There are initial questions such as: how many organisms need to be studied? Or: what kinds of organisms are optimal theoretically? Many neurophysiologists—perhaps the majority of those practicing today, express no, or very little, interest in invertebrate animals. Their goal is to understand the higher mammalian nervous system, period. Such an approach is their personal prerogative, but while some of their work will be incorporated into the general body of knowledge, by their narrow approach they deny themselves a position in the center of things. The invertebrates have nervous systems which will certainly be understood first in cellular and connectivity terms. Whether or not the cellular details of vertebrate nervous systems ultimately prove to be similar to those of invertebrates will have to await studies on both. But, by definition, the general principles must be those features which vertebrates possess in common with invertebrates.

For the time being it seems to me pointless to justify studies on invertebrates solely on the basis of their possibly providing models for understanding aspects of human nervous systems. They may not. But that should not detract, for each is in itself a living variation on the basic theme that we are trying to understand. The study of specifically human nervous system physiology is an applied, not a general, science. The same argument holds true for each particular invertebrate. The task set *Drosophila* genetics was not to understand the fruit fly, as such, but to discover the principles in operation in genetic segregation and reassembly. These emerged in large part from the *Drosophila* studies, in a manner which makes them acceptable as a basis for understanding human genetics.

While it can be argued that neuronal connectivity and multineuronal circuit operating principles are not qualitatively comparable to the mechanics of intracellular components, there is, nevertheless, no evidence that this is not so. If that level of "justification" seems to be needed by an individual to drive his personal operation, then he cannot deny it to himself on logical grounds.

Having, for whatever reason, decided to make studies on an invertebrate, he now faces the enormous task of selecting a choice of experimental subject. Unfortunately, neurobiology has so far not been blessed either by a T. H. Morgan, rallying the scattered workers to focus their efforts on one animal, or even by an uniquely favorable preparation.

No *Drosophila* of neurophysiology has emerged, in spite of strong claims on behalf of: the crayfish, *Aplysia, Tritonia,* the locust, the octopus, and nematodes. The reader may study the previous survey for a comparison of the relative merits of each preparation, but diverse human, emotional, geographical, and historical factors seem destined to play larger roles than logic in his choice. He may even be determined, as has often been the case, to find something nobody else is working on; with over a million species to choose from, this is not difficult. Professor T. H. Bullock declared that "a maximum of eclecticism is needed." I have personally preached a *Drosophila* approach, though being a Bullockist thus far in practice. Until some organism has come to the forefront for sound theoretical, as well as experimental, reasons, a diverse approach is clearly needed, just to find out what is there, and what the limitations of each preparation may be. But at some stage a concentrated attack by many workers on a single species is essential, for the task is gargantuan. Once a set of principles has become apparent for the selected animal, a return to eclecticism will become desirable again.

The next step in the investigator's odyssey must be a means of approach to the chosen animal unless, of course, this has already dictated itself. Some organisms present marvelous sensory preparations, but are well-nigh hopeless for motor-output analysis, and vice versa. Others have giant central neurons but possess unanalyzable peripheral nerve nets serving as largely autonomous relays.

The ideal organism would present the following features:

1. Behavior which includes both simple and complex forms, instinctive and also learned.
2. A nervous system containing: neurons which are large enough to study individually with intracellular electrodes; a relatively small total number of component neurons so that at least 10% of them eventually could be examined, with parts being analyzable almost 100%.
3. A short generation time, and a genetic system permitting Mendelian analysis.
4. Experimentally accessible developmental stage or stages.

All of these requirements are fulfilled only by some insect species.

Let us assume that the ideal organism has been chosen: The next task facing the investigator is to decide what to study first. Among the possibilities are the following, each of which has its adherents:

1. A minute analysis of the behavior, perhaps with gross intervention such as limb fixation and ablation. From the observations neurophysiological bases are inferred.

2. Sherringtonian or Pavlovian reflexological analysis.
3. A study of a particular sensory modality which influences behavior.
4. A direct examination of the central nervous system. This may be made by gross anatomy, by histology, by studies of degeneration and regeneration pathways following lesions, etc. Or it may be made by various physiological methods, with whole or partial isolation of a part of the nervous system such as a gangliaon.
5. Examination of the motor outputs underlying behavioral acts.

The present author prefers 5, as a starting point. For this approach, behavior is defined as sequences of muscular contractions. The motor output causing a behavior is recorded. The next step is to try to determine the way in which the observed pattern of motor output is determined by the central neurons. In an ideal preparation, behavior is obtained from a relatively intact animal while recordings are made from the neurons which cause behavior. With this approach, the last part to be analyzed is the sensory input. Those who start with the latter claim that they are working toward the motor output. But they seldom get that far. Few animals permit central unit analysis correlated with peripheral analysis, and none of those studied has yet been pursued very far in detail. But some invertebrate preparations permit this approach. These include locomotion of *Tritonia* (escape swimming), lobster (walking), and locust (walking, jumping, and flight), the singing of crickets, cicadas, and grasshoppers, and defensive and postural movements of the crayfish.

NEURONAL MACHINERY CONTROLLING BEHAVIOR

It has been realized for the last twenty years that behavior is only in small part due to reflex actions, simple and in chains, the mechanisms which had dominated the thinking about behavior for the first half of the century. An animal is endowed with genetically determined neural "programs" which come into operation under appropriate circumstances—or sometimes even under inappropriate ones, as "displacement" activities.

Reflexes versus Central Programs

All nervous systems include reflex arcs and a great deal of simple behavior consists solely of reflex actions. Since each such action requires a specific stimulus of sense organs, the most conspicuous of which are external to the animal, they are termed *exogeneous*. A reflex initiated from internal sense organs, say in the mesenteries or gut, is regarded as equivalent. The complimentary term, *endogenous,* is reserved for acts whose origin is *within central nerve cells* and not determined by the firing of sense organs. In the physiological literature of the late nineteenth and the first half of the twentieth century, all behavior was expressed in terms of reflex actions, *simple,* on the basis of inherited and directly developed neuronal circuitry, or *conditioned,* that is, "learned" associations made indirectly between sense organs and motor channels.

This view was challenged, first by ethologists, who found behavior to be inherited in the form of useful, that is, species-survivally-meaningful "packages" or fixed action patterns. Each package involves a complex sequence of stereotyped movements that follow each other in a fixed order. The package only appears at an appropriate stage of development, for example, sexual maturation, and it may appear spontaneously, in the absence of a specific stimulus, as an urge or "drive," or it may appear as a displacement activity when a state of conflict exists between opposed drives such as attack and flight. Later, invertebrate neurophysiologists found examples of behavior in which there is no requirement for immediate sensory input for patterned output to occur. In the mollusc *Tritonia,* in 1969, the isolated brain was shown to be able to produce the neural program of a characteristic fixed action pattern, associated in the intact animal with swimming escape behavior, after a single electric shock to a nerve. Conditioned reflexes have been demonstrated in a very few invertebrates. Only a few insects, notably bees and wasps, can learn to associate colors or objects with their home or food sources. Avoidance conditioning is, by contrast, easy to obtain.

Lability of Reflexes

Reflexes are easy to demonstrate in invertebrates, but they range from a classroom level of reliability to the embarrassingly fickle. I have seen a locust fry to death on a heated plate because for some reason its jump escape mechanism, which was being tested and had previously been found to be normal, was inhibited. Examples of reliable actions include the initiation of flight by combined loss of tarsal contact and air on the head or thorax, and the optomotor response—a turning tendency initiated by the movement of alternate black and white vertical stripes past the eyes.

Habituation of reflexes is a phenomenon sometimes classed as a simple form of learning, though not all investigators (including myself) so regard it. All invertebrate preparations provide material for the study of habituation and those which permit penetration of identified neurons, at the same time as the reflex is elicited, provide material for investigation of its cellular basis. Studies on reflexes of crayfish, cricket, cockroach, and others are all pointing to depletion of transmitters between the primary sense organs and the first integrating interneurons as being in part the basis for the phenomenon. However, this location has been ruled out for the gill-withdrawal reflex of *Aplysia,* and the synapses between an interneuron and motorneurons are implicated. Dishabituation appears to be caused by a facilitation process at the same synapses, but the physiological basis of the latter is not yet understood.

Endogenous Neural Programs

It is implicit in the concept of an endogenous neural mechanism that *specific* sensory input is not necessary for its pattern to be developed. General sensory input may be required in the form of background excitation, to depolarize a sufficient number of component neurons to cause them to fire. But from the moment of initiation of the action, central connectivity alone can determine the pattern. This is not to say that sensory feedback, when present, will not modify the precise timing and intensity of component discharges. These will occur, and may be essential for perfect performance, but it is not necessary for a basic program to appear. The simplest category of such behavior is the centrally determined rhythm. We shall consider these, and more complex actions, later. But first it is necessary to consider certain aspects of the motor systems through which the central nervous systems of invertebrates must operate. These are markedly different in a number of respects from those of vertebrates.

Muscle Activation, Synergism, and Antagonism

Behaviorally meaningful movements of invertebrates are, in a few instances, generated by a single neuron. For example, the rapid extension of the leg seen in a locust or grasshopper kick, hop, or jump is developed by a single motor neuron. But the jump is much more precise and powerful if extension is delayed by a prior, powerful flexion, so that there is time for maximum tension to be developed by the extensor muscle before takeoff. The moment of takeoff is then determined by flexor relaxation. Furthermore, if certain other leg muscles such as the depressor of the tarsus, the extensor of the trochanter and the promotor of the coxa, contract synchronously with the released tibial extensor, further enhancement of the movement will be promoted.

The extensor and the flexor must not contract at the same time when rapid movements occur, since they are antagonistic, so a central mechanism of reciprocal inhibition is needed between them. In many invertebrates, especially arthropods and annelids, a peripheral inhibitory mechanism is available to aid (it does not replace) central mechanisms in achieving reciprocity. Peripheral inhibitory action also accelerates relaxation, and in some instances promotes contraction when invoked to raise membrane potential immediately preceding excitation. Thus, there is a need for flexibility in the linkage between neurons. There will be other times when simultaneous activation of antagonists is called for; when, for example, a joint needs to be stiff, or when damping of a movement by rapid deceleration is required, and when a very delicate, relatively slow action is required of a powerful muscle.

When more than one motor neuron innervates a muscle, variation in the number of neurons active at the same time provides an important device for altering tension. Thus, excitatory coupling between the synergists is required, but a subtle kind, so that additional units fire exactly as needed. The additions of units will be different in principle, for each different act of behavior, implying complex mixing functions.

In most neuromuscular systems the tension developed by firing of a motor neuron is sensitive to the frequency of firing, and in most invertebrates variations in frequency provide the principal means of tension variation. Commonly, an arthropod muscle is innervated by two excitatory axons, one of which, the *slow* axon, produces slow, highly frequency-dependent tension development. The other axon, termed the *fast* axon, causes larger contractions at a low frequency of firing. The former are relatively unfatiguable, and are used for postural tonus and all slow movements. The latter fatigue rapidly when repeated and are used only for rapid movements. The grasshopper jump provides an extreme example which clearly illustrates the difference between the two. A single impulse in the fast axon leads to such a powerful twitch that it is of use only in defensive kicks and escape jumps. By contrast, usable tension is developed by the slow axon only at frequencies above 10 Hz, but all normal movements are determined by the single axon alone.

Tension development by muscles innervated by one or a small number of axons is highly sensitive not only to mean frequency but also to the pattern of nerve input. This sensitivity is enhanced by neuromuscular facilitation in slow axon transmission, and by intramuscle fiber facilitation phenomena associated with excitation-contraction coupling events. Further complications occur in some muscles as a result of still unexplained muscle phenomena, such as the Blaschko, Cattell, Kahn (1931) effect, discovered in certain crustacean muscles. In these a single extra impulse added to a regularly spaced train at low frequency leads to a large

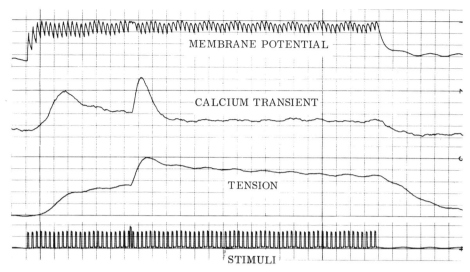

Figure 7. Illustrating the Blaschko *et al.* (1931) effect on a single crustacean muscle fiber from the giant barnacle *Balanus nubilus*. The upper trace registers intracellular membrane potential, the second trace the free calcium concentration (as light intensity by reaction with injected aequorin), the third trace is tension and the bottom trace denotes stimuli. The single extra pulse leads to a large intracellular calcium transient and tension increment which is maintained by a background rate that cannot itself develop so much tension. [From Hoyle and Ridgway, unpublished.]

tension increase (Figure 7). This increase is explained by a combination of the following events: neuromuscular junctional potential facilitation; the initiation by the single enhanced extra junctional potential of a large graded postsynaptic membrane response, or even a spike; and intramuscular facilitation of calcium release. But still unexplained is the fact that the increased tension is now maintained by the previous low background frequency of excitation, although by itself it could not cause development of so much tension.

The previously described phenomenon recalls the catch mechanism of certain molluscan muscles. According to Wilson and Larimer, hysteresis effects, indicative of partial catch properties, are to be seen in many different kinds of muscle when excitation levels are suddenly changed.

The "Size Principle"

In the lobster, somata of motorneurons supplying those muscles which are innervated by several motor neurons, there is a gradation in the sizes of the somata which is related both to their intrinsic biophysical properties and to the neuromuscular junctional transmission they effect.

If the neurons are placed in a series based upon the speed of contraction they cause in response to firing at similar rates, the following features are found. The faster neurons have physically larger somata and thicker axons, they show greater adaptation to applied current, less tendency to repetitive discharge and cause larger, less facilitating excitatory junctional potentials.

When these neurons are recruited in reflex actions, first the smaller slow motor neurons are excited, and later, progressively, the larger ones. Where several motorneurons innervate one muscle, of other crustaceans and of insects, the principle also appears to be applicable. This principle has also been shown to operate in vertebrate spinal cord.

All of the above phenomena point to the need, in examining neural mechanisms underlying behavior, to understand fully the neuromuscular mechanisms of the animal in question. There is clearly an enormous range of possibilities for an organism, with but few muscles and neurons, to have evolved subtleties in neural programing in order to take advantage of the great variety of contractile phenomena and neuromuscular mechanisms. One of the first expected payoffs of detailed comparative studies will be the discovery of the extent to which nervous systems have utilized their options. For example, do the relevant crustaceans know, as it were, that their muscles have the phenomenon discovered by Blaschko *et al.*, and can they utilize it by sending out an extra impulse during a train?

We already know that a behaviorally significant act can be determined by a single impulse, and that the difference in the movement caused by adding one, two, three, four, etc., impulses alters it sufficiently to be behaviorally identified and, therefore, subject to natural selection.

Motor Tape, Sensory Tape, and Mixed Control Mechanisms

Granted that a behavior has been determined to be endogenously generated, what kinds of programming mechanisms are there which might, in principle, be utilized? There are two logically distinct extremes. One is equivalent in mechanism to the Descartian automaton, with an energy source which is wound up like a spring, that causes sequential movements as it unwinds. The modern equivalent is a machine that responds to logic instructions punched in a tape or inscribed on a magnetic strip. When these instructions are "read out" they cause peripheral mechanical devices to perform subroutines. Such a device in its simplest form is termed *open-loop*, because there is no built-in device for detecting that the peripheral mechanisms are actually performing as they should and which can modify the output instructions if they are not. It is simply assumed that the actions will be satisfactory. The operation of this type of open-loop control system in an animal is termed *simple motor tape*.

An improvement in the quality of the controlled movement or product is achieved by the adoption of sensors which detect that the instructions are being followed. There must be added a second control stage, where the sensed information is compared with the taped instructions. By estimates made from the difference, final instructions are computed and sent to the operating part. Such a control device, in its machine-shop form, takes care of variations in hardness of a part being drilled, of the effects of local overheating and other variables, making a product that always conforms to some strict criteria as to final dimensions. Otherwise, the product is statistical; for example, expecting that a fixed number of rotations of a lathe at a programmed average speed will produce a cut of desired size. In fact there will be a range of product sizes having a Gaussian distribution about a mean, rather than precisely equivalent ones.

Animals do an equivalent job to the latter with the aid of peripheral proprioceptors as sensors. The lower motor centers receive the equivalent of taped instructions from higher ones, together with feedback from stress and position receptors in joints and in parallel with muscles. Taking into account the latter, the output is modified to compensate for load changes. Here the programming is principally for sequences of length changes, but it can also be used for sequences of force changes in which length variations are a major interfering variable. This second type of control system is of a type we may term, *motor tape with combined servo-loop*. It can function without feedback, but operates more precisely when it is present.

At the other extreme is a system which is totally dependent upon peripheral sensing devices, utilizing a transcription of their inputs. These are measurements of positions and their time derivatives calculated in advance or made during the performance of a perfect sequence of movements, as a standard source from which to calculate output. By comparing actual sensor input with the stored information sequence, motor output is continually computed. The output has to be that which will reduce the error, or difference between the requirement and the status quo. This type of control system we may term *sensory tape*.

Most investigators do not appear to have considered the sensory tape system as even a possibility, at least for invertebrates, and it is seldom seriously considered even by neurobiologists working on primates. In the human it is clear that tapes of sensory sequences exist because they can be recalled into conscious memory. Detecting whether or not complex movements are controlled by sensory (rather than motor) tapes requires first, that investigators raise in their own minds the possibility that a sensory tape as defined above might exist. Affirmation of sensory-tape control requires experimental demonstration: (*1*) that detailed proprioceptive input is essential for the performance; and (*2*) that in repetitions identical movements are produced by widely different detailed motor-output patterns. For simple motor tapes successive outputs are essentially identical. For motor tapes with servo-loop the outputs, though different, are recognizably similar. But in a system operated by sensory-tape control the outputs, when the same movement recurs, will be widely different in detail from each other. Cocontraction of antagonists is also likely to be used frequently during sensory-tape control, but infrequently during motor-tape control.

All animals with nervous systems, including coelenterates, probably possess motor-tape mechanisms, and they have been definitively demonstrated in molluscs. We cannot say which organisms might or might not have sensory-tape devices, but any animal with a nervous system in which a memory trace is retained can, in principle, develop a sensory-tape mechanism. There are indications that mammals learning of a non-inherent motor act is at first determined by a sensory tape. Later, a motor tape is prepared, and this progressively takes over the control function. Motor tapes are probably slicker in operation than sensory tapes and more reliable because there is less mixing of control signals with information from unrelated channels.

An animal which inherits a number of motor-tape programs may improve them continually or find it necessary to modify them, as it grows in size and changes in shape during maturation. This it can do on a trial and error basis, but it would be done most efficiently by reference to sensory input. The different tape-control concepts are illustrated in the diagrams shown in Figure 8.

The Locations of Motor and Sensory "Tapes"

Our view of neural organization, handed down from earlier generations of students of nervous systems—especially Sechenov, Pavlov, Sherrington, Herrick, Weiss, and Gray—is that, like the army, it is a hierarchy of centers. Motor neurons occupy the lowest level; then come centers controlling individual limbs; next are centers involving several limbs in a specific movement pattern such as swimming or walking; and finally, there are centers determining multiplexes, such as fighting or nest-building, working through the successive lower centers. The subjects on which these schemes were built, however, were all vertebrates, but it seemed probable that invertebrates would not be significantly different.

When F. Huber found, in 1960, that he could evoke characteristic behaviors of crickets, including flight, walking, and singing, by making specific brain lesions and by local electrical stimulation, there seemed no reason to suppose that even the insects were different. But the motor neurons of insects, after their activity had been recorded in behaving animals by M. Burrows and myself, were clearly seen to be themselves playing a major role in their own programming rather than simply following instructions. Also, in the *Tritonia* brain, the neurons, which cause discrete movements when stimulated with axons going directly out to the effectors, are also clearly concerned with generating their own output pattern. Here the motor tape is in the brain, but it will be recalled that complex movements are generated by single insect ganglia detached from the rest of the nervous system; and in 1972, W. Kutsch announced that he had observed normal songs of all types from crickets in which he had severed the connectives to the head.

Therefore, when brain lesions or local brain stimulation cause singing, it is not because a programming center has been hit and started to send out neural patterns, but merely because relevant neurons in thoracic ganglia have been indirectly excited or disinhibited. The motor tapes for singing lie within the second and third thoracic ganglia, where the wing motor neurons are located, and the motorneurons themselves determine the final pattern. The brain, or cerebral ganglion, is primarily a region for further integrating already integrated visual input and unanalyzed antennal input. It is not the supreme site where detailed instructions to lower centers are computed, or even where all of the decisions to act are made. Also, it must be emphasized that the motorneurons are not simple slavish followers of the tape instructions. They themselves play a major role in generating the pattern they put out, after instruction profiles have been received.

The sceptic will now feel more firmly convinced than ever that invertebrates are totally different from vertebrates. But he should, instead, question the prevailing vertebrate view, which is in any case no more than a working hypothesis. The alternative to a system of discrete hierarchies is a set of parallel tape and computing systems for each behavior. All of these would involve neurons through to the level of the motor neurons themselves. The currently available data for invertebrates suggests that parallel, not hierarchical, control is the rule. Furthermore, it suggests that relatively small numbers of neurons are utilized in computations and even fewer in decision-making.

What, then, are all those small neurons in nervous systems doing?

A

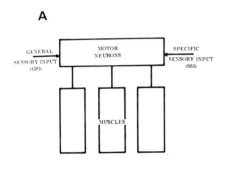

GENERAL
SENSORY INPUT
(GSI)

MOTOR
NEURONS

SPECIFIC
SENSORY INPUT
(SSI)

MUSCLES

B

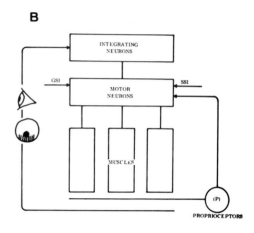

INTEGRATING
NEURONS

GSI

MOTOR
NEURONS

SSI

MUSCLES

(P)

PROPRIOCEPTORS

C

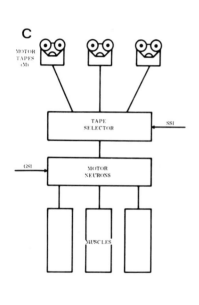

MOTOR
TAPES
(M)

TAPE
SELECTOR

SSI

GSI

MOTOR
NEURONS

MUSCLES

D

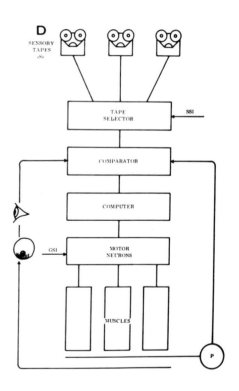

SENSORY
TAPES
(S)

TAPE
SELECTOR

SSI

COMPARATOR

COMPUTER

GSI

MOTOR
NEURONS

MUSCLES

P

28

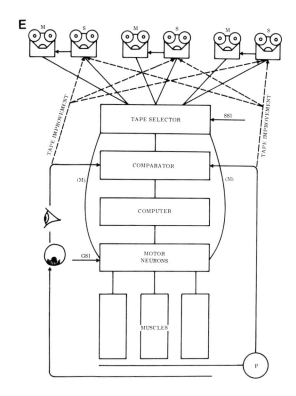

Figure 8. Diagrams illustrating the principal forms of behavioral control systems. A. Simple reflex. The system is open-loop, operating without either proprioceptive and visual feedback, or commands from nerve centers. B. Simple reflex with servo-loop additions. The motor output is continually modified during the reflex by feedback arising out of the animal's own movements. The latter is not only proprioceptive but also visual and kinesthetic. C. Motor tape. When the level of general sensory input (GSI) is sufficiently high, specific sensory input (SSI) leads to the turning on of an endogenous neural pattern generator symbolized by a magnetic tape of instructions. These are played into the motor output directly. D. Sensory tape. Sequences of stored sensory information are envisaged as controlling motor output. The actual sensory input during movement is compared with the memory trace of the input from previous performances. From the difference, corrected output is computed. E. Combined motor tape and sensory tape. The animal first uses the sensory tape as the movement is being learned and perfected by trial and error. As the movements become improved, the information is transferred to a parallel motor tape. This bypasses the comparator and computer elements and, hence, is more automatic and reliable than sensory-controlled movement.

Most of them are probably concerned solely to reduce the sensory input data, which is massive for all animal nervous systems, to a series of overall general excitations and inhibitions of motor neurons, which are relatively few in number.

We shall now summarize research progress in the experimentally amenable organisms mentioned above on the basis of selected topics of general interest rather than by phylogenetic position, paying special attention to the questions we listed previously.

Rhythmic Activities

A great many endogenously determined acts are repetitive. Some are more or less continuous, such as heart beats and respiratory movements. Others come in bursts of varying duration such as swimming, walking, flying, feeding, and singing. Some of these involve more than one rhythm. For example, in cricket song each sound chirp is composed of a rhythmic up-and-down wing movement at a relatively fixed, high frequency, while a complete song is composed of a rhythmic series of from three to seven chirps alternating with silence.

The late Donald Wilson introduced the notion that a means of generating these and perhaps other common behaviors is by temporary connection, achieved by central synapses equivalent to switches, of motor neurons to central nervous devices giving a constant rhythmic output irrespective of sensory input. He termed such devices "oscillators." Obviously, they could only be associated with behaviors which are themselves rhythmical, but a large number of behaviors are either totally rhythmic, or contain components which are rhythmic. In the simplest form, a single oscillator would be involved, but in more complex forms two or more oscillators, having markedly different periods, would be combined.

Reflecting the urge felt by investigators in this field quickly to discover general features of neural control mechanisms—and thereby establish a new scientific discipline—Wilson's concept of oscillators has been seized upon and developed to a point where for some behaviors the existence of such devices has become an accepted starting point. In fact, no such oscillators have been demonstrated conclusively! Serious attempts to locate neural circuits which are oscillating at the relevant behavioral frequency, when behavior is not occurring, have been made in several animals. But no direct evidence for them has yet been found. Also, there are few oscillatory behaviors which have a truly fixed period; most show an initial decrease, followed by a progressive increase, in period (slowing down) until the action ceases. We may take it that the property of oscillation is widespread in active neurons and neural networks.

The simplest forms of rhythmic pattern are seen in the cardiac pacemaker ganglia associated with neurogenic hearts, such as are found in crustaceans. The output takes the form of continuing rhythmic bursts, which are not arrested following isolation of the ganglion. The intervals between bursts, as well as the frequency of firing within a burst and the duration of a burst, vary according to synaptic inputs aimed at regulating the heart, but no input is necessary for a basic rhythmicity.

A similar rhythm occurs in respiratory centers, and whole nerve cords of insects completely isolated show spontaneous impulse firing at an average periodicity similar to that of natural respiration. Rhythms identical in timing to those of contractions of the stomach, occur in the totally isolated stomatogastric ganglion of spiny lobsters and of crabs.

In isolated molluscan ganglia certain cells are found which exhibit spontaneous rhythmic burst patterns (Figure 9). It has recently been shown, by long-term time-lapse cinematography, that some of these

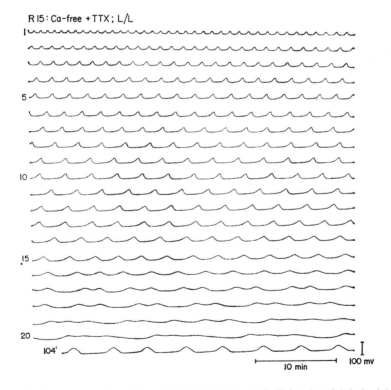

Figure 9. Continuous intracellular recording from the parabolic burster of *Aplysia* abdominal ganglion in the presence of calcium-free artificial sea water to block synaptic activity, and tetrodotoxin to block spikes. The last line shows the effect of hyperpolarizing the membrane in restoring the oscillation. The experiment shows that the bursts of impulse shown by this cell normally are secondary. The primary event is a membrane potential oscillation intrinsic to the cell. [From Strumwasser, 1971.]

rhythms correspond to naturally occurring movement rhythms, which they may therefore be generating. Slower rhythms also occur, at about a daily (circadian) interval, which is a rhythm definitely discernible in whole animals. A still slower rhythm may determine sexual activity. F. Strumwasser (1971) has shown that there are underlying biophysical changes—and presumed biochemical ones underlying these in turn—in individual neurons.

The basis of these rhythms is thereby seen to reside not within properties of circuits or interconnected nerve-cell clusters but within the biochemical program of an individual nerve cell. Coupling between cells must occur, in which case independent rhythms of coupled cells must be synchronized, and cells lacking intrinsic rhythms will be driven by those that have them. Can we not say, then, that they represent the neuronal oscillators? The answer to this question is that the question is unfair. The experiments show that the property of oscillation at a behavioral rhythm resides within some component cells. Whether or not this determines the behavior remains to be shown.

The available evidence for rhythms driving hearts and respiration in invertebrates, by contrast, does not implicate any cellular chemical cycles, but rather, suggests that properties of the circuits—connectivity, synaptic delays, excitations and inhibitions—between cells collectively determine that a rhythm shall be generated. There are only nine neurons in the lobster cardiac ganglion, and two or three are merely followers, so the remainder, in ways which are still very imperfectly understood, generate the rhythm by interacting.

In the spiny lobster stomatogastric ganglion there are about thirty nerve cells which between them generate four different, though related, rhythms. One is the rhythm of the muscular wall of the gastric mill, a second is the related rhythm of a central tooth in the mill, the third is that of two lateral teeth, and the fourth is of the pyloric sphincter regulating the exit. Several investigators—notably, D. M. Maynard, who introduced the preparation, M. Dando, A. Selverston, D. Hartline, and B. Mulloney—have combined to make detailed intracellular microelectrode studies of each neuron in the ganglion. They have recorded bioelectric activity when the ganglion is firing rhythmically and, using two electrodes, the details of interactions between pairs of cells. They have reached a point where they can now state just how many excitory, inhibitory and electronic synapses occur, and a surprising feature is the relatively large number of the latter and of inhibitory junctions. Most of the neurons have been categorized with confidence; 22 of them are motor, sending axons to stomach muscles, as well as being involved in the integration.

Intracellular recordings from key cell bodies involved in the rhythm are shown in Figure 10. The basic connections which form the wiring diagram of neurons involved in the pyloric cycle are illustrated in Figure 11. For the pyloric cycle Maynard was able to build a simple electronic model which simulated the principal feature of the membrane potential changes. The model required a total of nine elements in two parallel channels. Two sets are reciprocally connected, two sets are followers, while the ninth is a single common additive or mixer element. But while this unit simulated observed membrane potential shifts closely, it did not generate the basic rhythm and the model had to be driven by a superimposed rhythm. The stomatogastric ganglion preparation does not always fire, and no component cells are yet known which routinely possess an intrinsic oscillatory quality such as that demonstrated for *Aplysia*

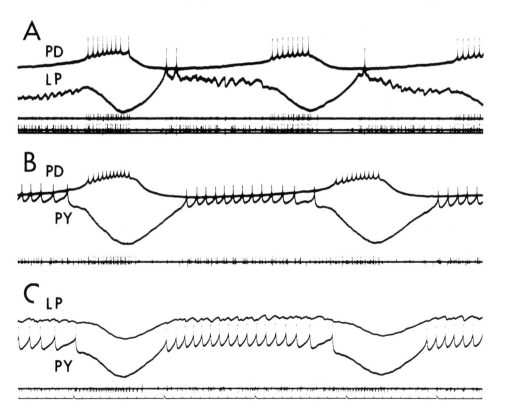

Figure 10. Intracellular recordings from the cell bodies of neurons of the stomatogastric ganglion of *Panulirus argus* which are concerned with the pyloric rhythm. The lower traces are extracellular records from nerves to stomach muscles: PD—pyloric dilator; LP—lateral pyloric; and PY—pyloric. Action potentials in PY always precede an ipsp in the LP neuron with a fixed latency. [From Maynard, 1972.]

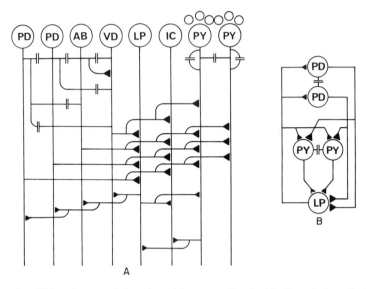

Figure 11. Wiring diagram of stomatogastric neurons involved in the pyloric cycle. A. Detailed diagram showing known connections between neurons (14) of the pyloric cycle: PD—pyloric dilator; AB—anterior burster; VD—ventral dilator; LP—lateral pyloric; IC—inferior cardiac; and PY—general pyloric. B. Simplified diagram reduced to five elements retaining critical features of the network. [From Maynard, 1972; modified to conform with symbols utilized in present article. Symbols used are explained in Figure 13.]

parabolic bursters. The prevalence of inhibition suggests that release from it may be a means of achieving excitation. Here the rhythm may be dependent upon network properties.

The complex connections between the neurons are clearly aimed at ensuring reciprocal actions between the synergistically acting groups while at the same time ensuring that the common level of excitability is maintained. The rhythm must be maintained and not allowed to die out, as can easily happen in principle in a system of neurons, through fatigue and accommodation. We shall meet this problem repeatedly in considering the question of generating behavior.

Specific Neural Programs for Which Unit Analysis Has Been Made

The Tritonia Escape Swim

The swimming escape behavior of *Tritonia diomedia* (Willows, Dorsett, & Hoyle, 1973) is just as readily elicited from animals whose brains have been exposed for intracellular recording, with six electrodes recording at the same time from different neurons, as it is in the intact animal

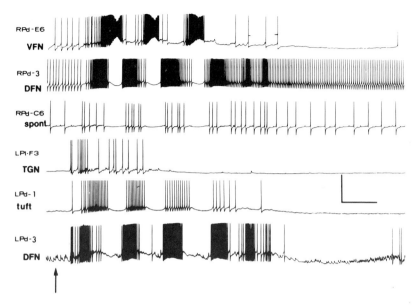

Figure 12. Simultaneous intracellular recordings from six different neurons of character-istic function from the brain of *Tritonia* during an escape swim. Trace 1—A ventral flexor neu-ron (VFN). The onset of inhibition which precedes termination is evidenced by the reduction of the third burst. Trace 2—A white neuron of the pedal ganglion which is presumed to be neurosecretory and is not involved in swimming. Trace 3—A neuron that is inhibited during swimming, especially during ventral flexions. Trace 4—A trigger group neuron (TGN) whose activity consists of a brief volley at the onset of swimming. Trace 5—A neuron envoking with-drawal of the left branchial tuft that fires in phase with DFN's. Trace 6—A dorsal flexion neu-ron (DFN). Onset of activity in this neuron is correlated with TGN activity. Note the depo-larization plateaus in the flexor neurons, that in the DFN outlasting that in the VFN. [From Willows *et al.*, 1973.]

(Figure 12). A touch from a starfish tentacle, or a drop of strong salt solution, invariably elicits the whole sequence, even in the isolated brain. Thus, it has been possible to find out which neurons on the exposed dorsal surface of the brain are involved in swimming. The activities of these known neurons appear to be adequate to account for most of the movements which occur. The movements are violent alternating flexions, ventral and dorsal, of which the first is always dorsal and the last ventral. There are commonly five complete cycles in a swim, followed by two or three weak dorsal flexions unaccompanied by ventral ones at termination. There are several neurons which drive each flexion and they are all located at various sites on the pedal ganglion surface. Ventral flexion units are intermingled with dorsal ones.

Only one other kind of neuron appears to be involved and it is one which is itself excited by both ventral flexion neurons (FN's) and dorsal flexion neurons (DFN's). In turn, this kind of cell, which Dennis Willows has termed a general excitatory neuron (GEN), also excites each type of flexion neuron.

An examination of the flexion neurons during swimming, and during the performance of similar activities in the isolated brain for a very large number of swims, showed that firing is always associated with a plateau of depolarization. The plateau can be eliminated with hyper-polarizing current, and without it the neuron does not fire. The purpose of the GEN cell, thus, may be simply to maintain the depolarization plateau. This possibility led to the proposal that the connectivity patterns of the three sets of neurons is itself sufficient to account for the pattern. All it needs to come into operation is a trigger. When the trigger group neurons (TGN's) were discovered, that aspect was fulfilled. One phenomenon remains unexplained, and that is termination, for no neurons have been found which are active at termination. Termination is preceded by a decline in the depolarization plateau, such as is most readily explained by a delayed inhibition. The sequence may be simply modeled by the scheme shown in Figure 13.

The Locust Jump

This escape behavior (Hoyle & Burrows, 1973) is readily evoked in a preparation in which the single fast extension tibiae (FETi) motor-neuron, and one antagonistic motorneuron or one interneuron are impaled at the same time by two microelectrodes (Figure 14). One sense organ mediating feedback is here essential, and that is the femoral chordotonal organ, which inhibits flexors, and releases the action.

The sequence is triggered by any sudden stimulus, when a high general level of excitation is also present, or it is generated spontaneously, at high body temperatures. First the flexors fire causing full cocking of the leg. This causes excitations of FETi, which has an autoexcitatory collateral that ensures a tetanic burst once the action is under way. At the same time reciprocal inhibitory input to the flexor occurs, mediated by chordotonal organ feedback. When the flexors relax, the energy stored in the extensor is released. Meantime, the extensor burst will have led to automatic flexion, in preparation for landing—and perhaps a further jump, because a collateral from the extensor excites all the flexors via an interneuron, which has been located and studied intracellularly.

The Crayfish Tail Flip

Another escape behavior, the tail flip of the crayfish, has been explained at the cellular level but it turns out to be little more than a

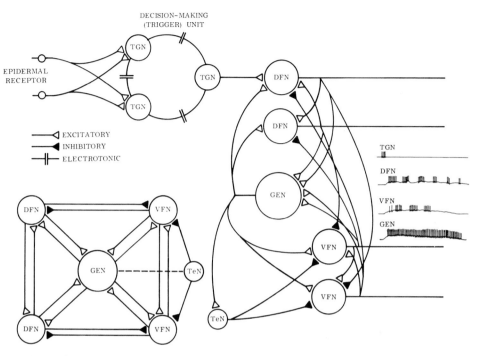

Figure 13. Neuronal circuit illustrating organization of neurons controlling the triggering, maintenance, and control of escape-swimming behavior in *Tritonia*. Reciprocal inhibition ensures alternation by antagonists, while common general excitatory neurons (GEN's), which are excited by both antagonists and in turn excite them both, ensure the maintenance of action by causing depolarization plateaus in all participating neurons. A delayed inhibition, not yet located, causes termination.

straightforward reflex (Kennedy, 1971). It involves a key single neuron and two small populations of neurons (Zucker *et al.*, 1971). The latter are the sensory elements, which are tactile receptors situated all over the animal, and synergistically acting, fast flexor motorneurons. The former feed onto four central elements, a tactile interneuron associated with a single segment, two tactile interneurons associated with several segments that are also interconnected, and the lateral giant neuron. The latter directly innervates both the fast flexor motorneurons and another giant neuron which also innervates the flexor muscles. Synapses from the fast flexor motorneurons onto the muscle fibers are facilitating, a property which would ordinarily cause them to be classified as slow, except that there is an additional set of neurons to the flexors causing still weaker contractions. By contrast, the motor giant is a truly fast axon which causes nonfacilitating junctional potentials and large twitch

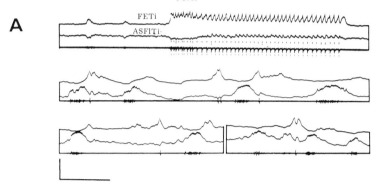

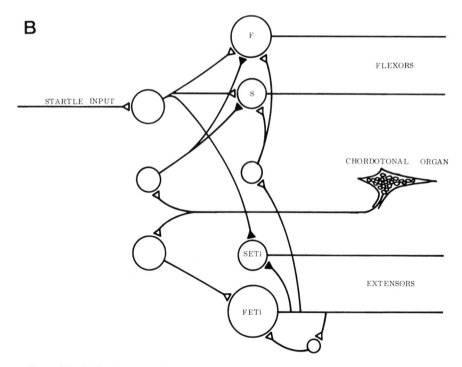

Figure 14. A. Simultaneous intracellular recordings from the fast extensor tibiae (FETi) motorneuron soma of the desert locust and one of its antagonists, during an evoked jump response. [From Hoyle and Burrows, 1973]. B. Diagram illustrating known neurons and connections involved in the locust jump.

contractions in the flexor muscles. Together, then, these motorneurons saturate the tail flexor muscles with excitation after the lateral giant fires. Hence, it is the latter which, as previously pointed out, makes the decision to flip.

Insect Flight

Insects in tethered positions produce flight movements of normal rate and amplitude after even quite extensive experimental dissection provided only that their feet cannot touch a firm base. This permits analysis of the neuronal mechanisms underlying flight, including intracellular recordings from slight motorneurons (Wilson, 1968; Burrows & Hoyle, 1973).

D. M. Wilson provided the first clear experimental demonstration of a behavior which can be centrally programmed, when he cut off all the proprioceptive feedback from locust wings and found that prolonged, normal-looking flight movements could still occur. The flight motor output consists of alternating bursts to elevators and depressors of the wings. Minimally, a burst comprises a single impulse, and maximally there are five, but since the increase above two is accomplished without an increase in duration of the burst, the frequency within the burst rises at the same time. The muscles respond to this by an increased power output per stroke.

Synergists have now been demonstrated to be cross-connected by chemically transmitting excitatory synapses. But although there is also marked reciprocity between antagonists, neither direct nor collaterally mediated inhibitory connections have been found. This means that the reciprocity is achieved at the level of interneurons which feed onto the final motorneurons. Nevertheless, antidromic impulses fed into the latter influence the timing of subsequent impulses and also those of the other neurons. This means that the motorneurons are not simply follower cells; they may generate their own rhythm, influenced by the input they receive. Each neuron has slightly different oscillatory characteristics and can operate independently of the others, but the cross-connections tend to pull synergists together and antagonists apart.

Intracellular work has established that flight requires input only as a trigger, for the initiation of sequences of a minute or so, but continuation for a longer time requires additional input. The membrane potential of each flight motor neuron oscillates by up to 20 mV peak to peak during activity (Figure 15A). The output spike pattern is determined by the rate of rise and instantaneous amplitude of the wave. A shallow wave produces no, or a single, impulse, while a steep wave produced two or more. The depolarizing phase, and also the repolarizing one,

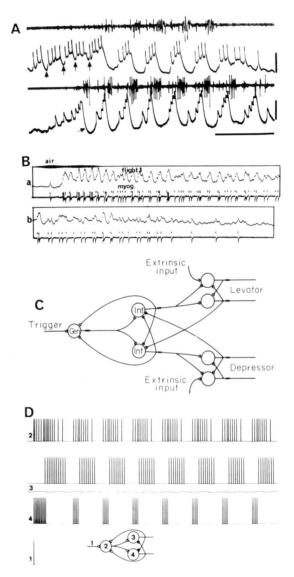

Figure 15. Intracellular recordings made from motorneuron cell bodies during tethered flight in: A. Cricket [from Bentley, 1969]. B. Locust [from Hoyle and Burrows, 1973]. C. Model of flight motor control system which is auto-oscillatory and requires only occasional, random input to maintain activity. D. Operation of a neuromime circuit set up according to the scheme illustrated in B.

represent the sum of intrinsic properties of the neurons, together with synaptic and ephaptic inputs to them.

The basic circuit generating flight may be of the type illustrated diagrammatically in Figure 15B.

Insect Song

Songs in insects are produced either by rubbing the wings together, by rubbing the hind legs together, by rubbing the hind legs against the wings, or by rubbing the wings against the body, except in cicadas, where a special muscle develops the sound by causing air vibrations in a tympanic chamber. But they all have one thing in common, that they use muscles which are, or have been derived from, flight muscles. The neural mechanisms which drive the singing are necessarily closely allied to those which operate in flight, and many of the sound patterns are determined by contractions recurring at the same rates as flight strokes. Characteristic songs are determined in part by the timing of silent periods interrupting impulse trains, and by the durations of active periods.

For example, crickets use about three pulses per chirp in calling song, and simply by increasing this number to five achieve a different sound with which a markedly different behavioral connotation is associated, namely, aggression. By reducing the amplitude of pulses—this is done by exciting a few fast synergists in each pulse—and continuous, rather than interrupted operation, they change to a courtship song. A different form of courtship sound is achieved by adding an occasional loud single pulse, giving a ticking sound, to a similar background. The detailed motor output to the muscles during singing has been analyzed for several species with the aid of implanted fine wires. In one instance intracellular recordings were also made by D. Bentley. The motorneurons showed the same kinds of large oscillations that occur during flight movements, and again the number of impulses produced was at least roughly dependent on the rate of rise of the depolarizing phase. The calling and aggression songs are, therefore, probably produced by a second periodic mechanism which interrupts the normal flight motor. There is a progressive increase in the intervals between successive chirps, however, which is not compatible with linkage to a second, low-frequency oscillator as has been suggested.

The pauses must be caused by inhibition, and the simplest mechanism which might be envisaged is a reciprocity between a general excitatory interneuron driving the antagonistic flight muscles and a second dummy interneuron. This kind of circuit would have the additional merit of permitting easy return to normal flight activities, for example, or to

continuous stridulation of the kind seen in the courtship pattern. All
that would be needed would be inhibition of the output of the dummy.

Command Interneurons and Behavior

In 1952 Professor Wiersma described interneurons he had found in
the crayfish circumesophageal connective which produce coordinated
movements immediately recognizable as identical to aspects of the normal behavior of the animal, such as the familiar startle response. Later
he termed them "command interneurons." Since then, several more such
neurons have been found, all in decapod crustaceans, causing tail movements, swimmeret beating, and walking, both forward and backward.

The early use of so powerful a term to denote these experimentally
defined neurons has been most unfortunate, however. For the term implies that the behaviors concerned are determined by a kind of push-
button operation in which the brain functions by selecting which button
from a panel to push in response to appropriate sensory clues, to achieve
the right behavior. This is not to say that we know that the brain does
not operate in this manner, but that this is what the concept implies.
Yet, at the present time we do not know how the brain controls these
behaviors, or if, indeed, it is the center from which instructions emerge.

However, these results do show that the thoracic ganglia possess the
neural machinery required to produce coordinated motor acts independently of detailed patterning from brain centers. The same can be observed readily in insects, by progressively removing the head, subesopha-
geal, thoracic, and abdominal ganglia. Without a head, insects can walk,
climb, fly (though crudely), jump, sing, learn leg position, and mate.
Mating not only continues normally but is enhanced, in the praying
mantis, as the female progressively devours the male, from the head
backward. All the operational instructions are contained in the last abdominal ganglion.

A less commital term for interneurons which evoke specific coordinated
movements is "driver" (Hoyle, 1964). There appear to be several kinds
of driver neuron. First, there are those which need to be active only
briefly, as triggers, like the *Tritonia* TGN's, to start a long-term activity
in motion. We may presume that in these cases GEN's intrinsic to the
neuronal "center" maintain the action. These constitute a second class,
in which the driver neuron must be continually active, and these I termed
general driver interneurons. Such neurons are functionally indistinguishable from GEN cells within a neuron cluster. They probably cause the
synergistically, reciprocally, and sequentially connected neurons in the
center to "do their thing" simply by lowering the membrane potentials

of a sufficient number of the components. A third class of interneuron must deliver rather precisely patterned input before the center goes into action. I earlier called such neurons *specific driver* neurons. N. Elsner has found examples in a grasshopper, but this class is more a speculation than reality at the present time. Obviously, any classification system is merely of value as a descriptive convenience. Terms with precise meanings can be utilized effectively only after a full description of the neuronal mechanism has been achieved, and we are still very far from achieving this objective for any behaviors above the reflex level.

Genetic Control of Motor Programs

A few of the principles governing the inheritance of motor programs have already been determined by studies at the gross behavioral level, and in one instance a correlation with neurophysiological data was made. The former shows that Mendelian rules are followed in the segregation of a behaviorally distinct subunit. The latter showed that genetic control determines the structure and physiological parameters of a circuit down to the ultimate possible level of refinement, the inclusion, or not, within a pulse train, of a single impulse.

In a clear behavioral example, Rothenbuhler crossed a strain of honey bees, termed "hygienic" because they remove dead larval bees from the comb, with a strain termed "unhygienic" which neither uncap cells nor remove dead larvae even if the cells are uncapped first. All the progeny were unhygenic, but a back-cross of the hybrids to the recessive, hygienic strain led to a simple segregation into four equal groups showing neat behavioral subunits. These were as follows: (*1*) bees which uncapped only and did not remove dead corpses; (*2*) bees which did not uncap but which removed dead larvae; (*3*) ones which did both— i.e., were hygienic; and (*4*) ones that did neither—i.e., were unhygienic. The combined genetic and neurophysiological work was done on the calling songs of male crickets of the genus *Teleogryllus* by D. Bentley and R. Hoy. Songs or chirps comprise sequences of trills each comprised of a number of pulses of high-frequency sound. The number of pulses in a trill is a species characteristic and one species, *T. Oceanicus,* has only two. *T. Commodus* has the most, with 14. The two species were mated by Bentley, and although F_1 females are sterile, the F_1 males are fertile and can be back-crossed to either parent type. Male crickets lack a Y chromosome, so genes on the X chromosome received from the maternal parent have no counterpart, and phenotypic differences in the songs can be attributed to X chromosome genes. Offspring from the interspecies cross were back-crossed with parents, producing a series of hybrid males. They could not be distinguished morphologically, but they did have different song patterns. In fact the songs could be placed in a series, related to the proportion of parental inheritance, with numbers of pulses in a trill of 3, 4, 5, or 6.

A pulse of the weakest strength is caused by a single impulse in the relevant wing motor neuron. Bentley had therefore shown that genetic information can specify neuronal output down to the level of a single impulse, in the output pattern of homologous neurons from different phenotypes. This it must do via control of factors which influence cell properties, probably, rather than circuits, in this instance.

How Decisions to Act Are Made by Invertebrates

At the level of reflexes it is usually the physiological state of an interneuron which determines whether or not an action shall occur. This is particularly the case where a giant fiber is involved. For example, the rapid tail flexion of the crayfish, which causes a backward escape movement, is determined by whether or not the sensory input to the lateral giant fiber exceeds its threshold. At that moment the decision is made to escape, for once the impulse occurs excitation of the fast flexor neurons must follow.

The decision to fire some rapid movements, including the strike of the mantis shrimp and the leap of the grasshopper, is made at the level of motorneurons which are antagonistic to the principal action neurons. Only if these cease to fire abruptly after there has been a prior buildup of tetanus tension in the extensor, does the rapid extension follow.

Only in one case has the cellular means by which the decision is made to initiate a fixed action pattern been determined. This is in the mollusc *Tritonia,* and the behavior is escape swimming. It is of the motor-tape control type and does not require any sensory input either for control of details or to maintain the excited state. Doing the latter is part of the program itself. The decision, to swim or not to swim, is made by two clusters of relatively small (for *Tritonia* brain, which has some neurons over 0.6 mm in diameter) neurons, one located in each half of the brain. They are situated in the waist between the cerebral and pleural ganglia. All the neurons are intimately coupled to each other by electrotonic junctions whose efficiency decreases with distance between somata. The neurons of the left and right sides are electrotonically coupled also across the commissure which joins the sides.

Sensory input from those sense organs which mediate escape-tactile and chemosensory, from all over the body surface but particularly from the oral veil, converges onto these neurons, which are termed *trigger group neurons* (TGN's). Some of this input is distributed in parallel to several neurons. The critical features are the total excitatory synaptic input and the number of spikes initiated at the same time. Because of the electrotonic connections a spike causes a depolarization in all the other TGN's with magnitude proportional to distance from the spiking cell. By itself, this potential will not initiate further spikes unless the general background depolarization is unusually large, as would happen

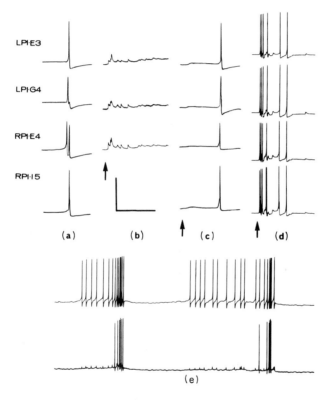

Figure 16. Illustrating the mode of operation of the decision-making, or trigger group neurons (TGN's) of the brain of *Tritonia*, which collectively determine whether or not an escape swim shall be initiated. Simultaneous intracellular recordings in each illustration. (a) A spontaneously occurring impulse in a TGN spread to others, exciting an impulse in a contralateral neuron. (b) Initiation of a TGN burst by application of a drop of strong salt solution to the oral veil (at arrow). The neurons received synchronous excitatory input. (c) A weak response initiated by pinching. (d) A strong response initiated by a small crystal of NaCl, which led to swimming. (e) The close coupling between two ipsilateral TGN's is illustrated in this spontaneous action. [From Willows *et al.*, 1973.]

only if the animal was thoroughly disturbed. But if a few spikes occur at about the same time in different TGN's they sum, causing more neurons to fire and leading to further excitation. Under these conditions, the whole group of neurons will be excited and fire together, inducing further reexcitation which leads to additional impulses, i.e., a cascading burst (Figure 16).

The cascading burst provides the stimulus which kicks off the "motor tape" neural machinery of the escape swim. Since weaker stimuli do not lead to a synchronous burst they are, in effect, filtered out. In this way the decision is made by the TGN's whether or not to initiate the swimming. Synergists are connected to each other via weak excitatory synaptic pathways while antagonists are cross-connected by somewhat stronger inhibitory connections.

CONCLUSIONS

We may conclude that a number of behaviors, including complex instinctive, and modifiable ones, shown by a wide range of invertebrates are open to in-depth analysis at the level of underlying neuronal phenomena and cellular interactions. Their diversity is such that the requirements of those who believe that it is necessary to show that a neural system principle must be demonstrated in more than one phylum to be considered general, is generously met.

A concentration of effort on an insect species suitable for combined genetic, developmental, and single neuron analysis is clearly highly desirable. The results should lead to the elucidation of principles whose generality may then be tested by studies on selected examples from other phyla.

The presently available data have revealed something of the way decisions to act are made in invertebrates. The common mechanism is exceeding the threshold for firing of either a key interneuron, or a cluster of interneurons that act as a single unit because of extensive electrotonic coupling among themselves.

Also revealed have been ways in which rhythms are produced. Both slow and medium rhythms have been shown to be produced by single neurons. In no case has rhythmicity definitively been shown to be a circuit property. Other, more complex, endogenous mechanisms have been disclosed in invertebrates and may be represented in model form by inherited motor tapes which are triggered to playback by specific sensory input. Once started, the program must run through its full sequence, in stereotyped order.

REFERENCES

Bentley, D. R. Intracellular activity in cricket neurons during the generation of behavior patterns. *Journal of Insect Physiology*, 1969, **15**, 677–699.

Bentley, D. R. Genetic control of an insect neuronal network. *Science*, 1971, **174**, 1139–1141.

Bentley, D. R., & Kutsch, W. The neuromuscular mechanism of stridulation in crickets (Orthoptera: Gryllidae). *Journal of Experimental Biology*, 1966, **45**, 151–164.

Blaschko, H., Cattell, M., & Kahn, J. L. On the nature of the two types of response in the neuromuscular system of the crustacean claw. *Journal of Physiology*, 1931, **73**, 25–35.

Bullock, T. H., & Horridge, G. A. *Structure and function in the nervous systems of invertebrates.* 2 vols. San Francisco: Freeman, 1965.

Burrows, M. The morphology of an elevator and a depressor motorneuron of the hindwing of a locust. *Journal of Comparative Physiology*, 1973, **83**, 165–178. (a)

Burrows, M. Physiological and morphological properties of the metathoracic common inhibitory neuron of the locust. *Journal of Comparative Physiology*, 1973, **82**, 59–78. (b)

Burrows, M., & Hoyle, G. Neural mechanisms underlying behavior in the locust *Schistocerca gregaria.* III. Topography of limb motorneurons in the metathoracic ganglion. *Journal of Neurobiology*, 1973, **4**, 167–186.

Castellucci, V., Pinsker, H., Kupferman, I., & Kandel, E. R. Neuronal mechanisms of habituation and dishabituation of the gill-withdrawal reflex in *Aplysia*. *Science*, 1970, **167**, 1745–1748.

Davis, W. J. Functional significance of motorneuron size and soma position in swimmeret system of the lobster. *Journal of Neurophysiology*, 1971, **34**, 274–288.

Dorsett, D. A., Willows, A. O. D., & Hoyle, G. The neuronal basis of behavior in *Tritonia.* IV. Neuronal mechanism of a fixed action pattern demonstrated in the isolated brain. *Journal of Neurobiology*, **4**, 287–300.

Elsner, N. The central nervous control of courtship behaviour in the grasshopper *Gomphocerippus rufus* L. In *Neurobiology of Invertebrates*. Budapest: Akádémia Kiadó, 1973.

Goldschmidt, R. Das Nervensystem von *Ascaris lumbricoides und megalocephala*. *Zeitschrift fur Wissenschaftliche Zoologie*, 1909, **92**, 306–357.

Hoyle, G. Exploration of neuronal behavior underlying behavior in insects. In R. F. Reiss (Ed.), *Neural theory and modelling*. Stanford, California: Stanford Univ. Press, 1964.

Hoyle, G., & Burrows, M. Neural mechanisms underlying behavior in the locust *Schistocerca gregaria.* I. Physiology of identified motorneurons in the metathoracic ganglion. *Journal of Neurobiology*, 1973, **4**, 3–41.

Hoyle, G., & Burrows, M. Neural mechanisms underlying behavior in the locust *Schistocerca gregaria.* II. Integrative activity in metathoracic neurons. *Journal of Neurobiology*, 1973, **4**, 43–67.

Huber, F. Untersuchungen uber die Funktion des Zentralnervensystems und insbesondere des Gerhirns bei der Forkbewegung und der Lauterzeugung der Grillen. *Zeitschrift fur Vergleichende Physiologie*, 1960, **44**, 60–132.

Kandel, E. R. The organization of subpopulations in the abdominal ganglion of *Aplysia*. In M. A. B. Brazier (Ed.), *The interneuron*. Berkeley: Univ. of California Press, 1969.

Kater, S. B., & Rowell, C. H. F. Integration of sensory and centrally programmed components in generation of cyclical feeding activity of *Helisoma trivolvis*. *Journal of Neurophysiology*, 1973, **36**, 142–155.

Kennedy, D. Nerve cells and behavior. *American Scientist*, 1971, **59**, 36–42.

Kutsch, W., & Otto, D. Evidence for spontaneous song production independent of head ganglia in *Gryllus campestris* L. *Journal of Comparative Physiology*, 1972, **81**, 115–119.

Maynard, D. M. Simpler networks. *Annals of the New York Academy of Sciences*, 1972, **193**, 59–72.

Ross, D. M., & Sutton, L. Swimming sea anemones of Puget Sound: Swimming of *Actinostola* new species in response to *Stomphia coccinea*. *Science*, 1967, **155**, 1419–1421.

Rothenbuhler, W. C. Behavior genetics of nest cleaning in honey-bees. *American Zoologist*, 1964, **4**, 111–123.

Smith, J. E. The nervous anatomy of body segments of nereid polychaetes. *Philosophical Transactions of the Royal Society of London. Series B*, 1957, **240**, 135–196.

Strumwasser, F. The cellular basis of behavior in *Aplysia*. *Journal of Psychiatric Research*, 1971, **8**, 237–257.

Stuart, A. E. Physiological and morphological properties of motorneurones in the central nervous system of the leech. *Journal of Physiology*, 1970, **209**, 627–646.

Waterman, T. H., Wiersma, C. A. G., & Bush, B. M. H. Afferent visual responses in the optic nerve of the crab, *Podophthalmus*. *Journal of Cellular and Comparative Physiology*, 1964, **63**, 135–155.

Wiersma, C. A. G. Neurons of arthropods. *Cold Spring Harbor Symposium on Quantitative Biology*, 1952, **17**, 155–163.

Wiersma, C. A. G. Regulative mechanisms for the discharges of specific interneurons. In M. A. B. Brazier (Ed.), *The interneuron*. Berkeley: Univ. of California Press, 1969.

Willows, A. O. D. Giant brain cells in mollusks. *Scientific American*, 1971, **224**, 68–75.

Willows, A. O. D., Dorsett, D. A., & Hoyle, G. The neuronal basis of behavior in *Tritonia*. III. Neuronal mechanism of a fixed action pattern. *Journal of Neurobiology*, 1973, **4**, 255–285.

Wilson, D. M. The nervous control of flight and related behavior. In J. E. Treherne & J. W. L. Beament (Eds.), *Recent advances in insect physiology*. New York: Academic Press, 1968.

Wilson, D. M., & Larimer, J. L. A catch mechanism in skeletal muscle. *Biophysical Journal*, 1969, **9**, A183.

Wine, J. J., & Krasne, F. B. The organization of escape behaviour in the crayfish. *Journal of Experimental Biology*, 1972, **56**, 1–18.

Zucker, R. S., Kennedy, D., & Selverston, A. I. Neuronal circuit mediating escape responses in crayfish. *Science*, 1971, **173**, 645–649.

Chapter 2

Similarities in the Visual Behavior of Arthropods and Men

MICHAEL F. LAND

University of Sussex, England

INTRODUCTION

Animals with ancestries as different as insects and men might well be expected to have evolved quite different ways of using their eyes to extract information from the environment, and the very different structures and optical principles of the compound eyes of arthropods and the simple eyes of vertebrates would seem to make this even more likely to be true. Nevertheless, all animals that rely on vision have to perform more or less the same kinds of tasks. For example, bees, racing pigeons, and men all navigate with precision over large tracts of countryside; swallows and dragonflies both catch insects on the wing; cats and jumping spiders both stalk and capture highly mobile prey; and many species of birds and fish, as well as insects and spiders, indulge in complicated species recognition displays concerned largely with mating the right kind of partner. This article is concerned with whether or not there are underlying similarities between the ways in which different animals use vision in performing similar kinds of behavior.

49

   Central to this discussion is the notion that an active animal is operating in a situation where the retinal image of the world is constantly changing, as a result of either the animal's own movements or the movements of objects in the world. This means that any action taken by an animal on the basis of visual information has visual consequences, so that, as von Holst and Mittelstaedt pointed out in their "Reafference Principle" (1950), an animal is inevitably part of a visual feedback loop. How an animal uses the visual feedback that is available to it is likely to depend on what kind of behavior it is engaged in: for example, if it is trying to remain still, or more specifically to keep its eyes still, one way would be to use any slip of the whole image across the retina to initiate movements of the eyes or body to minimize this slip, and such a mechanism forms the basis of the virtually ubiquitous "optomotor response," in which an animal in a moving striped drum will turn in the direction of the stripe movement. On the other hand, an animal turning "voluntarily," either simply to view another part of the environment or to orient toward a new stimulus, will in general need to do this unimpeded by a control system whose function is to prevent movement. Similarly, an animal may wish to follow a single rather small object which is moving relative to the background, and in this case it would be appropriate to minimize the slip of the "target" across the retina, but not the whole background, since a system minimizing background slip will actually be set in opposition to that minimizing target slip.
   Students of human eye-movements will recognize the three hypothetical examples just given as those which induce the three kinds of eye movements: optokinetic nystagmus (the involuntary following by the eyes of any large moving pattern), saccadic eye-movements (rapid relocations of the direction of gaze that are not controlled by visual feedback during their execution), and smooth tracking (the slow and accurate following of a target by the fovea). For a discussion of the control of human eye-movements consult Robinson (1968) or Stark (1971). It is generally thought that human eye-movements represent the highest development of visuomotor coordination in the animal kingdom, and in some ways this may be correct. However, all visual animals are faced with the same kinds of problems, and certainly among higher insects the means of dealing with them are not dissimilar in kind or complexity to those employed by man. One of the chief differences between ourselves and insects, which tends to obscure underlying similarities of mechanism, is simply that we have movable eyes on movable necks on movable bodies, and consequently our observable external behavior is largely emancipated from the needs of purely visual processes, which are taken care of by the oculomotor system. In an insect this may well not be true: many insects have some mobility of their necks (up to

15° each way in the blowfly, Land, 1973), but in general if they must move their eyes in performing a visual task, they must move their entire bodies also. The appearance of much insect behavior is, thus, very different from that of a higher vertebrate, but much of this difference disappears if one tries to imagine what one's behavior would be like if one's eyes were mechanically linked to one's feet!

In the remainder of this chapter I shall consider the kinds of control systems underlying four fairly simple kinds of visually mediated behavior: keeping still, "looking around" or visual search, locating specific targets, and tracking targets against a background. These are shown diagrammatically in Figure 1. There are, of course, much more complex kinds of visual behavior which could be considered (especially those concerned with navigation), but these four are sufficiently elemental and common for comparisons between species to be easily interpreted. This is not intended to be a review, and much of the material presented is derived from a study of a single species of insect, the hover fly *Syritta pipiens*, undertaken with Dr. Thomas Collett (Collett & Land, 1975). This insect was chosen in preference to others specifically because of its large repertoire of visual behavior: like other flies it uses vision to locate food sources (flowers), but in addition males stalk females in a very precise manner, and consequently it shares with man a need to locate small targets and an ability to track them—capabilities which may or may not be common to many other arthropods.

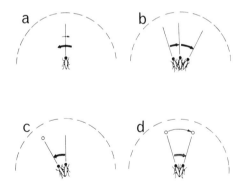

Figure 1. Diagram to illustrate the four types of visual behavior discussed. (a) Keeping still: any movement of the animal (small arrow) causes a slip of the image of the background across the retina, which evokes a counter-rotation (heavy arrow). (b) Looking around: voluntary turns made by the animal without any obvious stimulus, and in spite of stabilizing mechanisms like that indicated in (a). (c) Target location: here the presence of an important stimulus off the visual axis causes a turn of appropriate direction and amplitude, which results in the stimulus appearing directly ahead. (d) Tracking: movements which tend to keep the axis of the tracking animal closely locked to the direction of a particular stimulus. In each diagram the dashed semicircle is meant to represent a "background" at infinity, and open circles another animal acting as a stimulus for location or tracking. In each case a heavy arrow indicates the animal's response.

KEEPING STILL

It is presumably an advantage for good resolution to keep the image of the surroundings still on the retina, at least for short periods. In humans, a rapidly moving image becomes an uninterpretable blur, and it is a reasonable assumption that the confusion of temporal and spatial information that image movement produces will be minimized in other animals as well. Provided the rate of image slip across the retina can be measured, this information can be used to drive eye, head, or body movements in the appropriate direction to reduce the slip. Such a system will constitute a feedback loop which tends to keep movements of the retina relative to the outside world to a minimum (Figure 2). These responses are usually studied not by rotating the animal in a stationary world but by rotating a striped drum around the animal, and this sometimes gives the misleading impression that "optomotor responses" are to enable an animal to follow movement of the visual surrounding. This can scarcely be the case since the world itself never moves, and it is unreasonable to suppose that animals would be equipped to deal with something that does not happen. The alternative and more plausible

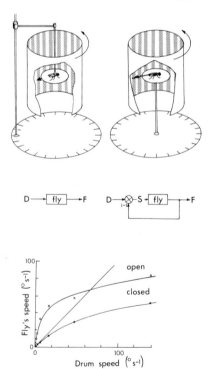

Figure 2. Mittelstaedt's technique (top) for investigating the "optomotor response" in normal (right) and open-loop (left) conditions. On the left the fly is fixed, but permitted to turn a light card disk, which it rotates in the opposite direction to the drum movement and at speed greater than that of the drum. On the right the more normal situation is shown, in which a free fly turns with the drum, but slightly slower. Middle: Control diagrams of the two situations. The left (open-loop) situation results in the drum speed (D) being converted by the fly into the fly speed (F) via some "gain" factor (g), so that $F = gD$. In the "closed-loop" situation (right) the fly only "sees" the slip speed of the image ($S = D - F$), the effect of which is to turn the fly into a follower, in which $F = (g/g + 1)D$. Bottom: Mittelstaedt's result using the open and closed-loop techniques. At low velocities the "open-loop" gain is very much higher than one, and the "closed-loop" gain approaches one, as predicted. At high velocities this relation breaks down presumably because higher rates of turning are limited by the rate at which the legs can step. [Modified from Mittelstaedt, 1964.]

explanation is that optomotor responses and their counterpart in humans—optokinetic nystagmus—are concerned with preventing image motion generated by the animal itself.

There have been several studies of these responses in arthropods. One of the earliest and most illuminating was that of Mittelstaedt (1951, 1964) (Figure 2) who showed that a fly (*Eristalis*) in a moving drum would rotate at a speed only slightly less than the drum speed; however, if the fly itself was prevented from turning, but allowed to turn a disk beneath its feet, it rotated the disk at many times the speed of the drum, and in the opposite direction. This is precisely what one would expect if the fly was measuring the rate of movement of the stripes across its retina and converting this, amplified, into leg movements tending to turn the animal in the direction of the stripes. When feedback is prevented in the second case the whole of the drum speed is amplified, but in the former case only the difference between the drum speed and the fly's speed (i.e., the slip speed). The performance of the fly can be expressed as a "gain" (g) which is the ratio of the fly's rate of turning to the rate of movement of the image across the retina, and g in this case is about 10 at reasonably low speeds. In the "closed loop" case, where the velocity of stripes across the retina is the drum speed *minus* the fly's speed, the "loop gain," i.e., the ratio of the fly's speed to the drum speed, should be equal to $g/(1+g)$, i.e., about 0.9, and this close to the ratio seen in practice.

A second example of a response tending to keep the eyes stationary in space is the visual control of eyestalk rotation in crabs. Horridge (1966) was able to show that the eyes would follow a pattern of stripes down to speeds as slow as $.001°sec^{-1}$, which is less than one revolution per day! By using one eye ("seeing," but prevented from moving) to drive the other (blinded, but free to move), Horridge was able to show that the open-loop velocity gain of the system (g) was between 5 and 25, giving a loop gain again very close to 1. An interesting feature of the crab system is that in addition to the visual input tending to prevent image slip the eyestalks also receive an input from the statocysts which compensates for body movements in very much the same way as vestibulo-ocular reflex in mammals (Sandeman & Okajima, 1973). A third similarity between crab and human eye movement control is the alternation, in both, of slow compensatory eye movements (induced visually or via the statocysts) with quick return phases that bring the eyes rapidly back to a position near the middle of their range of travel. The term "nystagmus," applied to this kind of follow-and-reset movement in humans, is equally appropriate in crabs.

Similar optomotor responses, involving the eyes or the whole body, have been demonstrated in flies, bees, locusts, and beetles and are prob-

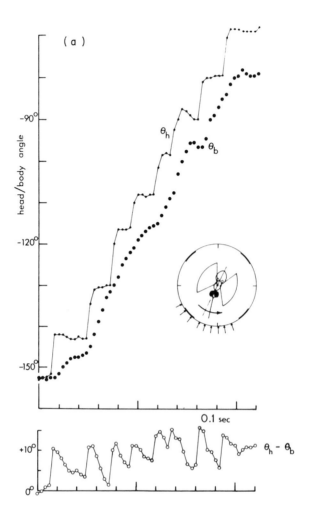

ably part of the behavioral repertoire of all arthropods. Although the image stabilizing aspect of this behavior has been emphasized here, a second function in animals without eyes that are movable independent of their bodies is inevitably going to be course stabilization—a tendency for a moving animal to adhere to a straight course. For example, Wilson and Hoy (1968) found that milkweed bugs (*Oncopeltus*) tend to follow straight paths, but show a directional bias when visual feedback is prevented. The inference is that the prevention of image slip by an optomotor response involving the legs also effectively prevents the animal from creating an angular slip across the retina by departing from a straight path.

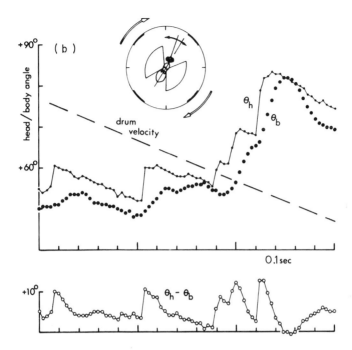

Figure 3. Saccadic head movements made by a blowfly *(Calliphora erythrocephala,* male) while flying at the center of a drum containing four 20° black stripes (indicated in the insets). The fly is fixed to a pivot and is free to rotate but not move forward or sideways. The direction of the body axis is shown as large dots, and that of the head axis by small dots; movements of the neck are shown as open circles below. (a) Fly turning between one stripe and another. Note that while the body turns almost smoothly, the head makes a series of sudden direction changes with steady periods between them. (b) As (a) except that the drum is rotating at 20° sec⁻¹ as well, as indicated by the dashed line. During the "steady" phases the head rotates at almost exactly the same speed as the drum, indicating that the stabilizing optomotor system is operating. [From Land, 1973.]

LOOKING AROUND

Clearly, no animal is constrained either to follow a straight course all the time or to keep its direction of gaze constant indefinitely. In fact, most animals spend much of the time in exploratory or food searching behavior which necessitates repeated changes of visual direction. The question this raises is how the feedback loop responsible for the maintenance of a constant direction is used, or overridden, when changes of visual direction are made.

Yarbus (1967) found that when the eye movements of human subjects were monitored while inspecting pictures or reading, the pattern of scanning was not smooth, but consisted of a succession of rapid (saccadic) relocations of gaze with periods of steady fixation between them.

Although there have been reports that some subjects are able to make smooth eye movements voluntarily, the normal human pattern of visual search involves the eyes being still for most of the time. Presumably, then, the stabilizing mechanisms discussed in the previous section are operating throughout this behavior, and are interrupted or overridden only during the 100 msec (or shorter) periods taken up by the saccades.

The alignment of the eyes during "voluntary" search behavior has not been much studied in arthropods, but two examples from flies can be given (Figures 3 and 4) which suggest that the way the eyes are used is not very different from man. Figure 3(a) shows the angular movements of the head (θ_h) and body (θ_b) of a male blowfly which is turning from one black stripe to another 90° to the left. The fly is tethered to a pivot which permits rotation, but not forward movement. Whereas the rotation of the body is fairly smooth, the head (of which the eyes are a part) moves in a series of small rapid (20–40 msec) angular changes, with periods of stability between them. The stability of the head angle (i.e., the direction of gaze) is achieved by two kinds of movements of the neck: the "saccades" themselves, and between these slower movements which accurately compensate for the movements of the body. Blowflies will also fixate stripes for long periods (Land, 1973), but while doing so they repeatedly change the direction of gaze from one position to another within and just outside the stripe, each new

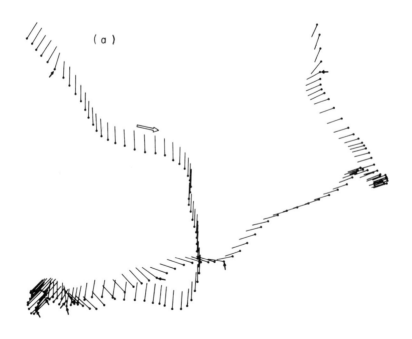

(a)

position being achieved by a saccade and thereafter kept steady in spite of body movements. Figure 3(b) demonstrates that the stability of the head between saccades is mediated by vision. The situation is the same as in Figure 3(a), except that the drum (containing four stripes) is rotated continuously at $20°$ sec^{-1}. The fly can still move its head saccadically around the field, but during the periods of stability the direction of gaze accurately follows the movement of the drum—even though for most of the time the fly is not fixating any particular stripe.

The second example is taken from a film of a hover fly, *Syritta*. When examined closely these flies are not seen to make head movements other than twisting movements of the head around the body axis during cleaning of the eyes. This is reflected in the flight path, in that body saccades are substituted for what in a blowfly would be head saccades. In the flight shown in Figure 4, which lasts about 3 sec, all the changes in

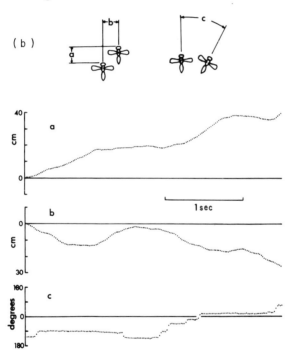

Figure 4. (a) Filmed record of the hover fly *Syritta pipiens* cruising around in a large transparent enclosure. The direction of the body axis and location of the head are indicated as they appeared on each frame (frame interval 20 msec). Arrows mark the instants at which saccadic turns occur. (b) Plot from the record in (a) of the cumulative forward, sideways and angular movements of the fly. Whereas forward and sideways movements can be performed smoothly, changes in the direction of the body axis are rapid and discontinuous (cf. Figure 3a). [From Collett and Land, 1975.]

direction of the body axis can be seen to be quite abrupt, usually occupy-ing no more than two frames of film (40 msec). Between these body direction changes, of which there are eight in all, the body is kept at a constant absolute angle, even though it is free to move forward, sideways, or backward. As in the previous example, when the fly hovers in a rotating striped drum, it turns with the drum during what ought to be the stabilized periods, again indicating the importance of the op-tomotor loop in maintaining directional stability.

In humans "looking around" involves eye movements, in blowflies head movements, and in hover flies body movements, but in all three cases the underlying strategy is the same: to retain stabilized vision for most of the time, and when rotations are to be made to make them as rapidly as possible. Presumably this permits a maximum of unblurred vision in all three cases. Probably some animals are less affected by image movement; tethered bees, for example, do not make obvious head saccades while rotating. It would be very interesting to know whether this is an odd exception to a good generalization or whether there really are two kinds of visual animal: those that can cope with a moving image of the environment and those that cannot.

LOCATING TARGETS

Some objects, especially ones that move, can have particular signifi-cance to an animal and require that they be attended to. In humans this involves transferring the image of the object from the periphery, where it appeared, to the fovea, where it can be fixated and examined. This is invariably accomplished with a saccadic eye movement of the correct size. These saccades have the interesting property that they are not affected by any movements of the target made during the course of the saccade, or a short period before it. This means that the control system responsible for the accuracy of the saccade is of the "open" type; it is precalibrated to give appropriate sized movements, given only the retinal location of the stimulating object, and operates without further visual feedback.

Systems like this are only likely to be encountered in animals which have a fovea, or at least some region of their eye or eyes which is specialized for high acuity vision; and foveas are most likely to be found in animals whose way of life requires them to locate, track, and identify other animals, for either sexual or predatory purposes. The hover fly *Syritta* again provides an excellent example of an insect with a fovea, and interestingly this fovea is present only in the males (Figure 5).

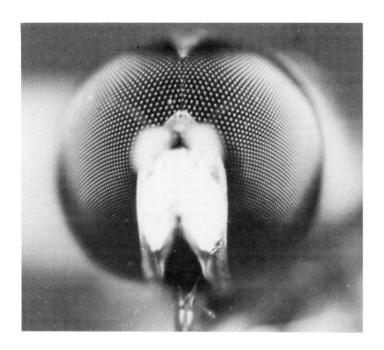

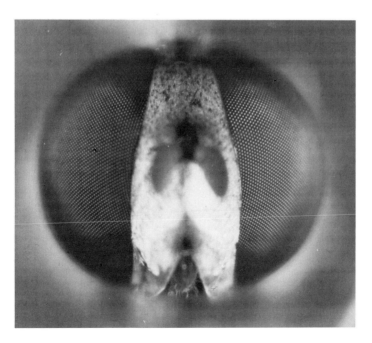

Figure 5. Eyes of male (upper) and female (lower) hover flies *(Syritta pipiens)*, showing the forward-pointing region of large facets, or "fovea," present in the male only.

The fovea consists of an axially situated region of large facets, each covering an angle of about .6° as opposed to about 1.4° which is typical for the rest of the eye, and for the entire eye in females. Both sexes feed on nectar from flowers, but in addition the males locate, stalk, and effectively rape the females by darting at them at high speed while they are feeding—or, occasionally in mid-air (Figure 9). Females do not track males. The first step is for the male to locate a female and turn so that she falls within his fovea (whose total extent in the horizontal plane is about 10°). These turns have all the characteristics of human directed saccades (Collett & Land, 1975). They occur when another fly comes within 5–10 cm; they are very brief (like the "voluntary" saccades discussed above), and both their duration and maximum angular velocity are proportional to the size of the turn. It can be shown from the absence of overshooting and the time course of the turns that they are made without visual feedback. All saccadic turns examined were accurate to within a few degrees, which is sufficient to bring the target into the foveal region. The conclusion is that the fly can accurately measure the angle between a target and its own body axis, and can translate this measurement automatically into a turn whose trajectory is such as to rotate the fly through the angle originally measured.

The second example of a directed saccade mechanism in an arthropod

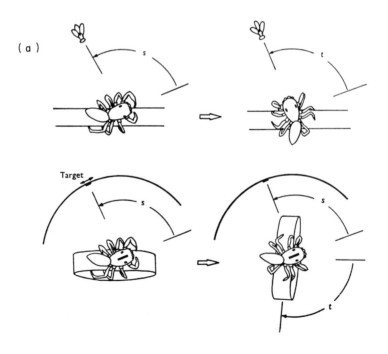

is the kind of turn made by a jumping spider (*Salticidae*) when it sees another object move to the side or behind (Figure 6). Jumping spiders are hunters, and do not build webs but stalk their prey—usually flies—using vision alone. Like *Syritta*, the first step is to turn to face anything that moves: this is done using information from one of the four lateral eyes, and the turn results in fixation of the stimulus by the anteriorly directed principal eyes, which are of higher resolution (10 min as opposed to 1°) and act as the "fovea" of the animal (Land, 1969, 1971). It can easily be shown that these turns are executed accurately without visual feedback by suspending the spider and giving it an artificial substrate to turn—a light card ring is ideal. Then, when a small spot is moved at a point to the side, the spider turns the ring

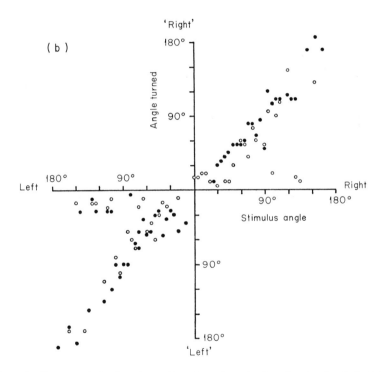

Figure 6. Turns made by jumping spiders toward moving targets, showing that accuracy does not depend on visual feedback. (a) Upper row illustrates a jumping spider *(Metaphidippus harfordi)* making a turn toward a fly, 100° to the spider's left. In the lower row the spider is fixed in space, but allowed to turn a card ring between its feet. Movement of a spot on a drum evokes a turn in which the ring is rotated through an (*t*) approximately equal to the stimulus angle (*s*). (b) Result of an experiment made under the "open-loop" conditions shown in (a). In nearly all cases the spider turned the ring accurately through the appropriate angle, even though the relative positions of the spider and stimulus were not affected by the turn. [From Land, 1971.]

through an angle which is equal to the angle it would have turned, if it had been free to move its own body. However, the stimulus is still in the same relative position as it was before the turn, which means that the turn, once initiated, continues and terminates accurately, without visual feedback. As in a human-eye saccade, the response is a preprogrammed response to movement at a given position. However, unlike both human saccades and the saccadic turns of *Syritta,* the exact time course of these turns is not specified—turns of the same size can vary in duration by a factor as great as ten. What is common to all turns of a given size is the number of steps the animal takes in accomplishing the turn, and presumably the "calibration" of this system resides in the accurate specification, execution, and counting of a particular number of steps, appropriate to the size of turn to be made (Land, 1972).

This kind of open-loop locating maneuver is by no means confined to visual systems. For example, Murphey has shown (1971, 1973) that the predatory aquatic bugs *Gerris* (a water-strider) and *Notonecta* (the backswimmer) can both turn accurately to a source of surface ripples using information from receptors in their legs and feet. Like a visual saccade, this is a single, preorganized action completed with surprising accuracy. What is startling about these animals is that the localization of a source of ripples is nothing like as easy as determining the direction of a ray of light on a resolved image, but nevertheless once the determination has been made the same kind of control system is employed.

The specially attractive feature of a saccadic system for locating targets is probably its speed. A closed-loop system in which the position of a stimulus continually specifies, for example, the velocity of turning toward it, is a theoretically possible alternative to an open system in which stimulus position specifies turn size in a once-for-all way. However, the main disadvantages of such a system would be, first, that given neural delays it would have to be quite slow in order to be stable, and second, that while the turn is being made the original stimulus may simply not be visible in the context of a moving background. Both these problems are obviated if, like a Sherringtonian scratch reflex, the target position of the response is related directly to the initial position of the stimulus. Certainly among visual systems there seem to be no exceptions to saccadic mechanisms like those discussed as the means of relocating seen stimuli (see Chapter 8 by Blakeman, for a discussion of the control of saccadic movements in monkeys).

TRACKING

While the last section dealt with how targets are transferred from the periphery to a fovea or forward pointing part of the retina, this section deals with the question of how they are kept there when they move. There is some dispute about the nature of the mechanisms responsible for tracking in humans (see Young, 1971), but, at least at low target velocities, tracking movements are continuous (not discrete like saccades) and are controlled to a large extent by the slip of the target across the eye.

In their smoothness and partial dependence on slip velocity, tracking movements resemble optokinetic nystagmus and optomotor responses (page 52); however, they differ sharply in being confined to the foveal region which is responsible for tracking small targets. If, as often occurs, there is a conflict between the foveal tracking system tending to move the eyes to follow a target, and the optokinetic system tending to prevent relative motion of the background and keep the eyes still, the foveal system "wins."

A difficulty in accepting a velocity feedback system as responsible for smooth tracking is that it will not on its own tend to keep the image on the fovea, unless the target starts on the fovea and the velocity match is perfect, which in a simple system is impossible because slip is needed to generate the tracking. Various ways have been suggested as to how this defect might be overcome. Saccades may be used to reposition the fovea (cf. page 58) and indeed at high tracking speeds ($>40°$ sec^{-1}) saccades take over as the principal mechanism of tracking. A second method would be to employ some position information in the production of smooth tracking, so that the "input" signal to the smooth tracking system would contain a measure of the retinal distance of the target from the center of the fovea. In the insect examples of tracking which follow, it is this type of "position error to velocity" system that mediates tracking.

Figure 7 shows a record of a male *Syritta* tracking another fly, probably a female. The tracking is very accurate in the sense that the leading fly is never permitted to stray further than a few degrees (i.e., a few facets of the eye) from the axis of the tracking fly. This is shown in more detail in Figure 8, where the angular movements of the tracking fly, the "error angle" and the angular velocity of tracking are compared. There are three points to notice: first, the fly rotates smoothly without any hint of the kind of saccadic jerkiness that characterizes the flight path during ordinary cruising (Figure 4, or the track of the followed fly in Figure 7). Second, the error angle (θ_e)—that is the angle between a line joining the leading fly to the following fly, and the following fly's axis—is always less than 5°, i.e., the leading fly is always located within the tracking fly's fovea (Figure 5). Furthermore, when the leading fly is moving right relative to the tracking fly this error angle is negative, which means that it appears in the right-hand part of the tracking fly's

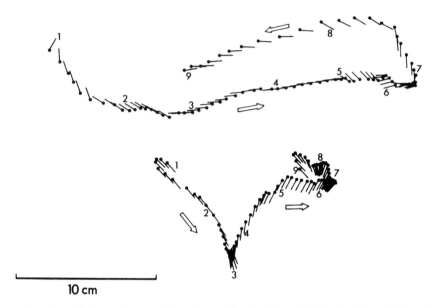

Figure 7. Male hover fly *(Syritta)* tracking another fly. Notice that the tracking fly (lower) keeps the upper fly directly in front at all times, and that while the angular movements of the tracking fly are smooth, those of the tracked fly are abrupt, as in Figure 4. The positions of the flies in every second frame are indicated (40 msec intervals). The positions of the two flies are numbered at the same instants, at intervals of .4 sec. [Collett and Land, 1975.]

fovea, and conversely, when the leading fly is moving left. Third, the angular velocity of the following fly—shown here as $\delta\theta_f$, the angle turned through in 20 msec—seems to vary in the same way as does the error angle. In fact, from this and other instances of tracking it is found that $\delta\theta_f$ correlates strongly with θ_e, after a lag of about 20 msec. What this suggests is that the position error—i.e., the dispacement of the leading fly relative to the center of the fovea of the following fly—is translated by the following fly into an angular velocity in the direction of the error after a total delay of 20 msec. The "gain" constant relating angular velocity to error angle is about $28°$ sec^{-1} per degree of error. Tracking can be modeled using these values for the gain and delay, and the results agree very well with what one sees—at least as far as strictly angular tracking is concerned; the fly also tracks in part by flying sideways. Interestingly, a system like this should show some instability, and if the leading fly makes sudden or rapid movements the tracking fly should show some damped oscillation. There are two instances in Figure 8 where this appears to occur, at about 0.7 sec and 1.4 sec after the beginning of the record.

An amusing example of tracking is shown in Figure 9. The fly on

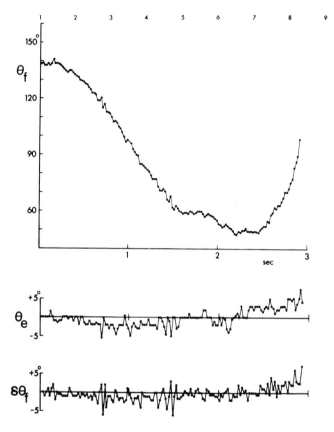

Figure 8. Changes in the direction of the body axis of the following fly (θ_f), the angular deviation of the center of the leading fly from the following fly's axis—or error angle (θ_e), and the angular velocity of the following fly— expressed as the change in direction ($\delta\theta_f$) of the body axis in 20 msec. The actual angular velocity in °sec^{-1} is $\delta\theta_f$ multiplied by 50. Numbers above refer to times in Figure 7.

The essential features are that tracking is very nearly smooth, that the "error" is kept within ±5° (within the fovea, Figure 5), and that an error angle in one direction, produced by a movement of the leading fly, leads to a change in direction of the following fly in the direction of the error. [From Collett and Land, 1975.]

the left had been hovering virtually motionless for several seconds about 10 cm from a second fly feeding on the flower. It was itself being tracked by a third fly (right). In the sequence shown the first fly suddenly accelerates between frames 1 and 10 (total duration of the flight is 200 msec) and lands on the fly on the flower in a typical "rape," and while doing this is tracked by the third fly. The movements of the tracking fly ("voyeur") are typical of other instances of smooth tracking: about half the required turn is achieved by sideways movement and

Figure 9. "Rapist and voyeur." For explanation see text. [Collett and Land, 1975.]

half by rotation, and the tracking ends in an overshoot predictable from the delay in the tracking process. In contrast to this, "saccadic" tracking (Figure 12) does not result in overshooting.

A rather more spectacular example of what is essentially the same kind of tracking system is shown in Figure 10, which is a plot of the flight paths of two houseflies (*Fannia canicularis*) engaged in a chase (Land and Collett, 1974). The chase takes place mainly in the horizontal plane, and it is possible to follow the maneuvers of the leading ($\bigcirc$) and chasing ($\bullet$) fly round the whole course, which only lasts $1\frac{1}{4}$ sec. Apart from the rapidity of the maneuvers, the striking thing to emerge from an analysis of this and similar chases is that the chasing fly is guided by the relative position of the leading fly in just the same way as in the hover fly example. Figure 11 shows a plot of the error angle (measured in this case as the angle between the leading fly, the following fly, and the direction of the latter's flight path) and beneath it a plot of the angular velocity of the chasing fly. Again there is an excellent correlation between the two after a lag of 30 msec. The "gain" in this case is rather less than in *Syritta* (about $20°$ sec^{-1} per degree error) and this improves the stability of the chasing fly. There is also a hint of some angular velocity error control in the region close to the chasing fly's axis which would further improve stability, as well as making it possible for the chasing fly to turn in the direction of motion of the leading fly, before the latter has crossed the chasing fly's line of flight.

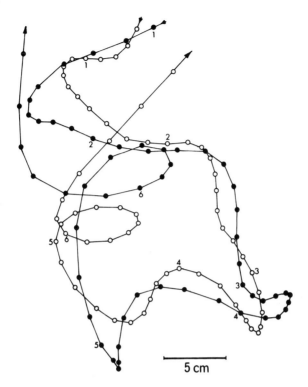

Figure 10. One lesser house fly *(Fannia canicularis)* chasing another. Leader ○, follower ●. The film was taken from vertically beneath a lampshade frequented by *Fannia*, and each point shows the position of a fly on one frame of film. The interval between points is 20 msec, and every tenth corresponding instant is numbered (.2 sec intervals). Although the whole sequence lasts less than $1\frac{1}{2}$ sec, each of the six maneuvers of the leading fly is followed by the chasing fly rapidly and accurately. [From Land and Collett, 1974.]

A good example of this can be seen in the left turn after 4 in Figure 10. Perhaps the most interesting feature of this tracking system, as opposed to that in *Syritta* or in man, is that here the whole eye is concerned, since the errors involved can be as great as 140°. Here, at least, the ability to track is not confined to a distinct fovea. There appears to be a continuous relation between the retinal location of the target, and angular velocity towards it, wherever on the eye the target is temporarily present.

Returning to the hover fly *Syritta,* it seems that in tracking there *is* a marked difference between fovea and periphery. *Syritta* can track objects that pass out of its fovea, but like humans it does so using a series of saccades, as indicated in Figures 12 and 13. Figure 12 shows a male attempting to track a very fast moving target, and it does so

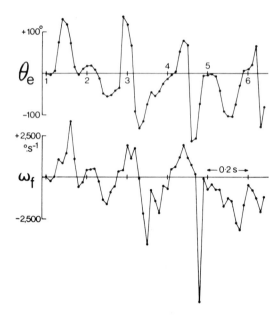

Figure 11. Plots of the angle between the leading fly, the following fly and the following fly's flight direction (θ_e), and the angular velocity of the following fly (ω_f) during the chase shown in Figure 9.

Like the following behavior of hover flies (Figures 7 and 8) chasing involves the continuous translation of stimulus position on the retina (approximately equal to θ_e) into an angular velocity (ω_f) of the following fly in the direction of the stimulus. [From Land and Collett, 1974.]

using three distinct saccadic movements (see inset for clarification) and there is no suggestion in the plot of angle versus time in Figure 13 of any smooth tracking of the kind seen in Figure 8. The error angles in this case reach 40–50° before being reduced to near zero by each saccade. There seem to be no differences in principle between these saccades and those discussed on page 58 as "locating" saccades: in both cases retinal location dictates the size of the turn to be made by specifying both the angular velocity of the movement (as in smooth tracking) and also its duration.

In man, one of the principal difficulties in trying to work out the nature of the smooth tracking system, is that the system has predictive capabilities, and will, for example, follow a target moving sinusoidally with much greater accuracy than one which moves unpredictably. An element of prediction can also be demonstrated in the tracking behavior of *Syritta* in the situation shown in Figure 14, namely, where the fly is flying around a stationary target. In this case the quality of tracking, as judged by the sizes of the angular errors that should be produced

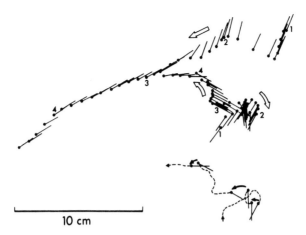

Figure 12. "Saccadic" tracking by the hover fly *Syritta*. As in Figure 7, the lower fly is tracking the upper fly, which in this case is moving very rapidly. There is no sign of smooth tracking, and the following fly makes three distinct turns—at 2, four frames before 3, and 4. Each turn realigns the following fly with the leading fly. Every frame is shown (20 msec intervals) and every tenth corresponding instant on the two tracks is numbered. The inset (below) indicates the path of the head of the following fly, and the three saccades. [From Collett and Land, 1975.]

Figure 13. Direction of the body axis of the following fly (θ_f), error angle (θ_e), and angular velocity of the following fly ($\delta\theta_f$ is the angle turned in 20 msec) during the tracking episode shown in Figure 11. As in Figure 4, the turns are rapid and discontinuous. Preceding each turn the error angle reaches 40–50°, in contrast to the situation in smooth tracking (Figure 8) where θ_e is less than 5°. Numbers above the figure refer to numbered instants in Figure 11. [From Collett and Land, 1975.]

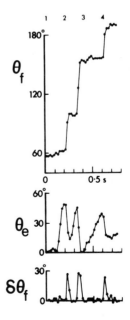

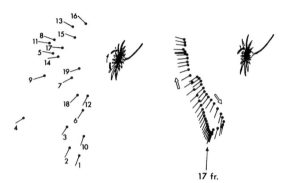

17 fr.

Figure 14. "Circling" tracking movements of the hover fly *Syritta* around a second individual on a flower. Left: different semistationary positions adopted by a male (in the sequence
shown) in a 25 sec period. Right: tracking during a rapid circling maneuver around the "target"
fly, including a brief period of steady fixation (20 msec interval between frames). Unlike
ordinary tracking (Figure 7) leftward movements are linked to clockwise rather than counterclockwise turning, and the angular tracking error is almost nonexistent. [From Collett and
Land, 1975.]

by the tracking fly's own sideways motion, is about 10 times better
than that of the ordinary tracking system discussed above. Furthermore,
no simple tracking system with a realistic delay could be this good,
because it would be permanently unstable. It is, therefore, necessary
to postulate that the fly knows how fast to rotate, for a given velocity
of self-induced sideways flight. In principle, all it needs to know is
its distance from the target since its angular velocity ($\dot{\theta}$, in radians
per second) should be equal to its sideways velocity ($\dot{s}$) divided by
the distance from the target, if the fly is to describe an arc of a circle
centered on the target. Since the flies are able to track stationary objects
in this way with great accuracy and at very different distances, this
provides good evidence both for their ability to judge distance, and
to use this knowledge to adjust the relation between angular and sideways
movement, i.e., to predict the amount of turning required.

CONCLUSIONS

I have tried to demonstrate in this chapter that there are basic similarities in the ways different animals use movements of the eyes in space
to organize visual perception and visually directed action. Specifically,
both flies and men possess visual control mechanisms (*1*) for preventing
the slip of the visual image across the retina, (*2*) for periodically altering
the direction of view using rapid saccadic movements, with only temporary suspension or overriding of the "slip-preventing" mechanism, (*3*)

for changing (again saccadically) the direction of view in order to face a new or important stimulus, and (4) for tracking a stimulus, even if this means permitting the slip of the image of the rest of the background across the retina.

These similarities are the more surprising because the requirements and functions of the visual systems of flies are different from those of men. A fly must, in general, point in the direction it intends to fly, and so movements of the eyes, mediated by the wings, are direct preparations for action in the immediate future. In man this is much less true, because of the independent movability of eyes, head, and body, and one might have expected human eye-movements to show rather different kinds of strategies. However, this does not really seem to be true, and there are two likely reasons. First, human eye-movement control has evolved from more "primitive" kinds of visual behavior in which there was a much closer connection between eye movements and "action directed" body movements (as, for example, in fish), and that control systems appropriate to such situations have survived—modified and refined, but not changed basically. Second, it seems likely that however the eyes are used in relation to behavior, visual processing itself has certain specific requirements, for example, a need to keep the image, or at least part of it, stationary on the retina if more complicated aspects of visual perception such as pattern recognition are to be possible.

It is always a surprise to find that unrelated animals have evolved similar engineering solutions to common problems. However, just as different kinds of kidney show basic similarities in construction and mechanism, presumably because filtration and reabsorption is the only efficient way to clean the blood, it is not hard to make the case that for a highly mobile animal there is only one efficient pattern of eye movement control.

ACKNOWLEDGMENTS

My thanks are due to Thomas Collett for reading the manuscript and for allowing me to use the results of unpublished joint research, and to the Science Research Council of the United Kingdom for support.

REFERENCES

Collett, T. S., & Land, M. F. Visual behaviour of the hoverfly *Syritta pipiens*. 1975 (in preparation).

Horridge, G. A. Study of a system, as illustrated by the optokinetic response. *Symposia of the Society of Experimental Biology*, 1966, **20**, 179–198.

Land, M. F. Structure of the principal eyes of jumping spiders (Salticidae: Dendryphantinae) in relation to visual optics. *Journal of Experimental Biology*, 1969, **51**, 443–470.

Land, M. F. Orientation by jumping spiders in the absence of visual feedback. *Journal of Experimental Biology*, 1971, **54**, 119–139.

Land, M. F. Stepping movements made by jumping spiders during turns mediated by the lateral eyes. *Journal of Experimental Biology*, 1972, **57**, 15–40.

Land, M. F. Head movements of flies during visually guided flight. *Nature*, 1973, **243**, 299–300.

Land, M. F., & Collett, T. S. Chasing behaviour of the housefly *Fannia canicularis:* A description and analysis. *Journal of Comparative Physiology*, 1974, **89**, 331–357.

Mittelstaedt, H. Zur Analyse physiologischer Regelungssysteme. *Verhandlungen deutscher zoologischer Gesellscheft: Wilhelmshaven*, 1951, 150–157.

Mittelstaedt, H. Basic control patterns of orientational homeostasis. *Symposia of the Society of Experimental Biology*, 1964, **18**, 365–385.

Murphey, R. K. Sensory aspects of the control of orientation to prey by the water-strider, *Gerris remigis*. *Zeitschrift für vergleichende Physiologie*, 1971, **72**, 168–185.

Murphey, R. K. Localization of receptors controlling orientation to prey by the back swimmer *Notonecta undulata*. *Journal of Comparative Physiology*, 1973, **84**, 19–30.

Robinson, D. A. The oculomotor system: A review. *Proc. I.E.E.E.*, 1968, **56**, 1032–1049.

Sandeman, D. C., & Okajima, A. Statocyst-induced eye movements in the crab *Scylla serrata*. *Journal of Experimental Biology*, 1973, **58**, 197–212.

Stark, L. The control system for versional eye movements. In P. Bach-y-Rita & C. C. Collins (Eds.), *The control of eye movements*. New York: Academic Press, 1971.

von Holst, E., & Mittelstaedt, H. Das Reafferenzprinzip. *Naturwissenschaften*, 1950, **37**, 464–476.

Wilson, D. M., & Hoy, R. R. Optomotor reaction, locomotory bias, and reactive inhibition in the milkweed bug *Oncopeltus* and the beetle *Zophobas*. *Zeitschrift für vergleichende Physiologie*, 1968, **58**, 136–152.

Yarbus, A. L. *Eye movements and vision*. New York: Plenum Press, 1967.

Young, L. R. Pursuit eye tracking movements. In P. Bach-y-Rita & C. C. Collins (Eds.), *The control of eye movements*. New York: Academic Press, 1971.

Chapter 3

Anatomical Methods for Tracing Connections in the Central Nervous System

LENNART HEIMER

University of Virginia

ANTHONY H. M. LOHMAN

University of Nijmegen, The Netherlands

INTRODUCTION

The function of the brain is dependent upon a highly organized interaction among billions of nerve cells, and one of the major objectives
in neuroanatomic research has always been to trace the neuron circuits
and the pathways that connect different brain regions with each other.
Increasingly more sophisticated methods have been introduced, and it
is now possible to study interneuronal connections in great detail. One
of the most important developments was undoubtedly the introduction
of the electron microscope as a tool for mapping interneuronal connections. The electron microscope, however, will not in itself permit us

to elucidate the organization of the brain. Considering the extremely limited amount of tissue present in an electron-microscopic section (less than .1 μm thick) it would be an exercise in endless futility to try to trace even the shortest pathway with the exclusive aid of the electron microscope. It is only with the guidance provided by light-microscopic techniques that the electron microscope can be used to full advantage in the tracing of pathways in the brain. Classic Golgi and reduced silver techniques are, therefore, as helpful as ever, and the recently developed fluorescence and autoradiographic techniques have proved to be of great value for the study of certain pathways.

The introduction of a new and powerful technique is always a celebrated event, and rightly so. There is, however, a tendency to overestimate the capacity of new techniques, and it seems appropriate to caution against too much enthusiasm in this regard. The introduction of the electron microscope in the field of experimental neuroanatomy may again serve as an illustration. In order to analyze the degenerating boutons of a specific pathway, it is obviously important that the tissue prepared for electron microscopy is representative. In other words, the distribution area of the fiber system under investigation must be fairly well outlined with light-microscopic techniques in order to avoid a time-consuming and highly irrational groping with the electron microscope in the search for degenerating boutons. Another example is provided by the autoradiographic method which is especially well-suited to answer questions related to the origin of certain fiber systems. The great expectations on behalf of the autoradiographic method are justified on the basis of observations that the isotopic label seems to be taken up only by the nerve cell bodies and not by fibers of passage that traverse the site of the injections. This very circumstance, however, also reveals one of the limitations of the autoradiographic tracing technique. Although well-defined and -circumscribed in its path, a fiber tract may have a widely dispersed origin, and the more widespread its origin the more difficult it becomes to inject differentially its cells of origin. With a simple surgical intervention somewhere along the path, however, the fiber system may be differentially marked by degeneration and subsequently visualized in its whole extent with the aid of experimental silver techniques.

PREPARATION OF THE MATERIAL

The quality of histological sections is to a large extent dependent on the preservation of the tissue during the essential steps of fixation, embedding and cutting prior to staining. The best results are nearly always achieved if the fixation of the brain can be initiated *in vivo* by vascular perfusion of the fixative, and since most neuroanatomic research is done on animals, fixation by perfusion is usually the method of choice. The technique of perfusion needs special attention and the main concern should be to get the fixative into the vascular system as rapidly as possible and under optimal conditions. The animal is anesthetized either intraperitoneally or intravenously with Nembutal. As the use of undiluted Nembutal may cause sudden death of the animal, it is advisable to give the chemical as a 30% solution in physiological saline. In the case of small animals, Equithesin* or inhalation anesthetics such as Metofane (methoxyflurane) and ether are also effective. When the level of deep anesthesia is reached, the chest is opened and a hemostat is placed on the descending aorta. The left ventricle of the heart is then incised and a canula is inserted through the heart cavity into the ascending aorta, where it is kept in place by a ligature. The blood is allowed to escape through an opening cut in the right auricle. In order to prevent the blood from clotting, the blood can be washed out with a small volume of physiological saline prior to the administration of the fixative. Another method to minimize the risk of coagulation of the blood by the fixative is to inject a small amount of heparin into the bloodstream before the perfusion.

Sufficient pressure for the perfusion can in general be achieved by placing the perfusion bottles at a height of about 4 ft above the heart (Koening, Groat, & Windle, 1945), whereas the speed of the flow can be regulated by a clamp on the canula fixation tube. The perfusion should start while the heart is still beating, and to make sure that no air bubbles enter the vascular system, the flow must be regulated so that physiological saline drips from the end of the canula while it is introduced into the aorta.

Although the perfusion technique is principally the same regardless of whether the tissue is going to be used for light or electron microscopy, the choice of fixative is dependent on the intended study. To obtain the excellent preservation needed for electron microscopic investigation of the nervous system, special attention must be given to the composition of the fixation solution as well as to the duration of the fixation (see Peters, 1970). For most light-microscopic procedures, however, 10% formalin has proved to be an excellent fixative. The formalin solution is prepared by diluting commercial (approximately 40%) formaldehyde solution 10 times with distilled water. To minimize shrinking of cells, Cammermeyer (1962) recommends the use of Heidenhain's Susa fixative,

---

* A combined sodium pentobarbital—chloral hydrate anesthetic manufactured by Jensen–Salbery Laboratories, Kansas City, Missouri.

which seems to be of special value if the cell structure is in the main focus of interest. Brains intended for light-microscopic study should as a rule be postfixed in the perfusion fluid for 1-4 weeks after they have been taken out of the skull.

The next step in the preparation is the embedding of tissue in a material that will give it mechanical support during the sectioning. The method of embedding is dependent upon the character of the investigation. Paraffin or celloidin embedding is often used in studies of normal cytoarchitecture. Investigators using experimental silver techniques, however, usually prefer frozen sections that have proved to be most suitable for silver stains. To prevent the formation of ice crystals in the tissue during the freezing stage, it is advisable to soak the formalin-fixed brain in 30% sucrose–formalin solution until the brain sinks to the bottom of the jar. This usually takes 2–3 days. To give the brain additional support during the cutting procedure, different investigators have developed techniques in which the brain is encaged in gelatin or egg yolk (Ebbesson, 1970).

STUDY OF PATHWAYS IN NORMAL BRAINS

The macroscopic dissection of the brain (Figure 1) was for many centuries the only method by which it was possible to define the position of the gray substance and the course of fiber pathways. However, the discovery of methods to stain nerve cells and their processes in histological sections in the second half of the nineteenth century, opened the possibility of studying the central nervous system with the light microscope. In 1858 Gerlach suggested the use of carmine to outline the position of cell masses and fiber paths. Of the numerous methods that have succeeded the carmine stain of Gerlach, the method developed by Franz Nissl in 1885 is one of the most well-known. The Nissl method stains selectively the cell bodies or perikarya of nerve cells, and is, therefore, widely used for the study of cytoarchitectural patterns in the central nervous system [Figure 2(a)]. There are many variants of the original Nissl method, but in our institutes the procedure published in 1950 by Powers and Clark (see Conn, Darrow, & Emmel, 1960, p. 93) has proven to be the most successful. The gallocyanin-chromalum technique of Einarson (1932) is recommended when subtle cytological characteristics are being studied.

The tracing of long fiber connections can be done reliably only in experimental material. It may be helpful, however, especially for the purpose of orientation, to study the normal fiber pattern with the aid

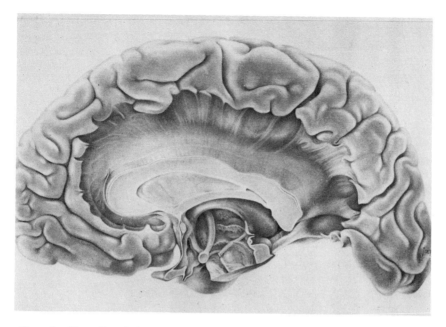

Figure 1. Dissection of a human brain from the medial surface according to the method by Ludwig and Klingler (1956). Large fiber tracts such as the corpus callosum, the fornix and the mammillo-thalamic tract, are well-displayed.

of a method that stains either the axons of nerve cells or the myelin sheaths around the axons. Although Fromman discovered the use of silver for the impregnation of axons in 1864, it was not until 1904 that Ramón y Cajal and Bielschowsky, independently of each other, developed the first silver methods that proved suitable for routine use. Especially the Bielschowsky procedure has proved to be valuable and modifications of the original method are now widely employed for both paraffin (Bodian, 1936; Romanes, 1950) and frozen sections (Schneider, 1969).

Ever since Weigert discovered that myelin treated in potassium dichromate stains bright red with acid fuchsin (Weigert, 1884; Pal, 1887), the staining of myelin sheaths around the axons of nerve cells has been of great value in anatomical and pathological studies. Modern variations of the Weigert–Pal technique include the procedures developed by Loyez (1920), Weil (1928), and Woelcke (1942). These methods, however, involve chromate treatment, and the sections can, therefore, not be counterstained with the Nissl method for the mapping of cell bodies. The discovery that Luxol fast blue stains the myelin without previous chromation of the sections was, therefore, of great value (Klüver & Barrera,

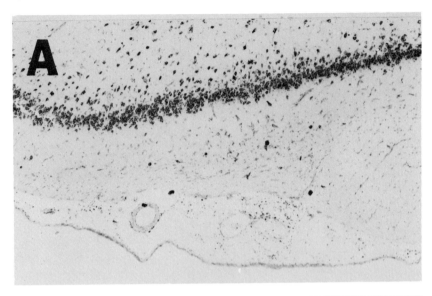

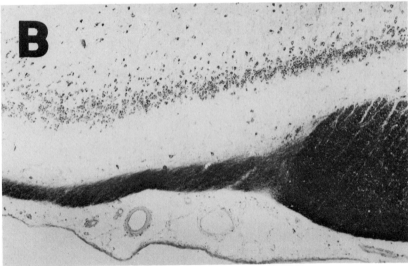

Figure 2. A. Nissl stain of the prepiriform cortex of the opossum. Whereas the characteristic distribution of cell bodies in the cortex is well demonstrated, the heavily myelinated olfactory tract at the surface remains unstained. B. Both cell bodies and myelinated fibers are well stained in a Klüver–Barerra preparation.

1953), and the Klüver–Barrera technique, in which the myelin stain is combined with a cresyl-violet staining of cell bodies, may rightfully be considered the method of choice for the simultaneous visualization of nerve cells and myelinated fibers [Figure 2(b)]. When it is desired to visualize the axons and myelin sheaths of the nerve fibers at the same time, the procedure of Häggqvist (see Busch, 1961) can be recommended. This method stains the axon and myelin sheath of individual fibers in sharply contrasting colors and can, therefore, be employed in the quantitative analysis of the caliber spectrum of fiber systems.

Recently, fluorescence microscopy for demonstrating monoamine-containing pathways has become extremely useful in neuroanatomic research (Falck, Hillarp, Thieme, & Thorp, 1962; Fuxe, Hökfelt, Jonsson, & Ungerstedt, 1970). The technique is based on the principle that biologically active monoamines condense with formaldehyde to yield strongly fluorescent products (Figure 3). The fluorescence method has not only increased our knowledge of the histochemistry of the peripheral and central nervous system, it has also revealed nervous connections not previously stained by the more traditional reduced silver techniques.

Golgi Technique

The Golgi technique, which is the method of choice for the study of neurons including their processes in normal preparations, plays such

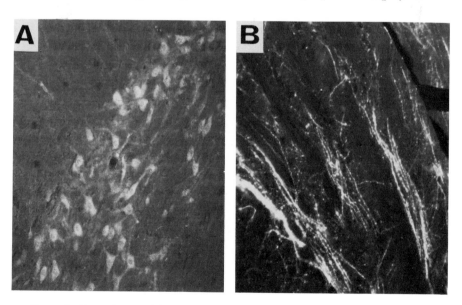

Figure 3. Dopamin containing cells in the rat brain demonstrated with the aid of the fluorescence method. The dopamin containing fibers (B), which have their cell bodies in substantia nigra (A), terminate in the striatum. [Courtesy of Dr. Kjell Fuxe, Karolinska Institutet, Stockholm.]

a paramount role in neurologic research that it rightly deserves to be set apart from the procedures just described. Immediately following its introduction in 1873 by the Italian scientist Camillo Golgi, the method became very popular and Golgi-studies by Cajal, van Gehuchten, Held, Retzius, and von Lenhossek, among others, laid the foundation of most of our present knowledge concerning the morphology of the brain.

The value of the Golgi method rests upon the fact that it demonstrates, for a still unknown reason, only a small number (5 to 10%) of the nerve cells in an area, but it impregnates these neurons with their axons and dendrites in full detail (Figure 4). The small number of cells impregnated allows for the cutting of relatively thick sections (80–300 μm),

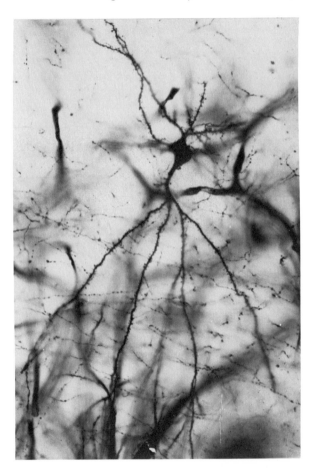

Figure 4. Rapid Golgi stain of a cell in the pallidum of the *axolotl.* Note the spines on the dendrites.

which in turn makes it possible to obtain a three-dimensional view of the nerve cells with their processes. Although our knowledge of the structure of neurons and the arrangement of short interneuronal connections is to a large extent based on Golgi studies, it is difficult and often impossible to trace longer axonal connections with the Golgi technique. The reason for this is that an axon seldom can be traced reliably from one section to another. The complete course of an axon can be followed only if the whole length of the axon is confined to one section.

Like many other staining procedures, the Golgi method has been the subject of frequent modifications. Problems related to the application and interpretation of the various Golgi procedures now in use were discussed by Valverde (1970) and Ramón–Moliner (1970) at a recent symposium on anatomical techniques (Nauta & Ebbesson, 1970).

EXPERIMENTAL METHODS FOR THE TRACING OF PATHWAYS

Reduced Silver Methods

It is only occasionally that the origin or the termination of the fibers can be determined in normal material, and we have to rely on animal experiments in order to reach certainty on these points. Following an experimentally induced lesion of a nerve fiber or its cell body, the fiber peripheral to the lesion undergoes anterograde degeneration, which in essence is a chemical and physical breakdown of the fiber into fragments. This type of degeneration is also called "Wallerian degeneration," after Augustus Waller, who noticed this form of reaction in 1850. Methods for staining of degenerating nerve fibers became available a few decades later, and the phenomenon of Wallerian degeneration has been successfully exploited in the tracing of pathways ever since. The axons that are "marked" by the Wallerian degeneration can be traced in serial sections through the brain or spinal cord.

The first method for the study of Wallerian degeneration was developed in 1886 by Marchi and Algeri, who noticed that degenerating myelin, that had been treated with a potassium dichromate solution, can be selectively stained by osmium tetroxide. Although the Marchi method has contributed considerably to our knowledge of pathways in the central nervous system, its usefulness is limited to the study of myelinated axons. Axon arborizations and synaptic end-structures, which are not covered by myelin, are beyond reach of the Marchi technique. Fortunately, the development and gradual refinement of the reduced silver technique, which impregnates degenerating axoplasm instead of degenerating myelin, has made it possible to demonstrate the course and the termination of both myelinated and unmyelinated fiber systems.

Many silver techniques were used for the study of the nervous system in the first half of this century. It remained for Glees (1946) and Nauta (1950), however, to demonstrate the extreme usefulness of silver methods for the tracing of degenerating nerve fibers in experimental material. Although the mechanisms involved in silver impregnation are poorly understood, the main procedural steps are rather simple. Following an initial impregnation in a silver nitrate solution, the section is transferred to an ammoniacal silver solution, and finally reduced in a dilute formaldehyde solution. One of the great advantages of the Nauta procedure, of which many variations are available (Ebbesson, 1970), is that it suppresses the argyrophilia (e.g., affinity for silver) of normal fibers, which in turn enhances the identification of the degenerating fibers. During the last few years the value of the reduced silver technique has increased considerably because of the introduction of more sensitive procedures (Fink & Heimer, 1967; Voneida & Trevarthen, 1969; Loewy, 1969; De Olmos, 1969; Eager, 1970; Velayos & Lizarraga, 1973; Wiitanen, 1969). These more recent modifications are capable of demonstrating the terminal axon ramifications and boutons of most fiber pathways in the brain and spinal cord.

The anterograde degeneration technique has been successfully exploited in determining the course and distribution area of many fiber tracts, and the value of using silver procedures is well-exemplified in the study of the olfactory bulb projection in the rat. Following an olfactory bulb lesion 5 days before sacrifice of the animal, it can be demonstrated that degenerated axons of the secondary olfactory neurons in the olfactory bulb reach the cerebral hemisphere by way of the lateral olfactory tract. The degenerating droplike fragments which are aligned in serial fashion are often connected to each other by thin bridges reminiscent of the morphology of disintegrating fibers [Figure 5(a)]. As the tract proceeds caudally, fibers deviate from it and spread out on the surface of practically the whole basal part of the forebrain, which indicates that these areas represent olfactory projection fields. This is confirmed by the fact that appropriately stained preparations show so-called terminal degeneration [Figure 5(b)] in the same areas. In an experimental preparation, such an area is identified by an accumulation of different-sized, more or less spherical argyrophilic particles some of which represent fragments of terminal axon branches, others degenerating synaptic end-structures (Heimer & Peters, 1968; Heimer, 1970a). The argyrophilic particles representing the terminal degeneration in the olfactory bulb

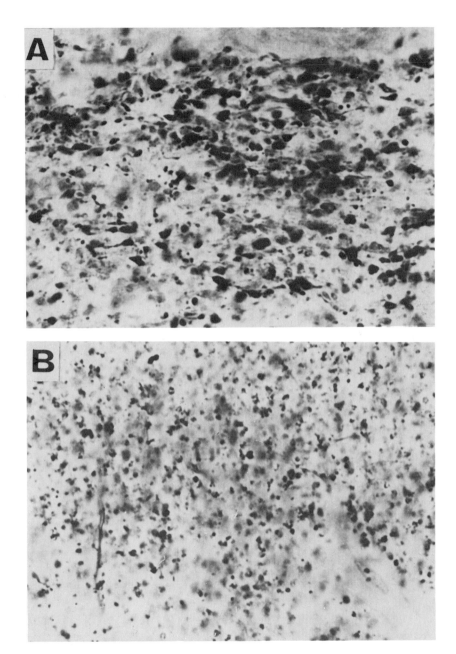

Figure 5. Silver impregnated degenerating olfactory bulb fibers in the prepiriform cortex of a rat, whose ipsilateral olfactory bulb was destroyed 5 days before sacrifice of the animal. Degenerating olfactory tract fibers are demonstrated in A, whereas B shows terminal degeneration in the plexiform layer.

projection are mainly confined to the superficial part of the cortex, where the distal arborizations of pyramidal cell dendrites are located (Figure 6). The inference might therefore be made that the synapses between the fibers of the lateral olfactory tract and the cells of the cortical mantle of the hemisphere are axodendritic.

The recent advances in experimental tracing of fiber connections is as much a result of increased awareness in regard to the importance

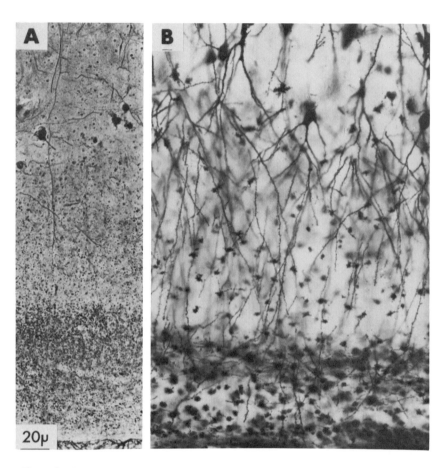

Figure 6. A. Low-power photograph of a silver-impregnated section from the prepiriform cortex of a rat whose ipsilateral olfactory bulb was removed 5 days before the sacrifice of the animal. The pial surface is seen at the bottom. B. A Golgi picture of the prepiriform cortex of a normal rat showing the distribution of olfactory bulb fibers in the superficial zone of the plexiform layer among the distal arborization of long dendrites emanating from the pyramidal cell layer. This superficial zone corresponds to the lamina of dense terminal degeneration shown in A. [Reproduced from *Journal of Anatomy*, 1968, **103**, 3, by courtesy of Cambridge University Press.]

of the survival time as it is a reflection of refined methods, and many of the mistakes in the light-microscopic tracing of fiber connections in the 1950s and 1960s were obviously due to a less suitable survival time of the animal rather than defective histotechnical methods. The optimal survival time for the staining of degenerating boutons does not necessarily correspond to the best survival time for the tracing of fibers of passage. Seven to ten days survival time, although suitable for the study of fiber degeneration, is often too long for determining the projection area of a pathway, at least in commonly used laboratory animals such as rats and guinea pigs. The argyrophilia of degenerating boutons and preterminal fibers of many fiber systems reaches an optimum after a couple of days following the surgical intervention and may disappear within five to seven days. The reason for this is quite simple. Degenerating preterminal fibers and boutons are in many areas rapidly engulfed by glia cells, which are able to remove most signs of terminal degeneration within less than a week (Colonnier, 1964; Heimer & Wall, 1968).

Combination of Light and Electron Microscopy

The recent modifications of the reduced silver method are highly sensitive in the sense that they are capable of impregnating, not only degenerating parent stem fibers but also axon arborizations and terminal end-structures. It is important, however, to recognize their limitations. It is not possible, for instance, to identify the single argyrophilic particles [Figure 5(b)], of which some represent degenerating boutons, and others fragments of preterminal fibers or perhaps even a fragmented impregnation of degenerating fibers. It remains for the electron microscope to identify and classify the synaptic end structures. Only with the high magnification provided with the electron microscope is it possible to identify synaptic vesicles and membranes, thereby clarifying the anatomical relations at the site of the synaptic contact (Figure 7).

Several techniques can be used for combined light- and electron-microscopic analysis. It is often possible to perform the light- and the electron-microscopic analysis on separate specimens, especially in the presence of easily recognizable topographic landmarks. In other cases, however, the sampling of tissue for electron-microscopic analysis of degenerating boutons might create difficulties that can be resolved only by performing the light- and electron-microscopic study on one and the same specimen. The usual way to "bridge the gap" between light and electron microscopy is to cut a "semithin" section from the plastic-embedded tissue prior to the cutting of the ultrathin sections. The "semithin" section, which can be successfully cut with a glass knife from blocks as large as 5×5

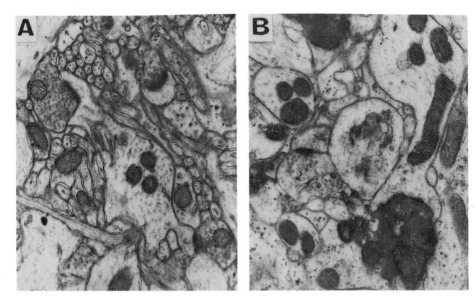

Figure 7. Electron micrographs from the superficial part of the prepiriform cortex of the rat. The micrograph in A, which is taken from a normal animal, shows a dendrite with two spines contacted by synaptic end structures. The electron micrograph in B is taken from a rat, whose ipsilateral olfactory bulb was destroyed 3 days before sacrifice of the animal. The degenerating bouton, that contacts the spine, belong to a fiber whose cell body was destroyed by the lesion in the bulb. Increased density, deformation, and glial reaction are important signs of degeneration.

mm, is then stained and used for orientation on the light microscopic level. The silver method used for the light microscopic mapping of degenerating fibers and boutons in frozen-cut sections can be successfully applied also on plastic embedded "semithin" sections (Heimer, 1969). Once the area of terminal degeneration has been determined in a semithin light microscopic section (Figure 8), it is easy to trim the original block so that ultrathin sections can be prepared from the desired region for subsequent electron microscopic analysis of synaptic contacts (Heimer, 1970b). It is important, however, to cut the "semithin" section thick enough (5–10 μm) so that it contains a fair amount of terminal degeneration.

Following the electron-microscopic identification of the synaptic complex, the classification of the postsynaptic neuron usually becomes a matter of prime concern. The identification of the postsynaptic neuron serves to illustrate a well-known dilemma in high magnification microscopy, namely, the gain of details at the expense of orientation. Although the electron microscope has made it possible to study synaptic interrela-

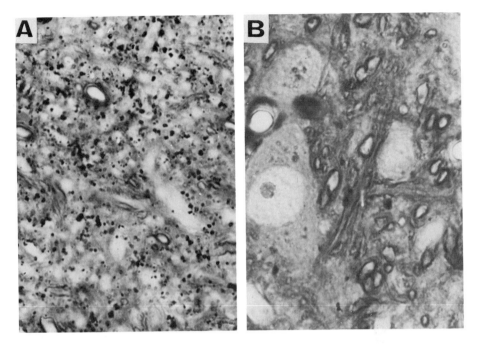

Figure 8. A. Light-microscopic picture of dense terminal degeneration in the ventral part of globus pallidus as a result of an olfactory tubercle lesion 2 days before sacrifice of the animal. B. Absence of terminal degeneration in a nearby region of the medial forebrain bundle. The photographs were taken from a semithin plastic-embedded section, which explains the good definition of cell bodies and neuronal processes.

tions in great detail, the limited amount of tissue used in electron-microscopic studies usually prevents us from identifying the different structures except in general terms like dendrite, axon, or glia profile. It is only occasionally that an electron micrograph or even a serial reconstruction will allow for the tracing of a dendrite back to its cell of origin, which would permit a classification of the neuron. A method developed independently by Stell (1965) and Blackstad (1965) seems, at least for the moment, to offer the greatest potential to overcome this difficulty. The Golgi method which impregnates the cell body and the process of a neuron is ideally suited for the classification of neurons, and Stell and Blackstad showed that meaningful electron-microscopic analysis can be made on previously Golgi-impregnated material. After a neuron had been classified in a successful Golgi preparation, it was possible to trim down a block that contained only one black-impregnated neuron into a small fragment, suitable for ultrathin sectioning. All black nerve processes in ultrathin sections prepared from this fragment, therefore, be-

longed to the neuron that had already been identified by the aid of
the Golgi method in the light microscope (Figure 9). Promising as
this procedure may be, its application is still hampered with technical
difficulties, and the method has been used only in a few instances. For
further details on the method see Blackstad (1970).

Autoradiographic Tracing Technique

When tracing experimentally induced axon degeneration, it is not
always possible to decide if the degenerating fibers belong to cell bodies

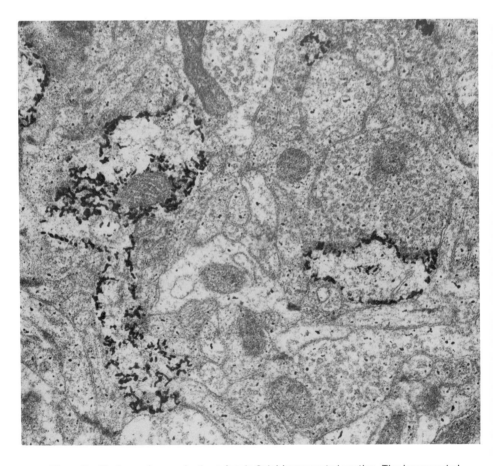

Figure 9. Electron micrograph of previously Golgi-impregnated section. The impregnated
profile to the left is a dendrite, whereas the black profile on the right side of the figure apparently is a spine in synaptic contact with a bouton. [Courtesy of Dr. Theodor Blackstad, Aarhus
Universitet, Aarhus.]

that were destroyed by the lesion or if they constitute the distal parts of fibers of passage destroyed by the lesion. In other words, it may not be possible to determine the origin of a fiber system in a fiber degeneration study. The autoradiographic method, however, makes use of the fact that tritiated amino acids such as proline and leucine injected into the nervous system are incorporated into proteins only by neuronal cell bodies and not by fibers passing through the site of the injection. The labeled protein is subsequently transported in centrifugal direction within the axons (Droz & LeBlond, 1963). By using animals with different survival times following the injection, the entire extent of the fiber system can be selectively demonstrated by light- (Figure 10) and electron-microscopic autoradiography (Lasek, Joseph, & Whitlock, 1968; Schonbach, Schonbach, & Cuenod, 1971; Cowan, Gottlieb, Hendrickson, Price, & Woolsey, 1972; Edwards, 1972; Hendrickson, 1972). Although there are certain difficulties in controlling the extent of the injection area, there is no indication that the usefulness of the autoradiographic technique is limited by extracellular or vascular spread of radioactive material.

Concluding Remarks

Silver impregnation methods and electron microscopy continue to be the methods of choice for many of the problems in the field of experimental neuroanatomy, especially when the interest is centered on the course and termination of fiber pathways. The great popularity of the experimental silver techniques seems justified on several grounds. Whatever the basis for the argyrophilic response may be, it is apparently dependent on neuronal properties of a general nature, which makes the silver methods potentially useful in all areas of the central nervous system. Another great advantage of the reduced silver method is its ability to portray rather accurately the morphological changes taking place in different parts of the degenerating axons. Although it is true that the final deposition of silver particles in the tissue is dependent on many factors and, therefore, can be somewhat erratic, it is also true that the "capriciousness" of the silver techniques is to a large extent a reflection of insufficient mastery. In order to use the increasingly more sophisticated silver methods to full advantage, special attention must be given, not only to the survival time of the animal but also to histotechnical details and problems of interpretation. A detailed and critical review of the neuroanatomic techniques, including the most commonly used silver procedures, was recently published by the Springer Verlag (Nauta & Ebbesson, 1970).

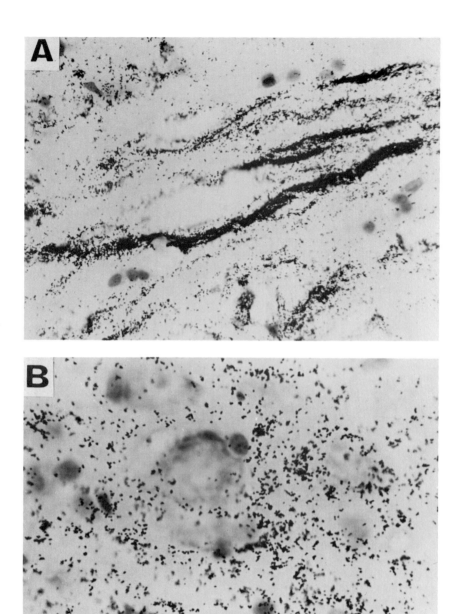

Figure 10. Photomicrographs from sections of the cat brain showing heavily labeled fibers in the brainstem (A), and an accumulation of grains indicating terminal end structures in the main sensory trigeminal nucleus (B). Tritiated leucine was injected into the red nucleus. [Courtesy of Dr. Stephen Edwards, University of Virginia, Charlottesville, Virginia. Reproduced from *Brain Research*, 1972, **48,** with permission from Elsevier Publishing Company, Amsterdam.]

When studying anterograde degeneration, the interest is primarily centered on the course and termination of fiber pathways. Depending on the extent and location of the experimental lesion, valuable data may also be obtained regarding the origin of a pathway. Exact information on this point, however, has in many cases been obtained only by studying the retrograde cellular changes, i.e., the changes occurring in the cell bodies of neurons whose axons have been sectioned. This is best done in newborn or very young animals, in which the neurons are especially sensitive to injury (Brodal, 1940). As indicated earlier, the autoradiographic technique also has a special value in studies designed to determine the exact origin of fiber systems.

Recently another histochemical method for the tracing of brain pathways has been introduced. This method, which is based on histochemical demonstration of horseradish peroxidase (HRP) in the neurons, was first introduced as a means of studying retrograde protein transport (Kristensson & Olsson, 1971; LaVail & LaVail, 1972). Following the injection of HRP into the termination area of a fiber system, the enzyme is apparently taken up by the axon terminals and transported in retrograde direction to the cell body, where it can be visualized in both light (Figure 11) and electron microscopy following relatively simple histochemical procedures. Lynch, Smith, Mensah, and Cotman (1973), on the other hand, recently demonstrated the projection of the mossy fibers following injection of the enzyme into the dentate gyrus of the hippocampus (Figure 12), and these investigators suggest the likelihood of an active uptake of HRP by the cell bodies and an axonal transport in anterograde direction by the aid of physiological mechanisms. Electronmicroscopic studies by Hansson (1973) as well as Nauta, Kaiserman–Abramof, and Lasek (personal communications) do indeed seem to suggest the possibility of an axonal anterograde transport of HRP in membrane-bound organelles following its uptake by the cell bodies. The accumulation of HRP in the axon terminals, however, seem to be granular and not necessarily massive enough to be detected in the light microscope. It is therefore doubtful if the results by Lynch and his co-workers can be explained on this basis. Adams and Warr as well as Scalia and Colman (personal communications), on the other hand, have recently shown that HRP can fill severed axons in both anterograde and retrograde direction. The filling in this case, however, is massive and therefore reminiscent of the results obtained by Lynch and his co-workers as shown in Figure 12. Although there are still many unresolved questions in regard to the uptake and transport of HRP, it is likely that the use of this enzyme or other protein markers will be of great help in future morphological studies of the nervous system.

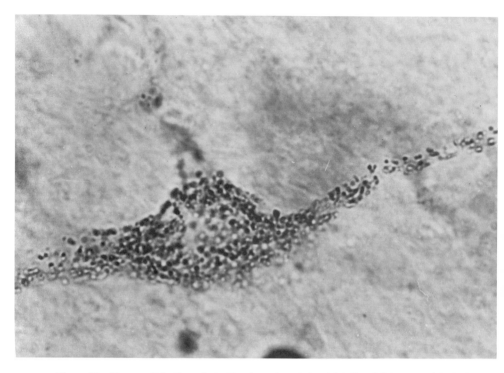

Figure 11. Nerve cell in the substantia nigra of a rat, in which the striatum was injected with horseradish peroxidase 24 hours before sacrifice of the animal. The cell body and the dendrites are filled with peroxidase granules. [Courtesy of Dr. H. J. W. Nauta.]

OPERATIVE PROCEDURES

The use of refined histological methods for the study of degenerative changes in neurons calls for operative techniques capable of inflicting small and well-defined lesions, in particular fiber tracts and cell groups, or in the case of laminar structures, in special cell layers. Surgical procedures such as ablation, suction, or transection do not usually meet these requirements, although they can be employed as a first approach, especially when superficial structures are involved. The need for accurate access to structures deep within the brain also holds true for electrical stimulation or recording and for the injection of chemical substances.

To reach this requirement, the *stereotaxic procedure* is the only method available. The relations between the exterior of the head and the brain can be established on the basis of a three-dimensional coordinate system. Stereotaxic atlases, equipped with these coordinates, are presently avail-

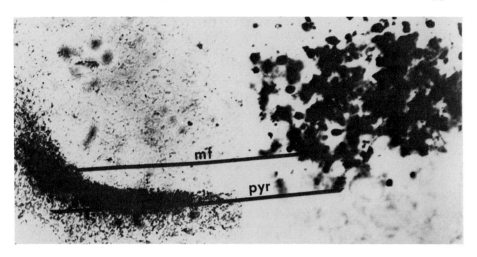

Figure 12. Micrograph showing labeled mossy fiber terminals in the hippocampus follow-ing an injection of horseradish peroxidase into the dentate gyrus of the rat. The panel to the left shows the labeled mossy fiber system (mf) at the base of the dendrites of CA3-pyramids (pyr). The panel to the right demonstrates mossy fiber terminals at higher magnification. [Courtesy of Dr. Gary Lynch, University of California, Irvine. Reproduced from Lynch et al., 1973.]

able for a wide range of animals (see page 97). The use of the stereo-taxic techniques requires an instrument in which the head of the animal can be oriented in the same way as indicated in the stereotaxic atlas. The stereotaxic instruments commercially available are all modifications of the original model developed by Horsley and Clarke in 1908. These instruments usually have the disadvantage that the electrodes must be inserted perpendicular to the horizontal plane, whereas oblique or hori-zontal approaches can only be achieved by the help of phantoms. The latter approaches, however, are desirable in order to avoid electrode tracks in structures that are located dorsal to the target area.

The deficiency in the design of most stereotaxic instruments prompted the development of a new stereotaxic apparatus, in which the electrodes can be inserted into any part of the brain from vertical as well as oblique or horizontal positions (Lohman & Peters, 1973). In this instrument (Figure 13) the electrode holder is carried by an arm, which can be rotated 90° laterally and along which the electrode holder itself can be shifted 30° toward its free end and 50° in the other direction. More-over, the arm with the electrode holder can be moved independently in frontal, lateral, and vertical directions. The apparatus is so constructed that, when all scales have zero readings and the electrode has been adjusted to a certain length, the tip of the electrode and the x and

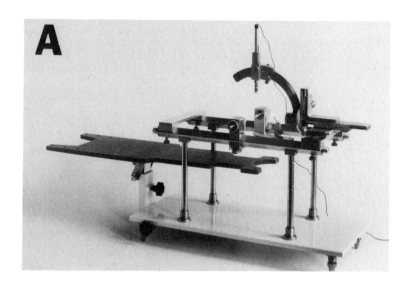

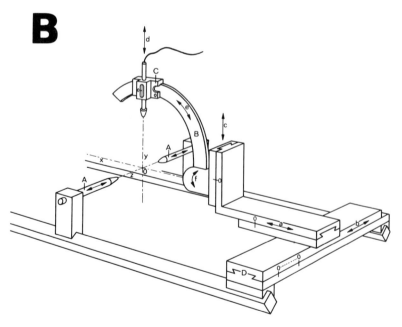

Figure 13. A. The universal stereotaxic instrument developed by Lohman and Peters (1973). B. Drawing of the same stereotaxic instrument showing the three axes of the instrument.

z axes of the instrument intersect each other at point 0. This is defined as the stereotaxic zero point, which is the center of a hemisphere in which the arm carrying the electrode holder forms part of its circumference. By means of the frontal, lateral, and vertical adjustments of the instrument the stereotaxic zero point can be transferred to any point of the brain. By rotating the arm laterally, and shifting the electrode holder along the arm, this point can be reached, not only vertically, but also by oblique or horizontal approaches without further calculations. Stereotaxic instruments based on a similar principle have been constructed by Stellar and Kraus (1954) and Minderhout (1968). The use of their instruments, however, is limited to small laboratory animals such as the rat.

The placement of lesions in stereotaxic experiments is mostly achieved by *unipolar electrolysis,* one pole being connected with an insulated needle electrode in the brain, the other attached to an indifferent electrode on the abdomen or in the anal canal of the animal. The electrode consists of a pointed metallic wire (stainless steel, platinum, nichrome, or tungsten) coated with insulated varnish such as epoxyresin or polysterene except at the tip. The size of the lesion depends not only on the electrode material and the length of the uninsulated electrode tip, but also on the type, quantity, and duration of the current (Carpenter & Whittier, 1952). In our own experimental studies, good results have been obtained by the use of stainless steel electrodes with diameters of .2 and .3 mm and exposed tips of .2–.5 mm. The electrodes have been supplied with either direct anodal current of .5–5 mA or low-power, continuous wave radio-frequency current[*] for 5–15 sec. The latter technique was developed by Hess (1932) and Wyss (1945) and is now routinely used in stereotaxic surgery of experimental animals and man (Sweet, Mark, & Hamlin, 1960).

Before each experiment the electrode should be tested to make sure that there are no bare regions other than at the tip. This can be done by dipping the electrode into a beaker filled with physiological saline. When a potential of a few volts is applied by means of a battery (with the electrode as the cathode), bubbles should be formed only at the tip of the electrode.

Considerable nuisances in stereotaxic surgery are the unintentional damage caused by the electrode tract and the interruption by the lesion of fibers of passage originating outside of the target area. As reported above, the first problem may be obviated by introducing electrodes

---

[*] Model LM4 lesion maker, manufactured by Gross Medical Instruments, Quincy, Massachusetts, U.S.A.

from various directions. Furthermore, the diameter of the electrodes should be as small as admitted by the increasing resistance. Recently, Leonard and Scott (1971) were successful in making lesions in the rat brain by the use of electrodes with diameters ranging from 60 to 90 μm.

To avoid altogether the disruption of intervening tissue by the electrode in the placement of subcortical lesions, the use of *focused ultrasound* has been investigated by Fry and Dunn (1962), Lele (1967), and Warwick and Pond (1968). The latter authors report that they were able to place sharply defined, reproducible lesions from .5 to 3 mm in diameter. Probably due to the high cost of equipment and the complicated technical procedures involved, this technique has not yet been generally applied in experimental neuroanatomical studies.

Structures that are located at the surface of the brain can be destroyed by the so-called *laminar thermocoagulation method* (Dusser de Barenne, 1933a,b), in which heat is applied to the exposed surface of the brain. The apparatus consists of an ordinary electric soldering iron in which the copper tip is replaced by a small flat silver plate of .5 mm thickness (Figure 14). By applying the proper voltage from a Variac, the silver plate is heated to 70° C. This temperature can be determined by placing a piece of Wood's metal (melting point between 68 and 72° C.) on the silver plate. When the plate has reached the desired temperature, it is brought in flat contact with the surface of the brain and, depending on the desired depth of the lesion, heat is applied for a varying number of seconds. A great advantage of the method is that the lesions are localized to the heated area without the added involvement of deeper structures by ischemic necrosis (Figure 15). This is of special value when surgical interference is attempted at structures on the base of the brain, where many blood vessels perforate the surface.

Figure 14. The heat lesion apparatus.

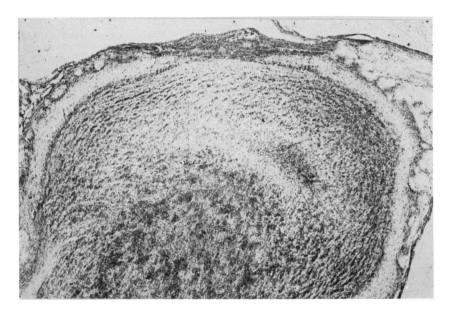

Figure 15.   Heat lesion of the outer layers of the olfactory bulb of a rat.

For a correct placing of heat lesions, the head of the experimental animal must be kept in a fixed position, which can be achieved by mounting the animal in a stereotaxic instrument. This procedure, however, may occasionally damage the soft tissue in the auditory canals or even destroy the eardrums with subsequent impairment of hearing. An alternative way of fixing the head is to tighten two needles to the posterior part of the occipital bone on both sides (Heimer, Kuikka, Larrson, & Nordström, 1971). This fixation technique offers the additional advantage of convenient subtemporal approach to structures on the basal side of the brain, especially if the head is tilted before it is fixed in position by the needles.

ACKNOWLEDGMENTS

This work has been supported by a Research grant (NS 08022- NEUB) from the NINDS and a Career Development Award (5 K2 MH 40, 51604) from the NIH. Drs. Theodor Blackstad, José deOlmos, Stephen Edwards, Kjell Fuxe, Gary Lynch, H. J. W. Nauta, and Bruce Warr have kindly provided figures and valuable suggestions.

STEREOTAXIC ATLASES

A bibliography of stereotaxic brain atlases arranged by class, order, and species (according to A classification of living animals, *by Lord Rothschild (London: Longmans, Green, 1961)).*

Amphibia

Kemali, M. & Braitenberg, V. *Atlas of the frog's brain.* Berlin: Springer-Verlag, 1969.

Birds

Karten, H. J. & Hodos, W. *A stereotaxic atlas of the pigeon (Columba livia).* Baltimore: Johns Hopkins Press, 1967.

Ralph, C. L. & Fraps, R. M. 1959. Long-term effects of diencephalic lesions on the ovary of the hen. *American Journal of Physiology,* 1959, **197,** 1279–1283.

Tienhoven, A. van & Juhásh, L. P. The chicken telencephalon, diencephalon and mensencephalon in stereotaxic coordinates. *Journal of Comparative Neurology,* 1962, **118,** 185–197.

Zweers, G. A. A stereotactic atlas of the brainstem of the Mallard (*Anas platyrhynchos* L.). A stereotactic apparatus for birds and an investigation of the individual variability of some headstructures. Assen: van Gorcum and Company, 1971.

Mammals—marsupials—opossum

Oswaldo-Cruz, E. & Rocha-Miranda, C. E. The diencephalon of the opossum in stereotaxic coordinates. I. The epithalamus and dorsal thalamus. *Journal of Comparative Neurology,* 1967, **129,** 1–38.

Oswaldo-Cruz, E. & Rocha-Miranda, C. E. The diencephalon of the opossum in stereotaxic coordinates. II. The ventral thalamus and hypothalamus. *Journal of Comparative Neurology,* 1967, **129,** 39–48.

Oswaldo-Cruz, E. & Rocha-Miranda, C. E. The brain of the opossum (*Didelphis marsupialis*) in stereotaxic coordinates. Rio de Janeiro: Instituto de Biophisica, Universidade Federal do Rio de Janeiro, 1968.

Salinas, M., G. A. Zeballos & Wang, M. B. A stereotaxic atlas of the opossum brain (*Didelphis virginiana*). *Brain, Behavior, and Evolution,* 1971, **4,** 114–150.

Mammals—primates—tree shrew

Marrocco, R. T., de Valois, R. L. & Boles, J. J. A stereotaxic atlas of the brain of the tree shrew (*Tupaia glis*). *Journal für Hirnforschung,* 1971, **12,** 307–312.

Tigges, J. & Shanta, T. R. *A stereotaxic atlas of the tree shrew (Tupaia glis).* Baltimore: Williams and Wilkins, 1969.

Mammals—primates—marmoset

Saavedra, J. P. & Mazzuchellu, A. L. A stereotaxic atlas of the brain of the marmoset (*Hapale jacchus*). *Journal für Hirnforschung,* 1969, **11,** 105–122.

Mammals—primates—cebus monkey

Eidelberg, E. & Saldias, C. A. A stereotaxic atlas for Cebus monkeys. *Journal of Comparative Neurology,* 1960, **115,** 103–123.

Manocha, S. L., Shanta, T. R. & Bourne, G. H. A stereotaxic atlas of the brain of the Cebus monkey (*Cebus apella*). Oxford: Clarendon Press, 1968.

Mammals—primates—squirrel monkey

Emmers, R. & Akert, K. *A stereotaxic atlas of the brain of the squirrel monkey* (*Saimiri sciureus*). Madison: Univ. of Wisconsin Press, 1963.

Gergen, J. A. & MacLean, P. D. *A stereotaxic atlas of the squirrel monkey's brain* (*Saimiri sciureus*). Bethesda, Maryland: U. S. Public Health Service Publ. No. 933, 1962.

Mammals—primates—macaque

Jackson, W. J. & Savarie, P. J. A comparison of stereotaxic atlases: coordinates for hippocampus in rhesus and stumptailed monkeys. *Lab. Primate Newsletter,* 1970. 9, 9–12.

Kusama, T. & Mabuchi, M. *Stereotaxic atlas of the brain of Macaca fuscata.* Baltimore: University Park Press, 1970.

Olszewski, J. *The thalamus of Macaca mulatta. An atlas for use with the stereotaxic instrument.* Basel: S. Karger, 1952.

Russel, G. V. Stereotaxic atlases. C. Hypothalamic, preoptic and septal regions of the monkey. In D. E. Sheer (Ed.), *Electrical stimulation of the brain.* Austin: Univ. of Texas Press, 1961.

Shanta, T. R., Manocha, S. L. & Bourne, G. H. *A stereotaxic atlas of the Java monkey brain* (*Macaca irus*). Basel: S. Karger, 1968.

Smith, O. A., Castella, K. G. & Randall, D. C. A stereotaxic atlas of the brainstem for Macaca mulatta in the sitting position. *Journal of Comparative Neurology,* 1972, 145, 1–24.

Snider, R. S. & Lee, J. C. *A stereotaxic atlas of the monkey brain* (*Macaca mulatta*). Chicago: Univ. of Chicago Press, 1961.

Winters, W. D., Kado, R. T. & Adey, W. R. *A stereotaxic brain atlas for Macaca nemestrina.* Berkeley: Univ. of California Press, 1969.

Mammals—primates—baboon

Davis, R. and Huffman, R. D. *A stereotaxic atlas of the brain of the baboon* (*Papio*). Austin: Univ. of Texas Press, 1968.

Riche, D., Christolomme, A., Bert, J. & Naquet, R. *Atlas stéréotaxique du cerveau de babouin* (*Papio papio*). Paris: Editions de Centre National de la Recherche Scientifique, 1968.

Mammals—primates—chimpanzee

De Lucchi, M. R., Dennis, B. J. & Adey, W. R. *A stereotaxic atlas of the chimpanzee brain* (*Pan satyrus*). Berkeley: Univ. of California Press, 1965.

Mammals—lagomorphs—rabbit

Fifková, E. & Marsala, J. Stereotaxic atlas for the rabbit. In J. Bures, M. Petrán & J. Zachar (Eds.), *Electrophysiological methods in biological research.* Prague: Publishing house of the Czechoslovak Academy of Sciences, 1960.

Matricali, B. A new stereotaxic coordinates system for the rabbit's brain stem. Thesis, University of Leyden, 1961.

McBride, R. L. & Klemm, W. R. Stereotaxic atlas of rabbit brain, based on the rapid method of photography of frozen unstained sections. *Comm. Behav. Biol.*, 1968, Part A, **2**, 179–215.

Monnier, M. & Gangloff, H. *Atlas for stereotaxic brain research on the conscious rabbit.* Amsterdam: Elsevier, 1961.

Sawyer, C. H., Everett, J. W. & Green, J. D. The rabbit diencephalon in stereotaxic coordinates. *Journal of Comparative Neurology*, 1954, **101**, 801–824.

Urban, J. & Richard, P. H. A stereotaxic atlas of the New Zealand rabbit's brain. Springfield, Illinois: Thomas, 1972.

Verley, R., & Siou, G. Relations spatiales de quelques structures diencéphaliques et mésencéphaliques chez le lapin noveau-né. *Journal de Physiologie*, 1957, **59**, 257–279.

Mammals—rodents—deermouse

Eleftheriou, B. E. & Zolovick, A. J. The forebrain of the deermouse in stereotaxic coordinates. Kansas. Technical bulletin 146 of the Kansas State Univ. of Agriculture & Applied Science, 1965.

Mammals—rodents—prairie dog

Carlson, R. H. & Kott, J. N. The forebrain of the prairie dog (*Gnomys ludovicianus*) in stereotaxic coordinates. *Acta Anatomic*, 1970, **77**, 321–340.

Mammals—rodents—golden hamster

Knigge, K. M. & Joseph, S. A. A stereotaxic atlas of the brain of the golden hamster. In R. A. Hoffman, P. F. Robinson & H. Magalhaes (Eds.), *The golden hamster. Its biology and use in medical research.* Ames: Iowa State Univ. Press, 1968.

Smith, O. A., Jr., & Bodemer, C. N. A stereotaxic atlas of the brain of the golden hamster (*Mesocricetus auratus*). *Journal of Comparative Neurology*, 1963, **120**, 53–63.

Mammals—rodents—rat

Albe-Fessard, D., Stutinsky, F. & Libouban, S. *Atlas stéréotaxique du diencephale du rat blanc.* Paris: Editions du Centre National de la Recherche Scientifique, 1966.

Bernardis, L. L. Stereotaxic localization of amygdaloid nuclei in rats from weaning to adulthood. *Experientia*, 1967, **23**, 158–160.

Bernardis, L. L. & Skelton, F. R. Stereotaxic localization of supraoptic, ventromedial and mamillary nuclei in the hypothalamus of weanling to mature rats. *American Journal of Anatomy*, 1965, **116**, 69–74.

Fifková, E. & Marsala, J. Stereotaxic atlas for the rat. In J. Bures, M. Petrán & J. Zachar (Eds.), *Electrophysiological methods in biological research.* Prague: Publishing House for the Czechoslovak Academy of Sciences, 1960.

De Groot, J. *The rat forebrain in stereotaxic coordinates.* Amsterdam: Noord-Hollandse Uitgevers Maatschappij, 1959.

De Groot, J. The rat hypothalamus in stereotaxic coordinates. *Journal of Comparative Neurology,* 1959, 113, 389–400.

Hurt, G. A., Hanaway, J. & Netsky, M. G. Stereotaxic atlas of the mesencephalon in the albino rat. *Confin. Neurol.,* 33, 93–115.

König, J. F. R. & Klippel, R. A. *The rat brain. A stereotaxic atlas of the forebrain and lower parts of the brain stem.* Baltimore: Williams and Wilkins, 1963.

Krieg, W. J. S. Accurate placement of minute lesions in the brain of the albino rat. *Quarterly Bulletin Northwestern University Medical School,* 1946, 20, 199–208.

Massopust, L. C. Jr. Stereotaxic atlases. A. Diencephalon of the rat. In D. Sheer (Ed.), *Electrical stimulation of the brain.* Austin: University of Texas Press, 1961.

Pellegrino, L. J. & Cushman, A. J. *A stereotaxic atlas of the rat brain.* New York: Appleton, 1967.

Sherwood, N. M. & Timiras, P. S. *A stereotaxic atlas of the developing rat brain.* Berkeley: Univ. of California Press, 1970.

Skinner, J. E. Stereotaxic atlas of the rat brain. In J. E. Skinner (Ed.), *Neuroscience: a laboratory manual.* Philadelphia: Saunders, 1971.

Szentágothai, J. Cytoarchitectonic atlas of the rat brain in Horsley-Clarke co-ordinates. In J. Szentagothai, B. Flerkó, B. Mess & B. Hálasz (Eds.), *Hypothalamic control of the anterior pituitary.* Budapest: Akadémiai Kiadó, 1962.

Wünscher, W., Schober, W. & Werner, L. *Architektonischer Atlas von Hirnstamm der Ratte.* Leipzig: S. Hirzel Verlag, 1965.

Mammals—rodents—mouse

Montemurro, D. G. & Dukelow, R. H. A stereotaxic atlas of the diencephalon and related structures of the mouse. Mount Kisco, New York: Futura Publishing Co., 1972.

Mammals—rodents—gerbil

Thiessen, D. D. & Goar, S. Stereotaxic atlas of the hypothalamus of the mongolian Gerbil (*Meriones unguiculatus*). *Journal of Comparative Neurology,* 1970, 140, 123–128.

Mammals—rodents—guinea pig

Benson, B. & Ward, J. W. The hypothalamus of the adult guinea pig in stereotaxic coordinates. *Journal of the Tennessee Academy of Science,* 1971, 463, 19–27.

Blobel, R., Gonzalo, L. & Schuchardt, E. Stereotaktisches Verfahren zur Lokalisation der Zwischenhirnkerne beim Meerschweinchen. *Endokrinologie,* 1960, 39, 167–172.

Luparello, T. J. *Stereotaxic atlas of the forebrain of the guinea pig.* Basel: S. Karger, 1967.

Luparello, T. J., Stein, M. & Park, C. D. A stereotaxic atlas of the hypothalamus of the guinea pig. *Journal of Comparative Neurology,* 1964, 122, 201–218.

Morrison, R. G. B. *A stereotaxic atlas of the forebrain of the guinea pig.* Adelaide: South Australian Education Department, 1972.

Rössner, W. *Stereotaktischer Hirnatlas vom Meerschweinchen.* Lochham bei München: Pallas Verlag, 1965.

Tindal, J. S. The forebrain of the guinea pig in stereotaxic coordinates. *Journal of Comparative Neurology,* 1956, **124,** 259–266.

Mammals—carnivores—dog

Adrianov, O. S. & Mering, T. A. *Atlas of the canine brain.* Ann Arbor: Univ. of Michigan Press, 1964.

Dua-Sharma, S., Sharma, K. N. & Jacobs, H. L. *The canine brain in stereotaxic coordinates.* Cambridge, Massachusetts: M. I. T. Press, 1970.

Lim, R. K. S., Liu, C. & Moffit, R. L. *A stereotaxic atlas of the dog's brain.* Springfield, Illinois: Thomas, 1960.

Vakolius, N. I. Stereotaxic coordinates of subcortical nuclei in the dog brain. *Neirofiziologiia,* 1969, **1,** 331–335.

Mammals—carnivores—cat

Berman, A. L. *The brain stem of the cat. A cytoarchitectonic atlas with stereotaxic coordinates.* Madison: Univ. of Wisconsin Press, 1968.

Bleier, R. *The hypothalamus of the cat. A cytoarchitectonic atlas with Horsley-Clarke co-ordinates.* Baltimore: Johns Hopkins Press, 1961.

Fifkova, E. & Marsala, J. Stereotaxic atlas of the cat. In J. Bures, M. Petrán & J. Zachar (Eds.), *Electrophysiological methods in biological research.* Prague: Publishing House of the Czechoslovak Academy of Sciences, 1960.

Ingram, W. R., Hannet, F. I. & Ranson, S. W. The topography of the nuclei of the diencephalon of the cat. *Journal of Comparative Neurology,* 1932, **55,** 333–394.

Jaspar, H. H. & Ajmone-Marsan, C. A. *A stereotaxic atlas of the diencephalon of the cat.* 2nd ed. Ottawa: National Research Council of Canada, 1960.

Jaspar, H. H. & Ajmone-Marsan, C. A. Stereotaxic atlases. B. Diencephalon of the cat. In D. E. Sheer, *Electrical stimulation of the brain.* Austin: Univ. of Texas Press, 1961.

Jiménez-Castellanos, J. Thalamus of the cat in Horsley-Clarke coordinates. *Journal of Comparative Neurology,* 1949, **91,** 307–330.

Reinoso-Suárez, F. *Topographischer Hirnatlas der Katze fur experimental-physiologische Untersuchungen.* Darmstadt: E. Merck A. G., 1961.

Rose, G. H. & E. F. Goodfellow. *A stereotaxic atlas of the kitten brain: Coordinates of 104 selected structures.* Los Angeles: UCLA Brain Information Service, 1972.

Snider, R. S. & Niemer, W. T. *A stereotaxic atlas of the cat brain.* Chicago: Univ. of Chicago Press, 1961.

Mammals—artiodactyls

Richard, P. *Atlas stéréotaxique du cerveau de brebis Préalpes-du-Sud.* Paris: Institut National de la Recherche Agronomique, 1967.

Sing, K., Singh, K., Soni, B. K. & Manchanda, S. K. Brain of the buffalo (*Bubalis babalis*) in stereotaxic coordinates. *Indian Journal of Physiology and Pharmacology*, 1972, **16**, 127–138.

Tindal, J. S., Knaggs, G. S. & Turvey, A. The forebrain of the goat in stereotaxic coordinates. *Journal of Anatomy* (Lond.), 1968, **103**, 457–469.

Welento, J., Szteyn, S. & Millart, Z. Observation on the stereotaxic configuration of the hypothalamus nuclei in the sheep. *Anatomischer Anzeiger*, 1969, 124, 1–27.

REFERENCES

Bielschowsky, M. Die Silberimprägnation der Neurofibrillen. *Journal für Psychologie und Neurologie*, 1904, 3, 169–188.

Blackstad, T. W. Mapping of experimental axon degeneration by electron microscopy of Golgi preparations. *Z. Zellforsch.*, 1965, **67**, 819–834.

Blackstad, T. W. Electron microscopy of Golgi preparations for the study of neuronal relations. In W. J. H. Nauta & S. O. E. Ebbesson (Eds.), *Comtemporary research methods in neuroanatomy*. Berlin: Springer-Verlag, 1970.

Bodian, D. A new method for staining nerve fibres and nerve endings in mounted paraffin sections. *Anatomical Record*, 1936, **65**, 89–97.

Brodal, A. Modification of Gudden method for study of cerebral localization. *Archives of Neurology and Psychiatry*, 1940, **43**, 46–58.

Busch, H. F. M. *An anatomical analysis of the white matter in the brain stem of the cat.* Thesis: Leiden, 1961.

Cajal, S. Ramón y. Quelques methods de coloration des cylindres axes, des neurofibrilles et des nids nerveux. *Travaux de Laboratoire de Recherches Biologiques*, 1904, 3, 1–7.

Cammermeyer, J. An evaluation of the significance of the "dark" neuron. *Ergebuisse der Anatomie und Entwicklungsgeschichte*, 1962, **36**, 1–61.

Carpenter, M. B. & Whittier, J. R. Study of methods for producing experimental lesions of the central nervous system with special reference to stereotaxic technique. *Journal of Comparative Neurology*, 1952, **93**, 73–132.

Colonnier, M. Experimental degeneration in the cerebral cortex. *Journal of Anatomy* (Lond.), 1964, **98**, 47–53.

Conn, H. J., Darrow, M. A. & Emmel, V. M. *Staining procedures.* 2nd ed. Baltimore: Williams and Wilkins, 1960.

Cowan, W. M., Gottlieb, D. I., Hendrickson, A. E., Price, J. L. & Woolsey, T. L. The autoradiographic demonstration of axonal connections in the central nervous system. *Brain Research*, 1972, **37**, 21–51.

De Olmos, J. S. A cupric-silver method for impregnation of terminal axon degeneration and its further use in staining granular argyrophilic neurons. *Brain, Behaviour and Evolution*, 1969, **2**, 213–237.

Dieckmann, G., Gabriel, E. & Hassler, R. Size, form and structural peculiarities of experimental brain lesions obtained by thermo-controlled radiofrequency. *Confinia Neurologica*, 1965, **26**, 134–142.

Droz, B. & Leblond, C. P. Axonal migration of proteins in the central nervous system and peripheral nerves as shown by radioautography. *Journal of Comparative Neurology*, 1963, **121**, 325–346.

Dusser de Barenne, J. G. Laminar destruction of the nerve cells of the cerebral cortex. *Science*, 1933, **77**, 546–547. (a)

Dusser de Barenne, J. G. Selektive abtötung der Nervenzellschichten zur Groszhirnrinde. Die Methode der laminaren Thermokoagulation der Rinde. *Zeitschrift für die gesamte Neurologie und Psychiatrie*, 1933, **147**, 280–290. (b)

Eager, R. P. Selective staining of degenerative axons in the central nervous system by a simplified silver method: spinal cord projections to extrenal cuneate and inferior olivery nuclei in the cat. *Brain Research*, 1970, **22**, 137–141.

Ebbesson, S. O. E. The selective silver-impregnation of degenerating axons and their synaptic endings in non-mammalian species. In W. J. H. Nauta & S. O. E. Ebbesson (Eds.), *Contemporary research methods in neuroanatomy*. Berlin: Springer-Verlag, 1970.

Edwards, S. B. The ascending and descending projections of the red nucleus in the cat: an experimental study using an autoradiographic tracing method. *Brain Research*, 1972, **48**, 45–63.

Einarson, L. A method for progressive selective staining of Nissl and nuclear substance in nerve cells. *American Journal of Pathology*, 1932, **8**, 295–307.

Falck, B., Hillarp, N. Å., Thieme, G. & Thorp, A. Fluorescence of catecholamines and related compounds condensed with formaldehyde. *Journal of Histochemistry and Cytochemistry*, 1962, **10**, 348–354.

Fink, R. P. & Heimer, L. Two methods for selective silver impregnation of degenerating axons and their synaptic endings in the central nervous system. *Brain Research*, 1967, **4**, 369–374.

Fromman, C. Uber die Farbung der Binde- und Nervensubstanz des Ruckenmarkes durch Argentum nitricum und über die Struktur der Nervenzellen. *Virchows Archiv für pathologische Anatomie*, 1864, **31**, 129–153.

Fry, W. J. and Dunn, F. Ultrasound: Analysis and experimental methods in biological research. In W. L. Nastuk (Ed.), *Physical techniques in biological research*. Vol. 4. New York: Academic Press, 1962.

Fuxe, K., Hökfelt, T., Jonsson, G. & Ungerstedt, U. Fluorescence microscopy in neuroanatomy. In W. J. H. Nauta & S. O. E. Ebbesson (Eds.), *Contemporary research methods in neuroanatomy*. Berlin: Springer-Verlag, 1970. Pp. 275–314.

Gerlach, J. *Mikroskopische Studien aus dem Gebiet der menschlichen Morphologie*. Erlangen, 1858.

Glees, P. Terminal degeneration within the central nervous system as studied by a new silver method. *Journal of Neuropathology and Experimental Neurology*, 1946, **5**, 54–59.

Golgi, C. Sultra struttura della sostanza grigia dell cervello. *Gazetta Medica Italiana Lombardia*, 1873, **33**, 244–246.

Hansson, H-A. Uptake and intracellular bidirectional transport of horseradish peroxidase in retinal ganglion cells. *Experimental Eye Research*, 1973, **16**, 377–388.

Heimer, L. Silver impregnation of degenerating axons and their terminals on Epon-Araldite sections. *Brain Research*, 1969, **12**, 246–249.

Heimer, L. Selective silver-impregnation of degenerating axoplasm. In W. J. H. Nauta & S. O. E. Ebbesson (Eds.), *Contemporary research method in neuroanatomy*. Berlin: Springer-Verlag. 1970. (a)

Heimer, L. Bridging the gap between light and electron microscopy in the experimental tracing of fiber connections. In W. J. H. Nauta & S. O. E. Ebbesson (Eds.), *Contemporary research method in neuroanatomy*. Berlin: Springer-Verlag, 1970. (b)

Heimer, L., Kuikka, V., Larsson, K. & Nordström, E. A headholder for stereotaxic operations of small laboratory animals. *Physiology and Behavior,* 1971, **7**, 263–264.

Heimer, L. & Peters, A. An electron-microscopic study of a silver stain for degenerating boutons. *Brain Research,* 1968, **8**, 337–346.

Heimer, L. & Wall, P. D. The dorsal root distribution to the substantia gelatinosa of the rat with a note on the distribution in the cat. *Experimental Brain Research,* 1968, **6**, 89–99.

Hendrickson, A. E. Electron microscopic distribution of axoplasmic transport. *Journal of Comparative Neurology,* 1972, **144**, 381–398.

Hess, W. R. *Beiträge zur Physiologie des Hirnstammes. i. Die Methodik der lokalisierten Reizung und Ausschaltung subkortikaler Hirnabschnitte.* Leipzig: G. Thieme, 1932.

Horsley, V. & Clarke, R. H. The structure and functions of the cerebellum examined by a new method. *Brain,* 1908, **31**, 45–124.

Klüver, H. & Barrera, E. A method for the combined staining of cells and fibers in the nervous system. *Journal of Neuropathology and Experimental Neurology,* 1953, **12**, 400–403.

Koening, H., Groat, R. A. & Windle, W. F. A physiological approach to perfusion-fixation of tissues with formalin. *Stain Technology,* 1945, **20**, 13–21.

Kristensson, K. & Olsson, Y. Retrograde axonal transport of protein. *Brain Research,* 1971, **29**, 363–365.

Lasek, R., Joseph, B. S. & Whitlock, D. G. Evaluation of a radioautographic neuroanatomical tracing method. *Brain Research,* 1968, **8**, 319–336.

LaVail, J. H. & LaVail, M. M. Retrograde axonal transport in the central nervous system. *Science,* 1972, **176**, 1415–1417.

Lele, P. P. Production of deep focal lesions by focused ultrasound-current status. *Ultrasonics,* 1967, **5**, 105–112.

Leonard, C. M. & Scott, J. W. Origin and distribution of the amygdalofugal pathways in the rat: an experimental neuroanatomical study. *Journal of Comparative Neurology,* 1971, **141**, 313–330.

Loewy, A. D. Ammoniacal silver staining of degenerating axons. *Acta Neuropathology,* 1969, **14**, 226–236.

Lohman, A. H. M. & Peters, K. A. A new stereotaxic instrument. *Brain Research,* 1973. (in press)

Loyez, M. Coloration des fibers nerveuses par la methode a l'hemotoxyline au fer apres inclusion à la celloidine. *Comptes Rendus des Seances de la Société de Biologie et de ses Filiales,* 1920, **62**, 511.

Ludwig, E. & Klingler, J. *Atlas cerebri humani.* Basel: S. Karger, 1956.

Lynch, G., Smith, R. L., Mensah, P. & Cotman, C. Tracing the Deutate gyrus–mossy fiber system with horseradish peroxidase histochemistry. *Experimental Neurology,* 1973.

Marchi, V. & Algeri, G. Sulle degenerazioni discendenti consecutive a lesioni sperimentale in diverse zone della corteccia cerebrale. *Rivista Sperimentale di Freniatria e di Medicina legale,* 1885, **11**, 492–494.

Minderhoud, J. M. An all-round apparatus for stereotaxic operations on small test animals (albino rat). *Brain Research,* 1968, **9**, 380–384.

Nauta, W. J. H. Uber die sogennante terminale Degeneration im Zentralnervensystem und ihre Darstellung durch Silberimprägnation. *Archives of Neurology and Psychiatry,* 1950, **66**, 353–376.

Nauta, W. J. H. & Ebbesson, S. O. E. (Eds.) *Contemporary research methods in neuroanatomy.* Berlin: Springer-Verlag, 1970.

Nissel, F. Uber die Untersuchungsmethoden der Grosshirnrinde. *Neurologisches Zentralblaft*, 1885, **4**, 500–501.

Pal, J. Ein Beitrag zur Nervenfarbetechnik. *Z. Wiss. Mikr.*, 1887, **4**, 92–96.

Peters, A. The fixation of central nervous tissue and the analysis of electron micrographs of the neurophil, with special reference to the cerebral cortex. In W. J. H. Nauta & S. O. E. Ebbesson (Eds.), *Contemporary research methods in neuroanatomy*. Berlin: Springer-Verlag, 1970.

Powers, M. M. & Clark, G. An evaluation of cresyl echt violet acetate as a Nissl stain. *Stain Technology*, 1950, **30**, 83–88.

Ramón-Moliner, E. The Golgi-Cox technique. In W. J. H. Nauta & S. O. E. Ebbesson (Eds.), *Contemporary research methods in neuroanatomy*. Berlin: Springer-Verlag, 1970.

Romanes, G. J. The staining of nerve fibres in paraffin sections with silver. *Journal of Anatomy* (Lond.), 1950, **84**, 104–115.

Rose, G. & Goodfellow, E. A stereotaxic atlas of the kitten brain. Los Angeles: UCLA Center for the Health Sciences, 1974.

Schneider, G. E. Two visual systems. *Science*, 1969, **163**, 895–902.

Schonbach, J., Schonbach, C. H. & Cuenod, M. Rapid phase of axoplasmic flow and synaptic proteins: an electron microscopical study. *Journal of Comparative Neurology*, 1971, **141**, 485–498.

Stell, W. K. Correlations of retinal cytoarchitecture and ultrastructure in Golgi preparations. *Anatomical Record*, 1965, **153**, 389–397.

Stellar, E. & Krause, N. P. New stereotaxic instrument for use with the rat. *Science*, 1954, **120**, 664–666.

Sweet, W. H., Mark, V. H. & Hamlin, H. Radiofrequency lesions in the central nervous system of man and cat. *Journal of Neurosurgery*, 1960, **17**, 213–225.

Valverde, F. The Golgi method. A tool for comparative structural analysis. In W. J. H. Nauta & S. O. E. Ebbesson (Eds.), *Contemporary research methods in neuroanatomy*. Berlin: Springer-Verlag, 1970.

Velayos, J. L. & Lizarraga, M. F. A simplified silver impregnation method. *Experientia*, 1973, **29**, 135–136.

Voneida, T. J. & Trevarthen, C. B. An experimental study of transcallosal connections between the proreus gyri of the cat. *Brain Research*, 1969, **12**, 384–395.

Waller, A. Experiments on the section of the glossopharyngeal and hypoglossal nerves of the frog, and observations of the alternations produced thereby in the structure of their primitive fibers. *Philosophical Transactions*, 1850, **140**, 423–469.

Warwick, R. & Pond, J. Trackless lesions in nervous tissues produced by high intensity focused ultrasound (high-frequency mechanical waves). *Journal of Anatomy* (Lond.), 1968, **102**, 387–405.

Weigert, K. Ausfurliche Beschreibung der in No. 2 dieser Zeitschrift erwähnten neuen färbungsmethode für das Zentralnervensystem. *Fortschrifte der Medizin*, 1884, **2**, 190–191.

Weil, A. A. A rapid method for staining myelin sheaths. *Archives of Neurology and Psychiatry* (Chicago), 1928, **20**, 392–393.

Wiitanen, J. T. Selective silver impregnation of degenerating axons and axon terminals in the central nervous system of the monkey (*Macaca mulatta*). *Brain Research*, 1969, **14**, 546–548.

Woelcke, M. Eine neue Methode der Markscheidenfärbung. *Journal of Physiology and Neurology*, 1942, **51**, 199–202.

Wyss, O. A. M. Ein Hochfrequenz-Koagulationsgerat zur reizlosen Ausschaltung. *Helvetica Physiologica et Pharmacologica Acta*, 1945, **3**, 437–443.

Chapter 4

The Perfectible Brain: Principles of Neuronal Development

HELMUT V. B. HIRSCH

State University of New York at Albany

MARCUS JACOBSON

University of Miami Medical School

One must remember, however, that it has yet to be demonstrated that the changes in circuit morphology affected by experience consist of more than an enhancement, maintenance, or neglect of connections that already are basically patterned by selective growth. [Sperry, 1971, p. 39]

It is possible that . . . the mode of differentiation of the postnatally formed microneurons is dependent upon stimulus input and feedback from the environment. [Altman, 1970, p. 236]

INTRODUCTION: INVARIANT AND VARIABLE
COMPONENTS OF THE NERVOUS SYSTEM

Some behavior patterns reflect the individual organism's past history, while other actions are highly stereotyped and characteristic of all members of the species. Developmental processes, thus, must provide for the formation of a system capable of guiding behavior with a high degree of constancy from one member of a species to the next, without sacrificing the flexibility of response in the individual that is necessary for survival in a capricious, constantly changing environment. In this chapter we attempt to outline how this dual requirement might be met.

We shall demonstrate that the nerve circuits found in adult animals are the product of a long program of development, initiated at the time of fertilization of the ovum and progressing through a series of carefully controlled steps. The end product in this sequence, in all species that have been studied anatomically, is a highly ordered and structured system. This evidence indicates that the nervous system contains a structural framework whose interconnections are highly invariant for all of the members of a given species. By invariant we will be referring to a range of situations. At one extreme, in simple organisms such as *Daphnia* (Macagno *et al.*, 1973) or the leech (Purves & McMahan, 1972; Stuart, 1970) the location of cell bodies and the main branching patterns and synaptic connections of nerve fibers are very similar from one individual to another, although even here the very fine structural details show variability. At the other end, in the vertebrate nervous system, comparable anatomical studies have not been carried out. Nonetheless, many of the main interconnections, for example, those linking the retina with visual centers such as the lateral geniculate body and visual cortex, appear to be precisely and invariantly established (Guillery & Kaas, 1971).

Contained in this invariant framework are a very large number of nerve cells which, by virtue of their diverse shapes and arrangements within the nervous system, suggest variability in their pattern of connectivity. These neurons typically do not have a unique shape; although they have recognizable features, the detailed pattern of their axons and dendrites is quite variable.

Now, hopefully without imposing an artificial construct on the nervous system, we will compare its invariant and variable anatomical and functional components. This will permit us to contrast the intrinsic genetic controls of neuronal development with the extrinsic and, thus, highly variable controls imposed by the individual organism's environment and personal experience. Establishing an antithesis between *nature* and *nurture* in this manner can serve as a useful point of departure. However, we should not forget that these are stereotypes which are but the endpoints of a continuum.

When we consider neuronal variability and invariance at different stages of development it will become evident that much remains to be learned. For example, we do not know when during ontogeny the cells which are destined to become the variable components of the nervous system are formed, or at what subsequent stage they are affected by conditions prevailing in the organism's environment. Some of these cells arise before any environmental influences can become active, others are known to develop postnatally. Environmental forces may thus affect either the formation of variable components of the nervous system, or, alternatively, modify or even delete existing neurons. Before any of these points can be clarified we must refine our research strategies, both to identify variable neurons and to quantify the extent to which the conditions in the animal's environment affect or modify them.

There is always a temptation in considering broad theoretical problems such as the nature–nurture controversy to generalize too freely, forgetting that what is applicable to the mouse may not necessarily be true for man. Morphological and physiological evidence make it likely that the invariant features of neuronal development and anatomy are the same in all vertebrate species. Thus, a structure such as the cerebellum has a developmental history and a final organization, consisting of the same cell types interconnected in similar fashion, in the mouse and in man. On the other hand, we are on much less certain grounds when we attempt to assess how universal is the role of *variable* components of the nervous system. This is because it is easier to observe the invariance of a given anatomical structure than it is to assay its variability. Furthermore, invariant components, particularly on a microscopic level, appear much more prominently and strike one more vividly than do variable ones. For that reason the latter, to a considerable extent, have been ignored. Finally, anatomical methodology itself has imposed what might be viewed as an artifact upon the subject; neurobiologists have been so intent upon the systemization of their subject that they have overlooked the variable aspects in favor of those which are invariant and, therefore, much easier to classify and place into rigid taxonomies.

Having entered the age of overassurance in which neurobiologists confidently assert that at least some components of the nervous system are invariantly constructed,* we should now be in a position to embark

---

* From various *obiter dicta* we select the following two: According to Brodal (1969, p. 684), "In the anatomical organization of the central nervous system there is an extremely high degree of order. This is true even with regard to the most minute structural features." Similarly, Palay (1967, p. 31) states that the organization of the nervous system "is highly specific, not merely in terms of the connections between particular neurons, but also in terms of the number, style, and location of terminals upon different parts of the same cell and the precise distribution of terminals arising from that cell."

on a critical reassessment of the evidence for invariance. In this chapter we will sketch out the evidence for the control of the invariant aspects of neuronal development and explore some of the evidence for variability. We then venture to outline how both of these together play a role in the functioning of the nervous system.

TWO NEURONAL ARCHETYPES: MACRONEURONS AND MICRONEURONS

Anatomists have long been struck by the observation that there appear to be two main classes of neurons (Golgi, 1886). In the first category are cells with apparently invariant morphology; these are the large principal neurons of each region of the brain, whose long axons constitute

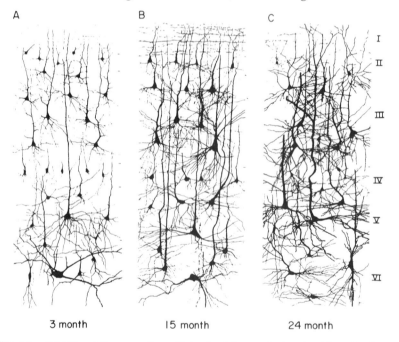

3 month   15 month   24 month

Figure 1. Golgi Type I Neurons: Pyramidal cells of the cerebral cortex. These tracings are made from Golgi-impregnated sections taken from the region of the superior temporal gyrus in human children aged 3, 15, and 24 months. Some cells in each of the cortical layers (I–VI) have dendritic processes extending up to the outermost layer (Layer I) of the cortex. Axons of these neurons enter white matter below the cortex, thus conducting activity outside the region containing the cell body. In motor cortex, some large pyramidal cells (Betz Cells) send their axons down into the spinal cord. These are the longest axons found in the mammalian nervous system. The characteristics of *Golgi Type I Neurons* make them ideally suited for *collecting* and *transmitting* information in the nervous system. Note also in this figure the changes in arborization of neurons and density of dendritic spines with increasing age. [From the work of Conel: (A) Conel, 1947; (B) Conel, 1955; (C) Conel, 1959.]

the main nerve-fiber tracts. They have been called Golgi Type I neurons (Golgi, 1886), macroneurons (Altman, 1967), or Class I neurons (Jacobson, 1970a,b, 1973) (Figure 1). The second category consists of highly variable interneurons. They are present in all parts of the nervous system, but are particularly prominent in areas of the brain such as the cerebral cortex that are known to be involved in modifiable behavior. Cells of this type have been called Golgi Type II neurons (Golgi, 1886), neurons with short axons (Ramón y Cajal, 1909–1911), microneurons (Altman, 1967), or Class II neurons (Jacobson, 1970a,b, 1973) (Figure 2).

These two classes of neurons differ not only in their morphology but also in their ontogeny. As a rule, the large neurons with long axons (macroneurons) are formed first in each part of the brain, and the small neurons with short axons are formed later. This generalization applies to all parts of the nervous system of vertebrates (Jacobson, 1970b) as well as to the parts of the invertebrate nervous system that are mainly concerned with variable or modifiable behavior, such as the corpora pedunculata of insects (Bullock and Horridge, 1965) and the vertical lobes and optic lobes of the octopus (Young, 1964; Giuditta et al., 1971).

The fact that these two classes of neurons arise at different times in development is further evidence for a functional difference between them. The large neurons originate and complete their development at a time when the embryo is still protected from the vagaries of the environment. Their development and differentiation thus appear to be largely or completely controlled by intrinsic ontogenetic mechanisms. In contrast, the small cells originate and undergo their differentiation at a later stage, often postnatally, when the embryo is exposed to environ-

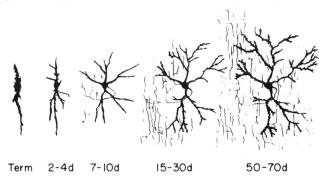

Term   2-4d   7-10d   15-30d   50-70d

Figure 2. Golgi Type II Neurons: Granule cells of the cerebral cortex. These tracings are made from Golgi-impregnated sections from the cortex of kittens of different ages. Note that the dendrites of these cells as well as their axons branch extensively within a fairly restricted neighborhood about the cell body. The geometry of these *Golgi Type II Neurons* would make them well-suited for the *processing* of information in the nervous system. [From Scheibel, 1962.]

mental influences. The suggestion has therefore been offered that the small variable cells may be responsive to environmental influences and thus responsible for the plastic or modifiable aspects of behavior (Ramón y Cajal, 1909–1911; Altman, 1967; Jacobson, 1969, 1973). In other words, the final step in postnatal differentiation of these cells may, at least in part, be regulated by conditions prevailing in the organism's environment. Such exogenous influences must operate within the constraints imposed by intrinsic ontogenetic mechanisms. Unfortunately there is little evidence to support the idea that environmental factors play a role in the ontogeny of these microneurons. Until variability can be assayed in a more quantitative fashion it will be difficult to determine whether experience alters the diversity of any nerve cell population. Even for the macroneurons, whose gross morphological invariance is easy to recognize, there is no reason why some parts of these cells could not also possess a degree of structural variability. Thus, in fact, there may be a range of various cell types in which variability is present in different proportions.

Comparison of these two classes of cells, using other than anatomical techniques, is very difficult. While electrophysiological procedures for studying the response characteristics of single neuronal elements are becoming extremely sophisticated, there is no reliable method for ascertaining whether the cell being studied is a macroneuron or a microneuron. Anatomical identification of large neurons can be made by intracellular injection of dye and subsequent histological examination. This procedure, however, is virtually impossible to implement in the case of small microneurons, making electrophysiological comparisons between the two classes of cells at the present time technically extremely difficult. [Recent results have, however, been promising (Van Essen & Kelly, 1973).] This greatly hinders any efforts to understand the functional differences between these two types of nerve cells. An alternative procedure for studying the functions of these two cell types—modifying their development and assaying the physiological and functional consequences—will be discussed in a later section.

DETERMINANTS OF NEURONAL INVARIANCE

Histogenesis of the Nervous System

Descriptive studies of the histogenesis and morphogenesis of the nervous system have delineated the main steps in its development. For expository purposes we shall divide the process into five components, realizing fully that these may overlap one another in time. They are: proliferation, migration, differentiation, growth, and cell death.

A very large number of mature nerve cells ultimately develop as a result of proliferation of a relatively small population of neuroepithelial germinal or stem cells. The mature nerve cells are not produced at random but in orderly sequences which have been well documented for many parts of the brain (summarized in Jacobson, 1970b). The nerve cells originate in germinal zones in which the neuronal stem cells reside. By a sequence of cell divisions these stem cells generate daughter cells, called young neurons, which then migrate from their sites of origin in the germinal zones to their final positions in various parts of the nervous system. From each region of the brain the newly formed cells migrate outward, passing by previously formed young neurons, so that the nervous system is formed in an "inside-out" fashion (Figure 3). These young nerve cells are destined never to divide again, and in fact, one of the hallmarks of the nerve cell is its incapacity to undergo mitosis.

The young nerve cell, if it survives, passes through a series of stages of differentiation. Morphologically, one sees the outgrowth of various cellular processes comprising the axons and dendrites. In the case of the large macroneurons the axons grow along well-circumscribed trajec-

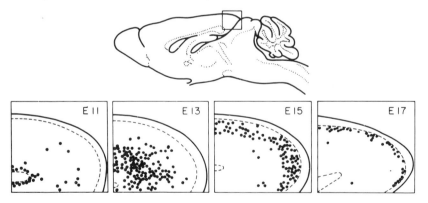

Figure 3. The "inside-out" sequence of time of origin of neurons in the cerebral isocortex of the mouse. Four different mice each received a single injection of tritiated thymidine on either the 11th, 13th, 15th, or 17th day of gestation (E11 to E17). This injection labels cells undergoing their final cell division, at which point they become young neurons. All animals were killed 10 days after birth when the neurons of the cerebral cortex had reached their final positions. The positions of the labeled cells are shown by dots in the autoradiographs of sections of the region of cerebral cortex outlined by the rectangle. Cell proliferation occurs along the walls of the ventricular lumen from which cells migrate to their final position. Note that cells formed on the 11th day of gestation travel only a short distance to their final position. Neurons differentiating later in development must migrate past them and so the cortex arises in an "inside-out" sequence. [From Angevine & Sidman, 1961.]

tories and eventually constitute the nerve-fiber tracts that are among the most invariant features of neuroanatomy. Their processes terminate at characteristic positions in the nervous system and in many instances appear to make synaptic connections with selected cells or even on particular parts of selected cells. In this way neuronal circuits develop by the formation of selective connections between different groups of neurons or between neurons and sensory receptors or muscle fibers. The formation of those nerve circuits, where it has been investigated, is not random, but occurs according to a fairly rigid timetable and in quite precise order. The basic circuits connecting the different parts of the nervous system, including the main input pathways and the main output pathways, are thus laid down. The mechanisms which are involved in the formation of selective interconnections between macroneurons as yet remain unknown, but are presumably under genetic control. In a later section we shall consider some of the possible mechanisms.

Neuronal differentiation of macroneurons, in addition to yielding an invariant circuitry, also results in another invariant feature of the nervous system; each region of the brain is characterized by nerve cells of distinctive types arranged in a particular pattern. Anatomists easily can identify hundreds of different types of nerve cells which are known to exist only in their typical positions. We do not yet understand the mechanisms which control differentiation of these specific neuronal phenotypes. It is the same problem that confronts the developmental biologist interested in explaining the origin of other cellular phenotypes, but the number of different neuronal phenotypes exceed all other types of cells in the body.

Differentiation of microneurons also involves the outgrowth of both axons and dendrites. However, rather than forming invariant interconnections between different parts of the nervous system, these seem to branch in some random fashion. This suggests that the interconnections involving these microneurons may be highly variable. To what extent the particular connectivity established is affected by environmental influences is not known. It is, however, clear that some intrinsically determined constraints must be operative since microneurons do have recognizable features despite the apparently random branching of their cellular processes.

Once the neurons have migrated to the appropriate regions of the nervous system and have undergone their particular differentiation, they continue their growth until adult size is eventually achieved. In some instances this process continues into the animal's postnatal life.

During the course of development a tremendous excess of neurons is produced. In some regions of the nervous system as many as 60%–80% of these cells die before the animal reaches adulthood (Jacobson,

1970b). This widespread cell death must, of necessity, have a profound—and often overlooked—effect on the final structure and function of the nervous system.

Cell death can play a role in determining the final shape of many of the cell groups or nuclei in the nervous system. For example, the dorsal horns of the spinal cord assume their characteristic adult shape as a result of the death of a large number of neurons. This occurs in a highly predictable and determined fashion, resulting in an important invariant feature of the nervous system, and is evidently subject to the control of intrinsic ontogenetic mechanisms. Cell death, in addition, may help to delete those neurons which, as a result of errors occurring during development, do not establish appropriate connections. This would explain why comparatively few "mistakes" or abnormalities are found in the course of anatomical investigations of the nervous system (Jacobson, 1970b).

Finally, selective cell death could provide an opportunity for modifying the development of the nervous system in accordance with conditions prevailing in the organism's environment. Thus, an overproduction of variable neuronal elements, followed by a "pruning away" of the excess might help to insure a match between the operational characteristics of the surviving neurons and the functional requirements of the animal as they are determined by its environment and past history (Jacobson, 1973). At least superficially, this bears some resemblance to the selective process whereby, during development, the fittest cells survive in the Darwinian struggle for existence (Roux, 1881).

To summarize, the developmental events leading up to the formation of the main interconnecting pathways of the nervous system are relatively unmodified by extrinsic agents which normally impinge upon the developing organism. Environmental factors, perhaps, may exert some influence on neuronal survival patterns and may also affect the differentiation of microneurons. Development of the basic interconnections can, of course, be upset by abnormal conditions such as malnutrition or exposure to injurious levels of radiation or poison.

Mechanisms in the Formation of Neuronal Connections

As emphasized, anatomical and physiological studies have demonstrated the existence of neuronal connections which are relatively invariant from one member of a species to the next. The mechanisms which make the establishment of this connectivity possible during the course of development are unknown. We shall, however, attempt to outline some of the problems and findings.

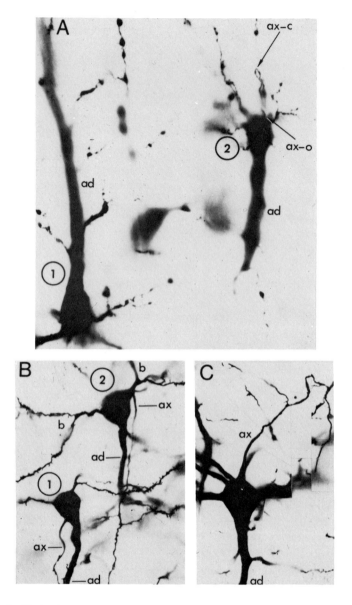

Figure 4. Photomicrographs of inverted pyramidal cells in the rabbit's cerebral cortex. (A) Two pyramidal cells in the cortex of a 5-day-old rabbit. *Cell 1* is properly oriented: its apical dendrite points toward the pial surface, while the axon emerges from the base of the cell body and grows downward toward the white matter underlying the cortex. Other correctly oriented pyramidal cells are illustrated in Figure 1. *Cell 2* is inverted. The cell body is oriented with its base toward the pial surface and its apical dendrite pointing away from the surface

The formation of neuronal interconnections can logically be divided into two phases: *guidance* of the outgrowing axon process to a general target area and *selection* of an appropriate target site, either another neuron or a sensory receptor or muscle fiber, within the region. Guiding the growing axon to its goal may seem to be a terribly difficult task if one considers the tortuous and complex pathways followed by fibers in the mature nervous system. We should bear in mind, however, that distances are shorter and pathways generally straight at the time when many of the main fiber tracts are laid down in the embryo. For example, the first peripheral nerve fibers to develop proceed a short distance along a straight course to reach their terminus. As the embryo grows and its shape is transformed these axons are presumably "towed along" (Weiss, 1941), thereby giving rise to the very complex fiber pathways found in the adult animal. This would imply that arrival of the outgrowing axon at its target region may depend largely upon the initial direction taken by the processes emerging from the cell body. Indeed, sprouting of the axonal process from the differentiating nerve cell occurs from the point on the cell body closest to the axon's eventual target. In the retina, for instance, axons emerge from the ganglion cells on the side nearest to the optic stalk toward which they must all grow (Ramón y Cajal, 1910, 1929).

If necessary, the axon's progress can be guided or corrected by factors operating within the cell's local environment. This is clearly demonstrated by the behavior of those few cells in the adult animal which are evidently incorrectly oriented within the nervous system (Van der Loos, 1965). Typically, all pyramidal cells in the mammalian cortex are oriented with their apices towards the surface of the brain, and the axons grow from the base of the cell towards deeper layers of the cortex (Figures 1 and 4). A small number of pyramidal cells, however, are inverted. Their axons either emerge from unusual parts of the cell body, or they grow out from the base and then abruptly change course (Figure 4). Initial

---

of the cortex. The axon emerges (ax-o) from the usual site at the base of the cell body, initially grows toward the pial surface, but then makes a hairpin turn (ax-c) to grow in the correct direction. (B) Two inverted pyramidal cells in the deeper half of the occipital cortex of an adult rabbit. In *Cell 1*, the axon (ax) grows out of the cell at an unusual point between the apical dendrite (ad) and the cell body itself. In *Cell 2,* the origin of the axon, at the base of the apical dendrite, is also abnormal. In both cases, the axon proceeds in the correct direction through the cortex while the shaft of the apical dendrite is oriented incorrectly. (C) Further example of a pyramidal cell taken from cerebral cortex of an adult rabbit in which the axon (ax) initially grows in the wrong direction and then curves to proceed in the correct direction. Again, the apical dendritic shaft (ad) is pointing in the wrong direction. The orientation of the apical dendritic shaft thus faithfully reflects the polarity of the cell body, while the course followed by the axon is apparently subject to correction by factors operating outside of the cell itself. [From Van der Loos, 1965.]

emergence of the axon thus reflects an intrinsic polarization of the cell, but there must be factors—chemical, mechanical, or electrical—operating outside the cell which can influence the axon's course. At present, nothing more is known about what these might be (Jacobson, 1970b).

Once the axonal process reaches its predetermined target region, it must make appropriate connections. As a rule the axon terminal undergoes extensive branching. Eventually, most of these branches will disappear, leaving behind those which were able to extablish permanent connections. This suggests that the axonal process makes contact at random within its target area, maintaining only those branches and tentative terminals which turn out to be functionally suitable. Recognition of a desirable postsynaptic surface may be based either upon genetically determined affinities between the cells or upon some measure of the functional adequacy of tentative connections.

Genetically determined affinities acquired by neurons during the course of ontogeny may well be based on a system of matching biochemical factors or properties. Quite possibly, these reside on the cell surface. Two neurons (or alternatively, a neuron and a sensory receptor or muscle fiber) might develop complementary affinities permitting permanent functional synapses to develop once contact between them has been established. The probability of such contact would be greatly enhanced by an overproduction of neurons and extensive branching of the axon terminals.

Functional validation of tentative connections could be based upon the pattern or quality of neuronal activity generated in the developing nerve circuits. Though we can only speculate on possible mechanisms, this could allow external factors dependent upon the animal's interaction with its environment to be involved in the selection of those neuronal interconnections which best serve the functional requirements of the nervous system. This would be a means of using information from the external environment to optimize the organism's ability to function within that environment. This validation of tentative connections would have to occur at the appropriate stage of development so it must be governed by an appropriate, presumably genetically controlled, timetable. In the case of mammals, for example, selection of the appropriate connections from among the excess that had been created would have to be delayed until the animal is first exposed to the external environment. One would, therefore, expect environmental factors to have their effect on the developing nervous system during some "critical period."

To complete the developing neuronal interconnections, the dendritic processes must also develop and establish contact with the proper axonal terminals. The dendrites constitute the major proportion of the input surfaces in the nervous system and many of the same questions that were asked about the outgrowth of axonal processes can be raised with reference to the development of dendritic structures. Intrinsic polarization of the cell also appears to control the point at which dendritic processes emerge from the cell. However, the dendritic processes growing out from misaligned pyramidal cells in the cortex were not observed to correct their direction (Figure 4). As a result, dendritic trees of such incorrectly oriented pyramidal cells failed to make proper connections with other neurons and remained stunted in the adult animal (Van der Loos, 1965).

Differentiation of dendritic processes follows rather precise timetables. In addition to spatio-temporal gradients in the sequence in which dendrites mature in the different regions of the brain, the differentiation of dendrites of large macroneurons occurs before that of smaller microneurons. Furthermore, there is a tendency for cells in the motor fields of the nervous system to develop their dendrites before those of cells located in sensory fields. Finally, the timing of dendritic differentiation, for example, the growth of dendritic spines, seems linked to the arrival of axonal terminals into a given region. For example, the dendrites of the large principal neurons of the lateral geniculate body complete their differentiation only when optic tract axons arrive at this structure (Morest, 1969). The smaller granule cells in the lateral geniculate body develop their dendrites much later. In summary, intrinsic polarization of nerve cells seems to determine the direction in which dendrites will grow out from the cell, although the stimulus setting off this event may well be the arrival of appropriate axon terminals.

Prior to establishing connections, dendritic processes undergo extensive branching. As one might expect from what has been said earlier, many of these branches disappear before the animal reaches adulthood. In some instances, sensory stimulation plays a role in determining the final shape and arrangement of the dendritic processes; this will be considered later.

The presence of structures typically associated with growing dendrites (filopodia and growth cones) on the dendritic trees of neurons in adult animals suggests that some plasticity may be maintained throughout the organism's lifetime (Morest, 1969). This would naturally constitute a further source of variability.

The Acquisition of Positional Information

If the formation of neuronal interconnections requires intercellular recognition, then nerve cells must acquire complementary affinities some time before they develop connections. This hypothetical process has been most extensively studied using systems with relatively simple but invariant patterns of organization. For example, pathways linking the retina with higher visual centers are arranged in a topographically ordered fashion. That is, electrophysiological and anatomical studies have shown that retinal ganglion cells located at adjacent positions on the retina connect with adjacent neurons in visual centers such as the tectum. Just how precise this arrangement is remains unknown; within the limits of resolution of the electrophysiological techniques used to map out the connections, however, they appear to be established retinotopically (Jacobson & Hunt, 1973).

Sperry was the first to demonstrate that this orderly retinotopic projection is not formed as a result of learning or experience (reviewed in Sperry, 1951). In adult frogs one eye was surgically inverted, and the optic nerve sectioned and allowed to regenerate. These animals recovered vision, but consistently mislocalized objects in visual space (Figure 5), even many years after the operation was performed. Electrophysiological assays of regenerating connections from the retina to the optic tectum confirmed that their organization was unaffected by visual experience (Gaze & Jacobson, 1963).

Thus, sensory stimulation does not play a role in the regeneration of orderly connections between retina and optic tectum of the frog. That visual experience does not affect the initial development of retinotectal connections in frog embryos is shown by the observation that visuomotor reflexes are permanently inverted after surgical inverson of the eye rudiment before the outgrowth of nerve fibers linking the eye and the brain (Stone, 1960; Jacobson, 1968b). Similarly, in the cat, connections between retina and visual cortex are established prenatally, before the animal can receive any visual stimulation (Hubel & Wiesel, 1963; Barlow & Pettigrew, 1971). We conclude that the information necessary for optic tract axons to connect with the appropriate part of the brain must be supplied by intrinsic developmental programs rather than by the animal's visual experience.

Some insights into the action of these developmental processes have been achieved. In amphibians the information needed to specify the connections between retina and tectum is present very early in embryonic development, as is known by the fact that the early eye rudiment may be isolated in tissue culture for up to two weeks, and when reimplanted in the eye socket will form a retinotectal projection that can be shown to be organized by information derived from the original donor embryo (Jacobson & Hunt, 1973). There is a time period, however, before which the retinal polarity, although present, can still be reversed by inverting the eye. Thus if the eye rudiment is rotated 180° at a very early stage

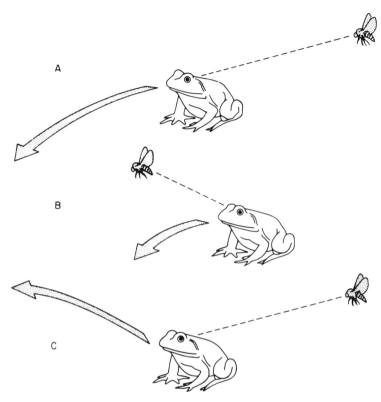

Figure 5. Typical errors in spatial localization of small objects made by animals in which one eye has been rotated or inverted. (A) If the eye is rotated by 180°, a frog will strike at points in the visual field diametrically opposite that point at which the target is actually presented. (B) Following dorsoventral inversion of one eye, a frog will strike correctly with respect to the nasotemporal axis of the visual field, but will invert responses with reference to the dorsoventral axis. (C) After nasotemporal inversion of an eye the frog will misdirect responses with respect to the nasotemporal axis, but will orient correctly with regard to the dorsoventral axis. [Redrawn from Sperry, 1951.]

of development, there is neither electrophysiological nor behavioral evidence that the eye was ever inverted (Figure 6). At an intermediate stage, rotation of the eye produces an inversion of the connections between retina and tectum, along one axis of the eye (Figure 6). Rotation at a subsequent stage leads to an inversion along both axes (Figure 6). Thus, the developmental program in the embryonic retina that gives rise to a biaxially organized retinotectal projection becomes unalterable in at least two stages during a critical period lasting 5 to 10 hr in the embryo (Jacobson, 1968b; Stone, 1960). The development of orderly retinotectal connections is based on positional information obtained by

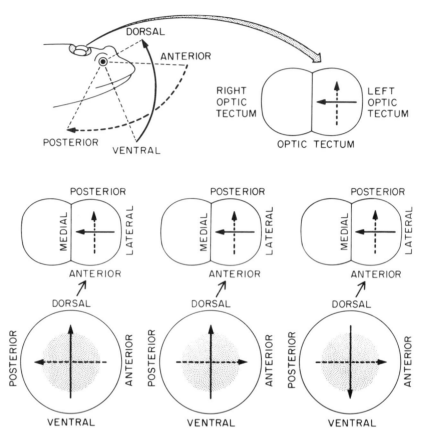

Figure 6. The effects of eye inversion at various stages of embryonic development on the retinotectal projection of the frog. As illustrated in Figure 5, if one eye is rotated 180° in an adult frog, the animal mislocalizes objects with respect to both axes of its visual field. The projection of retinal ganglion cell axons onto the tectum can be mapped electrophysiologically. An electrode is positioned successively at different points on the surface of the tectum. For each location of the electrode, a small stimulus spot is moved through the animal's visual field until a point is located from which the maximal neuronal discharge is obtained at the site of the electrode. From such a map, the distribution of retinal ganglion cell terminals can then be inferred. When the eye is rotated at different stages of embryonic development, the effects on the retinotectal projection are not always the same as after eye inversion in the adult animal. If the right eye is rotated 180° at embryonic stage 28, the retinotectal projection is normal *(left)*. When the eye is rotated slightly later, at stage 30, the map is inverted in one axis, the anteroposterior axis *(middle)*. Finally, if the eye is rotated at stage 31, or later, the map is inverted in both the anteroposterior and dorsoventral axis *(right)*. [After Jacobson and Hunt, 1973.]

the retinal cell population before the critical period of embryonic development. This positional information is required for the development of locus specificity in the retinal ganglion cells. Locus specificity is defined as the property each ganglion cell uses to send its axonal process to the proper position in the orderly array of ganglion cell axonal processes growing into the brain.

It is of some interest that the critical period during which the retinal locus specificities are irreversibly programmed in the retinal ganglion cells coincides with final cell division of the presumptive retinal ganglion cells (Jacobson, 1968a) (Figure 7). Before this time the retinal germinal cells are still dividing, and thus, as we have said earlier, do not qualify as neurons. Afterwards they will never again divide, but instead will differentiate into retinal ganglion cells which send out axonal processes to the tectum. We might thus say that when these cells become young

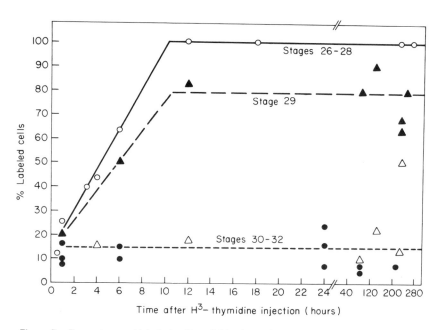

Figure 7. Percentages of labeled cell nuclei in the retina of *Xenopus* embryos after either one or several injections of tritiated thymidine. Cells undergoing mitosis will incorporate tritiated thymidine into the DNA being synthesized in the cell nucleus; those that have ceased mitosis fail to incorporate label into their nucleus. Thus, it is possible to determine the time at which groups of cells in the nervous system undergo their final cell division and become young neurons. Virtually all of the cell nuclei in the retina of *Xenopus* embryos were labeled if the initial injection was given between stage 26 and 28 (open circles), indicating that none of the cells had ceased dividing by this stage of development. On the other hand, by stage 29 (solid triangles) some retinal ganglion cells had undergone their final mitosis, and after stage 30 (open triangles) or 31–32 (solid circles) only a small proportion of dividing cells remained in the retina. When these data are compared to those presented in Figure 6, it becomes evident that many of the retinal ganglion cells are undergoing their final mitotic cycle at precisely that point in time when the specification of the retinal coordinate axis becomes unalterable. [From Jacobson, 1968a.]

neurons they no longer have the ability to acquire new positional information.

DETERMINANTS OF NEURONAL VARIABILITY

The most prominent neurons of the brain—the principal large neurons and their extensive systems of long axons linking the various parts of the nervous system—are laid down by developmental programs presumably operating under genetic control. These programs insure the development of an invariant framework which, it seems reasonable to assume, is a prerequisite for orderly nervous functioning.

Such an invariant structure, perhaps capable of a minimal degree of modification and variation as a result of learning, is able to use information from the environment to guide and control behavior. In some cases this must involve interpretation and transformation of sensory information presented to the central nervous system both by its numerous converging input pathways (see Chapter 8) as well as by the pathways conducting efferent outflow. The necessary flexibility in such information processing capabilities of the nervous system may well be provided by the variable connectivity of the countless small, short axon interneurons found throughout the brain, particularly where complex information-processing is known to occur. The apparently random branching pattern of their axons and dendrites, as well as their distribution in the nervous system, suggest that they might serve as variable components of the nervous system. Unfortunately, our knowledge about the function of these microneurons is so limited that their putative role as modifiable elements remains speculative.

We can, nonetheless, make some start in asking questions about the development and fate of these microneurons. First, to what extent do environmental factors play a role in determining the rate of formation and final number of these neurons? Second, once they have developed, what factors play a role in the formation of their connections with other neurons? Finally, what determines the survival and maintenance of the microneurons themselves and of their connections?

Anatomical studies on the influences of environmental factors on the development of the nervous system have provided limited information. Gross manipulations—either increasing or decreasing the overall sensory input to an organism—have been employed in an effort to determine what role sensory stimulation plays during the ontogeny of the nervous system. There is considerable evidence, for example, that an absence of visual stimulation leads to altered development of the visual cortex:

a decrease in cell size, loss of dendritic spines from pyramidal neurons, decreased length and branching of dendrites of stellate cells (Gyllenstein et al., 1965; Valverde, 1967; Globus & Scheibel, 1967; Coleman & Riesen, 1968; Fifkova, 1968). Such experiments, however, have failed to provide direct evidence that sensory stimulation affects the variability of neuronal interconnections. To obtain more direct and complete evidence for an effect of sensory stimulation we must, of necessity, turn to neurophysiological studies.

Development of the Cat's Visual Cortex

Neurons in the cat's visual cortex respond in a highly selective fashion to visual stimulation. Most of the cells that have been studied gave a maximal response to elongated stimuli at some particular orientation (Hubel & Wiesel, 1959, 1962), although other types of response patterns have also been reported (Spinelli & Barrett, 1969). The types of neurons generating these complex response characteristics are not known. One can, however, determine whether their development is guided by intrinsic ontogenetic mechanisms. With this in mind, Hubel and Wiesel (1963) studied cortical neurons in newborn, visually inexperienced kittens. Initial results suggested that neurons in the kitten's visual cortex were much like those present in adult animals, but their findings have recently been questioned (Barlow & Pettigrew, 1971; Pettigrew, 1974). It now appears that cells in the kitten's visual cortex begin to show an orientation sensitive response only during the second postnatal month (Figure 8); this is just when kittens first respond to visual stimuli. We must ask whether these changes occurring during the second month represent delayed maturation of cells in the visual system, or if sensory stimulation plays an active role in the final differentiation of these neurons.

To answer this question, cats were deprived of visual experience during the first few months of life (for a review, see Riesen, 1966). The visual cortex of such animals was quite abnormal; there was a sharp reduction in the number of cells responding in normal fashion to light stimuli (Chow & Stewart, 1972; Ganz et al., 1968; Wiesel & Hubel, 1963, 1965). The effect of the deprivation was increased when there was an imbalance in the exposure of the two eyes. Thus, after patterned stimulation was excluded from only one eye during the postnatal critical period while the other eye's visual exposure was not restricted, cells could be activated in normal fashion only through the nondeprived eye. On the other hand, contrary to what one might expect, after both eyes were covered during the critical period, about 40% of the neurons studied

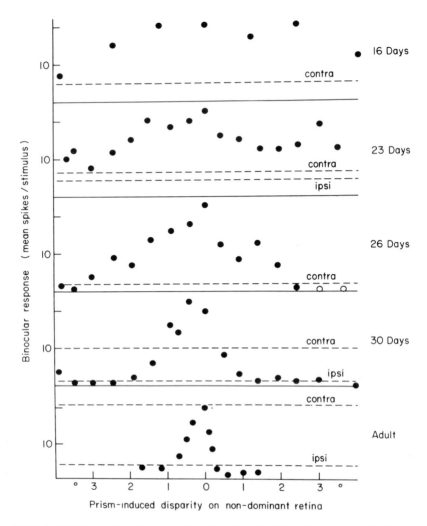

Figure 8. Developmental changes in the disparity selectivity of neurons in the kitten's striate cortex. The majority of neurons in the visual cortex respond to stimulation of either eye. When both eyes are stimulated simultaneously, there is often an optimal stimulus disparity to which the cell responds most strongly. This is determined by optimizing the stimulus position and orientation for the retina of one eye (dominant eye) and systematically shifting the location of the stimulus on the retina of the other eye (nondominant eye) with the aid of a variable prism. A sharply tuned peak in the cell's response is obtained when this disparity achieves its optimal value *(adult)*. Those neurons which were the most sharply tuned for disparity are illustrated for kittens of different ages. At 16 days of age, the response to stimulation of both eyes is high for a wide range of disparities; there is no response to stimulation of only the ipsilateral eye. In older animals, the response to stimulation of both eyes is reduced for all but a restricted set of stimulus disparities; the range of disparities to which cells give a strong response when stimulated binocularly is further reduced as the animals get older. It should be noted that the response at the optimal disparity is of approximately the

appeared to respond in a normal fashion to stimulation of one or both eyes (Wiesel & Hubel, 1965). Comparable results have recently been obtained for the frog, thereby extending the generality of this finding (Jacobson & Hirsch, 1973).

To obtain a different, though correlated assay of the visual system's development, the behavior of these deprived animals was studied. Excluding patterned stimulation from either one or both eyes resulted in deficits in visually guided behavior (Dews & Wiesel, 1970) and in the ability of the cats to discriminate between different patterns (Ganz & Fitch, 1968; Rizzolatti & Tradari, 1971). Ganz *et al.* (1972) also confirmed that unequal stimulation of the two eyes, for example, depriving only one eye, produced greater deficits in behavior than depriving both eyes to the same extent. The magnitude of the behavioral deficit was thus correlated with the extent of the physiological deficit.

Sensory stimulation is, therefore, necessary for the normal development of the cat's visual system. It is not known, however, whether the effect of the stimulation is to maintain connections that would otherwise succumb, or to promote the development and maintenance of functionally suitable connections.

Environmental Surgery of the Cat's Visual Cortex

As a first step toward understanding the role of visual stimulation, it must be demonstrated that sensory input can selectively affect functional properties, such as orientation sensitivity of cortical neurons. To this end a novel procedure called *environmental surgery* was developed (Hirsch, 1970, 1972; Hirsch & Spinelli, 1970, 1971). It involved raising kittens under conditions in which their visual experience was restricted and carefully controlled. For example, for some animals their only exposure to patterned visual stimuli consisted of viewing a field of vertical stripes with one eye, and simultaneously, a field of horizontal stripes with the other eye (Figure 9). As a consequence of this exposure history, elongated receptive fields of cells in the visual cortex were oriented either horizontally or vertically, in contrast to the full range of receptive field orientations characteristic of normal cats (Hirsch & Spinelli, 1970, 1971). All of these cells could be activated by only one eye, whereas normally

---

same magnitude as the response to binocular stimulation was in very young kittens. The response to stimulation of the contralateral eye gradually increases with age until it equals the response to stimulation of both eyes at the appropriate stimulus disparity. Response to stimulation of the ipsilateral eye also increases with age, but always remains well below the response to stimulation of both eyes or to stimulation of the contralateral eye. This figure illustrates very dramatically the increase in the stimulus selectivity of cortical neurons during the course of the kitten's early postnatal development. [From Pettigrew, 1974.]

128   Helmut V. B. Hirsch and Marcus Jacobson

Figure 9. Kitten wearing one of the masks used to provide selective visual exposure. Stimulus patterns are mounted on the inside surface of the black plastic sheet: one pattern in front of the left eye, a second in front of the right eye. A lens is mounted in the mask in front of each eye so that the patterns are located at the focal plane of the lens. When the kitten's eye is relaxed, the patterns should thus be in focus on the retina. Light entering through the white diffusing plastic illuminates the patterns. The animals wore these masks 8 hours a day from the time they were 3 or 4 weeks old until they were some 9 to 12 weeks old. Whenever the animals were not wearing these masks, they were kept in a dark room with their mother. Following this selective visual experience—a procedure that has been called *environmental surgery*—the receptive fields of single neurons in the visual cortex of these animals were studied. Examples of such data are illustrated in Figures 10 and 12. This particular kitten is viewing a field containing three black vertical stripes with its left eye, and one containing three back horizontal stripes with its right eye.

the overwhelming majority of cells can be driven by stimulation of either eye. Most important, cells with vertically oriented receptive fields were activated only by the eye that had viewed vertical lines while cells with horizontally oriented receptive fields were activated only by the eye that had viewed horizontal lines (Figure 10). The results of Hirsch and Spinelli (1970) were confirmed by Blakemore and Cooper (1970) and Blakemore and Mitchell (1973). Extending earlier work of Hubel and Wiesel (1970) they showed also that during a relatively brief critical period, between four and six weeks of age, very short exposure (possibly as little as one hour) to an environment containing lines of only one

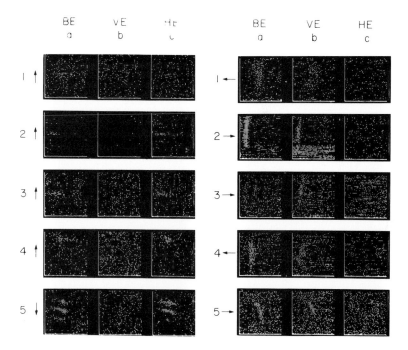

Figure 10. Receptive field maps for cells in the striate cortex of kittens given selective visual exposure during their early postnatal development. To make these maps, a black spot was moved under computer control across the cat's visual field. A cell's responses to this mapping stimulus were recorded with extracellular microelectrodes: Each response of the cell could be correlated with the position of the mapping spot when it occurred. In the figure, each black square represents a 25° by 25° portion of the cat's visual field: White spots indicate that the cell being studied fired at least once when the mapping spot crossed that particular point in the field. Cells are spontaneously active so there is often some background activity uncorrelated with the position of the spot. The presence of a receptive field is indicated by a clustering of the spots. Each row represents one cell: In the column labeled BE, the mapping was carried out with both of the cat's eyes opened; in the column labeled VE, the mapping was done with only the eye opened that had been exposed to vertical stripes; and in the column labeled HE, the mapping was carried out with only the eye exposed to horizontal bars open. The arrows to the left of each row indicate the direction in which the spot was moved during mapping. *Left:* Cortical cells with horizontally oriented receptive fields. Note that all receptive fields are oriented horizontally and that units are activated only by the eye that had been exposed to horizontal stripes. Rows 2 and 5 illustrate units having three parallel, elongated excitatory regions within their receptive field. *Right:* Cortical cells with vertically oriented receptive fields. Note that all receptive fields in this group are oriented vertically and activated only by the eye that had been exposed to vertical lines during the rearing. In row 2, there was a sudden change in the background firing of the cell during the mapping made with the eye exposed to vertical lines open. Although this produced a solid region of activity in the lower part of the field, the receptive field itself is quite clear and evidently vertically oriented. When this activity burst subsided, the cell's receptive field was remapped and now only the vertically oriented receptive field itself was present. These data indicate that, if cats are exposed with one eye to vertical bars and with the second eye to horizontal stimuli during early postnatal development, then elongated receptive fields are all oriented either vertically or horizontally. Cells with vertically oriented fields are activated only by the eye that had viewed vertical lines, while cells with horizontal fields were activated only by the eye that had been exposed to horizontal stimuli. During mapping, the spot moved at 10°/sec; incident light on the screen was 20 lm/m². [From Hirsch and Spinelli, 1971.]

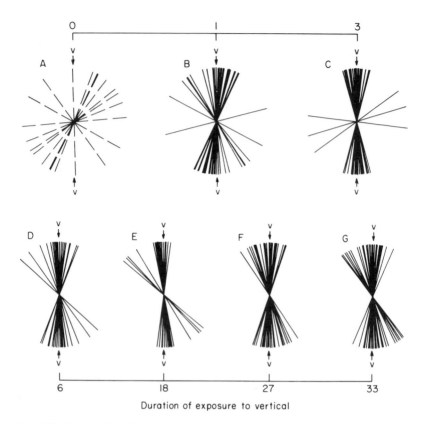

Figure 11. The cortical effect of short periods of selective visual exposure. The visual exposure of kittens was restricted by rearing them inside cylinders which had vertical stripes painted on their inside surfaces. An opaque neck ruff prevented the animals from seeing their own bodies. Four- to six-week-old kittens which had received no previous exposure to patterned stimuli were briefly exposed to such a vertically striped environment. In this figure, each line represents the optimal orientation for one cortical neuron. The number of hours of exposure inside the vertically striped drum for each kitten is indicated on the abscissa. In (A) the orientation preferences are indicated for cells recorded from a kitten that was deprived of all patterned visual stimulation. Interrupted lines were used, since cells in this cat showed only slight orientation preferences for the angles shown and also responded to all other orientations. In (B) through (G) it is clear that, in all animals after exposure to a vertical environment for as little as one hour, the majority of cells respond preferentially to vertical lines. Since the kittens were housed in a dark room for some weeks following their exposure to a patterned environment before orientation preferences of their cortical cells were determined, we know only that a few hours are sufficient to initiate dramatic changes in the striate cortex, but not the actual time course of these changes. [From Blakemore and Mitchell, 1973.]

orientation was sufficient to insure that all cells from which they recorded had receptive fields with an orientation matching that of the stimuli presented to the animals (Figure 11). These results all demonstrate that visual exposure can affect the functional characteristics of cells in the striate cortex of the cat in a highly selective and precise manner. Recent evidence indicates that these findings can be generalized to humans (Freeman & Thibos, 1973).

An animal's exposure history during its early postnatal development thus exerts a powerful influence on neuronal elements in the visual cortex. The effects are long lasting since they persist even after almost two years of normal visual experience (Spinelli et al., 1972; Pettigrew et al., 1973b) although the presence of some binocularly driven cells suggests that a partial recovery may have taken place (Figure 12).

Visually responsive neurons found in the cortex of cats reared in a restricted sensory environment all responded best to one of the patterns which the animals had seen. The most striking examples of this were cells whose receptive field contained three elongated excitatory regions—this very closely matched the lines to which these animals had been exposed (Figure 10) (Hirsch & Spinelli, 1971; Spinelli & Hirsch, 1971). We must ask how sensory stimulation might have such an effect on cortical neurons. This requires relating these physiological changes to anatomical ones to obtain a better understanding of the mechanisms by which environmental conditions can introduce variability into the developing nervous system.

Variable Components in an Invariant Framework

We have seen that the constancy of the long interconnecting pathways in the nervous system results from relatively tight developmental control of the morphology and connectivity of the large principal cells or macroneurons. In contrast, the shape and connections of microneurons appear less constant and more variable. They may thus provide a substrate for modifiable behavior or learning.

We suggest that these variable components of the nervous system exert their influence by modifying—either facilitating or inhibiting—the pattern of activity generated by neurons comprising the invariant framework of the nervous system. Indeed, the small, short axon microneurons seem ideally suited for introducing local changes in the activity patterns generated by larger macroneurons, and they might be thought of as local circuit neurons.

We have pointed out the dramatic effects which experience can have on the response properties of neurons in the cat's visual cortex. What the underlying mechanisms are remains an unresolved problem. One possibility is that some neurons require *functional validation* for the completion or maintenance of their intrinsically determined specificity. In that case, only those cells whose functional characteristics are appro-

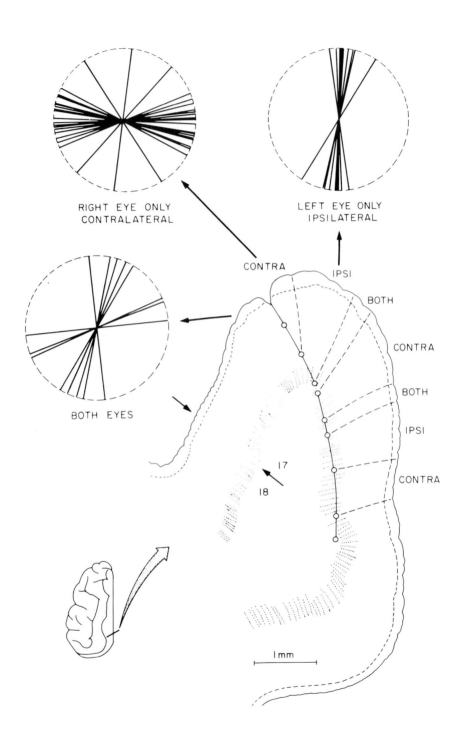

RIGHT EYE ONLY
CONTRALATERAL

LEFT EYE ONLY
IPSILATERAL

CONTRA

IPSI

BOTH

CONTRA

BOTH

IPSI

CONTRA

BOTH EYES

17

18

1 mm

priate to the animal's actual environment will survive. A second possibility is that development of the visual cortex requires *functional specification* of nonspecified, multi-potential neurons. This would mean that cortical cells initially lack the selective response properties characteristic of cells in the adult animal, and only develop them in response to visual input. In support of this, in visually inexperienced kittens sensory stimulation can induce small changes in the response properties of cortical cells in a matter of hours (Pettigrew et al., 1973a). One cannot tell from this whether longer periods of stimulation would produce larger, more permanent changes. Perhaps cells are capable of making small adjustments in their response properties if this will bring them into register with stimuli present in the visual world, but cease functioning when the necessary changes are too great. This hypothesis is difficult to verify since, on the one hand, "nonresponsive" neurons cannot be identified on anatomical grounds and, on the other hand, the available physiological recording techniques provide no means for reliably assaying the density and distribution of visually responsive neurons in the cortex.

While we do not understand the mechanisms by which sensory stimulation can affect cortical neurons, it is clear that environmentally induced changes do not occur in a haphazard fashion. Cells in the cat's striate cortex are grouped together according to their functional characteristics (see Chapter 8) so we must postulate that they do not develop their response properties independently of one another. Rather, it seems that the fate of an individual neuron is, in some way, related to that of

Figure 12. Recovery from the effects of selective visual exposure during early postnatal development. After viewing only vertical stripes with one eye and only horizontal stripes with the other eye during the first 3 months of life, cats were permitted unrestricted visual experience for some 18–24 months. At this time, receptive fields of neurons in their visual cortex were studied. The majority of cells recorded from were still activated by only one eye. Those activated by the left eye preferred vertically oriented stimuli, while those activated by the right eye for the most part responded best to horizontal lines. Such orientation preferences were consistent with the animals early selective visual exposure. Some neurons, however, could be activated by either eye: Their preferred orientations were not as clearly vertical or horizontal as were those of monocularly activated cells. Binocularly activated cells were clustered in groups at the borders of regions in which all of the cells could be activated only by one eye. This is illustrated by the reconstruction of an electrode tract: Groups of cells activated exclusively by the contralateral or ipsilateral eye alternated with smaller groups of cells that responded to stimulation of either eye. Binocularly activated cells located near a column of cells activated by the ipsilateral eye had orientation preferences near the vertical, while those located in the neighborhood of a cell cluster activated by the contralateral eye had orientation preferences nearer horizontal than vertical. Clearly, the distribution of receptive field orientations of cells activated by both eyes does not resemble that found either in a normal cat or in a binocularly deprived cat (see Figure 11). The insert at the bottom lefthand corner of the figure indicates the region of cortex in which the electrode track was located. [Modified from Pettigrew et al., 1973b.]

its neighbors. The functional properties of individual neurons thus develop within the context of a set or group of neurons.

In conclusion, then, we have returned to the beginning, to ask how we can at one and the same time account for both the flexibility of the nervous system as well as for its invariant structure. It is clear there must be means for orchestrating the variability of individual neuronal components so that the overall organization of the nervous system remains constant and predictable. It is hard to see how the nervous system could function properly without this.

ACKNOWLEDGMENTS

We thank Andrea J. Elberger, Richard Gordon, Barry Skarf, and S. A. B. Tieman for reading and criticizing the manuscript.

REFERENCES

Altman, J. Postnatal growth and differentiation of the mammalian brain, with implications for a morphological theory of memory. In G. Quarton, T. Melnechuk, & F. O. Schmitt (Eds.), *The neurosciences: A study program.* New York: Rockefeller Univ. Press, 1967. Pp. 723–743.

Altman, J. Postnatal neurogenesis and the problem of neural plasticity. In W. A. Himwich (Ed.), *Developmental neurobiology.* Springfield, Ill.: Thomas, 1970. Pp. 197–237.

Angevine, J. B., Jr., & Sidman, R. L. Autoradiographic study of cell migration during histogenesis of cerebral cortex in the mouse. *Nature,* 1961, **192,** 766–768.

Barlow, H. B., & Pettigrew, J. D. Lack of specificity of neurons in the visual cortex of young kittens. *Journal of Physiology (London),* 1971, 99P–101P.

Blakemore, C., & Cooper, G. F. Development of the brain depends on visual environment. *Nature (London),* 1970, **288,** 477–478.

Blakemore, C., & Mitchell, D. E. Environmental modification of the visual cortex and the neural basis of learning and memory. *Nature (London),* 1973, **241,** 467–468.

Brodal, A. *Neurological anatomy.* 2nd ed. Oxford: Oxford Univ. Press, 1969.

Bullock, T. H., & Horridge, G. A. *Structure and function in the nervous systems of invertebrates.* 2 Vols. San Francisco: Freeman, 1965.

Chow, K. L., & Stewart, D. L. Structural and functional effects of long-term visual deprivation in cats. *Experimental Neurology,* 1972, 34, 409–433.

Coleman, P. D., & Riesen, A. H. Environmental effects on cortical dendritic fields. I. Rearing in the dark. *Journal of Anatomy (London),* 1968, **102,** 363–374.

Conel, J. L. *Postnatal development of the human cerebral cortex.* Vols. 1–6. Cambridge, Mass.: Harvard Univ. Press, 1939–1963.

Dews, P. B., & Wiesel, T. N. Consequences of monocular deprivation on visual behavior in kittens. *Journal of Physiology (London),* 1970, **206,** 437–455.

Fifkova, E. Changes in the visual cortex of rats after unilateral deprivation. *Nature (London),* 1968, **220,** 379–381.

Freeman, R. D., & Thibos, L. N. Electrophysiological evidence that abnormal early visual experience can modify the human brain. *Science,* 1973, **180,** 876–878.

Ganz, L., & Fitch, M. The effect of visual deprivation on perceptual behavior. *Experimental Neurology*, 1968, **22**, 638–660.

Ganz, L., Fitch, M., & Satterberg, J. A. The selective effect of visual deprivation on receptive field shape determined neurophysiologically. *Experimental Neurology*, 1968, **22**, 614–637.

Ganz, L., Hirsch, H. V. B., & Bliss Tieman, S. The nature of perceptual deficits in visually deprived cats. *Brain Research*, 1972, **44**, 547–568.

Gaze, R. M., & Jacobson, M. A study of the retinotectal projection during regeneration of the optic nerve in the frog. *Proceedings of the Royal Society (London)*, *Series B*, 1963, **157**, 420–448.

Giuditta, A., Libonati, M., Packard, A., & Prozzo, N. Nuclear counts in the brain lobes of *Octopus* Vulgaris as a function of body size. *Brain Research*, 1971, **25**, 55–62.

Globus, A., & Scheibel, A. B. The effect of visual deprivation on cortical neurons: A Golgi study. *Experimental Neurology*, 1967, **19**, 331–345.

Golgi, C. *Sulla fina Anatomia degli organi centrali del sistema nervosa*. Pavia, 1886.

Guillery, R. W., & Kaas, J. H. A study of normal and congenitally abnormal retinogeniculate projections in cats. *Journal of Comparative Neurology*, 1971, **143**, 73–100.

Gyllenstein, L., Malmfors, T., & Norrlin, M. L. Effect of visual deprivation on the optic centers of growing and adult mice. *Journal of Comparative Neurology*, 1965, **124**, 149–160.

Hirsch, H. V. B. The modification of receptive field orientation and visual discrimination by selective exposure during development. Ph.D. Thesis, Stanford University, 1970.

Hirsch, H. V. B. Visual perception in cats after environmental surgery. *Experimental Brain Research*, 1972, **15**, 405–423.

Hirsch, H. V. B., & Spinelli, D. N. Visual experience modifies distribution of horizontally and vertically oriented receptive fields in cats. *Science*, 1970, **168**, 869–871.

Hirsch, H. V. B., & Spinelli, D. N. Modification of the distribution of receptive field orientation in cats by selective visual exposure during development. *Experimental Brain Research*, 1971, **12**, 509–527.

Hubel, D. H., & Wiesel, T. N. Receptive fields of single neurons in the cat's striate cortex. *Journal of Physiology (London)*, 1959, **148**, 574–591.

Hubel, D. H., & Wiesel, T. N. Receptive fields, binocular interaction and functional architecture in the cat's visual cortex. *Journal of Physiology (London)*, 1962, **160**, 106–154.

Hubel, D. H., & Wiesel, T. N. Receptive fields of cells in striate cortex of very young, visually inexperienced kittens. *Journal of Neurophysiology*, 1963, **26**, 994–1002.

Hubel, D. H., & Wiesel, T. N. The period of susceptibility to the physiological effects of unilateral eye closure in kittens. *Journal of Physiology (London)*, 1970, **206**, 419–436.

Jacobson, M. Cessation of DNA synthesis in retinal ganglion cells correlated with the time of specification of their central connections. *Developmental Biology*, 1968, **17**, 219–232. (a)

Jacobson, M. Development of neuronal specificity in retinal ganglion cells of *Xenopus*. *Developmental Biology*, 1968, **17**, 202–218. (b)

Jacobson, M. Development of specific neuronal connections. *Science*, 1969, **163**, 543–547.

Jacobson, M. Development, specification and diversification of neuronal circuits. In F. O. Schmidt, (Ed.) *The neurosciences: Second study program.* New York: Rockefeller Univ. Press, 1970. Pp. 116–129. (a)

Jacobson, M. *Developmental neurobiology.* New York: Holt, 1970. (b)

Jacobson, M. A plenitude of neurons. In G. Gottlieb (Ed.), *Studies on the development of behavior and the nervous system.* Vol. 2. New York: Academic Press, 1973. Pp. 151–166.

Jacobson, M., & Hirsch, H. V. B. Development and maintenance of connectivity in the visual system of the frog. I. The effect of eye rotation and visual deprivation. *Brain Research,* 1973, 49, 47–65.

Jacobson, M., & Hunt, R. K. Origins of neuronal specificity. *Scientific American,* 1973, 228, 26–35.

Macagno, E. R., Lopresti, V., & Levinthal, C. Structure and development of neuronal connections in isogenic organisms: Variations and similarities in the optic system of *Daphnia magna. Proceedings of the National Academy of Science,* 1973, 70, 57–61.

Morest, D. K. The growth of dendrites in the mammalian brain. *Zeitschrift für Anatomie und Entwicklungsgeschichte,* 1969, 128, 290–317.

Palay, S. L. Principles of cellular organization in the nervous system. In G. Quarton, T. Melnechuk, & F. O. Schmitt (Eds.), *The Neurosciences: A study program.* New York: Rockefeller Univ. Press, 1967. Pp. 24–31.

Pettigrew, J. D. The effect of visual experience on the development of stimulus specificity by kitten cortical neurons. *Journal of Physiology (London),* 1974, 237, 49–74.

Pettigrew, J. D., Olson, C., & Barlow, H. B. Kitten visual cortex: Short-term stimulus-induced changes in connectivity. *Science,* 1973, 180, 1202–1203. (a)

Pettigrew, J. D., Olson, C., & Hirsch, H. V. B. Cortical effect of selective visual experience: Degeneration or reorganization? *Brain Research,* 1973, 51, 345–351. (b)

Purves, D., & McMahan, U. J. The distribution of synapses on a physiologically identified motor neuron in the central nervous system of the leech. *Journal of Cell Biology,* 1972, 55, 205–220.

Ramón y Cajal, S. *Histologie du Systeme Nerveux de l'Homme et des Vertébrés* 2 Vols. Trans. by L. Azoulay. 1909–1911. Reprinted by Instituto Ramón y Cajal del C.S.I.C., Madrid, 1952–1955.

Ramón y Cajal, S. Algunas observaciones favorables á la hipótesis neurotrópica. *Trabajos del Laboratorio del Investegaciones Biologicas de la Universidad de Madrid,* 1910, 8, 63–134.

Ramón y Cajal, S. (1929). Étude sur la neurogenèse de quelques vertébrés. In *Studies on vertebrate neurogenesis.* Trans. by L. Guth. Springfield, Ill.: Thomas, 1960.

Riesen, A. H. Sensory deprivation. In E. Stellar & J. M. Sprague (Eds.), *Progress in physiological psychology.* New York: Academic Press, 1966. Pp. 117–147.

Rizzolatti, G., & Tradardi, V. Pattern discrimination in monocularly reared cats. *Experimental Neurology,* 1971, 33, 181–194.

Roux, W. *Der Kampf der Theile im Organismus.* Leipzig: Wilhelm Engelmann, Verlag, 1881.

Scheibel, A. Neural correlation of psychophysiological development in young organisms. In J. Wortis (Ed.), *Recent advances in biological psychiatry.* Vol. 4. New York: Plenum, 1962. Pp. 313–327.

Sperry, R. W. Mechanisms of neural maturation. In S. S. Stevens (Ed.), *Handbook of experimental psychology.* New York: Wiley, 1951. Pp. 236–280.

Sperry, R. W. How a developing brain gets itself properly wired for adaptive function. In E. Tobach, L. A. Aronson, & E. Shaw, (Eds.), *The Biopsychology of development.* New York: Academic Press, 1971. Pp. 27–44.

Spinelli, D. N., & Barrett, T. W. Visual receptive field organization of single units in the cat's visual cortex. *Experimental Neurology,* 1969, **24**, 76–98.

Spinelli, D. N., & Hirsch, H. V. B. Genesis of receptive field shapes in single units of cat's visual cortex. *Federation Proceedings,* 1971, **30**, 615.

Spinelli, D. N., Hirsch, H. V. B., Phelps, R. W., & Metzler, J. Visual experience as a determinant of the response characteristics of cortical receptive fields in cats. *Experimental Brain Research,* 1972, **15**, 289–304.

Stone, L. S. Polarization of the retina and development of vision. *Journal of Experimental Zoology,* 1960, **145**, 85–93.

Stuart, A. Physiological and morphological properties of motoneurons in the central nervous system of the leech. *Journal of Physiology (London),* 1970, **209**, 627–646.

Valverde, F. Apical dendritic spines of the visual cortex and light deprivation in the mouse. *Experimental Brain Research,* 1967, 3, 337–352.

Van der Loos, H. The "improperly" oriented pyramidal cell in the cerebral cortex and its possible bearing on problems of growth and cell proliferation. *Bulletin of Johns Hopkins Hospital,* 1965, **117**, 228–250.

Van Essen, D., & Kelly, J. Correlation of cell shape and function in the visual cortex of the cat. *Nature,* 1973, **241**, 403–405.

Weiss, P. Nerve patterns: The mechanics of nerve growth. In *Third Growth Symposium.* Vol. 5, 1941. Pp. 163–203.

Wiesel, T. N., & Hubel, D. H. Single-cell responses in striate cortex of kittens deprived of vision in one eye. *Journal of Neurophysiology,* 1963, **26**, 1003–1017.

Wiesel, T. N., & Hubel, D. H. Comparison of the effects of unilateral and bilateral eye closure on cortical unit responses in kittens. *Journal of Neurophysiology,* 1965, **28**, 1029–1040.

Young, J. Z. *A model of the brain.* Oxford: Oxford Univ. Press, 1964.

Part II

THE CHEMISTRY
OF BEHAVIOR

Chapter 5

Chemical Pathways in the Brain

SUSAN D. IVERSEN
LESLIE L. IVERSEN
University of Cambridge, England

INTRODUCTION

The concept of chemical transmission in the central nervous system (CNS) is a relatively new one. We are only beginning to define the "chemical pathways" in the CNS, in the sense of describing neuroanatomical pathways of neurones with known transmitters. In this chapter this information, in so far as it is currently available, will be summarized for the transmitters noradrenaline, dopamine, 5-hydroxytryptamine, acetylcholine, and the amino acids GABA and glycine.

In some instances our understanding of such chemical pathways has progressed rapidly following the development of histochemical techniques for selectively staining neurones that contain particular transmitters. A fluorescence technique pioneered by Hillarp and his colleagues has been used successfully for staining NA, DA, and 5-HT containing pathways. With such methods the neurones and fibers with their terminals containing vesicular stores of specific chemical transmitters can be visualized and their distribution plotted through the CNS.

There is still no direct staining technique for ACh and knowledge of its distribution rests largely on the indirect approach of staining the degrading enzyme acetylcholinesterase. In the case of GABA the evidence for distribution is based mainly on the method of autoradiography. Brain tissue is exposed to radioactively labeled GABA which is selectively taken up by GABA-containing neurones. The density of radioactivity in autoradiograms of the exposed tissue indicates the identity of such neurones. There are, however, at present no complete maps of the distribution of GABA or glycine-containing neurones in the CNS.

DISTRIBUTION OF NA, DA, AND 5-HT

Fluorescence Histochemical Technique

Aqueous formaldehyde-condensation was first used by Eranko to visualize catecholamines in the adrenal medulla but Falck and Hillarp saw the tremendous potential in the technique, using the highly fluorescent condensation products to plot amine pathways in the brain. The technique depends for its success on rapid freezing of the brain tissue followed by freeze drying at −40°C and subsequent exposure to formaldehyde vapor at 80°C for one hour.

The first reaction involves condensation of catecholamines with formaldehyde to give nonfluorescent derivatives (6,7-dihydroxy-1,2,3,4,-tetrahydroisoquinolines). Protein in the tissue then catalyses the subsequent dehydrogenation of these compounds to yield the corresponding 6,7-dihydroxy-3,4,-dihydroisoquinoline compounds. These products are in a pH-dependent equilibrium with their tautomeric quinone structures, which are responsible for the strong fluorescence. Although NA and DA both form compounds that yield green fluorescence on exposure to ultraviolet light, they are difficult to distinguish by eye. However, the NA fluorophore has a labile hydroxyl group in position 4, and after treatment with hydrochloric acid its emission peak shifts slightly. This small change can be detected with microspectrofluorimetric methods. Alternatively pharmacological treatments which interfere with one of the catecholamines can be combined with fluorescence histochemistry to visualize NA and DA independently.

Indolamines such as 5-HT also form fluorescent condensation products when exposed to formaldehyde gas, but in this case the yellow fluorescence color is easy to distinguish from the greenish products produced from NA and DA.

Dahlström and Fuxe (1964) were the first to use these techniques to the distribution of cell bodies, axons, and terminals in the CNS containing NA, DA, and 5-HT (Figure 1). Ungerstedt (1971) more recently has extended such studies and added more details of the terminal distribution of the NA and DA pathways (Figure 2).

Figure 1. Formaldehyde-induced fluorescence in catecholamine-containing nerve terminals in rat hypothalamus. A section of tissue was dried and treated with formaldehyde vapor before examination in a fluorescence microscope. The left border of the figure passes through the third ventricle (asterisks), limited at the base by the median eminence (M). A high density of catecholamine terminals is present in the paraventricular nucleus (P) and basal hypothalamus (B), with lower densities in the anterior (A) and ventromedial (N VM) hypothalamic nuclei. Magnification × 70. [From Hokfelt & Ljungdahl, 1972.]

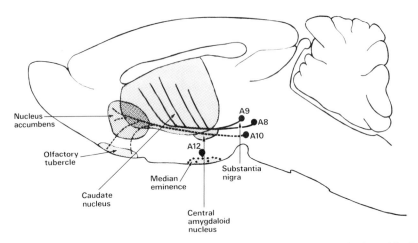

Figure 2. Diagram of the distribution of dopamine-containing neurones in the rat brain. Cell bodies of the dopaminergic neurones are located in groups A8–A12, with axon projections to areas indicated by cross-hatching. [From Livett (1973), modified from Ungerstedt (1971).]

Distribution of NA Pathways

The NA terminals of the forebrain originate from neurone cell body groups located in the pons and medulla, designated cell groups A1, 2, 4, 5, 6, and 7. The neurones of the locus coeruleus (A6) have multipolar axons, one branch innervating the Purkinje cells of the cerebellum and others forming a "dorsal bundle" of NA fibers which terminate in the cortex and hippocampus. The remaining cell groups contribute axons mainly to a "ventral bundle" which innervates the hypothalamus, basal forebrain, and parts of the limbic system. Some of the cell groups, especially A1, also produce descending fibers to the spinal cord where they terminate mainly in the lateral column in the region of the preganglion sympathetic neurones. Recently Olson and Fuxe (1972) have reexamined the organization of cell groups A5, 6, and 7, and the fibers arising from them. The dorsal part of A6 is still designated the locus coeruleus and innervates hippocampus and cortex; the ventral part of A6 together with A7 form the "subcoeruleus" area and gives rise to a fairly thick periventricular plexus along the third ventricle of the hypothalamus and preoptic areas. The NA cell bodies A1 and 2 together with A5 give rise to fibers that innervate the basal and lateral parts of the hypothalamus, preoptic area, and the ventral stria terminalis area This kind of anatomical detail may seem academic but when, for example, suggestions are made about the role of the locus coeruleus, it is important to know that in addition to its innervation of the cortex and hippocampus, this nucleus also innervates the medial hypothalamus.

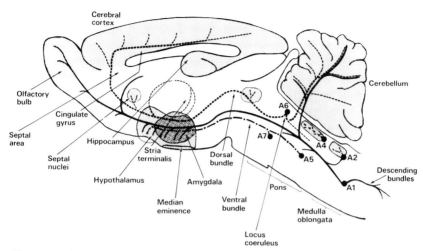

Figure 3. Distribution of noradrenaline-containing neurones in rat brain as in Figure 2. [From Livett (1973), modified from Ungerstedt (1971).]

At the level of the diencephalon the ascending NA fibers in the dorsal and ventral bundles merge in the medial forebrain bundle, and the fibers then separate to make their discrete terminal innervations. Additional methods have been employed to verify the observations made with normal fluorescence histochemistry. Originally pharmacological agents that interact with NA were used to decrease or increase the amine content of NA-containing neurone systems, and the resulting changes in the intensity of fluorescence observed were used to confirm the identity of these neurones.

The value of such manipulations is limited because few drugs interact specifically with only one of the amine systems. Of more value has been the method of local injection of 6-hydroxydopamine, pioneered by Ungerstedt. If 6-hydroxydopamine is injected in the vicinity of cate-cholamine-containing neurone cell bodies, or their axons or terminals, it is taken up and destroys the amine containing neurones. 6-hydroxy-dopamine can be used to verify the distribution of a particular part of an amine-containing neurone system if it is injected locally to destroy a discrete part of the system. Standard histochemical fluorescence methods do not give good resolution of the fine terminals of NA-containing neurones. Recently an immunofluorescence technique for staining the enzyme dopamine-β-hydroxylase (that is exclusively localized in NA neurones) has proved very successful in demonstrating the terminal plexus. Dopamine-β-hydroxylase is purified from the adrenal medulla and a specific antiserum to it produced in rabbits. Frozen sections of brain are incubated with the antiserum and the binding of enzyme antibodies is visualized by coupling to a fluorescent compound. The NA terminal plexus in the cortex has been elegantly demonstrated in this way, and some further details of the forebrain innervation pattern added to the Ungerstedt map (see Livett, 1973). Another recently developed method which gives good resolution of fine amine-containing nerve terminals utilizes glyoxylic acid instead of formaldehyde as a condensation reagent giving rise to fluorescent amine derivatives.

Dopamine Systems

The DA-containing neurone systems lie principally in the midbrain anterior to the NA cell bodies. Cell groups A8 and 9 are located in the substantia nigra and their fibers form the nigrostriatal pathway which projects together with the NA fibers in the medial forebrain bundle at the diencephalic level. The dopamine fibers leave the medial forebrain bundle more anteriorly to innervate the corpus striatum (caudate nucleus and putamen) and globus pallidus. Cell group A10, which lies medial to the substantia nigra, produces an ascending fiber system innervating the nucleus accumbens, amygdala, and the olfactory tubercle. Cell group A12 is contained within the arcuate nucleus of the hypothalamus, and the short axons of these neurones innervate the median eminence. Dopamine-containing neurones are also found in the amacrine cell layer of the retina, which contains no NA or 5-HT neurones.

The contribution of the fluorescence histochemical method to our knowledge of neuroanatomy is well illustrated by the discovery of the nigrostriatal DA pathway. Although this neurone system contains about three-quarters of all DA in the brain the fibers are extremely small and their existence had not been detected with classic silver staining histological methods. After the histochemical demonstration of this pathway with the fluorescence method, more refined modifications of the Nauta silver staining technique were used, and, relatively recently, fibers from the substantia nigra to the striatum were demonstrated with such methods. Experiments are also indicating that this major chemical pathway has an important role to play in motor integration (Ungerstedt); it is doubtful whether its existence would have been known without the application of this sophisticated chemical mapping approach.

Distribution of 5-HT

The location of 5-HT neurones and their fiber and terminal systems in the rat brain were described initially by Dahlström and Fuxe (1964). In the succeeding years the 5-HT system has received less anatomical attention than the NA and DA systems.

In the rat the 5-HT containing neurones are located in the nuclei of the raphe system situated dorsally near the midline of the brain stem. They extend from nucleus raphe pallidus in the caudal medulla to n. raphe dorsalis in the caudal mesencephalon. Some 5-HT cell groups are located more laterally in the n. paragiganto cellularis and in the ventral part of the area postrema. At least some of the ascending 5-HT fibers to the forebrain travel in the medial forebrain bundle. The terminals of the 5-HT axons innervate the pontomesencephalic reticular formation, hypothalamus, lateral geniculate nuclei, the amygdala, pallidum system, hippocampus, anterior hypothalamus and preoptic area, and cortex. The raphe 5-HT neurones also send descending fibers to the spinal cord.

The 5-HT neurone system of the cat brain has been studied in detail, and is essentially similar to that of the rat. It is in the cat that functional evidence was obtained of a crucial role of the raphe 5-HT system in sleep processes (See Jouvet, Chapter 17).

Limitations of the Fluorescence and Immunofluorescence Techniques

These techniques have been invaluable for plotting the distribution of amine pathways in the CNS. But they are essentially qualitative or at best semiquantitative techniques and can indicate only whether or not a given structure contains amine terminals.

They are not strictly quantitative, although many workers have used the density of fluorescence as an index of the amount of functional innervation present. Functional studies of amine transmitter systems suggest that very small changes in the amount of functional transmitter can cause changes in behavior. Differences of this magnitude cannot be detected by fluorescence histochemistry. Biochemical assay provides the complimentary technique for quantifying transmitter concentration in localized brain areas. For example Ungerstedt (1971) reported that the nigrostriatal fibers innervate the striatum (caudate and putamen) and globus pallidus and more recent biochemical assay of these segments of the basal ganglia reveal a much higher concentration of DA in the striatum than in the globus pallidus (Broch & Marsden, 1972).

More Recent Modifications of Fluorescence Techniques to Improve the Visualization of Catecholamine Innervation in the Forebrain

The original description by Dahlström and Fuxe of the catecholamine and indolamine cell groups in the mid and hind brain and their principal fiber projections has stood well the test of time and scrutiny. Recent technical innovations have mainly improved the resolution of the terminal distribution in the forebrain. The immunofluorescence techniques have already been referred to but there are also recent modifications of the Falck–Hillarp technique itself.

With the original technique a premium was placed on rapid freezing and drying of the brain tissue before sections were cut for processing. This procedure results in tissue which tends to fragment very rapidly during the physical manipulation of sectioning, and it was, therefore, difficult to obtain complete brain sections.

Hokfelt and Ljungdahl (1972) have recently used a novel sectioning device, the "vibratome," to obtain very thin (10 μm) sections of fresh or formalin fixed brain tissue. With this method the intact section is obtained on a slide; the thinness of the sections ensures very rapid drying which is a prerequisite to prevent amine leakage and diffusion from amine neurones prior to the formaldehyde reaction. Both the physical integrity of the tissue and the thin sections improve the visualization of the amine terminals. An added advantage is that the tissue can be initially fixed with formalin *in vivo* and thus some of the vibratome sections can be stained by other conventional histological procedures, to demonstrate other histological or histochemical properties of the tissue, e.g., ACh esterase or the Fink–Heimer silver impregnation method to demonstrate fine unmyelinated terminal axons.

Lindvall, Björklund, Hökfelt, and Ljungdahl (1973) have reported that perfusion of the brain with glyoxylic acid before vibratome sectioning followed by condensation with glyoxylic acid vapor improves yet further the terminal resolution, at least in certain brain areas. For example, the original studies reported sparse NA innervation of the thalamus, but with the glyoxlic acid method fine NA terminals innervating particularly the anterior, lateral, and ventral nuclear groups have been found. These thalamic axons were more delicate and less fluorescent than those in the hypothalamus, which are readily demonstrated with the standard techniques.

DISTRIBUTION OF ACh ESTERASE

Lewis and Shute made use of the acetylcholinesterase staining technique for the detailed mapping of cholinergic pathways in the rat CNS. Their original maps (Shute & Lewis, 1967) remain the most comprehensive description of the acetylcholine systems, if one accepts that the enzyme stain is a reliable indicator of cholinergic neurones (Figure 4).

In this technique the brain is perfused *in vivo* with formaldehyde and sectioned on a freezing microtome. Alternate sections are incubated with butyrylthiocholine as a substrate for nonspecific pseudocholinesterase (ChE) or acetylthiocholine together with a specific pseudocholinesterase inhibitor to demonstrate true acetylcholinesterase (AChE). The sections are then treated with sodium sulfide which gives a black staining at sites of enzyme activity. As AChE is distributed throughout the neurone it is impossible to determine from any given section whether stained axons are efferent or afferent to a given brain structure. Lewis and Shute, however, discovered that AChE activity changed in different ways on the two sides of a cut axon. The enzyme accumulated on the cell body side of a cut, and disappeared on the terminal side. The combination of selective lesions with the enzyme staining technique helped in plotting the neuronal pathways. So did biochemical assays for the enzyme

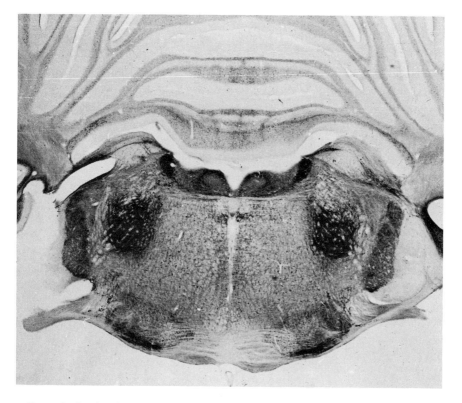

Figure 4. Section through rat brain stem stained histochemically to demonstrate localization of acetylcholinesterase activity. Note prominent staining of the cholinergic neurones in the motor nucleus of the trigeminal nerve (black areas in lateral brain stem) and weak staining of the overlying cerebellar structures. [From Dr. P. Lewis (unpublished material).]

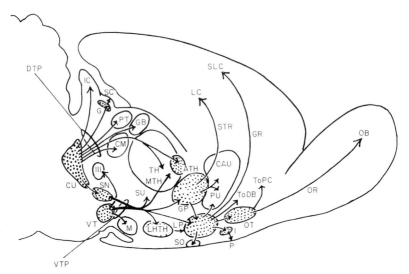

Figure 5. Diagram of the distribution of acetylcholinesterase staining (presumaptive cholinergic) neurones in rat forebrain. Cell bodies with positive reaction are present in nuclei indicated by stippling, and the direction and distribution of the axons of these neurones is shown by the arrows. [From Shute & Lewis, 1967.]

choline acetyltransferase, the enzyme concerned in the synthesis of the ACh which is an exclusive marker for cholinergic neurones. After lesions to a cholinergic pathway the activity of both AChE and choline acetyltransferase fell in a parallel manner. With this combination of techniques, for example, the cholinergic input to the hippocampus via the fornix was demonstrated. There are two main cholinesterase-containing pathways projecting to the rat forebrain (Figure 5). The dorsal tegmental pathway which arises mainly from the nucleus cunciformis situated in the dorsolateral part of the mesencephalic reticular formation, and ascends to innervate the tectum, pretectal area, geniculate bodies, and the thalamus. The fibers of the ventral tegmental pathway arise from the substantia nigra and ventral tegmental area of the midbrain and innervate the basal forebrain areas from which additional AChE containing neurones project to all regions of the cerebral cortex and olfactory bulb. These pathways are considered part of the ascending reticular formation and are implicated in arousal.

Unfortunately the enzyme cholinesterase is not exclusively limited to the synaptic region of cholinergic neurones. It is present in the cell body, the axons and synaptic terminals of cholinergic neurones, but may also be present in the cell membranes of neurones that receive a cholinergic input especially in the cell bodies and dendrites of such cells. AChE staining of cell bodies cannot, therefore, be used as the sole criterion that a neurone is cholinergic, and a misleading picture could be obtained. For example, the noradrenergic cell bodies of the locus coeruleus stain intensely for acetylcholinesterase, as do the noradrenergic neurones of peripheral sympathetic ganglia, yet these are almost certainly not cholinergic neurones.

It is likely that reliable histochemical or immunochemical methods to localize the enzyme choline acetyltransferase will be developed in the future, and these should provide a less ambiguous method for mapping cholinergic pathways since there is little doubt that this biosynthetic enzyme is strictly localized to cholinergic neurones.

DISTRIBUTION OF GABA AND GLYCINE

These two amino acids seem to be the most important inhibitory transmitter substances in the mammalian CNS. Little is known, however, of the pathways in which they are used. Glycine acts only in the spinal cord and lower brain stem as an inhibitory transmitter. It seems to be localized in the spinal cord in small inhibitory interneurones located in the medial gray and projecting to motor neurones and other cells in the ventral horn. There is, however, no histochemical procedure that allows such neurones to be visualized directly. Glycine neurones possess a specific high-affinity uptake process for glycine and this may allow the development of autoradiographic techniques to visualize such neurones after labeling them with radioactive glycine.

The identity and distribution of GABA neurones is similarly unknown at the present. In spinal cord biochemical estimations and neuropharmacological evidence suggest that GABA is largely confined to nerve terminals in the dorsal horn. In supraspinal regions GABA is present in substantial concentrations in nearly all gray matter, where it is probably the major inhibitory substance. Few long-axon pathways involving this transmitter, however, have been described. Although histochemical procedures for visualizing sites of GABA metabolism have been developed, these are not considered reliable indicators for GABA-containing neurones. Autoradiographic techniques are also under development, based on the existence of specific high-affinity uptake sites for labeled GABA in neurones using this transmitter. Such autoradiographic studies at the electron-microscope level have indicated that in many regions of the brain GABA may be the transmitter in up to a third of all the synaptic terminals. Biochemical findings after various experimental lesions suggest that some long-axon pathways containing GABA may exist. There is good evidence that the Purkinje cells and their axons, terminating in the various cerebellar nuclei use GABA as their transmitter. Lesion studies also indicate the existence of a pathway of GABA containing fibers from the globus pallidus (and possibly other striatal regions) terminating in the substantia nigra, which contains the highest density of GABA terminals in the brain (Robinson & Wells, 1973).

Until reliable autoradiographic methods, or immunochemical techniques based on localizing the GABA biosynthetic enzyme glutamic decarboxylase, are available little progress can be made in understanding the detailed mapping of GABA pathways.

REFERENCES

Broch, O. J., & Marsden, C. A. Regional distribution of monoamines in the corpus striatum of the rat. *Brain Research,* 1972, 38, 425–428.

Dahlström, A., & Fuxe, K. A method for the demonstration of monoamine containing nerve fibers in the central nervous system. *Acta Physiologica Scandinavica,* 1964, 60, 293–295.

Hokfelt, T., & Ljungdahl, A. Modification of the Falck-Hillarp formaldehyde fluorescence method using the vibratome (R): Simple, rapid and sensitive localization of catecholamine in sections of unfixed or formalin fixed brain tissue. *Histochimie,* 1972, 29, 325–339.

Lindvall, O., Björklund, A., Hökfelt, T., & Ljungdahl, A. Application of the glyoxylic acid method to vibratome sections for the improved visualization of central catecholamine neurons. *Histochimie,* 1973, 35, 31–38.

Livett, B. G. Histochemical visualization of adrenergic neurones, peripherally and in the central nervous system. *British Medical Bulletin,* 1973, 29, 93–99.

Olson, L., & Fuxe, K. Further mapping out of central noradrenaline neuron systems: Projections of the subcoeruleus area. *Brain Research,* 1972, 43, 289–295.

Robinson, N., & Wells, F. Distribution and localization of sites of gamma aminobutyric acid metabolism in the adult rat brain. *Journal of Anatomy,* 1973, 114, 365–378.

Shute, C. C. D., & Lewis, P. R. The ascending cholinergic reticular system: Neocortical olfactory and subcortical projections. *Brain,* 1967, 90, 497–520.

Ungerstedt, U. On the anatomy, pharmacology and function of the nigro-striatal dopamine system. *Acta Physiologica Scandinavica,* 1971, Suppl. 367.

Chapter 6

Central Neurotransmitters and the Regulation of Behavior

SUSAN D. IVERSEN
LESLIE L. IVERSEN

University of Cambridge, England

INTRODUCTION

Chemical Transmission in the Central Nervous System

There is a growing interest in the neurochemical basis of behavior, and in this field many investigators have focused on the role of chemical transmitter substances in the CNS. This is not because other aspects of brain chemistry are not important in influencing behavior but rather that the study of chemical transmission offers a convenient way of exploring the interrelationships between chemical events in the nervous system and behavior. The use of chemical transmitters is one way in which the biochemical functions of the brain differ from those of most other tissues. The nature of the biochemical mechanisms involved in chemical transmission and the way in which they can be manipulated by drugs are already understood because of the work that has been done on chemical transmission in various parts of the peripheral nervous system.

Fortunately, several of the substances used as transmitters in the peripheral system are also involved in synaptic transmission in CNS.

Our knowledge of chemical transmission in CNS, however, is far from complete. Although it is generally recognized that synaptic transmission between neurons in the nervous system usually involves the release of a chemical transmitter substance from the presynaptic terminal, the identity of the particular transmitters used at various points in particular neural pathways in CNS is obscure. We do not even yet have a complete list of all the substances involved as transmitters in the CNS. The criteria that have to be met before a substance can be established to be a transmitter can be summarized briefly as follows: (1) There are specific receptors on postsynaptic membranes with which the substance interacts to cause inhibition or excitation of neuronal activity, usually through specific ionic permeability changes in the postsynaptic membrane; (2) the substance is present in presynaptic nerve terminals in the appropriate neuronal pathways; and (3) special mechanisms exist in the presynaptic terminal for the storage of the substance—usually in synaptic vesicles—and for its release in response to the arrival of action potentials at the presynaptic nerve terminals.

The present list of CNS transmitters is as follows:

Acetylcholine   (ACh)
Noradrenaline   (NA)
Dopamine   (DA)
5-Hydroxytryptamine   (5-HT)
α-Aminobutyric acid   (GABA)
Glycine   (spinal cord only)
Glutamic acid   status doubtful
Aspartic acid   status doubtful

This is probably by no means the complete catalogue of CNS transmitters; for example, the transmitter(s) used by primary and secondary neurons in most sensory pathways do not appear to be any of those just listed.

Chemical Pathways in the Brain

Our understanding of the functional importance of the various different chemical transmitters in CNS has been increased greatly in the last decade by the availability of histochemical and cytochemical methods which have allowed at least a preliminary mapping of the occurrence of these chemicals in particular neuronal pathways in the brain and spinal cord (Livett, 1973). This information is reviewed in our accompanying Chapter 5. It is clear that substances such as ACh, NA, DA, and 5-HT are present in many distinct pathways, and it seems a priori unlikely that each of these substances has only one role to play in cerebral functions. In the periphery, for example, ACh mediates neuromuscular transmission in all skeletal muscles, but it also acts as transmitter in peripheral ganglia, and in tissues innervated by the parasympathetic nervous system. We should clearly think of the functional importance of a particular transmitter in a specific neuronal pathway when discussing the functions of CNS transmitters. This, unfortunately, is advice which is extremely difficult to follow in practice.

For example, the mapping of CNS transmitters does not yet extend to the amino acid transmitters, such as GABA, glycine, glutamate, and aspartate—for which histochemical mapping techniques do not exist. How are we to investigate the functional importance of such chemicals without this information? It is undoubtedly because so much is known of the distribution of the adrenergic, cholinergic, and tryptaminergic (5-HT) neurons in CNS, and because an armory of drugs exist for the manipulation of these systems, that almost all attention has focused on these three transmitters so far. Even with these substances, however, we must admit that although the distribution of the pathways using them is fairly well known, the neurophysiological functions served by these pathways remain largely obscure.

Experimental Approaches

Administration of Transmitters or Their Precursors

An obvious way in which to determine the functional importance of a particular transmitter is to administer the substance to an experimental animal and observe the effects. Most transmitter substances, however, do not penetrate at all readily into the brain from the blood, so that systemic administrations are usually quite ineffective. One way of circumventing this problem has been to administer instead some substance that does penetrate into the CNS from the blood, and which can be converted into one or other of the transmitters by metabolic conversion in the CNS. The amino acid L-DOPA, for example, can be administered systemically and enters the brain where it is converted by the normal biosynthetic pathways into DA and NA (Figure 1). Similarly the amino acid 5-hydroxytrytophan is used as a precursor for brain 5-HT (Figure 1). No analagous procedures exist, however, for introducing precursors for any other transmitters in CNS. The use of L-DOPA

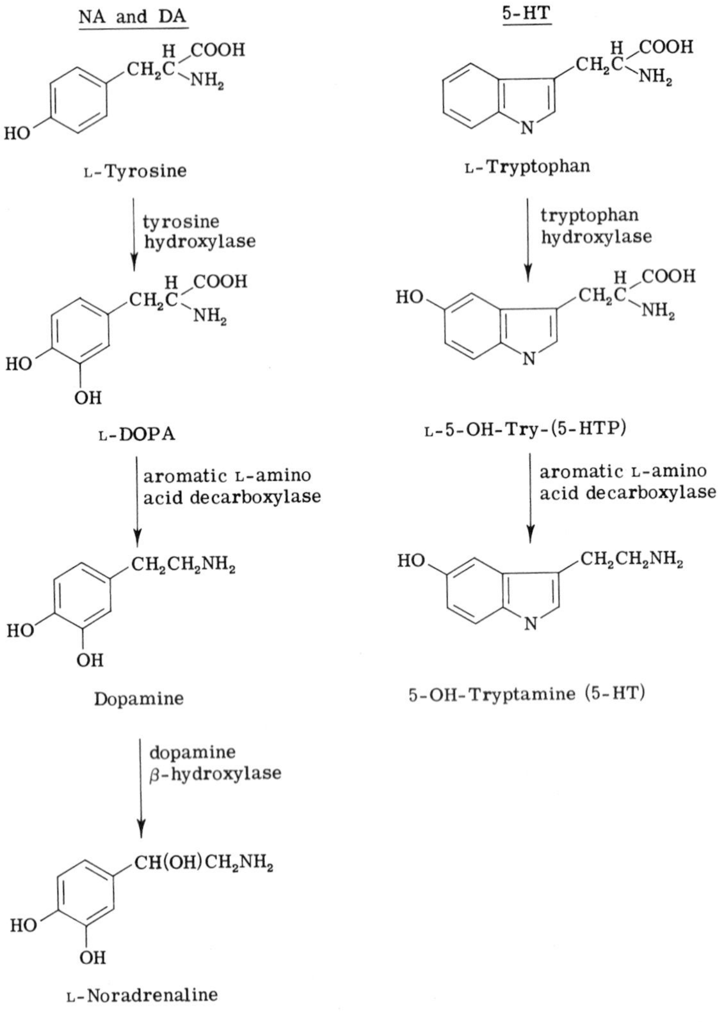

Figure 1. Biosynthesis of the monoamine transmitters dopamine (DA), noradrenaline (NA), and 5-hydroxytryptamine (5-HT).

and 5-hydroxytryptophan should also be viewed with some caution. Although these amino acids do penetrate into neurons in the brain, they are also decarboxylated to the corresponding amines (DA and 5-HT) in the walls of cerebral blood vessels, and it is possible that DA and 5-HT may be formed after administration of these substances in both adrenergic and tryptaminergic neurons, since both types of neurons appear to share a common decarboxylase enzyme, capable of using either amino acid as substrate. Thus although L-DOPA and 5-hydroxytryptophan do lead to a formation of the corresponding amines in brain, at least some of the amines produced may be in entirely inappropriate cellular locations, which do not normally contain the amine.

Another solution to the problems raised by the blood–brain barrier has been to use animals in which this barrier is defective or incomplete. In newly hatched chicks, for example, the blood–brain barrier does not appear to exist, and substances such as the catecholamines, 5-HT, and GABA, penetrate readily into CNS after systemic administrations. On the other hand, the neurochemical and neuropharmacological substrates for transmitter action are also not fully developed in young animals.

A direct way of circumventing the blood–brain barrier is to administer the substances directly into the brain or cerebrospinal fluid (CSF). This approach has been widely used, involving injections or infusions into CSF—often through permanently implanted microcannulae, which permit injection in unanesthetized animals. Alternatively, microinjections or implants of solid crystals of chemical may be made through microcannulae previously implanted using stereotaxic coordinates into various particular areas of the brain. The latter approach has the advantage that a particular area of CNS may be selectively stimulated by the chemical, whereas with CSF injections the entire brain is flooded with the substance. By perfusing various parts of the ventricular system, however, it is also possible to expose restricted areas of CNS to administered substances. The administration of solutions or solids directly into brain tissue inevitably leads to tissue damage at the site of administration and along the track of the cannula. Such microinjections or applications also lead to a spread of the administered substance for an indeterminate distance from the site of application. Unless very small injection volumes are used (less than 1 μliter) this spread is likely to be as much as 1 mm from the site of application, and since most studies of this type are performed in animals with small brains, the neuroanatomical precision with which such applications can be made is only poor. However, it is unlikely that extremely localized chemical stimulation of small numbers of neurons in the brain would be effective in causing sufficiently large changes in brain function to be reflected in a change in behavior. Extremely localized application of transmitters and drugs can be made through microelectrodes using the technique of expelling drugs by iontophoresis. This technique has been a powerful tool for studies of the pharmacological and neurophysiological properties of transmitter receptors in CNS, but it seems unlikely to be useful for behavioral studies. This approach is also largely restricted, for technical reasons, to fully anesthetized or immobilized animal preparations.

A puzzling feature of most of the studies involving direct injections of transmitter amines such as ACh, NA, DA, or 5-HT into CSF or brain is that rather high doses of amine must usually be administered before any behavioral changes are elicited. Thus, the whole rat brain contains only about 1 μg each of the amines NA, DA, and 5-HT, but injections of 5–10 μg of these amines are commonly used, and much larger amounts have sometimes been used. There have been disappointingly few attempts to take such considerations into account; in most studies involving intracerebral injections, for example, no attempt has been made to determine the precise spread of injected amine from the site of injection.

Experimental Lesions of Transmitter-Specific Pathways

With the improved understanding of the anatomical distribution of transmitters in specific pathways in CNS a promising approach to studies of the functions of these pathways is to produce selective lesions in them and then to examine the functional and behavioral consequences. This is an approach which has already contributed importantly to our understanding, for example, of the role of the 5-HT pathways in the control of sleep patterns (see chapter by Jouvet). Surgical lesions of the various catecholamine-containing pathways have also been widely used.

A new technique has also recently become available for the production of experimental lesions in catecholamine-containing neurones by use of a selective chemical neurotoxic agent. The compound 6-hydroxydopamine (6-OHDA) (Figure 2) is a close relative of NA and DA structurally, and it is selectively taken up and concentrated in NA- and DA-containing neurons. The substance is also chemically labile and causes degenerative changes in these neurons, leading, in the CNS, to an irreversible lesion of the nerve terminals. The 6-OHDA can be administered either into the CSF, or by local microinjections into specific brain regions. The latter approach allows the production of chemical lesions in particular adrenergic pathways, and these may be placed bilaterally or unilaterally. For example, injection of a few microgrames of 6-OHDA in a very small volume (2–4 μliter) into the substantia nigra of rat brain causes a permanent destruction of virtually all the fibers in the nigro-

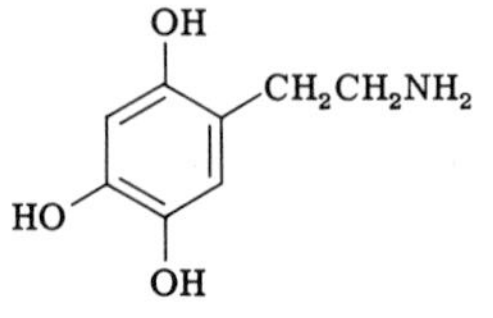

Figure 2. 6-Hydroxydopamine.

striatal dopaminergic system. The use of 6-OHDA has several advantages over more conventional surgical lesions: it requires a less accurate placement of the lesion, since the chemically selective nature of the 6-OHDA lesion is effective for some distance from the site of injection, furthermore, although the injection of 6-OHDA causes some unspecific tissue damage at the site of injection, it is far more selective than any other lesion technique, in which a large amount of damage is inevitably done to structures other than those intended.

Correlation of Behavioral Changes with
Alterations in Transmitter Turnover or Release

Direct electrical recordings of the activity of neurons in various transmitter-specific pathways during different behavioral states has so far proved technically difficult or impossible. An indirect approach to this goal, however, is offered by biochemical measurements of the rate of turnover or release of various amine transmitters in animals in different behavioral states. Much earlier work has involved attempts to correlate changes in functional activity in transmitter-specific pathways with changes in the tissue content of amines such as NA, DA, or 5-HT. It is generally recognized, however, that such changes in the steady-state levels of amines can only reflect very extreme changes in the underlying state of activity of the neurons containing them. Changes in neuronal activity are associated with changes in the rate of release of transmitter, which in turn are compensated for by an increased rate of transmitter biosynthesis. Thus, under most conditions there are changes in the rate of turnover of transmitters in brain, without any changes in their steady-state concentrations in the tissue. Biochemical measurements can give an index of the dynamic state of various transmitters, which in turn can indicate which pathways are activated during various behavioral manipulations. Such experiments usually involve the injection of a small amount of radioactivity labeled amines (NA, DA, or 5-HT) into the CSF or brain. The various transmitter-specific neurons each possess a special "amine pump," a specific uptake mechanism for their own particular amine (Iversen, 1971). The labeled amines are thus selectively concentrated in the appropriate neurons, although NA- and DA-containing neurons are capable of taking up both catecholamines, so some overlap occurs here. The rate of turnover of the transmitter pool in these neurons can then be estimated by following the rate of disappearance of the radioactive amine transmitters from brain, or from particular brain regions. The rate of release of labeled amines from nerve terminals can also be estimated more directly by perfusing the ventricular system of the brain or the surface of the brain, and measuring the amounts of radioactive amine released into the perfusing fluid from the surrounding brain tissue.

Neuropharmacological Manipulation of Transmitter Functions in the Brain

One of the most important experimental approaches has been the study of the behavioral actions of drugs which more or less specifically alter the properties of one or another of the transmitter systems in the brain. Since virtually nothing is yet known of the pharmacology of other transmitter systems, this approach has in practice been restricted to drugs which interact with NA, DA, 5-HT, or ACh. Most of the drugs used were discovered and studied first on peripheral adrenergic or cholinergic junctions. Not all agents that are active in the periphery, however, are suitable for CNS studies, because of the inability of many drugs to penetrate into the brain from the blood. Drugs that affect synaptic transmission can be in any of the following categories:

Receptor Agonists and Antagonists

These are compounds that mimic or antagonize the actions of the naturally occurring transmitter at postsynaptic receptor sites.

Inhibitors of Transmitter Synthesis

These are substances that depress synaptic transmission by blocking the normal replacement of the transmitter in the presynaptic terminal by biosynthesis.

False Transmitters

These are substances that are taken up and stored in presynaptic terminals and released in place of the naturally occurring transmitter. Such substances are usually less effective in stimulating the postsynaptic receptors and thus effectively depress synaptic transmission. This group also includes chemicals that can be converted by the normal biosynthetic enzymes into false transmitters.

Inhibitors of Transmitter Inactivation

These are compounds that potentiate and prolong the actions of the naturally occurring transmitter by inhibiting the mechanisms normally responsible for terminating the actions of the transmitter at postsynaptic receptor sites. This may involve either a degradative enzyme (e.g., acetylcholinesterase) or an "amine pump" mechanism, or both. For example, in monoaminergic neurons the degradative enzyme monoamine oxidase

(MAO) is present in the nerve terminals using the catecholamines NA and DA or 5-HT and is important in regulating the storage level of these amines, but these nerve terminals also possess specific uptake mechanisms, which appear to be responsible for terminating the actions of the released amine transmitters. Inhibition of either mechanism will tend to potentiate synaptic transmission in such neurons.

Depleting Agents

The main example is the alkaloid reserpine, which blocks the normal storage of NA, DA, and 5-HT in synaptic vesicles and, thus, leads to a profound and long-lasting depletion of these three amines from the brain, and a block in synaptic transmission in neurons using these transmitters.

Displacing Agents

These are compounds that cause a release of the naturally occurring transmitter onto receptor sites by displacing it from neuronal storage sites. Such substances are themselves usually without direct affects on postsynaptic receptors, but cause an indirect stimulation of these receptors by releasing the endogenous transmitter.

Examples of these different types of drug are summarized in Table 1. Although the neuropharmacological approach has been very widely used, it too suffers from several weaknesses. Probably the most important problem concerns the lack of pharmacological specificity in the drugs available for this type of study. Reserpine, for example, is a powerful tool with a unique neuropharmacological action and profound behavioral effects—but it does not distinguish between systems using the monoamines, NA, DA, and 5-HT, all of which are affected. Indeed very few drugs exist whose actions can be stated confidently to be completely specific for one particular transmitter system; unfortunately many studies in this area have been performed with a quite unwarranted degree of faith in the specificity of the drugs used. Even within a particular transmitter system, drugs tend to have multiple sites of action; for example, amphetamine is a displacing agent for NA (and DA), but it is also an inhibitor of the enzyme MAO, and it inhibits the uptake of NA and DA. Since transmitters must interact with various different biochemical mechanisms, including the receptor sites, the inactivating enzymes, storage mechanisms, and uptake sites, we should not be surprised to find that drugs—which generally resemble the transmitter chemically—also interact with more than one target. Thus, while pharmacology has provided a battery of "magic bullets," on closer examination these are often found to be quite blunt instruments.

TABLE 1 Some Commonly Used Drugs and Their Actions on Transmitter Systems

Type of action	Transmitter system			
	Noradrenaline	Dopamine	5-Hydroxytryptamine	Acetylcholine
Receptor stimulant (+)[a]	Isoprenaline Clonidine	Apomorphine ET 495[c]	Tryptamine	Oxotremorine Arecoline Carbachol
Receptor antagonist (−)[b]	Propranolol Chlorpromazine Phenoxybenzamine	Chlorpromazine Pimozide Clozapine Haloperidol	Lysergic acid-diethylamide Methysergide	Scopolamine Atropine Benztropine
Inhibitor of uptake (+)	Cocaine Amitriptyline Desipramine	Cocaine Amphetamine	Chlorimipramine Imipramine	—
Inhibitor of metabolic breakdown (+)	MAO inhibitors: pheniprazine, iproniazid, pargyline, tranylcypromine, phenelzine			Di-isopropylfluorophosphate (DFP) Physostigmine
Inhibitors of biosynthesis (−)	α-Methyl-p-tyrosine Disulphiram, FLA-63[d]	α-Methyl-p-tyrosine	p-Chlorophenylalanine	Hemicholinium
Displacing agent (+)	d-Amphetamine	d-Amphetamine l-Amphetamine	—	—
Precursor (+)	L-DOPA	L-DOPA	L-5-Hydroxytryptophan	—
False transmitter (±)	α-Methyl-m-tyrosine α-Methyl DOPA, Metaraminol			
Depleting agent (−)	reserpine, tetrabenazine, and related substances			

[a] (+) = stimulates or enhances actions of transmitter.
[b] (−) = antagonizes or decreases actions of transmitter.
[c] ET 495 = 1-3,4-methylenedioxybenzyl-4(2-pyrimidye) piperazine.
[d] FLA 63 = bis (4 Methyl-1-homopiperazinyl thiocarbonyl) disulphide.

DOPAMINE

Introduction

In this and succeeding sections we will consider how the experimental approaches previously outlined have been applied to studies of the functional significance of specific transmitter systems in the brain. This treatment is restricted to the amines DA, NA, and ACh, because so far these are the only transmitters that have been studied in sufficient detail.

Until recently little attention has been paid to the transmitter DA, but once it was appreciated that discrete neuronal pathways containing this substance exist in the brain, and that an abnormality in DA systems underlies Parkinsonism in man (Hornykiewicz, 1973), it became important to study the functional role of these systems.

The accompanying chapter describes in detail the anatomical distribution of NA- and DA-containing neurons in the brain (see Chapter 5, Figure 2). Groups of DA neurons in the midbrain send axons in the medial forebrain bundle to the tuberculum olfactorium, nucleus accumbens and corpus striatum (caudate and putamen). The system of DA neurons projecting from the substantia nigra to the corpus striatum, the nigro-striatal pathway, is the one that has been most extensively studied. The fibers in this pathway are of very small diameter (less than .5 μm) and unmyelinated, and until the application of the specific fluorescence histochemical technique for catecholamines their existence had not been revealed by traditional anatomical silver staining techniques.

Dopamine and Parkinsonism

The substantia nigra and corpus striatum are extrapyramidal motor structures, and patients with Parkinson's disease, which is associated with extrapyramidal motor symptoms, including akinesia, rigidity, and tremor, show degeneration and loss of DA from the nigro-striatal pathway. These clinical findings, together with improved understanding of the biosynthesis of catecholamines in brain, provided the impetus for the successful introduction of L-DOPA for the treatment of Parkinson's disease. Remarkable improvements in motor performance are seen in many cases of Parkinsonism after L-DOPA treatment. This compound is a precursor for DA synthesis in the brain, and presumably increases the functional efficiency of the inadequate dopaminergic mechanism of the striatum by replacing some of the missing amine. The tremor of Parkinsonism, which is thought to be of thalamic origin, does not, however, respond as well as to L-DOPA therapy. In man the facts, thus, suggest a role for DA in certain forms of motor control (Papeschi, 1972).

These clinical findings have occasioned much experimental work in animals, seeking to define the importance of endogenous DA for normal motor control. In the rat, electrolytic lesions of the substantia nigra result in hypokinesia and hind-limb rigidity. More recently 6-OHDA has been successfully used to induce more selective damage to the nigro-striatal pathway and has resulted in the same motor symptoms. However, the monkey provides a more relevant experimental model for studies of Parkinsonism. Lesions to the substantia nigra again result in hypo-kinesia and rigidity associated with loss of striatal DA. But if the lesions extend to include the ventral tegmental region of the midbrain, the Parkinson tremor symptom is also obtained. Poirier, Bedard, Boucher, Boavier, Larochelle, Olivier, Parent, and Singh (1969) suggest that 5-HT depletion is a corollary of the tegmental damage, and that this amine may be responsible for the tremor symptoms. As in humans the hypo-kinesia and rigidity produced by experimental lesions in monkeys re-spond well to L-DOPA, the tremor does not, and can be abolished by ventrolateral thalamic lesions. However, despite the progress in under-standing and treating Parkinsonism it should not be assumed that our knowledge is complete. The DA terminals are found in several sites in addition to the caudate nucleus, although the nucleus has been the focus for most experimental work. Parkinsonism shows many different forms, and it is possible that the different etiologies and courses of devel-opment of the disease result from different patterns of local pathology. In developing the animal model system, Poirier has noted that damage to other parts of the extrapyramidal motor circuitry, in particular the rubrocerebeller circuit, seems to contribute in some unspecified way to the motor disturbance.

Furthermore, while the rationale for treating Parkinson patients with L-DOPA is sound, the outcome of the treatment is not successful in all patients. An inverse correlation appears to exist between the extent of damage to the DA systems, as measured by decreased concentrations of the DA metabolite homovanillic acid in CSF before DOPA treatment starts, and the clinical efficacy of the drug. This seems to suggest that the more complete the DA loss the more successful the treatment may be.

Actions of Drugs That Interact with Dopamine Neurons

Perhaps the clearest evidence for a dopaminergic involvement in nor-mal motor mechanism comes from studies of the consequences of the blockade of DA receptors in the brain with specific antagonist drugs such as chlorpromazine. At high doses chlorpromazine depresses unconditioned behavior, such as locomotor activity, and also conditioned responding in operant situations. At such doses of chlorpromazine extrapyramidal motor symptoms are often seen. Indeed, extrapyramidal Parkinson-like symptoms constitute one of the major undesirable side effects encountered in the use of phenothiazines and other neuroleptics, including the butyrophenones, to treat psychosis in man. Repetitive chewing, involuntary jaw movements, and neck tension are often reported as side effects with these drugs.

Just as pharmacological blockade of DA receptors results in abnormal extrapyramidal activity, so does overstimulation of these receptors. Excessive doses of L-DOPA have such an effect in normal animals and induce intense extrapyramidal activity in monkeys who have motor disturbances after midbrain ventral tegmental lesions. The movements may be slight, such as lip-smacking or tongue-rolling, or extreme with grimacing, associated with choreiform movements of the head and upper and lower extremities. These are very similar to the undesirable choreiform involuntary movements seen in Parkinson patients after large doses of L-DOPA. A common practice is to gradually increase the dose of L-DOPA to achieve maximum relief from the akinesia and rigidity of Parkinsonism until extrapyramidal side effects are encountered.

Intense stimulation of DA receptors with the drug apomorphine, which mimics the action of DA at such receptors (Ernst, 1967), demonstrates most clearly the normal role of striatal DA mechanisms in holding motor mechanisms in balance. The striatum is the most anterior part of an extensive system of neural circuitry controlling involuntary motor acts which are normally emitted in the balanced arrangement required for a particular behavior. The administration of apomorphine to experimental animals, such as the rat, results in an abnormal type of behavior called stereotypy. The rat normally generates a specific sequence of acts in an environment, but during stereotypy one element of the behavioral repertoire is repeated to the exclusion of all other forms of behavior. When placed in a novel environment, for example, instead of walking around, rearing, sniffing the walls, etc., the rat may stand in one place and make a small repeated movement of the neck for periods of up to two hours. Any element of the normal behavioral repertoire may be involved in stereotypy, although neck movements, sniffing, and compulsive gnawing are most often seen in the rat. This type of behavior has been extensively studied in the laboratory of Randrup and his colleagues (Munkvad, Pakkenberg, & Randrup, 1968). Stereotypy can also be produced in animals by drugs that induce a release of endogenous DA in the striatum. The stimulant drug *d*-amphetamine has several effects on the amine transmitter systems, but it is generally accepted that the induction of stereotypy by this compound is related to its ability to release endogenous DA in the brain. The result is again the repetition of an element from the normal behavioral repertoire, or if the rat has been trained to make a certain response, e.g., bar pressing for food, a stimulation of that response to the exclusion of all other behavior. For example, a rat trained to press a bar on a fixed ratio operant schedule for food may press repeatedly without accepting reinforcement. Elevation of brain DA content by the administration of large doses of the precursor L-DOPA can produce similar effects, and, in social situations, rats treated in this way may maintain bizarre catatonic social postures for long periods of time. A form of stereotyped social interaction is often seen when apomorphine treated rats are put together. The animals show elements of normal aggressive behavior, rearing on hind legs, vocalization, and boxing with the forepaws, but the rats remain as it were transfixed in this behavior pattern usually without fighting. The cholinergic system plays a modulatory role in these stereotyped behaviors, as it has been shown that atropine blocks the social aggression and anticholinergics potentiate the licking, biting, and sniffing stereotypy.

Randrup has stressed the value of drug-induced stereotypy as a potentially useful animal model system for evaluating antipsychotic drugs. Pretreatment with DA receptor blocking agents such as chlorpromazine and haloperidol prevents amphetamine-induced stereotypy. All of the drugs which are effective neuroleptics have been found to have dopamine-blocking activity in animal tests. Such results have encouraged the speculation that psychotic states in man may be related to some abnormality in dopaminergic systems. This hypothesis is strengthened by the fact that large doses of amphetamine can consistently induce a psychotic state in human volunteers or in addicts, which is difficult to distinguish clinically from paranoid schizophrenia (Ellinwood, Sudilovsky, & Nelson, 1972). However, caution must be exercised in making such correlations purely on the basis of neuropharmacological evidence. Phenothiazines and other neuroleptics block NA as well as DA receptors in the brain. The L-DOPA increases brain NA as well as DA amphetamine influences NA, 5-HT, and the enzyme monoamine oxidase, for example.

More convincing evidence for the existence of a correlation between stereotypy and brain DA has come from studies which combine neuropharmacological and anatomical methods. For example, using cannulae implanted in the rat striatum for the injection of drugs, Fog has demonstrated directly many of the neuropharmacological effects of drugs interacting with DA mechanisms which had been inferred, after systemic administration of these drugs. Parahydroxyamphetamine or DA, for example, injected bilaterally into the striatum induced hyperactivity and stereotypy. Intraperitonal or intrastriatal injections of neuroleptics prevented amphetamine-induced stereotypy. Large surgical lesions of the striatum inhibited amphetamine stereotypy, and also prevented the catalepsy normally induced by the injection of neuroleptics.

The behavioral effects mediated by DA release can also be abolished if the terminals in the corpus striatum are selectively damaged by lesions to the substantia nigra. The results obtained with electrolytic lesions are difficult to interpret, as the lesion inevitably damages all ascending fibers in the vicinity of the substantia nigra, as well as descending connections. However, by using stereotaxically placed microinjections of 6-OHDA a selective lesion can be made to the DA synthesizing cells in substantia nigra, and such lesions also block amphetamine-induced stereotypy.

These results, together with those of Fog (1972) on striatal lesions, strongly suggest that the nigro-striatal pathway is crucial for the induction of stereotypy. However, it should be borne in mind that other DA containing neurone groups in the vicinity of the substantia nigra also give rise to fibers that innervate the tuberculum olfactorium and the nucleus accumbens. Certainly 6-OHDA lesions of the substantia nigra would be expected to damage DA fibers in all these areas in addition to the striatum. It has been reported, for example, that tuberculum olfactorium lesions can prevent apomorphine-induced stereotypy. This may suggest that DA activation in the olfactorium is a necessary prerequisite for stereotypy to emerge, and that without it an intact striatum is inadequate. The use of striatal injections of apomorphine or DA may activate DA receptors quite widely in the forebrain to induce stereotypy.

Chemical lesions of the DA containing pathways are known to result in denervation supersensitivity of the postsynaptic membranes. If the DA receptors of the striatum are stimulated directly with apomorphine, with DA or by L-DOPA treatment an unusually intense stereotypy is seen in 6-OHDA treated animals, and such animals respond to lower doses of these drugs.

Ungerstedt (1971) has demonstrated these phenomena elegantly by making unilateral 6-OHDA lesions of the substantia nigra. After such treatment the DA-containing terminals are intact on one side, and the postsynaptic membranes are supersensitive to DA on the other side, consequent upon degeneration of the nigro-striatal pathway. The rats are then treated with drugs which act either by releasing endogenous

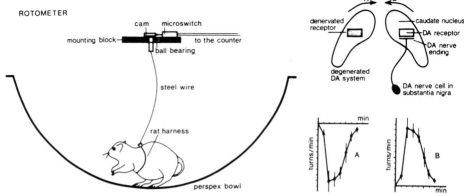

Figure 3a. Left: Schematic drawing of the rotometer. The movements of the rat are transferred by the steel wire to the microswitch arrangement. Upper right: Principal outline of the experimental situation shown in a horizontal projection of the nigro-striatal DA system. When the stimulation of the denervated receptor dominates, the animal rotates in direction A. When stimulation of the innervated receptor dominates, the animal rotates in direction B. Lower right: The rotational behavior is presented as turns per minute versus time. The curves are given negative y-values when stimulation of the denervated receptor dominates, the positive y-values when stimulation of the innervated receptor dominates. Each point represents the mean ± SEM of a certain number of animals. [Reproduced from Ungerstedt (1971).]

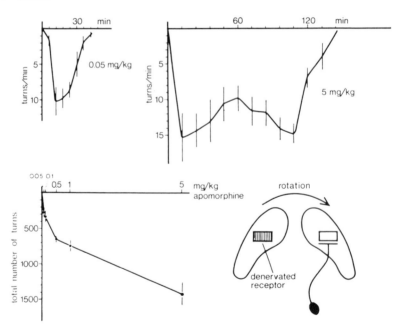

Figure 3b. Apomorphine-induced rotation in rats unilaterally denervated with 6-OH-DA. Rotation curves are shown for the smallest and the largest dose. Each point represents the mean ± SEM for 6 animals. Lower left: dose-response. Lower right: Horizontal outline of the experimental situation in the nigro-striatal DA system [see Figure 4(a)]. [Reproduced from Ungerstedt (1971).]

amines or by direct stimulation of DA receptors. After systemic injections either group of drug will influence both sides of the brain, but the amine levels and receptor sensitivity are asymmetric in the preparation, and, thus, after either treatment one side of the brain will be more strongly activated than the other.

The net result of unequal neurochemical activation is asymmetric motor behavior, or turning. This is measured in a specially designed rotometer; each time the rat turns the rotation is registered (Figure 3). Amphetamine causes a release of DA from the terminals on the normal side and produces turning toward the lesioned side, and apomorphine preferentially stimulates DA receptors on the supersensitive lesioned side, to produce turning in the opposite direction, toward the normal side (Figure 3b).

Dopamine and Spontaneous Motor Activity

The neurochemical basis of some of the symptoms of Parkinson's disease, amphetamine psychosis, the side effects of neuroleptic treatment and of amphetamine-induced stereotypy is well-documented. However, the neurochemical basis of a closely related behavior pattern has proved much more difficult to determine. When placed in a novel environment a rat investigates it. The behavior includes locomotion, rearing, sniffing, and the duration, intensity, and probability of these behaviors vary with the environment into which the animal is placed and how many times it has been there before. An instinctive sequence of integrated motor acts of this kind is likely to be controlled by, or at least modulated by, the extrapyramidal motor system and the associated DA system. It has been demonstrated that catecholamines are involved in these behaviors since reserpine, α-methyltyrosine and dopamine-receptor blocking agents depress or abolish spontaneous motor activity in the many different environments that have been investigated. The L-DOPA restores such behavior after α-methyltyrosine, but it is not clear whether the restoration of brain NA or DA by L-DOPA is more important in the reversal of psychomotor depression. Several workers, on the basis of quite different experimental evidence, have concluded that both NA and DA may be equally important for the control of spontaneous locomotor activity.

Taylor and Snyder (1971), for example, have compared the potency of the *d*- and *l*-isomers of amphetamine in stimulating locomotor activity, in inducing stereotypy, and in inhibiting the uptake of NA and DA by nerve terminals prepared from the hypothalamus and striatum. *d*-amphetamine was approximately ten times more potent that *l*-amphetamine in inhibiting NA uptake sites in the hypothalamus and in stimulating motor activity. The *d*- and *l*-isomers were approximately equally potent as inhibitors of DA uptake in the nerve terminals of the striatum and in inducing stereotyped behavior in the rat. They, therefore, conclude that amphetamine-induced stereotypy may be primarily related to DA, and locomotor activity to NA.

Another approach to this problem has been to use biochemical manipulations that selectively alter brain NA *or* DA, in an effort to determine which of these amines is most closely correlated with changes in locomotor activity (Svensson & Waldeck, 1970). For example, drugs have been used that inhibit the synthesis of NA by blocking the enzyme dopamine-β-hydroxylase. After such treatment DA neurons still contain normal amounts of DA, whereas NA neurons are depleted of NA. After such drugs, however, mobility fell only very slightly. On the other hand, if α-methyltyrosine is used to inhibit the synthesis of both NA and DA, motor activity is depressed and the time course of this behavioral effect correlates closely with brain DA rather than with the slower fall in brain NA.

These experiments, however, are fraught with difficulties. Enzyme inhibitors are rarely completely specific, and may have independent behavioral effects which are difficult to distinguish from the specific behavior being studied. Thus, the apparently simple question of correlating brain catecholamines with spontaneous locomotor activity remains unsolved.

The fact that 6-OHDA lesions to SN which abolish stereotypy do not abolish the locomotor response, also weakens the case for DA playing a primary role in this particular behavior. Injections of 6-OHDA into the CSF of the infant rat brain produces virtually total striatal DA loss, an equally complete loss of NA from cortex and limbic system and a 70% loss of hypothalamic NA. These animals, when adult, show a loss of locomotor activity and stereotypy responses to amphetamine, but it is again not clear whether this is related to the depletion of DA or of NA to the simultaneous depletion of both amines (Iversen, 1974). In all such interpretations there is a strong tendency to think about transmitters in the brain or in particular pathways as existing in unitary functional pools. We talk of striatal DA in this way, and we are puzzled that 10% of the total striatal DA seems to be sufficient to maintain one behavior and insufficient for another. It may be, however, that the lesion techniques used spare a particular part of the nigro-striatal DA projection that is crucial for locomotor activity. The important thing may be not that 10% of striatal DA remains, but that one particular part of the system remains intact. Experiments in which surgical lesions are made in various parts of the basal ganglia are beginning to reveal a localization of function of this type within the basal ganglia.

Conclusions

Can one on the basis of the various experiments discussed suggest a general role in DA behavior? Ungerstedt (1971) has attempted to do this, and suggests that the behavioral features seen after destruction of DA systems result from an inability of normal sensory arousing stimuli to activate the striatum. He considers the role of DA pathways is to sensitize the animal, and, thus, induce appropriate responses. Normal behavior may occur only when the striatal activity is set within normal limits by the modulating influence of the nigro-striatal DA system.

Central to this hypothesis are his observations on aphagia and adipsia after lesions of dopaminergic pathways. After 6-OHDA lesions of the substantia nigra, rats become severely aphagic and adipsic, and die unless tube-fed. Spontaneous feeding gradually recovers, and the pattern of behavioral deficits is indistinguishable from that reported many years ago by Teitelbaum and Epstein after surgical lesions in the lateral hypothalamus (Milner, 1970). The nigro-striatal DA pathway happens to project to the forebrain in the medial forebrain bundle which passes through the lateral hypothalamus. Ungerstedt believes that the earlier lesion and stimulation studies involving this brain region were interfering with the ascending DA system essential for the arousal of feeding behavior, rather than with an efferent pathway involving in eating. This idea has been received enthusiastically (Marshaum & Teitelbaum, 1973) because the traditional views of reciprocal centers in the hypothalamus controlling food intake and satiety was already under pressure for other reasons (see pp. 178–182).

Reconciling all the data, Ungerstedt says "when considering the curious hypokinesia, lack of exploratory behavior and difficulty to initiate activity that occurs after selective lesions of the nigro-striatal DA system it is more probable that the DA system and the striatum control a general arousal or drive level that is necessary for performing a number of activities, where eating and drinking deficits are noticed only because they are easily measured by the observer and disastrous to the animal."

Ungerstedt's suggestion that the lack of responsiveness associated with DA depletion reflects a sensory loss is supported by the observation that unilateral electrolytic lateral hypothalamic lesions result in orientation deficits to visual, olfactory, and somaesthetic stimuli presented to the contralateral half of the body after hypothalamic lesions. However lack of motivation or depression of the motor response mechanism itself could equally well produce a lack of responsiveness. Methods have been developed to study changes in responsiveness in humans, and this field of work together with the associated mathematical theories for analysing performance is signal detection theory. These are potentially powerful methods for the analysis of behavior in animals as well as in man, and can provide a direct means of dissociating changes in motivation, sensory awareness and motor readiness.

Such techniques are currently in use in this laboratory to study drug induced changes in behavior in the rat (Robbins & Iversen, 1973). The animal is trained on a schedule of reinforcement which requires a low, steady rate of bar pressing (DRL schedule). The d-amphetamine disrupts this performance, the rat responds at a faster rate and loses reinforcement. Signal detection analysis shows that with low doses of amphetamine which do not overtly disrupt response control, a change in response bias can be detected. This increases with dosage. If amphetamine induced disruption of motor behavior is mediated by the DA system these results suggest that activation of the nigro-striatal system induces response arousal. Many factors such as motivation state, degree of familiarity with the environment, strength of condition versus unconditioned behavior in given situations would interact to determine the degree of activity in the DA system. On this model different responses would require different levels of activity in the dopaminergic neurones for activation. Highly probable responses would only require low DA activity, while improbable ones would have higher "DA thresholds."

The results of studies of drug-induced stereotypy and locomotor activity are consistent with a dopaminergic mechanism acting like a motor probability generator, determining at any one time which motor response shall have priority. With impairment of such a mechanism, i.e., after DA receptor blocking drugs which induced catalepsy, or substantia nigra lesions, there is no motor arousal. Overstimulation of the DA mechanism, on the other hand, may induce intense motor behavior but, lacking the normal balance and including a distorted repertoire of responses, such behavior becomes repetitive and purposeless, as in stereotypy.

It is also our impression that the aphagia seen after electrolytic lesions of the substantia nigra may be explained in a similar way. When presented with highly palatable food these animals are not unresponsive. A great deal of behavior is generated around the food. Intensive gnawing of the edge of the bowl is often seen, but not the appropriate ingestive responses. It seems as though the responses necessary for feeding become highly improbable, in a situation which would normally elicit them.

Such a proposed role for DA in behavior is entirely consistent with the limited evidence from microiontophoretic studies involving DA application to single neurons. Such studies have proved technically difficult in the striatal dopaminergic pathway. However, most of the successful studies of this type have suggested that DA exerts a predominantly inhibitory effect on striatal neurons (York, 1972). Stimulation of the substantia nigra also results in a depression of firing in caudate neurons, presumably as a result of DA release. It is thus suggested that a certain level of activity of the DA neurons in the striatum is required for normal responsiveness. The net result of increased DA activity—and, thus, greater inhibition of caudate neurons is a greater responsiveness as, for example, after amphetamine treatment. In contrast, if the striatal DA neurons are rendered functionally inactive by blocking the DA receptors with neuroleptics, then normal behavioral arousal is abolished.

However, as with most simple models of complex neural functions, doubt is already being cast on the idea that nigro-striatal connections are purely inhibitory in function (Feltz & de Champlain, 1972). For example, it has been shown that there are cells in the caudate nucleus that are excited by nigral stimulation, and that do not respond to iontophoretically applied DA while other cells are inhibited by DA. The transmitter involved in the excitatory nigro-striatal pathway has not been identified, although ACh is highly concentrated in the striatum and is one probable candidate. Cholinergic drugs do affect motor behavior, and it is likely that a balance between cholinergic and dopaminergic activity determines the excitability of the corpus striatum (see pp. 193–195).

NORADRENALINE

Role of NA in Behavioral Control

It has frequently been reported that NA applied by microiontophoresis depresses the firing of neurons in the CNS. Excitatory responses, however, have also been found, although it has been suggested that these may be artifacts secondary to the localized vasoconstriction caused by NA applications. The inhibitory actions of NA could explain why flooding the brain with NA, for example, by intraventricular injection *in vivo*, produces a general behavioral dampening, or sleep. Marley and Dewhurst have reported that the systemic administration of NA induces sleep in the chick, during the first week after hatching, until the blood brain barrier to NA develops.

However, the fact that a widespread activation of NA receptors leads to behavioral depression does not prove that NA is necessarily an inhibitory transmitter at all sites in the CNS (Vogt, 1973). An excitation of inhibitory neurons by NA, for example, might produce the same effect. It is also possible that some neuronal systems using NA as transmitter are involved in behavioral arousal, rather than depression.

For example, the increase in brain NA concentration induced by monoamine oxidase inhibitors is accompanied by behavioral or EEG arousal. Increased levels of waking induced by stress are accompanied by an increased rate of catecholamine turnover. There is no evidence for an involvement of DA in these mechanisms, as a virtually complete loss of striatal DA, while resulting in behavioral unresponsiveness, does not alter sleepwalking patterns. Inhibition of CA synthesis by α-methyltyrosine, on the other hand, decreases EEG arousal, and lesion studies indicate that it is the dorsal bundle of NA fibers which is directly concerned with cortical arousal.

Furthermore, intraventricular injections of NA do not invariably produce sedation; small doses can induce behavioral activation. It is difficult to interpret all the conflicting results obtained after systemic and ventricular injections of NA, and perhaps the availability of techniques for the localized manipulation of specific NA pathways in any case make such a blunderbuss approach no longer rewarding.

Behavior Conditioned with Positive or Negative Reinforcers

Before it was known that NA was localized in specific anatomical pathways in the CNS, there was growing evidence of the involvement of NA in behavioral control. In many independent studies it was reported that disruptions of the NA and DA systems in the brain impaired both spontaneous and learned behaviors.

Reserpine, for example, abolishes conditioned avoidance of electric shock, and response patterns controlled by food or water reinforcement; α-methyltyrosine has similar effects on conditioned behavior, and depresses spontaneous locomotor activity. Chlorpromazine has similar effects, and abolishes what one might call behavioral "tone." The neuropharmacological properties of these classes of drug have the common feature of interfering with adrenergic transmission. The correlation between adrenergic functions and these behaviors is further strengthened by the finding that the depressing effects of reserpine and α-methyltyrosine on behavior can be prevented or reversed by treatment with precursors of the catecholamines such as L-DOPA (Andén, Carlson, & Haggendal, 1969). Many workers have reported such effects for reserpine and α-methyltyrosine on spontaneous locomotor activity, conditioned avoidance, and operant behavior for positive reinforcement.

Reserpine and α-methyltyrosine reduce both brain DA and NA, and L-DOPA can replace, at least temporarily, both of the amines. It, therefore, remains unclear whether both of these amines are involved, or whether one or the other is most important for the depression and subsequent reinstatement of these forms of behavior. This problem is difficult to solve with existing neuropharmacological techniques, NA and DA share a common biosynthetic pathway, and receptor structures which must be very similar since considerable overlap exists in the actions of blocking agents and agonists.

Efforts have, nevertheless, been made to correlate behavior independently with brain levels of NA and DA. For example, attempts have been made to determine whether the disruption of NA or DA mechanisms was responsible for the reserpine-induced depression of avoidance behavior and whether reinstatement by L-DOPA was more closely correlated with the brain content of one or other of the amines (Ahlenius & Engel, 1971). After L-DOPA injections (combined with inhibition of peripheral DOPA-decarboxylase to ensure that brain levels of L-DOPA are maximal) the reinstatement of avoidance behavior appears to be more closely correlated with a rise in brain DA than with increased NA. Potentially useful pharmacological tools have become available to investigate such problems. The DA and NA share a common biosynthetic pathway except for the final step, in which DA is transformed to NA. The enzyme catalyzing this reaction is dopamine-β-hydroxylase, and inhibitors of this enzyme are available, disulfiram is one such compound, although this is relatively nonspecific and toxic agent. The newer drug FLA-63 (Table 1) appears to be a more potent and specific inhibitor of NA synthesis. If the conversion of DA to NA is prevented by inhibition of dopamine-β-hydroxylase, then L-DOPA does not reinstate avoidance behavior so effectively. After such drug treatment, the L-DOPA-induced increase in brain DA is unaffected, suggesting that NA rather than DA may be central for this behavioral effect. The conclusion that NA plays a central role in avoidance behavior is in agreement with other studies in which disulfiram was reported to reduce the effects of L-DOPA on reserpine suppression of avoidance behavior. Rather different evidence supporting this view comes from histochemical studies (Fuxe & Hanson, 1967). Rats were trained to criterion of 85% correct responses on an avoidance task. On the next test day they were treated with α-methyltyrosine to prevent the synthesis of both DA and NA and were then tested for 4 hr on the avoidance task. The brains were then processed histochemically, and NA terminals in the forebrain were found to be more severely depleted of transmitter than the DA-containing ones. A correlation was found between the degree of depletion of NA terminals and the extent of disruption of avoidance behavior after α-methyltyrosine. This finding is also consistent with the many lesion studies that have shown that damage to areas of the ventral mid brain, medial forebrain bundle and limbic areas impair avoidance; the sites of these lesions conform closely to the route of the noradrenergic neuronal systems of the brain.

Avoidance is a complex form of conditioned behavior; it differs from other forms of learning only in that the reinforcement is the removal of an unpleasant stimulus. Learning induced by positive reinforcers also appears to involve noradrenergic mechanisms (Lewy & Seiden, 1972). Lever pressing for water in rats, for example, is depressed after α-methyltyrosine treatment, and now it has been shown that this behavior is normally associated with an increased role of NA turnover in the brain. Rats learned to press a lever for water reinforcement which became available on average every 30 sec (VI 30 sec). On the 15th day of training they were injected intracerebrally with tritiated NA, and after the training session the amount of ^{3}H-NA in their brains was compared to that of similarly water deprived rats who had not responded in the skinner box. The "behaving" group had lower specific activity indicative of a high turnover of NA. Sparber and Tilson (1972) have used ventricular push–pull cannulas in the rat. They report that after radioactive pulsing of the brain with ^{3}H-NA, there is a greater release of radioactivity into the perfusate when the rat is responding in a skinner box on FR30 schedule of reinforcement than when it is responding on a FR of 10 (Figure 4).

The independent variable in both of these experiments was the behavior of the animal; and it is interesting to note that not only may the neurochemical state of the CNS determine behavior, but that the converse may also occur. This finding has important implications for understanding drug interactions with behavior and Kelleher and Morse (1966) have pointed out that the ongoing behavior is a major determinant of drug actions on behavior. The ability, for example, of amphetamine to increase rates of responding is related to the rate of ongoing behavior. Low rates of responding at the beginning of the intervals of a fixed interval schedule are increased enormously, whereas the high rates of responding on a fixed ratio schedule are barely increased. As amphetamine is thought to produce its stimulatory effects by a release of catecholamines, different levels of activity in pathways using these transmitters in the two behavioral situations could provide the substrates for these differential amphetamine effects.

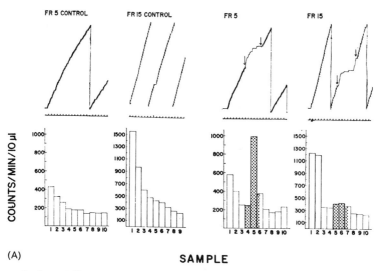

(A)   **SAMPLE**

Figure 4. Cumulative records and radioactivity histograms for control and extinction (arrows) session for rat A26(A). Tracings of event pens beneath the cumulative records indicate a 1-min time base and each bar of the histogram represents the radioactivity, in 10 μl of ventricular perfusate, from NE-7-^{3}H injected 1 hr prior to the sessions. Perfusate rates were 20 μliter/min^{-1}, using artificial CSF as the medium. Shaded areas indicate the extinction-perfusion samples. [Reproduced from Sparber and Tilson (1972).]

Response to Environmental Stress

The above experiments point to a role of NA in determining the response of the animal to events in its environment. Experiments have been described in which changes in the behavior (in many cases learning) indicate the responsiveness of the animal. But there is also a large literature demonstrating that the turnover of NA in the brain can be changed in response to environmental stimuli irrespective of whether the animal develops complex changes in its behavior. The application of various types of stressful stimuli falls into this category. Repeated electrical foot-shocks increases turnover of NA in the hind brain, electroconvulsive treatments increases forebrain NA turnover, overcrowded living conditions and forced activity depletes forebrain NA. Trauma in general, therefore, induces activity in NA systems. Acute stress is also known to make rats more susceptible to the lethal stimulant effects of large doses of amphetamine, and the lability of NA mechanisms may provide the substrate for this amphetamine effect.

It would be misleading to imply that the effects of stress on brain NA are clear. Certain apparently stressful conditions seem to lower rather than increase activity in NA pathways. The behavioral evaluation of different stress conditions is inadequate, and in most studies regional assays of changes in NA turnover have not usually been performed. There is no reason to suppose that a transmitter system with such varied projections to different parts of the brain will operate as a single entity.

Feeding, Drinking, and Satiety

The results of lesion studies and their effects on eating appeared to demonstrate the existence of a reciprocally interacting pair of nuclei in the hypothalamus which control food intake. The ventromedial nucleus (VMM) apparently acting as a satiety mechanism, by exerting a controlling influence on the lateral hypothalamus, initiates or stops eating. Food deprivation and reinforcing stimuli, operating finally through the ventromedial nucleus, would be the reciprocal modulators of this mechanism.

Grossman discovered a role for NA in the control of feeding before the distribution of NA pathways to the hypothalamus was described. He reported that application of crystalline NA to the lateral hypothalamus induced eating in satiated animals and that carbachol induced drinking. These results were followed up in Miller's laboratory where liquid rather than crystalline injections were used in order to specify more precisely the effective dose of NA. Slangen and Miller (1969) reported that 22 μg of NA produced a good feeding response in their experiment. They went on to explore the pharmacological specificity of the system and found that the α-adrenoceptor antagonist, phentolamine, prevented NA-induced eating, suggesting that an α-adrenergic eating system existed in the hypothalamus.

At this time the NA-induced eating seemed to be compatible with the neural model of feeding based on lesion and stimulation studies. Lesions to the lateral hypothalamus abolish feeding and electrical stimulation induces eating. It was, therefore, assumed that NA, like electrical current, was initiating eating via terminals in the lateral hypothalamus. This interpretation was challenged by two lines of evidence. Firstly, Booth (1967) used small electrodes and cannulas to explore in detail the effects elicited at different hypothalamic sites, and was able to show that electrically induced and NA-induced eating were elicited at different sites. The critical focus for NA effects was in the perifornical medial forebrain brain bundle.

The results of Slangen and Miller suggested an eating system mediated by α-adrenergic mechanisms. The brain, like the peripheral adrenergic system contains both α- and β-adrenoceptors, and agonist and antagonist drugs are available for the selective manipulation of these receptors.

Liebowitz (1971), using the basic measures of dry food intake or water intake in appropriately satiated rats, and the perifornical cannula placements as described by Booth, has made an extensive and systematic study of the pharmacological specificity of NA-induced eating. The α-adrenergic nature of the eating response was confirmed, and she has gone on to show that satiety in her system is probably mediated by a β-adrenoceptor. The β-agonist, isoprenaline, depressed food intake, and the β-antagonist drug, propranolol, resulted in an increase in food intake. By varying the placement of the cannulae, Liebowitz also concluded that the α-adrenergic effects are localized medially in the hypothalamus; probably, therefore, their action relates to the ventromedial nucleus, whereas the β-adrenergic satiety effects are localized in the lateral hypothalamus, probably in the classical "feeding" area. Liebowitz concludes that the NA effect is due to inhibition of the normal inhibitory effects of the ventromedial nucleus on the lateral feeding area.

In parallel studies she measured water intake after these adrenergic manipulations and found a set of effects perfectly reciprocal with the effects of the same drugs on food intake. She, therefore, concluded that food and water intake are held in balance by the activity in adrenoreceptors and presumably the involvement of adrenoreceptors in water control must work in parallel to the cholinergic receptors which are clearly also involved in the control of water uptake.

Margules (1970a,b) recently explored this problem in a similar pharmacological way and was influenced by the wide range of existing knowledge about the various sensory factors controlling food intake. He returned to the classic findings that normal feeding is controlled by both intero- and exteroceptive satiety signals. Rats can clearly sense and adjust their food intake appropriately to both these kinds of information independently. If fed diets adulterated with bulky nonnutrients, such as kaolin, or food sweetened with noncalorific substances like saccharin, they adjust their food intake to achieve the desired calorie input. Any neural or pharmacological model of satiety should, therefore, consider the possibility that these two kinds of information are mediated by different neural mechanisms. Using a liquid food, milk, which is known to be highly palatable for the rat, Margules finds that NA depresses food intake, whereas α-adrenoceptor blocking agents, like phentolamine, increase milk intake. This observation leads him to propose that there is an α-adrenergic satiety mechanism in the perifornical region. Feeding is normally induced when the synaptic release of NA is low, and the satiety system is not inhibited; after food has been ingested NA may then be released to activate satiety mechanisms and stop feeding. However, if unpleasant tasting milk (quinine adulterated) is used, β-adrenoceptor antagonists increase intake, and the β-adrenoceptor agonists and α-adrenoceptor antagonists suppress it. Thus a β-adrenergic receptor system mediates satiety concerned with the exteroceptive stimuli associated with food.

Margules suggests that the ventromedial nucleus is the source of the interoceptive satiety information. Stimulation of the ventromedial nucleus suppresses feeding. Electrophysiological recordings of ventromedial nucleus neurones show increased neural activity in response to stomach distension and glucose utilization. Lesions to the ventromedial nucleus produce hyperphagia. The animals are said to be insensitive to the glucostatic and lipostatic metabolic stimuli which reflect available glucose and fatty acid levels, and are thus unable to adjust food intake to their metabolic needs. Despite their voracious appetite, such animals are known to be finicky. They overeat palatable food but are abnormally sensitive to unpleasant foods. On the Margules model, it is the β-adrenergic receptors that are concerned with such enteroceptive cues, and it could be suggested that removal of the α-adrenoreceptor system after ventromedial nucleus lesions leaves the β-adrenoceptor system without its normal reciprocal interaction, and in some way abnormally active. Finally there is anatomical evidence of projections from neurones in the ventromedial nucleus to the perifornical region, where an abundance of NA-containing terminals are seen with the fluorescence histochemical technique and of connections to the ventromedial nucleus from the anterior hypothalamus and amygdala.

As cannulae in the perifornical region produce both α- and β-adrenergic effects, the two kinds of receptors presumably coexist. The origin of the β-adrenoceptor system is uncertain. Perhaps it too arises from neurons in the vicinity of the ventromedial nucleus. But it should be noted that the ventral NA bundle arising from the hindbrain neuron groups passes through the lateral hypothalamus and innervates the anterior hypothalamus. This pathway yields high rates of intracranial self-stimulation and its anatomical origin near the termination of the gustatory afferent system suggests that it may be important in conveying exteroceptive sensory information about food and water.

These two pharmacological models of food intake control are based on sets of results (Table 2) which appear at first sight to be very different, but it is probable that these results can be reconciled. The Margules model is intuitively more appealing: as Hoebel comments, "It begins to cope with the problem of homeostasis versus hedonism at the neurochemical level." It proposes that food intake is controlled, on a kind of hydrostatic model, by the amount of endogenous NA released at any time. NA levels fall gradually after a meal and when a certain

TABLE 2   Food and Water Intake after Pharmacological Manipulation of α and β Adrenoceptors in the Perifornical Region of the Hypothalamus[a]

Adrenoceptor	Dry food	Water	Milk	Quinine milk
α-Agonist Noradrenaline	↑	↓	↓	↓
α-Antagonist Phentolamine	↓	↑	↑	↓
β-Agonist Isoproterenol	↓	↑	Normal	↓
β-Antagonist Propranolol (Liebowitz) LB 46 (Margules)	↑	↓		↑

[a] Comparison of the results of Liebowitz and Margules.

level is reached the behavior is released. The sensory events associated with food intake release NA which gradually builds up until a level of inhibition is reached and feeding stops. The threshold levels could vary with the experience of the animal; for example, 24 hr deprivation reduces NA mediated satiety. This model is compatible with iontophoretic studies which show that NA inhibits neurones, and explains how amphetamine which has been shown to release endogenous NA from the hypothalamus, depresses food intake.

Margules also claims that he can explain Liebowitz's results with phentolamine, isoprenaline, and propranolol on dry food intake. The pattern of results are identical to his when using quinine-adulterated milk. Dry food, he maintains, is known to be unpalatable for rats. The one finding he cannot satisfactorily explain is the NA-induced eating of dry food reported by Grossman and many others. Margules says that NA-induced eating is a weak effect with puzzling features. Rats eat 3–4 gm food in the hour after NA administration but may consume more than 100 gm when eating is elicited electrically. The NA-induced eating has a long latency and indeed in the dry food studies, intake may initially be depressed. Even in his own study with milk, NA produced an increase in intake some time after the injection. The NA-induced eating is also abnormally sensitive to the exteroceptive stimuli associated with food, and also, unlike the effects of natural deprivation, will not induce complex operant behavior. He resorts to special neuropharmacological pleading to explain the NA-effect and suggests that flooding the NA axons and terminals with NA saturates the presynaptic membrane uptake mechanisms as well as stimulates the postsynaptic receptors. After some time, the uptake system is restored to normal function and NA is taken up from the extracellular space, such a reduction in the extracellular level of NA may be sufficient to induce a weak eating response. Margules presumably would also have difficulty in explaining why 6-OHDA injected into the perifornical region induces intake of dry food. It had been assumed that the endogenous NA released by 6-OHDA stimulated the α-adrenergic feeding system.

These apparently contradictory results on the ability of NA to induce eating or satiety have been explained in several ways. For example, Liebowitz suggests that Margules was studying liquid intake rather than food and that his results with milk agree with her results on water intake. This parsimonious interpretation is, however, weakened by Margules's studies with quinine-adulterated milk which give a different pattern of pharmacological sensitivity. And different interpretation is afforded by the recent findings of Margules, Lewis, Dragovich, and Margules (1972), who report that hypothalamically injected NA can induce eating or satiety depending on the time of day it is injected. If injected during daylight hours, when rats sleep and endogenous NA levels are low, it induces eating. However, if injected during darkness when rats are normally active, eat, and have higher endogenous NA levels, then satiety is induced. They argue that eating is associated with a "threshold" synaptic NA level. In Liebowitz's studies, the exogenously added NA brings the synaptic NA level to threshold, and eating is seen; while in Margules' study, the exogenous amines result in the threshold being surpassed, and satiety ensues. Under natural conditions, eating results in the release of synaptic NA to mediate satiety.

On neurological considerations, the Liebowitz model, depending as it does on the interaction of α-adrenergic mechanisms in the ventromedial nucleus and β-adrenergic mechanisms in the lateral hypothalamus, is weakened by recent suggestions that there may be no lateral hypothalamic feeding center. Ungerstedt (1971) suggests that the feeding deficits seen after lateral hypothalamic lesions are due to severance of the DA-containing nigro-striatal pathway which happens to pass through the lateral hypothalamus. If this is the case, the neural activity involved in the interaction of satiety with the parts of the other brain concerned in feeding behavior will require reappraisal.

Intracranial Self-Stimulation and Reinforcement

Animals who require food (or any other basic need) will learn new responses to obtain it, and the extero- and interoceptive cues associated with food as well as ultimately mediating satiety act, it is believed, as reinforcers. Reinforcement is an event and if it follows a response, the probability of that response increases. Reinforcement theory is a central part of any theory of learning and there have been many attempts over the years to define the significance of reinforcement for behavior.

As satiety signals at least contribute to reinforcement, and NA mechanisms mediate satiety, it is not surprising to find that other lines of evidence implicate NA in the reinforcement mechanism. Intracranial self-stimulation (ICS) is a remarkable phenomenon whereby a rat will press a lever to obtain a small electrical current through an electrode implanted in certain parts of the brain (Milner, 1970). The phenomenon was originally obtained from the medial forebrain bundle but it is now known that sites in a wide circuitry from the hindbrain to the forebrain will sustain ICS. Animals will press for brain stimulation until they collapse from exhaustion. These results immediately suggested that the current was mimicking reinforcement. Valenstein (1966) pointed out that exteroceptive cues associated with food were immediate whereas the interoceptive cues occur only after digestion and absorption of food has occurred. Behavior theory demands a reinforcing event be in close temporal contiguity with the response. He suggested that ICS mimicked this immediate reinforcement. This view is also supported by the finding that ICS rates vary with the level of food deprivation or prior priming with reinforcing stimuli. For example, insulin, which lowers blood glucose, increased ICS rates, whereas forced feeding or gastric distension had an inhibitory effect. Brain catecholamines have long been implicated in the mediation of ICS.

The distribution of positive ICS sites in rat brain also corresponds quite closely to the distribution of the NA and DA systems. Poschel and Ninteman (1963) reported that α-methyl-p tyrosine abolishes ICS and Stein confirmed this. He also implicated NA rather than DA or 5-HT by demonstrating that ICS suppressed by α-methyltyrosine treatment could be reinstated with the active l-isomer (Figure 5) of NA but not with the d-isomer, or with DA or 5-HT (Stein, 1968). ICS rates can also be predictably changed by drugs that alter adrenergic transmission. Beta-adrenoceptor blocking agents abolish ICS, and amphetamine, which releases endogenous catecholamines, increases ICS. This latter effect is potentiated by antidepressants, such as imipramine or iproniazid, which potentiate the actions of catecholamines at the synapse.

More recently Stein and Wise have reported that ICS in the hypothalamus and amygdala is associated with release of NA (Stein, 1968). Rats were implanted with push–pull cannulas in the ventricle and ICS electrodes. The brain was pulsed with radioactive NA and during ICS responding the radioactivity of the ventricular perfusate was assayed.

Further studies of ICS as a phenomenon have revealed characteristics

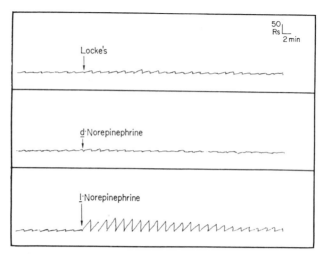

Figure 5. Facilitation of self-stimulation of the lateral hypothalamus in the rat (at the level of the ventromedial nucleus) by an intraventricular injection of 1-noradrenaline HC1 (10 μg). Control injections of Ringer–Locke solution (10 μl) or d-noradrenaline HC1 (10 μg) had negligible effects. Pen cumulates selfstimulation responses and resets automatically every 2 min (see key). [Reproduced from Stein (1971).]

which indicate its relationship to satiety signals. Rats still press more avidly for ICS if allowed to eat highly palatable food in the testing situation. ICS will not, however, sustain complex conditioned behavior, which is a feature of the hyperphagic ventromedial-nucleus-lesioned rat. But there are features of ICS, for example, its failure to show satiation, which cannot obviously be fitted into a NA satiety theory. If ICS releases NA this should gradually build up and suppress the reinforcement seeking behavior (ICS) just as it does feeding. Antidepressants which result in increased availability of synaptic NA increase rather than decrease ICS. If the ICS results are to be correlated with the NA α-adrenergic satiety mechanisms special pleading is again required to explain how NA release maintains a behavior. The other alternative is that the released NA occupies the α-adrenergic satiety system, leaving the β-adrenergic satiety system overactive, as after a lesion of the ventromedial nucleus. The animal then becomes exceptionally sensitive to exteroceptive cues. The β-adrenergic satiety signals in isolation are not sufficient to produce satiation and terminate ICS.

More recently small electrodes have been used to plot the ICS circuitry, and Crow (1972) has suggested that the DA-containing nigro-striatal pathway and the dorsal NA bundle mediate two forms of ICS, whereas the ventral NA bundle does not seem to be involved. This finding is compatible with the original observation that the medial forebrain bundle was a strong positive site for ICS because this bundle carries, at the level of the hypothalamus, all the catecholamine pathways traveling to the forebrain. These results await evaluation, however. ICS responding may be a corollary of several different neural functions, including reinforcement mediated by NA terminals of the perifornical region.

ACETYLCHOLINE

Arousal

It seems that NA and DA control behavioral activity largely by inhibiting otherwise over-excited parts of the nervous system. By contrast, ACh seems to be important as an excitant of neural activity, maintaining sufficient activity to generate behavior. The best index of this general excitation is the EEG, and ACh injected intravenously or topically applied to the cortex increases the arousal level as indicated by EEG. Anticholinergic drugs impair cortical arousal in the absence of effects on behavioral arousal. Interference with NA actions, on the other hand, impairs both aspects of arousal, and this difference suggests that while ACh contributes to the quality of arousal during the waking phase, its action may be modulatory rather than directive. The ACh does not seem to be involved in the organization of sleep–waking patterns, but via the locus coeruleus may be important for triggering paradoxical sleep.

The localization of cholinergic neurones in the ascending reticular formation, thalamic centers, and cerebral cortex is consistent with such a role for ACh. In addition, however, ACh operates at local sites in the brain to control more specific aspects of behavior.

Drinking

The regulation of water intake appears to involve cholinergic mechanisms (Fisher 1973). Grossman originally showed that carbachol injections into the hypothalamus induced drinking in water-sated rats. It was subsequently shown that an extensive neural circuitry in the limbic system shared this cholinergic sensitivity. The concept of a cholinergic drinking system has been challenged on many fronts, not in an effort to disprove the original findings but to demonstrate that the control of drinking was more complex than originally envisaged. It now appears that cellular dehydration, as a stimulus for drinking, is mediated by a cholinoceptive system of neurons in the preoptic region of the hypothalamus. Water intake is also controlled by a mechanism sensitive to changes in the volume of the extracellular space produced by loss of *isotonic* body fluids rather than dehydration.

This control mechanism results in the release of renin from the kidneys, which subsequently leads to an increased formation of the peptide angiotensin, and this in turn stimulates neurons in the preoptic region, leading to increased fluid and salt intake. The neurons in the preoptic area are sensitive to direct applications of angiotensin; endogenous catecholamine systems are required for this action (Settler, 1973). Angiotensin-induced drinking, for example, is unaffected by atropine but prevented by application of haloperidol, or by local injection of 6-OHDA.

A challenge to the cholinergic theory came from the β-adrenergic mechanism proposed by Lehr, Mallow, and Krukowski (1967) and Liebowitz (1971). They proposed that drinking was mediated by a β-adrenergic system and satiation by an α-adrenergic system. The β-agonist drug isoprenaline is reported to induce drinking when injected systematically. A similar result was obtained by injecting isoprenaline into the perifornical region. Leibowitz interpreted this as evidence of a hypothalamic β-adrenergic drinking site. However, she injected 40 μg of isoprenaline centrally, and since Lehr *et al.* reported drinking with systemic injections of 5 μg of isoprenaline, it is probable that leakage of isoprenaline from brain to the peripheral circulation may have occurred. The more recent finding that nephrectomy abolishes isoprenaline-induced drinking suggests that a release of renin and angiotensin production may have been the mediating factor. The evidence that β-adrenoceptor antagonists such as propranolol suppressed naturally induced thirst can also be criticized as the dose of propranolol used may have reached concentrations giving nonspecific local anesthetic-like effects. It is, therefore, not surprising that propranolol blocks both natural thirst and ACh-induced thirst. The local anesthetic hypothesis is supported by the finding that 40–80 μg of propranolol blocked both angiotensin and ACh-induced drinking, and that the *d*- and *l*-isomers of propranolol were equally potent; the *l*-isomer is, however, at least 100 times more potent than the *d*-isomer as a β-adrenoceptor antagonist.

Sham Rage and Behavior Attack

There is also a growing interest in the possibility that cholinergic mechanisms are involved in emotional responses to unpleasant stimuli. It is fashionable to use the term "aggression," but it is probably more satisfactory to describe the behaviors and the stimuli which elicit them. Bard and Mountcastle in the 1930s established that emotional stability depended on the integrity of the hypothalamus. If the hypothalamus with its efferent connections was intact, rage behavior with its associated patterns of autonomic activity could be induced by provoking stimuli. In the absence of the forebrain the range was undirected "sham rage." Areas in the limbic system such as the amygdala and septum were found to be especially important for the control of lower centers. Sham rage is induced by cholinergic stimulation of the amygdala or septum but NA mechanisms also appear to be involved in this behavior, although the interaction between the transmitters is unknown. Increased NA turnover is associated with sham rage induced by stress or amygdala stimulation. The clearest correlation has been found by Reis and Fuxe (1969) who induced sham rage by decerebration and found a correlation between the intensity of sham rage and the extent of depletion of brain stem NA, especially in the sympathetic centers.

In other studies electrical stimulation of the midbrain ventral tegmental region induced directed attack behavior, which is often described as aggression. Mouse or frog killing by the rat, or rat killing by the cat, is one such example which is being studied. It is far from clear, however, if this induced behavior is aggression in the normal sense of the word, or merely a component of the species-specific feeding behavior. Prey are killed before they are eaten. Fighting itself, without necessarily killing, is more commonly associated with territory defense. There is evidence that the pathways involved in killing attack in the rat are cholinergically organized (Bandler, 1971): cholinergic stimulation in the amygdala, lateral hypothalamus, and midbrain tegmental regions induces biting attack in the cat and rat, and these sites are blocked with atropine.

Fighting in response to territorial stress, e.g., in mice put together after social isolation, is not associated with ACh but rather with NA mechanisms.

Punishment

The ACh plays a role in some of the general aspects of behavioral control. The aversion system described by Stein and Margules is one such example (Stein, 1968). Just as reinforcing stimuli increase the probability of responding, so punishing stimuli decrease it. The punishment system has been plotted with intracranial self-stimulation methods, and a circuitry has been found where ICS is punishing. The system is called "periventricular," lying as it does in a chain of structures and fiber tracts from the hind brain to the forebrain close to the ventricles. The ventromedial nucleus of the hypothalamus is part of the punishment system which has been intensively studied using a behavioral procedure described by Geller and Seifter (1960). Rats are trained to respond for food on a variable interval schedule and subsequently when every response is followed by an electric shock severe depression of the behavior is seen. Responding is released by lesions of the ventromedial necleus, (VMN), and can be suppressed again by the injection of the cholinomimetic agents carbachol and physostigmine in the medial hypothalamus.

Anticholinergic agents, such as atropine, have the same effects as lesions to the ventromedial nucleus and disinhibit punished responding. The minor tranquilizers, such as oxazepam, act like anticholinergic agents to release punished responding, when injected systemically or into the VMN. Oxazepam releases behavior suppressed by a variety of aversive stimuli, including footshock, extinction, punishing ICS, satiation, and the bitter taste of quinine. On pp. 182–185, the role of the ascending NA system to the hypothalamus and limbic system in coding reinforcing stimuli was discussed. Reinforcing stimuli eventually mediate satiation which is in a sense a punishing event, in so far as it decreases the probability of responding. Anatomical connections from anterior hypothalamus and the amygdala to the VMN exist and the results with cholinergic drugs suggest that a cholinomimetic mechanism in the VMN may be responsible for the interaction between reinforcing and punishing contingencies (Figure 6). Such a neural interaction could explain how sensitively behavior is able to reflect the constantly changing influence of reinforcement and punishment. The VMN is an obvious site for this mechanism; it is known to respond to reinforcing stimuli and to be involved in punishment mechanisms. It is also interesting to note that it is one of the few structures which will sustain both positive and negative ICS.

Behavior Inhibition

The effects of anticholinesterases, cholinomimetic drugs, and cholinergic receptor blocking agents on conditioned response patterns have been intensively studied. As early as 1963 Carlton proposed a general theory that ACh mediated behavioral inhibition. If animals are trained in behavioral tasks requiring a response to be withheld in some circumstance and not in others, e.g., a discriminated "go–no go" task, anticholinergic drugs abolished such inhibition, and the animals responded when they should not have. Russell and his collaborators (1969) developed these early studies to reach similar conclusions. Recently Warburton (1972) has applied ingenious behavioral techniques to determine if the loss of behavioral control induced by cholinergic blockers is due to sensory

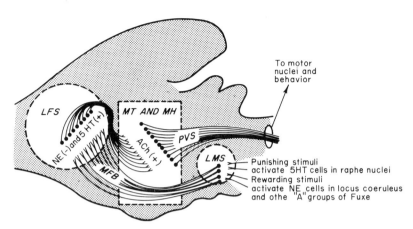

Figure 6. Diagram representing hypothetical relationships between reward and punish-ment mechanisms and behavior. Signals of positive reinforcement release behavior from periventricular system (PVS) suppression by the following sequences of events: (a) Acti-vation of noradrenaline containing cells in lower brain stem (LMS) by stimuli previously associated with reward (or the avoidance of punishment) causes release of noradrenaline (NE) into amygdala and other forebrain suppressor areas (LFS) via the medial forebrain bundle (MFB). (b) Inhibitory (−) action of noradrenaline suppresses activity of the LFS, thus reducing its cholinergically mediated excitation (+) of medial thalamus and hypothala-mus (MT and MH). (c) Decreased cholinergic (ACh) transmission at synapses in MT and MH lessens the activity of the periventricular system, thereby reducing its inhibitory influ-ence over motor nuclei of the brain stem. Signals of failure or punishment increase be-havioral suppression by the release of 5-hydroxytryptamine (5-HT), which either excites suppressor cells in the LFS or disinhibits them by antagonizing the inhibitory action of nor-adrenaline. [Reproduced from Stein (1971).]

disinhibition or loss of response control, i.e., do the animals fail because they do not notice the cues in the environment which indicate when responding is required, or because they simply cannot control their re-sponses to the stimuli? Rats were trained on a discrete trial procedure to respond in the presence of a briefly presented light cue which was three times above background illumination (signal-to-noise ratio of 3). The intertrial intervals (ITI) were variable (mean 15 sec) and responses within the last 9 sec of the (ITI) reset the interval. The rat was required to attend in order to detect the change in illumination which signalled the availability of responding. Inappropriate responses delayed further the next occurrence of reinforcement. Physostigmine improved discrimi-nation in this task. By contrast, anticholinergics such as scopolamine reduced the accuracy of detection, but an analysis of the patterning of errors and false alarm responses using the mathematical methods of signal detention theory showed that the deficit was due to sensory inattention and not to a loss of response control. This result may be compared with the change in sensitivity rather than response criterion reported by Broadbent and Gregory when humans shift attention in detection tasks.

The ventral hippocampus is the focus for this particular cholinergically mediated aspect of behavioral control. Using a behavioral task in which the rat is trained to press a lever on alternate trials, it is reported that intrahippocampal injections of cholinergic blocking agents, such as atropine, disrupted "no–go" responding controlled both by interoceptive cues (cue about previous trial which indicates whether right response should be "go" or "no–go") and exteroceptive cues (the blackout during the ITI).

Shute and Lewis (1967) report a cholinergic pathway from the reticular formation to the septum and hippocampus which is an obvious candidate for this function. Electrophysiological studies suggest that this pathway controls the theta rhythm of the hippocampus, thought by many to be correlated with the coding of meaningful or reinforcing stimuli.

Conclusions

The involvement of a hippocampal cholinergic mechanism in behavioral inhibition can be reconciled with a general proposal concerning the role of ACh in sensory behavioral arousal. The ACh pathways essentially form a dorsal tegmental system, a ventral tegmental system and a hippocampal circuit. The dorsal system innervates the thalamus and corresponds to the thalamic reticular activating system.

The ventral system projects to the cortex, striatum- and hypothalamus and is also the origin of the hippocampal circuit which projects back to the tegmentum.

Nauta (1963) has suggested that these circuits maintain tone in the control systems for behavior, the cortical influences maintaining electrocortical arousal and the hypothalamic projections homeostatic tone. The term "tone" refers to a background on which more specific mechanisms operate to determine details of behavior. The pathway from the ventral tegmentum passes to the hippocampus via a septal relay. Cholinergic stimulation of the septum produces changes in hippocampus theta rhythm similar to that produced by sensory stimulation and decreases *responding*, while septal application of atropine increases it.

On this model the hippocampal circuit could operate as a safety feedback loop cutting out irrelevant stimuli and preventing them from inducing unnecessary arousal. Interference with the cholinergic system at their tegmental origin produce consistent results.

Disruptive electrical stimulation in the tegmental regions impairs behavioral inhibition, or for example passive avoidance tasks and Fuster reported that at lower levels such stimulation could enhance discrimination performance in monkeys. If cholinergic mechanisms are blocked in the ventral tegmentum with atropine, behavioral inhibition is seen on the Warburton single alternation task. Tegmental stimulation activates a spectrum of ascending systems but it is likely that improved discrimination performance relates to cholinergic activation in the cortical arousal system as well as in the hippocampal circuit. Tegmental stimulation induces release of ACh from the cortex. Cells in the sensory cortex are excited by iontopheretic application of ACh and blocked by atropine, and visual stimulation induces a release of ACh from the visual cortex. Using intracellular recording methods it has been found that ACh changes membrane conductance of cortical neurons to potassium, enhancing and prolonging excitatory responses to subsequent synaptic inputs. The consequence of such changes might be an improved detection of sensory stimuli.

TRANSMITTER INTERACTIONS

Introduction

Some cases in which chemical neurotransmitters mediate specific functions at localized sites in the CNS have been discussed, but in many more instances it seems that we should be thinking in terms of interacting neurochemical systems as the substrates for behavioral control. To illustrate this point some particular cases where such interaction has been demonstrated will be described.

5-Hydroxytryptamine and Noradrenaline in Sleeping and Waking

It will not be appropriate to describe this particular example in detail, although it is probably one of the most important areas of study relating neurotransmitters to behavior, because this area is reviewed elsewhere in this volume by Jouvet (Chapter 17).

The 5-HT is localized in pathways whose anatomical distribution is not unlike that of the NA and DA systems. The cell bodies of the 5-HT neurons lie in the tegmental nuclei, as do the NA cell bodies. A series of classic lesion studies in the 1930s demonstrated that ascending tegmental pathways to the forebrain were necessary for normal waking and sleeping patterns. Subsequent experiments have shown that intraventricular injections of NA induce EEG arousal, whereas 5-HT induces sleep. Inhibition of the synthesis of 5-HT with the compound p-chloro-phenylalanine reduces the duration of sleep in cats. By making lesions in the cell body groups of the 5-HT neurons, which lie in the raphe nuclei, Jouvet and his colleagues were also able to disrupt sleeping patterns in cats. The reduction in sleep was highly correlated with the loss of 5-HT from the forebrain.

In contrast, α-methyltyrosine, which inhibits the synthesis of catechol-amines, reduces arousal and results in sedation. Consistent with a role of NA in arousal are the reports that increased arousal, induced, for example, by repeated foot shock in the rat, results in increased NA turnover in the brain. It is thought that the dorsal bundle of NA fibers arising from the locus coeruleus, rather than the ventral bundle, is impor-tant for EEG arousal. The dorsal NA bundle innervates the cortex, and the locus coeruleus is in the general region reported many years ago by Moruzzi and Magoun to be an excellent site from which to elicit EEG arousal with electrical stimulation. Chu and Bloom (1973) have recently made unit recordings from locus coeruleus neurones during the various stages of the sleep–waking cycle in the cat. They find that during attentive waking and paradoxical sleep there is EEG arousal and increased activity in the neurons. During slow-wave sleep the neuronal activity falls and becomes irregular. Although DA is important for the behavioral arousal of the response mechanisms of the extrapyrami-dal motor system it does not seem to be involved in EEG arousal, as lesions and stimulation in the substantia nigra do not produce such effects.

Presumably 5-HT and NA interact in the forebrain to trigger and modulate sleep–waking patterns. But in addition an interaction between them is thought to occur at the tegmental level, which may be important in the control of paradoxical sleep (PS). During normal slow-wave sleep, periods of EEG arousal occur although the animal is in a state of behav-ioral sleep, i.e., insensitive to environmental stimuli. This paradoxical decrease in cortical synchronization is accompanied by rapid eye move-ments, much as in normal arousal, despite the general lack of tone in axial musculature. A caudal region of the raphe nuclei has been found which apparently triggers PS. This is thought to activate the rostral raphe nuclei, which controls the normal synchronization of slow-wave sleep, and the caudal locus coeruleus, which has been shown to control the muscular activity and eye movements associated with PS. It is not known which part of this system is ultimately responsible for the cortical desyn-chronization of PS. It does not appear to be the dorsal NA bundle from the locus coeruleus, although this pathway is involved in normal waking. There are reciprocal connections between the locus coeruleus and the raphe nuclei and it is possible that a feedback from the locus coeruleus to the anterior raphe control the EEG synchronization associated with PS.

Noradrenaline and Acetylcholine in Arousal

The sleep–waking research attributes a central role in arousal to NA, but there is a great deal of evidence from lesion and drug studies which credits ACh with a similar function. Cholinergic stimulation is associated with high levels of sensory arousal and efficient behavioral performance on tasks requiring inhibition of responding (Carlton). Conversely a lack of ACh is associated with disruption of behavior on such tasks and a disinhibition of previously controlled responses. It seems possible that NA systems produce a general arousal, and that cholinergic pathways modulate this to provide the optimum level of arousal for behavioral efficiency. The cortex is an obvious site for such an interaction and there are ontogenetic studies which suggest that the NA and ACh systems become functionally developed at different times (Fibiger, Lytle, & Campbell, 1970). In adult rats drugs that stimulate adrenergic systems, such as amphetamine, activate behavior, whereas drugs such as pilocarpine that increase cholinergic activity depress behavior. Pilocarpine will also antagonize amphetamine-induced activity in adult rats, but this does not occur until the rats are at least 15 days of age. Before this time, amphetamine increases activity, just as in adults, but because the cholinergic innervation of the cortex has not developed, no antagonism by pilocarpine can be demonstrated. Cholinergic antagonist drugs like scopolamine also induce behavioral excitation in adult rats and potentiate the effects of amphetamine but this is again not seen until about 20 days of age (Figure 7). The emergence of cholinergic responsiveness coincides with deposition of myelin in the forebrain, the appearance of synaptic junctions, and the emergence of adult EEG patterns.

Acetylcholine and Dopamine in the Corpus Striatum

According to the high activity of acetylcholinesterase and of choline acetyl transferase in the striatum, this brain region contains an abundance of cholinergic neurons. Histochemical studies show intense acetylcholinesterase activity in the substantia nigra suggesting the possible existence of cholinergic striatal or pallidal projections to the substantia nigra. Such efferent pathways could function in a feedback loop from the striatum on to the neurons of the dopaminergic nigro-striatal pathway. The

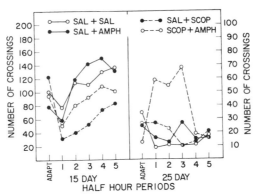

Figure 7. The effect of subthreshold doses of scopolamine and amphetamine, administered separately or in combination, on behavioral arousal in 15- and 25-day-old rats measured as increased number of crossings in an open field test. (Note different scales on ordinates due to different base lines.) [Reproduced from Fibiger et al., "Cholinergic modulation of adrenergic arousal in the developing rat," Journal of Comparative Physiological Psychology, 1970, 72, 384–389. Copyright 1970 by The American Psychological Association and reproduced by permission.]

modulation of DA metabolism would result in altered DA activity in the striatum. Because the methods for localizing and mapping cholinergic neurons are all indirect, it is difficult to determine precisely where the cell bodies and the terminals of the cholinergic systems in the basal ganglia are located. However, it is likely that DA and ACh operate in a delicately balanced system to control certain aspects of motor behavior.

Stereotypy is one form of behavior which illustrates this interaction well. Anti-cholinergic drugs potentiate stereotypy induced by the intracerebral microinjection of dopamine, amphetamine, or apomorphine. Cholinomimetic drugs, on the other hand, like neuroleptics, induce catalepsy and act synergistically with the neuroleptics. This suggests that the normal function of the cholinergic mechanisms in the striatum may be inhibitory to the DA mechanism. It seems that this inhibitory influence of ACh may contribute to the pattern of behavioral disruption induced by neuroleptic drugs, which block DA receptors. Neuroleptics induce catalepsy, antagonize the behavioral effects of amphetamine, and disrupt avoidance behavior, and all of these effects are antagonized by anti-cholinergic drugs.

Anti-cholinergic drugs may achieve these behavioral effects in part by inhibiting DA reuptake, which is one of the secondary neuropharmacological properties of many compounds of this type. Cole and Snyder reported a strong correlation between the ability of anti-cholinergics to inhibit DA uptake and their clinical efficiency in relieving Parkinson symptoms. Other workers suggest that the potentiation of amphetamine-induced stereotypy by anticholinergics is a more relevant model system for evaluating anti-Parkinson drugs. Using this behavioral measure and inhibition of DA uptake as a biochemical measure they report a good correlation between clinical efficiency and potentiation of stereotypy and a poor correlation with DA uptake inhibition.

Another example of ACh–DA interactions in the basal ganglia has been revealed by studies of the tremor associated with Parkinson's disease. The akinesia and rigidity characteristic of this disease seem to be associated with the loss of DA since these symptoms usually improve after L-DOPA therapy. These symptoms are often accompanied by tremor, which is not as readily improved with L-DOPA. It is suggested that the tremor may be due to overactivity in the cholinergic systems of the striatum, since anti-cholinergics are often effective in relieving Parkinsonian tremor.

Injection of anti-cholinesterases into the striatum induced a tremor in cats that was antagonized if ACh synthesis was inhibited or ACh receptors blocked with scopolamine. The neural circuitry activated by ACh to produce tremors probably involves the striatal projections to the thalamic nuclei which in turn are connected with the motor cortex. In electrophysiological studies electrical activity corresponding to the tremor is recorded in the thalamic nuclei, and this may explain why surgical lesions of such nuclei is thought by some clinicians to remain the most satisfactory treatment for the tremor symptom of Parkinsonism.

Interaction of Noradrenaline and Dopamine

As we have seen both ACh and DA are involved in the operation of the motor circuits mediating stereotypy. However, when one form of behavior is happening in an animal it is not an isolated event but part of a sequence of related acts to achieve a goal. It should not be surprising to find, therefore, that the potential of the DA system at any time is determined partly by what is going on in other neural circuits responsible for other aspects of behavioral control.

For example, if reserpinized rats are treated with L-DOPA, avoidance behavior and activity are reinstated. If, however, the animals are also pretreated with FLA-63, which prevents NA but not DA from being formed from L-DOPA, the behavior is clearly more stereotyped in nature. This suggests that although NA plays no direct role in the organization of striatal mechanisms it is operating an inhibitory function which prevents behavior becoming repetitive and purposeless when dopaminergic mechanisms are active.

Where these kind of interactions occur is quite unknown and these are the challenges for further studies in which drugs are applied to discrete sites in the brain, rather than given systematically. Only with such methods will the neural circuitries involved in these kind of subtle interactions be revealed.

PROBLEMS FOR FUTURE STUDIES

Functional Compensation in Transmitter Systems

The dogma that central nervous tissue, unlike peripheral nerves, once damaged could not regenerate is generally accepted. However, some recent experiments with transmitter-specific pathways indicates that this is not necessarily true. It seems that functional compensation and even anatomical regeneration can occur in CNS and this, therefore, complicates experimental efforts to correlate changes in neurochemical substrates with behavior. Denervation supersensitivity has long been known to occur in the peripheral nervous system. Presynaptic denervation results in a gradually increasing sensitivity of the postsynaptic receptors to normal agonists. A similar phenomenon has now been observed in the aminergic receptors of the brain after presynaptic denervation with 6-OHDA in either the neonate or adult rat (Creese and Iversen, 1973). After 6-OHDA treatment, for example, rats showed an enhanced locomotor response to low doses of L-DOPA, and more intense stereotypy if the DA receptors are stimulated directly with low doses of apomorphine.

Another remarkable finding is that if amine containing fiber pathways are surgically interrupted, the cut ends show sprouting, and morphologically recognizable terminals with storage vesicles can form. Adrenergic pathways do not have to be cut in order to sprout. A fornical cut, for example, abolishes the hippocampal septal fibers, and the NA terminals of the medial forebrain bundle which innervates the septum sprout profusely after such lesions (Moore, Bjorklund, & Stenevi, 1971). The functional significance of such mechanisms is obscure.

It has been reported that lateral hypothalamic lesions produce a less severe aphagia and adipsia if the rats have previously sustained frontal lesions (Glick & Greenstein, 1972). The frontal cortex projects to the hypothalamus and striatum, and removal of these connections may induce hypertrophy of amine terminals, thus strengthening these systems against subsequent assault by lesions in the lateral hypothalamus. Similarly α-methyl-tyrosine pretreatment is reported to alleviate symptoms of the lateral hypothalamic syndrome, supposedly because massive depletion increases postsynaptic sensitivity and the functional efficacy of synapses which survive the lesion (Glick, Greenstein, & Zimmerberg, 1972). It has, in an independent study, been reported that although recovered lateral hypothalamic animals have a lower body weight than controls, they appear in some respects to be more highly motivated for food. They begin to eat with a shorter latency in both familiar and novel environments and acquire food-rewarded discrimination tasks more quickly. Such results undermine any simple idea of a lateral hypothalamic eating center. If one result of the lateral hypothalamic lesion is denervation of aminergically inervated synapses due to severing of the pathways in the medial forebrain bundle, denervation supersensitivity may ultimately develop. An increase in motivation lability could well be the functional consequence of such changes in the brain.

Amino Acid Transmitters

The amino acid transmitters have not yet been sufficiently studied, nor is the neuropharmacology of these systems adequately developed, for meaningful speculation concerning their role in behavioral functions. It may be that studies of these systems will in any case prove more difficult, if only because substances such as GABA or glutamate are probably involved at far larger numbers of synaptic junctions in CNS than are the amine transmitters which numerically are probably of quite minor importance. Blockade of GABA receptors, for example, with the drug bicucueline, removes the major inhibitory influence on neuronal excitation in CNS, and the result is convulsive behavior, which is dramatic but not of great interest neuropsychologically. It may be possible to devise more subtle pharmacolical tools to modify the effectiveness of GABA or other amino acids as transmitters, but such tools are not yet available.

REFERENCES

Ahlenius, S., & Engel, J. Behavioural and biochemical effects of L-DOPA after inhibition of dopamine-β-hydroxylase in reserpine pretreated rats. *Naunyn-Schmiedebergs Archives of Pharmacology*, 1971, **270**, 349–360.
Andén, N-E., Carlson, A., & Haggendal, J. Adrenergic mechanisms. *Annual Review of Pharmacology*, 1969, 9, 119–134.
Bandler, R. J. Chemical stimulation of the rat midbrain and aggressive behavior. *Nature, New Biology*, 1971, **229**, 222–223.

Booth, D. A. Localization of the adrenergic feeding system in the rat diencephalon. *Science*, 1967, **158**, 515–517.

Carlton, P. L. Cholinergic mechanisms in the control of behavior by the brain. *Psychological Review*, 1963, **70**, 19–39.

Chu, N-S., & Bloom, F. E. Norepinephrine-containing neurons: Changes in spontaneous discharge patterns during sleeping and waking. *Science*, 1973, **179**, 908–910.

Creese, I. N. R., & Iversen, S. D. Blockage of amphetamine induced motor stimulation & stereotypy in the adult rat following neonatal treatment with 6-hydroxydopamine. *Brain Research*, 1973, **55**, 369–382.

Crow, T. J. A map of the rat mesencephalon for electrical self-stimulation. *Brain Research*, 1972, **36**, 265–273.

Ellinwood, E. H., Sudilovsky, A., & Nelson, L. Behavioral analysis of chronic amphetamine intoxication. *Biological Psychiatry*, 1972, **4**, 215–230.

Ernst, A. M. Mode of action of apomorphine and dexamphetamine on gnawing compulsion in rats. *Psychopharmacologia (Berlin)*, 1967, **10**, 316–323.

Feltz, P., & de Champlain, J. Persistence of caudate unitary responses to nigral stimulation after destruction and functional impairment of the striatal dopaminergic terminals. *Brain Research*, 1972, **43**, 595–600.

Fibiger, H. C., Lytle, L. D., & Campbell, B. A. Cholinergic modulation of adrenergic arousal in the developing rat. *Journal of Comparative Physiological Psychology*, 1970, **72**, 384–389.

Fisher, A. E. Relationships between cholinergic and other dipsogens in the central mediation of thirst. In E. Steller, A. N. Epstein, & H. Kissileff (Eds.), *The neuropsychology of thirst*. New York: Winston, 1973.

Fog, R. On stereotypy and catalepsy: Studies on the effect of amphetamines and neuroleptics in rats. *Acta Neurologica Scandinavica*, 1972, **48**, suppl. 50.

Fuxe, K., & Hanson, L. C. F. Central catecholamine neurons and conditioned avoidance behavior. *Psychopharmacologia*, 1967, **11**, 439–447.

Geller, I., & Seifter, J. The effects of meprobromate, barbiturate, d-amphetamine and promazine on experimentally induced conflict in the rat. *Psychopharmacologia*, 1960, **1**, 482–492.

Glick, S. D., & Greenstein, S. Facilitation of recovery after lateral hypothalamic damage by prior ablation of frontal cortex. *Nature, New Biology*, 1972, **239**, 187–188.

Glick, S. D., Greenstein, S., & Zimmerberg, B. Facilitation of recovery by α-Methyl-p-tyrosine after lateral hypothalamic damage. *Science*, 1972, **177**, 534–535.

Hoebel, B. G. Feeding: Neural control of intake. *Annual Review of Physiology*, 1971, **33**, 533–568.

Hornykiewicz, O. Dopamine in the basal ganglia. *British Medical Bulletin*, 1973, **29**, 172–178.

Iversen, L. L. Third Gaddum Memorial Lecture. Role of uptake processes for amines and amino acids in synaptic neurotransmission. *British Journal of Pharmacology*, 1971, **41**, 571–591.

Iversen, S. D. 6-hydroxydopamine: A chemical lesion technique for studying the role of amine neurotransmitters in behavior. *Third Neurosciences Programme Meeting*, Boulder, 1974.

Jouvet, M. The role of monoamines and acetylcholine-containing neurons in the regulation of the sleep–waking cycle. *Ergebnisse der Physiologie*, 1972, **64**, 166–307.

Kelleher, K. T., & Morse, W. H. Determinants of the specificity of behavioral effects of drugs. *Ergebnisse der Physiologie*, 1966, **60**, 1–56.

Lehr, D., Mallow, J., and Krukowski, M. Copious drinking and simultaneous inhibition of urine flow elicited by beta-adrenergic stimulation and contrary effect of alpha-adrenergic stimulation. *Journal of Pharmacology and Experimental Therapeutics*, 1967, **158**, 150–163.

Leibowitz, S. F. Hypothalamic alpha and beta-adrenergic systems regulate both thirst and hunger in the cat. *Proceedings of the National Academy of Science*, 1971, **68**, 332–334.

Lewy, A. J., & Seiden, L. S. Operant behaviour changes norepinephrine metabolism in rat brain. *Science*, 1972, **175**, 454–455.

Livett, B. G. Histochemical visualization of adrenergic neurones. *British Medical Bulletin*, 1973, **29**, 93–99.

Margules, D. L. Alpha-adrenergic receptors in hypothalamus for the suppression of feeding behaviour by satiety. *Journal of Comparative Physiological Psychology*, 1970, **73**, 1–12. (a)

Margules, D. L. Beta-adrenergic receptors in the hypothalamus for learned and unlearned task aversions. *Journal of Comparative Physiological Psychology*, 1970, **73**, 13–21. (b)

Margules, D. L., Lewis, M. J., Dragovich, J. A., & Margules, A. S. Hypothalamic norepinephrine: Circadian rhythms and the control of feeding behaviour. *Science*, 1972, **178**, 640–643.

Marshall, J. F., & Teitelbaum, P. A comparison of the eating in response to hypothermic and glucoprivic challenges after nigral 6-hydroxydopamine and lateral hypothalamic electrolytic lesions in rats. *Brain Research*, 1973, **55**, 229–233.

Milner, P. *Physiological psychology*. New York: Holt, 1970.

Moore, R. Y., Bjorklund, A., & Stenevi, V. Plastic changes in the adrenergic innervation of the rat septal area in response to denervation. *Brain Research*, 1971, **33**, 13–35.

Munkvad, I., Pakkenberg, H., & Randrup, A. Aminergic systems in basal ganglia associated with stereotyped hyperactive behavior and catalepsy. *Brain Behavior and Evolution*, 1968, **1**, 89–100.

Nauta, W. J. H. Central neurons organization and the endocrine motor system. In A. V. Nalbandor (Ed.), *Advances in neuroendocrinology*. Urbana, Illinois: Univ. of Illinois Press, 1963. Pp. 5–21.

Papeschi, R. Dopamine, extrapyramidal system and psychomotor function. *Psychiatria, Neurologia, Neurochirurgia*, 1972, **75**, 13–48.

Poirier, L. J., Bedard, P., Boucher, B., Bouvier, B., Larochelle, L., Olivier, A., Parent, A., & Singh, P. The origin of different striato- and thalamopetal neurochemical pathways and their relationships to motor activity. In F. J. Gillingham & I. M. L. Donaldson (Eds.), *Third symposium on Parkinson's disease*. Edinburgh: Livingstone, 1969. Pp. 60–66.

Poschel, B. P. H., & Ninteman, F. W. Norepinephrine: A possible excitatory neurohormone of the reward system. *Life Sciences*, 1963, **3**, 782.

Reis, D. J., & Fuxe, K. Brain norepinephrine: Evidence that neuronal release is essential for sham rage behaviour following brainstem transection in cat. *Proceedings of the National Academy of Science*, 1969, **64**, 108–112.

Robbins, T. W., & Iversen, S. D. Amphetamine induced disruption of temporal discrimination by response disinhibition. *Nature, New Biology*, 1973, **245**, 191–192.

Russell, R. W. Behavioural aspect of cholinergic transmission. *Federation Proceedings*, 1969, **28**, 121–131.

Settler, R. E. The role of catecholamines in thirst. In E. Stellar, A. N. Epstein, & H. Kissileff (Eds.), *The neuropsychology of thirst*. New York: Winston, 1973.

Shute, C. C. D., & Lewis, P. R. The ascending cholinergic reticular system: Neocortical, olfactory and subcortical projections. *Brain*, 1967, **90**, 497–520.

Slangen, J., & Miller, N. E. Pharmacological tests for the function of hypothalamic norepinephrine in eating behaviour. *Physiology and Behaviour*, 1969, **4**, 543–552.

Sparber, S. B., & Tilson, H. A. Schedule controlled and drug induced release of norepinephrine-7-^{3}H into the lateral ventricle of rats. *Neuropharmacology*, 1972, **11**, 453–464.

Stein, L. Chemistry of reward and punishment. In D. H. Efron (Ed.), *Psychopharmacology: A review of progress, 1957–67*. Washington: U.S. Government Publishing Office, 1968.

Stein, L., & Wise, C. D. Release of norepinephrine from hypothalamus and amygdala by rewarding medial forebrain bundle stimulation and amphetamine. *Journal of Comparative Physiological Psychology*, 1969, **67**, 189–198.

Svensson, T. H., & Waldeck, B. On the role of brain catecholamines in motor activity: Experiments with inhibitors of synthesis and of monoamine oxidase. *Psychopharmacologia*, 1970, **18**, 357–365.

Taylor, K. M., & Snyder, J. H. Differential effects of D- and L-amphetamine on behaviour and on catecholamine disposition in dopamine and norepinephrine containing neurones of rat brain. *Brain Research*, 1971, **28**, 295–309.

Ungerstedt, U. On the anatomy, pharmacology and function of the nigro-striatal dopamine system. *Acta Physiologica Scandinavica*, 1971, Suppl. 367.

Valenstein, E. S. The anatomical locus of reinforcement. *Progress in physiological psychology*, 1966, **1**, 149–190.

Vogt, M. Functional aspects of catecholamines in central nervous system. *British Medical Bulletin*, 1973, **29**, 168–172.

Warburton, D. M. The cholinergic control of internal inhibition. In R. Boakes & M. S. Halliday (Eds.), *Inhibition and learning*. London: Academic Press, 1972. Pp. 431–460.

York, D. H. Dopamine receptor blockade—essential action of chlorpromazine on striatal neurones. *Brain Research*, 1972, **37**, 91–100.

Chapter 7

Biochemical Approaches to the Biological Basis of Memory

DAN ENTINGH, ADRIAN DUNN, EDWARD GLASSMAN,
JOHN ERIC WILSON, EDWARD HOGAN, and
TERRI DAMSTRA

University of North Carolina

INTRODUCTION

The formation of memory requires chemical changes in the brain. The empirical and inferential bases for this assertion have been reviewed adequately (Booth, 1967; Bogoch, 1968; Glassman, 1969; Barondes, 1970; McGaugh & Dawson, 1971; Jarvik, 1972). The purposes of this chapter are to discuss some current approaches to identifying the biochemical processes that are involved in memory formation and storage, and to suggest future applications of those approaches.

The search for the molecular basis of memory has generated much excitement over the past decade. Much of this resulted from an analogy drawn between behavioral memory and genetic memory, coupled with the assumption that both processes might be mediated by similar biochemical mechanisms. Suggestions that memory might be stored by changes in the primary structure of RNA and proteins (Hydén, 1960;

Landauer, 1964) have received little experimental support (Hydén, 1967; Glassman, 1969).

Investigations into the chemical basis of memory are now more critical and cautious. Instead of searching for a unique memory molecule, many investigators deem it more useful to detect the ways in which the metabolism of neurons responds to various kinds of stimulation, and then to analyze the roles that such metabolic changes have in biological processes in the cells in which they occur. Much more must be known about the biology of the neuron before it will be possible to understand the role that any single biochemical change of state plays in memory storage.

Chemical changes of state can be related to physiological and anatomical changes by the working assumption that memory is mediated by changes in the connectivity (both functional and anatomical) between various neurons or sets of neurons. The remembered *information* is assumed to be stored in altered functional neuroanatomical pathways that are probably formed by changes in the biochemistry of individual neurons or other brain cells, affecting either synaptic function or excitation thresholds (but see John, 1972, for an alternate view). The behavioral neurochemist may, thus, concentrate on specific chemical changes in specific cells, leaving aside for the present the nature of the intercellular physiological events that initiate the intracellular basis of plastic changes.

The task of defining "learning" and "memory" precisely has confounded many authors and will not be attempted here. Both are complex events, mediated by processes that occur in poorly defined substrates. For theoretical convenience, the present discussion employs the consolidation model of memory formation (McGaugh & Dawson, 1971), which states that the storage of information by vertebrate brains involves sequential processes. The earlier stages of these processes are more easily disrupted by amnestic treatments than are the later stages.

We assume that this decreasing susceptibility of memory to amnestic treatments with time occurs because the physical basis of a given memory shifts from initial neurobiological processes with short lifetimes (e.g., electrical activity, conformational changes in protein structure) to processes with long lifetimes (e.g., changed membrane constituents, altered synaptic anatomy). Thus, the physical basis of memory is likely to be comprised of a complex chain of metabolic reactions, with initial chemical changes occurring in close temporal proximity to the development of short-term memory (STM) and the first signs of adaptive performance in a learning situation. It seems likely that some of these earlier events may include changes in which gene expression is modified in individual brain cells. The genomic changes may culminate in changes in metabolism that alter the functional relationships between brain cells, to mediate long-term memory (LTM).

Four major questions are addressed in this chapter: (1) Which neurochemical systems are sensitive to environmental stimulation? (2) When during the course of stimulation do changes occur? (3) Where in the brain do these events happen? and (4) What aspects of stimulation episodes are critical for inducing these changes?

CURRENT STATUS OF KNOWLEDGE

Which Neurochemical Systems Are Sensitive to Environmental Stimulation?

A variety of neurochemical responses have been elicited by environmental stimulation ranging from flickering lights to a wide variety of training paradigms. In spite of conflicting reports and differing opinions, especially on the interpretation of the data, the nervous system is now known to be capable of an impressive array of such responses.

Reactivity in Macromolecules

A convenient scheme for interrelating some of these changes is depicted in Figure 1. This scheme is certainly oversimplified and portions of it will prove false. Nonetheless, it is helpful for visualizing how the known central neurochemical reactions might be related to various states of neuronal activity, to the metabolic phenomena that follow physiological forms of stimulation, and to the microanatomical changes in brain mass and synaptic organization that follow prolonged environmental stimulation.

It seems likely that some of the known neurochemical changes depicted in Figure 1 participate in information storage mechanisms, but this has not been demonstrated for any of the changes. As discussed on pp. 212, 221, and 229, interpretation of the biological role of a detected neurochemical change is made difficult by the high likelihood that many chemical reactions participate in the formation of any engram.

The final chemical changes of state that store LTM must be stable and longlasting. It is not clear whether this implies that some of the molecules directly involved in a memory must themselves be stable, or that the stability resides in a self-regenerative system in which the longevities of the particular molecules are unimportant. For example, if a specific memory involved the formation of a new synapse, all of the molecules involved in that synapse could be replaced periodically, but, provided that the synapse itself were retained, the memory would remain. Since it has been difficult to conceive of small molecules being the *exclusive* basis of stable changes, macromolecules, with their generally longer half-lives, have been favored for study.

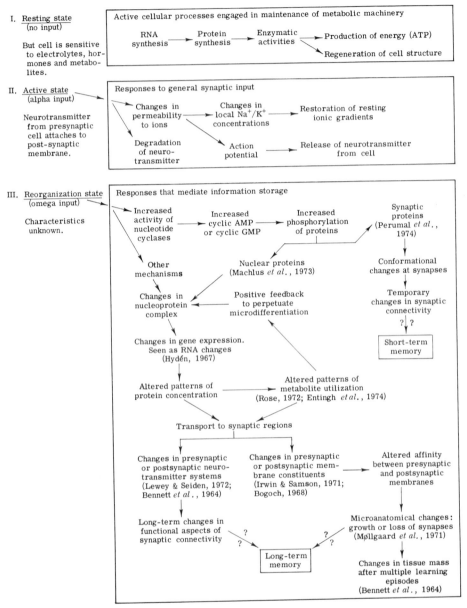

Figure 1. A working hypothesis of how the metabolism of a neuron might respond to the physiological effects of environmental stimulation.

This scheme is an attempt to place in a time-ordered perspective some of the changes in brain metabolism that have been reported to occur after stimulation. Many of the individual metabolic steps are discussed in the text. The scheme presumes that *information* retained in memory is stored in the form of altered relationships between neurons, whereas the

Macromolecules (e.g., RNA, proteins, glycoproteins) are likely candidates for major roles in the biological processes that underlie memory storage because these molecules are involved in genetic control mechanisms and cellular differentiation, and because they are the principal components of the anatomical structures of cells. Thus, macromolecules serve as the logical constructs that enable biologists to envision a variety of ways in which the ionic fluxes of action potentials and the various actions of neurotransmitters can be related to the observable structures of neurons.

The greatest emphasis has been placed on the metabolism of nucleic acids and of proteins. Some studies have attempted to link memory formation with brain DNA because of its great stability and genetic role in the control of metabolism (Reinis, 1972), but there is little evidence that the function of brain DNA is any way distinct from that of any other organ. Inhibitors of brain DNA synthesis apparently do not impair memory consolidation (Casola, Lim, Davis, & Agranoff, 1969).

The work of Hydén (Hydén, 1967) among many others has indicated that neural RNA is extraordinarily sensitive to changes in amount and composition as a result of external stimuli. While technical problems plague this area, a large mass of data strongly indicates that changes of RNA metabolism are associated with certain types of neural stimulation (Pevzner, 1966) and behavioral stimulation (Glassman, 1969, 1974).

Inhibitors of RNA synthesis probably affect long-term memory formation. While inhibiting RNA synthesis through different modes of action, both actinomycin D (Agranoff, 1968) and camptothecin (Neale, Klinger, & Agranoff, 1973) impaired LTM but had minimal effects on STM. Amnesia is rarely complete after blockade of RNA metabolism, and actinomycin D has complicating side effects. Nonetheless these are interesting though not clearly interpretable findings.

It is worth noting that concepts of RNA metabolism in eukaryotic

---

chemicophysical changes responsible for initiating and maintaining such alterations involve the reorganization of the internal metabolism of individual neurons. The scheme neglects possible roles of glial cells only for reasons of simplicity.

It is convenient to consider any given process as being related mainly to one of three major metabolic conditions within the neuron, as depicted along the left margin. Processes associated with the Resting state occur continuously. Processes associated with the Active state can occur at any time, even when the processes involved in the Reorganization state are operating.

The Reorganization state is conceived to include all the metabolic activities of the cell that may participate in a process of microscale cellular differentiation. Such differentiation has as its outcome an altered functional relationship between the neuron in question and its synaptic neighbors. The pericellular conditions that activate such reorganization are not known, but two of the more likely candidates for this role are (1) large local ionic imbalances within the neuron following prolonged operation of the Active state, or (2) activation of neuronal adenyl or guanyl cyclase following synaptic reception of appropriate neurotransmitters.

cells are undergoing constant revision. Past assumptions that increases in radioactivity in nuclear RNA reflect the synthesis of new messenger RNAs, eventually to be translated into new proteins in the cytoplasm, may be oversimplified and be misleading. In particular the relationship between nuclear and cytoplasmic messenger RNAs is not clear, because it is now thought that most nuclear RNAs never leave the nucleus (Soiero, Vaughan, Warner, & Darnell, 1968). Further, there is evidence that the synthesis of new or new types of proteins in eukaryotic cells often does not require changes of RNA metabolism within the nucleus (Tomkins, Levinson, Baxter, & Dethlefsen, 1972).

Reports of changes in protein metabolism induced by training have been plentiful (Hydén & Lange, 1970a; Beach, Emmens, Kimble, & Lickey, 1969; Horn, Rose, & Bateson, 1973; Rees, Brogan, Entingh, Dunn, Shinkman, Damstra-Entingh, Wilson, & Glassman, 1974). Most of the reported changes have not yet been analyzed in biochemical or behavioral detail, but there is broad agreement that cerebral protein metabolism can be altered by behavioral input.

There is no doubt that inhibitors of protein synthesis interfere with memory in several species tested on various tasks (Barondes, 1970). However, it is not clear whether such effects are due to the inhibition of protein synthesis or to other actions of the drugs, nor whether the effects are caused by inhibition of the synthesis of proteins that have some specificity for memory processes. Protein synthesis must be severely inhibited by the antibiotics (puromycin, cycloheximide, and acetoxycycloheximide) in order for amnestic defects to be observed. Because the brain has a comparatively high rate of protein synthesis, it seems unlikely that near-total inhibition of protein synthesis in the brain would not result in major impairments.

Two other types of large molecules have been reported to be related metabolically to behavior, namely glycoproteins (Bogoch, 1968; Damstra-Entingh, Entingh, Wilson, & Glassman, 1973) and gangliosides (Irwin & Samson, 1971; Dunn, Entingh, Damstra-Entingh, Gispen, Machlus, Perumal, Rees, & Brogan, 1974). Both compounds are of great interest because they are components of membranes, particularly of synaptic membranes.

The potential significance for memory mechanisms of the chemical modification of preexisting molecules during learning is great. Minor chemical modification of membrane components can greatly alter the cohesiveness between cells. Conformational changes in macromolecules induced by minor chemical modifications may also control the transcription of DNA into RNA, the translation of RNA into protein, or alter the metabolism of neurotransmitters. Possible modifications of proteins include the addition or removal of disulfide bonds, alkyl groups, acyl groups, phosphate groups, or glycosides. Similar possibilities exist for other macromolecules and lipids. Hydén and Lange (1970b) have reported that training induces conformational changes in the neural-tissue protein S-100. Machlus, Wilson, and Glassman (1974) found that the phosphoserine content of the nonhistone acid-extractable protein (NAEP) of rat brain nuclei is increased with training. This class of proteins has been implicated in the control of gene transcription in non-neural tissues, but further work is needed to determine if this neurochemical response is indeed related to gene expression. In the same laboratory, Perumal (1973) has shown that there is increased radioactivity associated with phosphoserine of synaptic membrane proteins in mice given avoidance training.

Radiochemical Tracer Methods

Radiochemical tracer methods have been particularly valuable in detecting and quantifying stimulation-induced neurochemical responses. The central role that such methods play in these investigations makes it imperative that one understand their limitations.

Radiochemical methods are used because they can measure changes in rates of ongoing reactions. They thus allow the study of neurochemical changes that occur during brief time periods, and are capable of elevating relatively small changes above the background noise of ongoing metabolism and preexisting concentrations of metabolites. An undetectable increase of 1% in the concentration of a brain protein during a 30-min period might, for example, be detected as a large increase in the amount of radioactive amino acid incorporated into that protein. Other benefits of the methods are that they can be used to measure the longevities of compounds and of changes. Double isotope methods, in which the same chemical precursor labeled with two different radioactive isotopes (e.g., ^3H and ^{14}C, or ^{32}P and ^{33}P) is administered to two different animals, can also be used to reduce experimental errors caused by nonquantitative recoveries of reaction products, and certain other chemical artifacts.

Figure 2 diagrams a hypothetical experiment addressed at the question: "What effect does 15 min of avoidance training have on the rate of protein synthesis in rat brain?" After it is injected, radioactive lysine mixes with the nonradioactive lysine already present in the body. Some of the radioactive compound penetrates from the blood stream into the brain (and other organs), where it is incorporated into proteins and also metabolized in other ways. The primary indication of an increased rate of protein synthesis is an increase in the ratio of the proportion

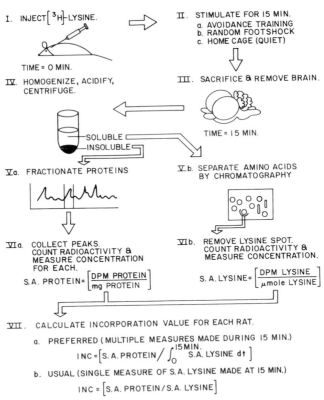

Figure 2. The plan of an experiment using a radioactive amino acid to measure the effects of training on cerebral protein synthesis.

The goal is to determine if avoidance training enhances the rate at which specific protein groups (peaks in Va) incorporate lysine. The calculations (VII) are made for each protein group. Treatment group IIb (Random Footshock) is included to detect effects induced by the general stimuli used in the training task.

of lysine molecules in protein that are radioactive to the average proportion across time of lysine molecules not in protein that are radioactive. This value is called the ratio of the specific activities of protein and free lysine (relative specific activity).

A major problem is the accurate measurement within tissue samples of the radioactivity in the precursor compound. Before the radioactive compound is incorporated into the product molecule it is affected by a series of factors, each of which influences the concentration of radioactive precursor at sites of macromolecular synthesis, and each of which might be affected by behavioral stimulation. Such factors include: (1) uptake of precursor at the site of injection, (2) penetration from blood to brain, in the case of systemic injections, (3) penetration from extracellular fluid into brain cells, (4) conversion of the tracer from the injected compound into the form that participates in macromolecular synthesis ([³H]-uridine triphosphate when [³H]-uridine is used as a tracer for RNA synthesis), (5) metabolism of the tracer into compounds that are not incorporated into the product of interest, (6) metabolism of the tracer in peripheral organs, and (7) distribution of the tracer into various intracellular compartments.

Accurate measurement of the radioactivity in the precursor pool requires that: (a) the proportion of precursor molecules that are radioactive (specific activity) must be measured, (b) such measurement must be made at the site where synthesis of the macromolecules occurs, and (c) such measurements must be made at multiple time points during the course of the incorporation period. If these stringent criteria are met, then measures of synthesis rates are independent of the confounding factors described in the paragraph above (Lajtha & Marks, 1971). However, such measurements can rarely be made in living vertebrate systems, particularly because it is difficult to isolate the compartment where synthesis occurs (e.g., the chromatin of the cell nucleus, in the case of RNA synthesis) from other metabolic compartments. Thus, for brain, measurements of synthesis rates can be made only indirectly.

Thus approximations are usually used in assaying neurochemical responses to stimulation. The total amount of radioactivity in all low-molecular-weight compounds (e.g., "acid-soluble fraction") in brain is used to approximate the specific activity of the precursor pool, and measurements are made only at the end of an incorporation period of fixed length. Especially questionable is the common practice of assuming that measurement at a single point provides an adequate description of the rate at which the tracer has entered the immediate precursor pool. Because such measures do not correct for any of the confounding factors cited above, and, least of all, for fluctuations in specific activity of the precursor due to stimulus-induced alterations in the metabolism of the precursor compound, the reported changes in incorporation values are as likely to be caused by changes in metabolism of precursor as they are by changes in synthesis rates.

The analysis of cerebral RNA metabolism by the use of radioactive tracers provides examples of some of these difficulties. The precursors used for these studies (uridine and cytidine) become incorporated into many different biochemical compounds in all tissues (Figure 3). Besides their roles in RNA and DNA metabolism, uridine and cytidine nucleotides are extensively involved as cofactors in the metabolism of sugars that are related to the biosynthesis of polysaccharides, glycoproteins

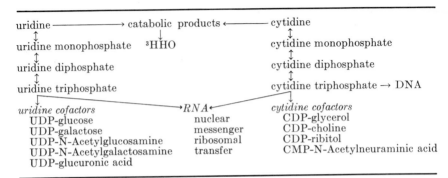

Figure 3. Metabolic fates of radioactivity injected into an animal in the form [5-³H]uridine. The compounds listed on this chart become radioactive in mammals when [5-³H]uridine or [5-³H]cytidine is injected either cerebrally or systemically. The large number of compounds involved makes it difficult to specify experimentally exactly what happens to the metabolism of these compounds in the brains of stimulated animals.

and glycolipids, and of alcohols involved in the biosynthesis of lipids. Changes in the metabolism of any of these compounds may affect nucleotide metabolism and hence the radioactivity in the precursor pools for RNA synthesis.

Glassman and Wilson and their coworkers (Glassman & Wilson, 1970) have reported consistent increases in the incorporation of radioactive uridine into brain RNA associated with avoidance training. A double isotope procedure was used to control for variability in the recovery of RNA, and the incorporation of radioactivity into uridine-5′-monophosphate (UMP) was used to measure uptake of uridine by the tissue. More recent results from the same laboratory, where some of the problems of quantitation in recovery of nucleotides have now been solved, indicate that the amount of radioactivity in UMP is lowered by training (Entingh, Damstra-Entingh, Dunn, Wilson, & Glassman, 1974). Those changes are large enough to account for some of the apparent changes in RNA metabolism, particularly those associated with nuclear RNA. In a similar vein, Baskin, Masiarz, and Agranoff (1972) have recently suggested that changes in RNA radioactivity patterns detected by Shashoua (1970) in goldfish brain during a float-balancing learning task may be due to stimulation-induced alterations in rates of conversion of cytidine to uridine.

Different routes of administration of precursors into the body each raise their own special problems. When peripheral injections are used, many compounds enter the brain from the blood stream only poorly. Stimulation-induced changes in the rate of transfer between blood and brain have been documented (Sokoloff, 1961), and these must be distinguished from intracellular effects. On the other hand, intracranial injections damage part of the brain, and the damage or any anesthesia involved may alter the behavior. Furthermore, the reproducibility of intracranial injections is usually poor, the distribution of the injectate over the brain may be variable, and concentration gradients from the ventricles may occur. There is also the danger of direct pharmacological effects of the injected compound, its solvent or impurities in either. Kottler, Bowman, and Haasch (1972) have reported that an increased incorporation of cytidine into brain RNA consequent to spatial reversal training is detected if radioactive cytidine is injected intravenously in rats, but not if it is injected directly into the brain ventricles. Hydén and Lange have reported that *increased* incorporation of leucine into brain proteins occurs during transfer-of-handedness training in rats if the precursor is injected into the brain (Hydén & Lange, 1968), but that *decreased* incorporation is apparent when the precursor is delivered through a peripheral route (Hydén & Lange, 1972).

Thus the results of incorporation studies, as published to date, are difficult to interpret, and at best can be only *suggestive* of changes in the rates of synthesis of macromolecules. Conclusions from tracer studies can be made clearer if (*a*) the radioactivity used for precursor measurements is derived from intracellular compartments, (*b*) the immediate precursor form is isolated from other compounds, and (*c*) measurements are made at progressive intervals within the period during which changes in incorporation were first detected.

At least three different approaches have been useful in circumventing the ambiguities of radiochemical tracer methods with regard to demonstrating conclusively that changes in rates of synthesis of macromolecules have occurred. (*1*) Demonstration *in vitro* that there is altered activity of the enzyme that is necessary for the observed *in vivo* changes. Haywood, Rose, and Bateson (1970) showed that, after imprinting, the *in vitro* activity of RNA polymerase was increased concomitant to increased *in vivo* incorporation of uridine into RNA. Since the activity of enzymes in the brain may be affected by subtle changes (e.g., membrane conformation, metabolite concentration) that may not be preserved when tissue extracts are made for the *in vitro* assays, results from this approach are meaningful only when the *in vitro* assays actually detect changes. (*2*) Demonstration *in vivo* that other chemical reactions in the same general metabolic sequence are affected by the experience in the same direction as was the step for which the changes were originally detected. Horn *et al.* (1973, pp. 509–510) and their associates have found sequentially that protein metabolism *in vivo*, RNA metabolism *in vivo*, and RNA metabolism *in vitro* are all enhanced by an imprinting experience. (*3*)

Demonstration of changes in the chemical concentrations of the macro-molecule's products, without the use of radioactivity. Machlus *et al.* (1974b) have shown that the phosphoserine content of the NAEP of rat brain nuclei is increased by avoidance training, having originally discovered the phenomenon by using radioactive phosphate.

A final problem is the detection of the biological functions of metabolic changes, and the elucidation of their neurobiological function. To dis-cover that, in the brains of trained animals, there occurs an enhanced synthesis of a protein that has been characterized only by physicochemi-cal separation methods (centrifugation, chromatography, electrophoresis, etc.) is not particularly enlightening until it is known what enzymatic or functional role that protein serves. At the moment, the biological roles of the known neurochemical responses to stimulation are only poorly understood.

When during Learning Do Chemical Changes Occur?

The dynamic aspects of memory-related metabolic phenomena have been approached in two ways. Following the injection of metabolic inhibitors at various times in relation to training, measurements of the retention of memory are made to determine when the formation of LTM is susceptible to metabolic interference. Alternatively, direct measure-ments are made of the magnitude of metabolic changes at different times during periods of learning and consolidation.

Interpretation of the temporal changes detected by either approach is complicated by the high likelihood that sequential chemical steps are necessary to form LTM. Effects of amnestic agents suggest that at least three sequential physiological processes are involved in the forma-tion of LTM. The last two of these are probably mediated by chemical processs, since they seem to be merely halted, and not reversed or eradi-cated, by application of electroconvulsive shock (McGaugh & Dawson, 1971). Biochemical considerations also suggest that multiple reaction steps are necessary, beginning with short-lived reactions of membrane-coupled compounds at presynaptic sites, and ending with slower and more permanent changes in molecules related to synaptic function or structure (see Figure 1).

If multiple chemical reactions are involved in consolidation, it is un-likely that any single chemical change will exhibit exactly the same time course as behavioral memory. Some of the initial reactions in the total process may come to an end long before consolidation is complete, and changes that participate in long-term aspects of memory storage may not be initiated until late phases of consolidation. This lack of any necessary simultaneity between behavioral memory and any single contributing biochemical change of state makes it difficult to draw conclusive inferences based on correlations between behavioral and biochemical time courses.

Further complications in interpreting such correlations arise from suggestions that the consolidation of LTM does not necessarily begin as soon as STM is established in the training episode. Leaving animals in the training apparatus can delay the onset of consolidation (Davis & Agranoff, 1966; Robustelli, Gellar, & Jarvik, 1968), and consolidative processes can seemingly be restarted after being disrupted by various treatments (Albert, 1966; Barondes & Cohen, 1968a). Although they complicate the current interpretive picture, such phenomena eventually should prove to be valuable tools for establishing causal links between neurochemical changes and the effects of amnestic agents.

Another barrier to relating the data from susceptibility studies to direct measurement of neurochemical responses is that the two approaches historically have concentrated on different types of training paradigms. The dominant paradigm for consolidation studies has been the one-trial inhibitory avoidance task, important because it can separate certain procative and retroactive effects of amnestic treatments (McGaugh & Dawson, 1971; but see Spevack & Suboski, 1969), and because the effects of amnestic treatments diminish in proportion to the number of original training trials (Barondes & Cohen, 1967). On the other hand, the direct studies of chemical changes have used large numbers of training trials, both because of the apparent necessity to demonstrate that each trained subject has learned something before it is sacrificed for neurochemical analysis, and because of a general belief that neurochemical effects of single training trials might be difficult to detect or interpret. Hydén and his colleagues (Hydén, 1967; Hydén & Lange, 1970a) have reported changes in cerebral RNA and protein metabolism after rats were exposed for three to five days to a rope-climbing task or a change of handedness task. Other workers (Zemp, Wilson, Schlesinger, Boggan, & Glassman, 1966; Kottler et al., 1972) have detected changes in RNA or nucleotide metabolism at the end of briefer (15 to 60 min) single episodes of active avoidance or position-habit reversal training. We know of no biochemical changes detected using the one-trial inhibitory avoidance paradigm as the stimulating experience.

Dynamic studies of consolidative processes have revealed particularly fascinating results when inhibitors of protein synthesis have been used to manipulate retention. Severe inhibition of protein synthesis seems to have only minimal effects on response acquisition and the formation of STM, but prevents LTM from developing in the normal fashion.

For instance, Barondes and Cohen (1968b) found that acetoxycyclohexi-mide injected shortly before training had no effects on the retention by mice of a T-maze habit for as long as 4 hr after training, but seriously disrupted retention measured at 6 or 24 hr or at 7 days (Figure 4). The decay of retention in such situations may reflect the decay of STM, but it is possible that the presence of the metabolic inhibitor increases the rate at which STM decays (McGaugh & Dawson, 1971).

Barondes and Cohen (1968b) also found that protein synthesis had to be inhibited *during* the period of training in order to prevent the formation of LTM. This suggests that LTM is dependent on a critical brief period of protein formation that occurs during or very soon after the learning experience. Changes in protein metabolism at early times after brief experiences have been observed (Rees *et al.*, 1974), but the proteins involved have not been definitively analyzed, nor is it at all clear what relationship, if any, those changes have to memory.

Some agents, particularly puromycin and diisopropylfluorophosphate (which inhibits cholinesterase, as well as other enzymes) have effects on retention if injected many hours or days after training (Flexner, Flexner, & Stellar, 1963; Deutsch, Hamburg, & Dahl, 1966). Although it is common practice to ascribe such late effects to interference with retrieval mechanisms, these phenomena also are suggestive of a continued development of the final form of memory storage over a period of several days. A primary example of this is that bilateral injections of puromycin into the temporal region of the mouse brain block the expression of memory for a maze habit if given within three days of training, but six days after training injections must be made at multiple sites in the brain if the memory is to be disrupted (Flexner *et al.*, 1963). This implies that the final storage form of memory is not achieved for at least three days.

Studies of the time course of metabolic changes during the training period are few in number. Perumal (1973) has recently generated the

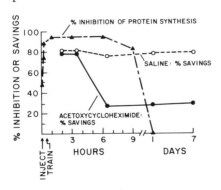

Figure 4. Effects of acetoxycycloheximide on cerebral protein synthesis and retention.

Mice were injected with the inhibitor 30 min before they were trained to escape shock by choosing the lighted limb of a T-maze to a criterion of five out of six consecutive correct responses. Different groups were tested for retention at the indicated times. For details, see Barondes and Cohen, 1968a. [From Barondes, 1970. Redrawn by permission of Dr. S. H. Barondes and The Rockefeller University Press.]

time curve for an increased phosphorylation of mouse brain synaptosomal proteins during avoidance training (Figure 5). The onset of this neurochemical change parallels the same time course as the avoidance-learning curve but occurs after significant changes in performance take place. The change persists for at least 30 min after the mouse is removed from the training apparatus. Although it has not been demonstrated that this chemical response is necessary for learning or consolidation, the data strongly support the notion that important neurochemical changes occur early in the training process.

Measurements of the rates at which stimulation-induced neurochemical changes decay are also important, for they are needed to answer the question of whether some of the chemical events that mediate LTM storage have longevities on the same order of magnitude as the observable behavior changes. The only long-term change in a cerebral compound following training that has been reported is an increased amount of hexose sugar glycosidically bound to a particular fraction of brain proteins in pigeons that were killed three to eleven months after their last exposure to discrimination training (Bogoch, 1968). As suggested above, chemical changes with short half-lives may be related to initiating events in the consolidation sequence. Many of the changes reported

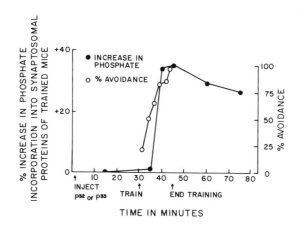

Figure 5. Comparison of avoidance behavior and changes in the amount of radioactive phosphate bound to the synaptosomal proteins of learning mice.

In this experiment, mice were trained for either 5, 10, or 15 min, starting 30 min after they had been injected with radioactive phosphate. The biochemical changes are against the baseline figures for mice that were held quietly in individual cages for the appropriate interval. Mice from groups 1, 2, and 3 were sacrificed immediately after training had ended. Other mice were kept in individual cages for either 15 min (Group 4) or 30 min (Group 5) after 15 min of training were completed, and then sacrificed. The biochemical response is not evident at 5 min of training, but appears at maximal amplitude by the end of 10 min of training (about 20 trials of avoidance learning).

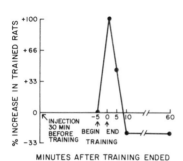

MINUTES AFTER TRAINING ENDED

Figure 6. Effect of avoidance training on the incorporation of radioactive phosphate into brain non-histone acid-extractable nuclear proteins of naive rats, and decay of that effect after training has ended.

Rats were given 10 trials during 5 min, on a one-way step-up avoidance task. Different groups were sacrificed at 0, 5, 10, and 60 min after training was completed. Changes in the incorporation of phosphate into protein in trained rats are measured against the baseline from rats that were held quietly in individual cages, with 8 pairs of rats per point.

This neurochemical response to stimulation decays quite rapidly, in marked contrast to the effects of similar avoidance training of mice on the incorporation of radioactive phosphate into synaptic proteins (see Figure 5).

by Hydén are not apparent 24 hr after the final training session. An increased incorporation of radioactive phosphate into certain nuclear proteins of brain (Machlus *et al.*, 1974b) is apparent from 5 min of avoidance training, but has decayed to the baseline within 5 min after the rat is removed from the training situation (Figure 6). Much work on the temporal aspects of known metabolic changes remains to be done.

Where in the Brain Do Changes Occur?

The search for the brain regions where amnestic agents have the greatest effects, and where intrinsic biochemical processes show the greatest degree of lability to behavioral stimulation is important for at least three reasons. First, the determination of whether or not a given neurochemical change occurs in only a few brain regions may provide insight into the mechanisms that contribute to the change. Certain treatments might be expected to have similar effects on many kinds of neurons, while other treatments will probably directly alter the activity and metabolism of only few types of brain cells. Second, the localization of specific activated regions allows further investigations of more complex biochemical and biological phenomena at those sites. Third, the localization of biochemical responses allows the interdisciplinary comparison of memory-related phenomena against the common background of neuroanatomical and neurophysiological knowledge.

The belief that the biochemical analysis of brain regions will contribute to the understanding of cerebral functioning arises from reports that specific physiological stimulation of the nervous system produces biochemical changes in the anatomically appropriate cells. For example, rotation of an animal produces elevated RNA content in appropriate nuclei in the brain stem and cerebellum as analyzed by a variety of neurochemical techniques (Jarlstedt, 1966; Watson, 1965). Another excellent example of the precise localization of metabolic effects is the production of specific geometrical bands of elevated protein synthesis in the optic tectum of fish and frog upon stimulation of the eye by fixed bands of flashing light (Rensch, Rahmann, & Skrizipek, 1968; Wegener, 1970). The maps of this increased incorporation of radioactive amino acid into the proteins of the tectum correspond to the known topography of the anatomical projections of the retina to this region.

A major difficulty in planning and interpreting anatomical studies is that the accuracy of the biochemical information obtained tends to be directly proportional to the size of the tissue sample used. Thus, one can obtain accurate biochemical measurements, or accurate anatomical localization, but not both, in a given experiment. Since most experiments have analyzed cells, even when studied on an individual basis, as *representative* of the cells of a particular neuroanatomical nucleus, data from them have to be interpreted as changes in that cell type for the entire nucleus, and are more properly thought of as tissue changes than as single cell changes. Tissue responsiveness may be related only secondarily to the responses of individual neurons and glia cells.

Refined neuroanatomical analysis ultimately requires that biochemical events in single cells be measured. Notable efforts have been made in this direction by Hydén (1967), Lowry (1962), and Osborne and Neuhoff (1973). However, most biochemical analyses require quite large tissue samples, from milligrams to grams in size. Consequently, most published studies have focused on overall reactions in blocks of tissue that contain many thousands or millions of cells of different types. Because of the heterogeneity of these samples, the interpretation of metabolic changes with respect to cellular biology is almost impossible. For example, Hydén has attempted to separate the large Deiter's neurons of the vestibular nucleus from their companion glial cells by microdissection techniques, and to analyze the RNA content and base composition of the two fractions. The results indicated that RNA was lost from neurons and later replenished from glial cells, during both the acquisition of a rope-climbing skill and during transfer-of-handedness training (Hydén, 1967). Rose (1968) has suggested that the microdissection methods used by Hydén do not permit distinction between glial and neuropil RNA; thus the data of Hydén may indicate a redistribution of RNA *within* neurons.

Micromethods that have been used with some success include histochemistry, microspectrophotometry, and autoradiography. Particularly plaguing is the difficulty of measuring precursor uptake in autoradiographic studies of macromolecular metabolism. Reasonable approaches toward partially circumventing this problem are exemplified by studies of Watson (1965), Carreres-Quevedo, Gomez Bosque, Rodriguez Andres, and Coca Garcia (1972), and Rensch and his associates (Rensch *et al.*, 1968; Wegener, 1970). Promising starts have also been made in analyzing the biochemistry of identifiable large neurons in invertebrate preparations (Wilson & Berry, 1972). In any of these methods, an unsolved key problem is the compartmentation of chemicals within the cell, particularly in the outlying regions of dendrites and synapses.

Some information concerning the localization of memory-related processes has been gained by injecting puromycin into the brain at various intervals after training (Flexner *et al.*, 1963). After three days, bitemporal injections are sufficient to destroy evidence of memory on a later retest, but after six days, puromycin must be injected at multiple sites in the brain to have amnestic effects. These results indicate that the anatomical distribution of processes necessary for information storage may change with time after training.

The anatomical distribution of stimulation-induced changes that have been detected by analytical chemical means are most conveniently described under headings of general categories of behavioral stimulation.

Environmental Enrichment

When rats are exposed to a diverse range of stimuli and given opportunities for manipulating those stimuli for a period of some weeks or months, there occurs a general growth of the posterior neocortex, both in absolute terms and relative to other brain regions (Bennett, Diamond, Krech, & Rosenzweig, 1964). There are also changes in the enzyme content of those areas. Recent microscopic studies suggest that pronounced changes in synaptic arrangements occur in this region as a result of environmental enrichment (Møllgaard, Diamond, Bennett, Rosenzweig, & Lindner, 1971).

Wire Climbing

When rats were trained to climb a wire to obtain food, the large Deiter's cells of the vestibular nucleus and the cells of the nucleus gigantocellularis of the reticular formation (the only cells examined) both showed increased RNA content per cell (Hydén, 1967). The microchemical methods also detected changes in the base-composition of the RNA of the Deiter's cells. The information available with respect to this behavior is insufficient to decide if localized changes exist or not.

Reversal Situations

Hydén and his associates have also studied neurochemical changes that occur when rats, over a series of days, learn to use their nonpreferred paw to get food pellets. When groups of cells were isolated from a region of motor cortex known to be important for this behavior, changes in ratios among the RNA bases were detected contralateral to the non-preferred (now used) paw, but not on the other side of the brain (Hydén, 1967). The incorporation of intracranially injected radioactive leucine into protein in the hippocampus of rats trained on this task was examined (Hydén & Lange, 1968). No clear changes were seen in region CA1, while CA3 showed elevations in this measure after 4 days or 14 days of training and CA4 showed a significant elevation after 4 days of training but not after 15 days. Region CA2 was not studied. These phenomena have been reexamined recently using periph-eral administration of radioactive leucine. With that route of injection leucine incorporation into proteins was *depressed* in trained animals in neocortex, dorsomedial nucleus of the thalamus, entorhinal cortex, septal nuclei, mammillary bodies, hippocampus CA3, dentate gyrus, and the reticular formation (Hydén & Lange, 1972). The data presented to date do not offer unequivocal evidence that reversal of paw preference leads to changed rates of cerebral protein synthesis (see pp. 210–211 and Bowman & Harding, 1969).

Although the task used by Hydén is usually thought of as being related specifically to the transfer of hand preference to the initially nonpreferred paw, as a task it can be included in a broad category of procedures called reversal learning, wherein the subject is made to switch its domi-nant mode of response from one form to another, usually an opposite form. Kottler *et al.* (1972), studying a situation where rats were made to repeatedly reverse the side to which they ran in a Y-maze for water reward, found that the incorporation of cytidine into total and nuclear RNA was significantly elevated in the total hippocampus, but not in anterior neocortex, posterior neocortex, pyriform cortex, nor the remain-ing subcortical regions anterior to a midcollicular coronal section; the posterior part of the brain was not analyzed. Elias and Eleftheriou (1972) reported that reversals in a water maze increase the incorporation of radioactive uridine into total RNA in mice in all regions studied (frontal cortex, hypothalamus, hippocampus, amygdala, and cerebellum) when incorporation was monitored during a final training session. The changes were reduced and more variable when studied after the training was completed.

Thus reversal training of various sorts seems to biochemically activate the hippocampus in rats and mice, but also has detectable effects on many other brain structures.

Active Avoidance

Metabolic changes have been reported in many areas of the brain when rodents are trained to avoid painful footshock by making some response such as jumping to a shelf or running down a given limb of a maze. Using dissection methods, Zemp, Wilson, and Glassman (1967) reported that the incorporation of uridine into RNA, during training of mice to jump to a ledge, was depressed in neocortex but elevated in a large piece of tissue consisting of the anterior forebrain; no statistical analysis of those changes was given. Later work (Entingh *et al.*, 1974) suggests that nuclear RNA changes observed in the subcortical forebrain may have been due to alterations in precursor metabolism, and that incorporation of uridine into RNA of neocortex is enhanced by this kind of training. Also using the jump up task, Uphouse, MacInnes, and Schlesinger (1972a) have reported that the incorporation of uridine into polysomal RNA is elevated in the neocortex and the hippocampus, but not in some other regions of the brain studied.

Three autoradiographic studies of effects of avoidance training, all using intracranial injections of precursor, have reported increased incorporation of [^{3}H]uridine into limbic system structures. Carreres-Quevedo *et al.* (1971) analyzed the ratio of radioactivity in the cytoplasm to that in the nucleus for individual neurons, and found this measure elevated in rat hippocampus. This is the only published autoradiographic study of training effects to have made any attempt to correct for precursor uptake. Pohle and Matthies (1971) reported that incorporation was increased in all regions of the hippocampus (CA1, CA2, CA3, CA4, and dentate gyrus), in visual cortex and cingulate gyrus, but not largely altered in the ventral cortex, habenular nucleus or hypothalamus. Kahan, Krigman, Wilson, and Glassman (1970) reported that increased uridine incorporation occurred in many subcortical regions, especially those classically associated with the limbic system, and that incorporation in neocortex was depressed. Unfortunately, the results of this last study were poorly quantitated.

Increased incorporation of amino acids into protein during avoidance training was detected by Beach *et al.* (1969) in an autoradiographic study. Changes were reported to occur in hippocampus CA1 and the medial septal nucleus, and possibly in the entorhinal cortex, but not in the lateral septal nuclei, superior colliculus, or lateral geniculate. No attempt to correct for pool variations was made; thus these results must await confirmation by more sophisticated techniques. A recent study that dissected the brain into six parts (Rees *et al.*, 1974) found that exposure to footshock alone or to loud noises or handling produced an increase of some 20 to 30% in the incorporation of various amino acids into protein in almost all regions except dorsal cortex. This suggests that a general humoral agent or brain state that promotes general protein synthesis may be induced by stressful stimuli. Such phenomena confound attempts to detect any synthesis of new proteins at specific loci during learning experiences through the use of general tracer methods.

Imprinting and Novel Exposure to Light

Horn *et al.* (1973) have reported that increased RNA synthesis occurs in the optic tectum, but not other regions of the chick brain, during exposure to an imprinting stimulus, a flashing light. Since the concept of imprinting requires that the learning experience is identical to the first exposure to a novel strong stimulus (a rotating light), it is impossible to determine whether this localized effect has anything to do with learning per se, or is related to general stimulation or novel stimulation. This problem is discussed at length later.

It appears that the work of tracing neurochemical responses to stimulation onto neuroanatomical structures has barely begun. The data are not sufficient to localize mnemonic systems onto a specific set of brain loci. Nor are they sufficient to distinguish between diffusely represented and highly localized memory storage systems. Mnemonic systems might be redundant either at the tissue level, with multiple neuronal networks participating within the same region of the brain, or in a manner that involves different neuroanatomical nuclei. Experiments designed to detect redundant memory systems are rare.

What Aspects of Experience Are Critical for the Observed Biochemical Changes?

Relationships of Biochemical Changes to Behavioral States

In most papers about neurochemical responses to stimulation the question is raised: Are the observed biochemical changes *specific* for the learning process? When posed in this form, the inquiry concerning the relationship of neurochemical responses to behavioral states becomes confused, for "specific" can mean a number of things. For example:

1. Is the neurochemical response *necessary* for the formation of LTM?
2. Will behavioral states or episodes other than learning trigger the neurochemical response? If so, what is the significance of such events?
3. Does the process of memory formation *always* produce the neurochemical response?
4. Given that the chemical response is related to memory, does the particular change directly code for the memory, or is it an intermediate in the chain of events that results in the chemical engram?

Thus, what appears to be a simple question regarding specificity is comprised of a series of more precise questions aimed at the complex interface of biochemistry, neurophysiology, and behavioral analysis.

The only unequivocal answers to question 1 will come from experiments in which precise metabolic inhibition of the chemical process in question is applied to animals that are given adequate opportunity to learn and to exhibit long-term retention. A demonstration that simple stimulation induces a neurochemical change *does not* imply logically that the change is unnecessary for learning or consolidation.

Question 2 has been raised most often in attempts to demonstrate that the learning aspects of a training experience are alone responsible for a given neurochemical response. The question is, however, more directly related to the biochemical change than to the particular training situation, for it is the chemical change that may be said to be specific to training, consolidation, recall, arousal, performance, etc. Thus the control groups that are built into experiments on the biochemistry of behavior, ostensibly to determine whether the observed chemical response is directly related to learning, provide information of great value for understanding the ways in which a particular chemical process within the brain responds to environmental input.

A positive answer to question 3 would suggest that all information storage in brain is mediated by a single mechanism, but logically does not imply that the neurochemical response under study is necessary for memory. Adequate answers to question 4 are most likely to come from studies that compare the longevities of neurochemical responses to the longevities of the memory to which they are related. Answering this apparently biochemical problem will probably require more skill at behavioral analysis than at biochemistry.

From the available data, three classes of biochemical responses can be defined:

Stimulation-related

The occurrence of modally adequate stimuli, not contingent upon the animal's behavior, is sufficient to induce the chemical response.

Acquisition-related

The biochemical response occurs when the organism is in a training situation in which its behavior becomes adapted to the contingencies of the situation (i.e., the animal learns and performs as the experimenter expects it to). The neurochemical response does not occur, or is much reduced, when the organism is exposed to the training stimuli while unable to learn the adequate behavioral response (e.g., yoked condition), or when a previously trained animal performs the response in the acquisition apparatus.

Experience-related

The neurochemical response occurs during both acquisition and performance sessions, and also upon exposure to any of the significant stimuli related to the training procedure, once the organism is overtrained. Both this category and its appellation are tentative, since they are suggested by the behavioral characteristics of only one known neurochemical response.

These three categories are attempts to define concepts that describe the existing data, rather than predefined concepts to which data have to be fitted. Some of the data from our own laboratories that have contributed to this classification are summarized in Table 1.

Two of these changes (A and B) fit into the first category, since they are induced by simple stimuli. The incorporation of lysine and other amino acids into protein has been found to be increased from 10 to 30% by a wide variety of stimuli, including handling by the experimenter, loud sounds, and repeated presentation of footshocks of moderate intensity (Rees et al., 1974). This chemical response occurs in animals that have been adrenalectomized, and is reduced in magnitude in mice that have been habituated to the test stimuli over a course of days. To categorize cerebral protein synthesis as being labile to almost any type of novel peripheral stimulation appears to be consistent with most previous reports.

Alteration in the rate at which ^{3}H-fucose is incorporated into brain glycoproteins seems also to be a function of simple stimulation (Damstra-Entingh et al., 1974). In this case, the relationship between the magnitude of the biochemical effects and the type and magnitude of stimulation appears to be somewhat more complex than in the case of the incorporation of amino acid into protein (Figure 7). A tentative explanation for the shape of the curves in Figure 7 is that exposure to the urine from shocked mice increases the rate of incorporation of fucose into glycoproteins, but that intensive stimulation by footshock decreases the incor-

TABLE 1 Magnitudes of Changes in Metabolism as a Result of Subjecting Rats or Mice to Different Conditions of Environmental Stimulation[a]

	Brain						Liver	
	Trained	Yoked	Apparatus exposed	Footshock	Extin- guished	Per- forming	Trained	Yoked
A. Incorporation of lysine into proteins[b]	+17*	+12*	+7	+12*	—	+5	+62*	+61*
B. Incorporation of fucose into glycoproteins[c]	+28*	+25*	+63*	variable[c]	—	—	+17*	+17*
C. Incorporation of uridine into polysome- associated RNA[d]	+38*	0	0	+1	+37*	-1	0	—
D. Phosphorylation of synaptic proteins[e]	+34*	+1	—	—	+39*	-3*	—	—
E. Phosphorylation of nonhistone basic nuclear proteins[f]	+102*	-3	—	-10	+104*	+95*	+100*	+118*

[a] Changes are expressed as the percentage difference between stimulated animals and controls. These five measures of macromolecular metabolism are affected by stimulation conditions in different manners. Changes A and B are stimulation-related; whether they have any role in information processing is unclear. Changes C and D seem to be related to the acquisition of information. Changes in E appear to be related to experience but are not limited to episodes of acquisition.
Asterisk (*) indicates $p < .05$; dash (—) indicates data are not available.
[b] Rees et al., 1974.
[c] Damstra-Entingh et al., 1974; see also Figure 7.
[d] Adair et al., 1968.
[e] Perumal, 1973; see also Figure 4.
[f] Machlus et al., 1974a; see also Figure 5.

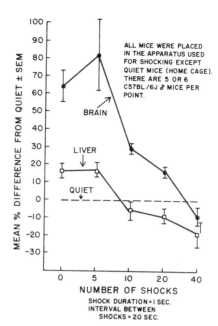

Figure 7. Effects of exposure to apparatus and of footshocks on the incorporation of fucose into brain glycoproteins.

Vertical bars represent the standard error of the mean. Holding mice for 15 min in the apparatus used for delivering footshock induces an elevation in the incorporation of fucose into glycoproteins of the brain, and a smaller elevation of similar processes in the liver. The stimulus that is responsible for the elevations may be the odor of urine released by other mice that received footshock in the apparatus. Delivering 10 or more footshocks drives the level of incorporation of fucose down again, in both brain and liver, in a manner proportional to the amount of shock received.

ALL MICE WERE PLACED IN THE APPARATUS USED FOR SHOCKING EXCEPT QUIET MICE (HOME CAGE). THERE ARE 5 OR 6 C57BL/6J ♂ MICE PER POINT.

NUMBER OF SHOCKS

SHOCK DURATION = I SEC.
INTERVAL BETWEEN SHOCKS = 20 SEC.

poration rate. If exposure to urine or the receipt of but a few footshocks constitutes mild stress, whereas receipt of many footshocks constitutes strong stress, then the behavioral characteristics of this neurochemical response fit the common pattern of increased neurochemical reaction on mild stimulation and decreased reaction on strong stimulation (see Pevzner, 1966).

Biochemical processes C and D listed in Table 1 have characteristics that fit into the second category, acquisition-related changes. This category is suggested because both the polyribosome and the synaptic protein phosphorylation occur during training and extinction, but not during yoked stimulation nor during performance of the avoidance response after extensive overtraining. Confirming evidence for the polysome effect has been reported (Uphouse, MacInnes, & Schlesinger, 1972b). The synaptic protein phosphorylation effect has been found so recently that its inclusion in this category is tentative.

Biochemical response E listed in Table 1, an increased incorporation of phosphate into a particular class of nuclear proteins (NAEP), stands alone in terms of its input contingencies. It is not induced by simple stimuli, and while it occurs in the acquisition states of original training or extinction, it also occurs in other circumstances. The original training consists of giving rats, over a period of 5 min, ten training trials in a straight alley in which the rat can avoid footshock by running from the grid-floor end of the box onto a platform. Each trial is started by manually placing the rat on the grid floor. Thus the proximal stimulus for the avoidance response might be handling, touching the grid floor, or seeing the walls of the apparatus. Rats sacrificed at the end of this training session, compared to rats held in the home cage, show an increase of about 100% in the amount of radioactive phosphate bound to NAEP. This increase does not occur in rats that are yoked in stimuli (handling, sequence and duration of footshock, etc.) to trained rats but given no opportunity to climb onto the safety platform, and thus does not fit into the stimulation-related category of changes.

The distinguishing aspect of this chemical response is that, if animals have been trained for 1 to 6 days previously, then the chemical response appears in the absence of further shocks or performance of the learned response, if the rat is (*a*) placed for 5 min on the grid floor, (*b*) placed for 5 min on the safety platform, (*c*) placed on the grid floor for 30 sec and then returned to its home cage for 5 min, or (*d*) merely picked up once and returned immediately to its home cage 5 min prior to sacrifice (Machlus, Entingh, Wilson, & Glassman, 1974). If a rat has been only randomly shocked previously (and not allowed to learn an avoidance response) none of these treatments activates the NAEP response. Since this change has a life of 5 min or less (see Figure 5), we tentatively conclude that the response, as observed, is initiated each time the trained rat is placed in a stimulus situation that prepared the animal to make an avoidance response.

This pattern of reactivity of NAEP looks as though it might be linked to the initiation of recall of information that is related to the performance of an instrumental response. As such, some general emotional disturbance might be suspected to be the immediate cause of the response, but if so, that emotional response does not occur in yoked rats. The yoked rats would seem to have had ample opportunity to develop a diffuse conditioned emotional response to these stimuli. Moreover, the NAEP response during the first training session occurs in the absence of either the adrenal glands, or the pituitary, suggesting that hormones liberated by stress from either of those glands are not necessary for the chemical change.

Stimulus Novelty or Performance Changes?

A major interpretive problem is beginning to arise from the repeated findings that stimulation in the absence of overt behavior changes can induce many biochemical changes in the brain. Such changes are often detected through treatments described as "controls" for aspects of training experiences, and have frequently been thought to be of minor importance in relationship to the chemical basis of memory. These changes may, however, represent an integral part of the memory storing apparatus or process, even though they are not uniquely evoked only during acquisition.

A large body of behavioral and physiological evidence suggests that information is registered and/or stored in the brain whenever any stimulus engages the attention of a mammal (Kimble, 1961, pp. 226–234; Pribram, 1971, pp. 48–54; Grossman, 1967, pp. 621–670). Sokolov (1960) has suggested that habituation to a novel stimulus requires that the brain form and store a representation of that stimulus; novelty is then ascribed to any stimulus for which the brain has stored no representation. Since the experimental situations and stimuli that induce cerebral chemical changes are usually novel to the animal, it is possible that many of the known stimulation-related neurochemical changes are related to the storage of information necessary for habituation to novel stimuli. And, in those cases where the neurochemical response is found to habituate to the evoking stimulus, a strong suspicion that chemical change is related to information storage must be entertained.

The habituation of stimulation-related central neurochemical responses is an important but largely unexplored area of experimentation, and a number of interesting questions remain.

1. When such habituation occurs, is it specific for the stimulus to which the animal is repeatedly exposed? The detection of stimulus-specific habituation of neurochemical responses would suggest that the chemical changes are related to the information content to that stimulus.

2. Do different brain tissues react to different degrees and then habituate at different rates? Changes in some brain regions (perhaps primary sensory nuclei) might not habituate, while changes in other regions might display varying time courses of habituation. Such phenomena will not be easy to interpret, but they might aid in identifying brain regions that are involved in the mediation of long-term habituation and learning.

3. Are acquisition-related neurochemical responses obscured, at the empirical level, by the stimulation-related responses to a given training situation? Habituation to all stimuli prior to the actual training procedure might increase the signal (acquisition-related effect) to noise (stimulation-related effect) ratio in the search for learning-related phenomena. The study by Beach et al. (1969) used such an approach.

Metabolic Responses to the Use versus the Reorganization of Tissue

Besides the distinctions made concerning behavioral states or conditions, particular neurochemical responses may be related to different biological functions within cells. This is suggested in Figure 1. The accurate identification of biochemical changes with information storage mechanisms is made particularly difficult because at least two different types of departure from the ground-state metabolic equilibrium are expected to occur in brain cells. The first of these includes reversible changes that accompany general neurophysiological activity (the Active state, Figure 1), for example, the influx of sodium ions into neurons during the action potential and their later expulsion to maintain a fixed internal concentration of that ion. The second consists of long-term, possibly irreversible, shifts in the levels at which various chemical concentrations and metabolic rates are set, to establish a different ground state. These are related to changes in internal cellular functions and the relationship of the affected cell to others (the Reorganization State, Figure 1). For ease of reference, the first kind of change may be called "use" and the second, "reorganization." It will prove very difficult to distinguish between these two roles for any particular cerebral molecular alteration.

A good example of this comes from studies of functions of the hippocampus. Changes in RNA and/or protein metabolism in this structure have been detected during or shortly after the training of rodents on two different kinds of tasks: One-way active avoidance learning, and position-habit reversal learning (see pp. 219–221).

A number of lines of evidence suggest that the hippocampus is functionally active during learning. (1) Damage to the hippocampus can seriously retard, but does not obliterate, learning in the two situations mentioned above (Kimble & Kimble, 1968; Olton & Isaacson, 1968). (2) Human patients with bilateral hippocampal-complex damage suffer from an apparent inability to form any but the simplest long-term memory (Penfield & Milner, 1958). (3) Electrical activation in the form of an increased amount of theta-wave activity occurs in the hippocampus during the orienting response (Grastyán, Lissak, Madarasz, & Donhoffer, 1959). (4) Elevated amounts of theta-wave activity appear in cortical regions of rats during periods of memory consolidation (Lanfield, McGaugh, & Tusa, 1972). (5) Finally, it has been reported that driving the hippocampus at theta-frequencies, by electrically stimulating the fornix, produces large, frequency-specific, alterations in RNA metabolism in this brain area (Marichich & Izquierdo, 1970). Thus there is a web of evidence that the hippocampus may be an important locus for the storage of memory.

But to conclude this would be premature. All of these data are fully consistent with the alternative concept that the hippocampus may play a role in the immediate organization of new information, to perhaps have ultimate effects on the storage of long-term memory in cells elsewhere in the brain. Thus, given the available evidence, it seems just as likely that the neurochemical responses detected in the hippocampus are related to tissue use rather than the reorganization of tissue functions. Indeed, it seems that the distinction between "vegetative" use and the reorganization of tissue may prove to be most troublesome in the future, and this limitation seems to set the maximum physiological role that can be assigned to any of the known neurochemical responses at the moment.

TENTATIVE GUIDELINES FOR RECOGNIZING LEARNING-RELATED CEREBRAL NEUROCHEMICAL EVENTS

The question of how the chemical basis of the engram will be initially recognized looms large for present-day investigators. The previous discussion has made it clear that this will not be easy. So far, correlative studies have not provided very much solid information about the relationship of learning to neurochemical processes.

The ultimate proof that a particular neurochemical process is involved in memory storage will require that the *information* contained in memory (as defined in stimulus, or stimulus–response contingency terms) can be deciphered by the examination of chemical changes of state in the appropriate sets of neurons. Clearly, this is presently unobtainable, and may never be achieved in mammalian systems. But guidelines for getting closer to this goal can be formulated and continuously refined, in the form of interim criteria for establishing by correlative means that a given neurochemical response is probably related to memory storage.

The specific guidelines have been stimulated by two considerations. First, the analysis of learning behavior suggests that the storage of specific bits of information are events that can be fixed empirically in space and time, so that the important neurochemical events must be initiated when and where learning occurs. Second, the concept that learning is mediated by changes in functional connectivity between brain cells imposes constraints upon the process of relating neurochemical changes to the biology of brain cells and physiological processes.

Before defining the guidelines, it is important to emphasize again the potential complexities of the chemical processes of consolidation. Given what is known of the biochemistry of cells and of consolidation processes, it is likely that these processes may require many chemical steps to achieve the final state of memory storage. Until the roles of individual chemical reactions can be related to overall biochemical processes, guidelines can be drawn only with respect to certain aspects of processes as a whole.

Prevention of the chemical change should prevent the formation of memory. The only way to demonstrate that a neurochemical change is *necessary* for the formation of memory is to show that the blockade of that reaction prevents the formation of memory. A great limitation to the practical application of this criterion is that the commonly used metabolic inhibitors have only limited specificities for particular biochemical processes (see Glassman, 1969). If all available inhibitors of a particular process are effective in blocking retention, the case for the involvement of that process will be strengthened, especially if their mechanisms of action are distinct. The only conclusive data obtainable from an experiment in which only one inhibitor is used are those where the total disruption of a central neurochemical process does *not* affect memory formation. In like manner, the agent(s) must act with behavioral specificity, blocking memory in the demonstrated absence of effects on sensory or motor performance, or on recall.

The first stages of the chemical process should occur at the beginning of memory consolidation. At present, this guideline is of limited help, for estimates of consolidation time range from less than a second to more than 24 hours. Although most amnestic agents apparently must be acting during the initial stage of learning in order to block consolidation, it is possible that the biochemical reactions that they affect are only the first of many reactions in a chain of metabolic events that requires many days to reach a stable endpoint.

The duration of the final chemical change(s) of state should be comparable to the duration of memory. It is clear that changed molecules themselves need not be as long-lived as the memory, but the changes of state with which the molecules are associated should be long-lived. A good example is the production of a new intercellular contact, the chemical components of which may be continuously renewed without changing the functioning of the contact. Since the only nervous system compounds with sufficiently long known lives are laid down mainly during growth and do not change much thereafter (Lajtha & Marks, 1971), this guideline suggests that metabolic control loops within cells must be permanently modified to allow the critical cells to continuously replace the newly important compounds. The long-lived compounds are neuronal DNA, and the components of myelin and connective tissue, but it is possible that hitherto unidentified compounds with long lives exist.

The neurochemical change should be related to the learning aspects of the experience; it should not occur in a brain to which the experience is no longer a novelty. Both instrumental learning and the habituation to novel stimuli require the storage of information, and probably require that neurochemical changes occur. Given the proviso that different types of learning may eventually be found to be mediated by different neurochemical processes, it is to be expected that chemical changes be observable after any experience that induces long term anterograde changes in behavioral responsiveness, but not after experiences that have no detectable effects on behavior.

If learning can be shown to occur at various strengths or intensities, then it is to be expected that the magnitude of the final chemical changes of state will reflect the strength of the memory. Correlations between strength of learning (manipulable by varying the number of training trials, etc.) and magnitude of chemical response is one way to get at the information content of both experiences and cerebral changes. Note should be taken of the difficulties associated with measuring the strength of learning (usually by means of extinction curves) and chemical changes in the same animal. Stronger memories may be found to be associated with larger chemical changes in a fixed number of neurons, or from involvement of a larger number of neurons in a given memory.

The detected neurochemical change should be demonstrably related to concomitant changes in either structural macromolecules or the activity of some enzyme. The changes that will be most readily interpretable will be those in the anatomical fine structure of neurons, or changes in the activities of enzymes associated with the metabolism of neurotransmitter substances. No isolated chemical change is meaningful until it can be integrated into the general knowledge of how cells function internally, and in relationship to each other.

It should be demonstrated that the chemical change indeed alters the functional connectivity between neurons. Cells where the chemical changes occur should exhibit altered responsive or response-inducing properties. Electrophysiological analysis of the properties of single cells will be of value for such demonstrations.

Since changes in functional connectivities are involved, some consequences of the altered neurochemical process should probably, but not necessarily, be detectable in the vicinity of synapses. Some possibilities of synaptic changes are: changes in the biochemical constitution or metabolism of presynaptic or postsynaptic membranes; changes in amount of neurotransmitter or number of postsynaptic receptor sites; alterations of morphology of either presynaptic boutons, synaptic vesicles, or postsynaptic membrane specializations; or the appearance of new synapses or loss of old ones.

Chemical changes observed after environmental stimulation should be demonstrable after appropriate electrical and chemical stimulation of the brain. The immediate physiological antecedents to the chemical changes should be demonstrable. This requires the detection of the critical presynaptic inputs to the cells that demonstrate chemical responses.

The neurochemical changes should map onto the structure of the brain in a manner consistent with the neuroanatomical loci of memory-related electrical events and the effects of localized injections of amnestic drugs. Chemical findings will not be interpretable with respect to cerebral function until they are correlated neuroanatomically with other information about memory mechanisms.

FUTURE DIRECTIONS FOR RESEARCH IN THIS AREA

Besides the improvements of techniques and the extensions of "classic" approaches just discussed, other exciting trends of great promise are apparent.

The biochemical emphasis in this field, once so restricted that the entire endeavor could be named "RNA and Memory," is becoming more general. Major attention is focusing on two specific areas: the control of gene expression in neurons, and biochemical reactions that are associated with synapses, particularly reactions that are associated with neurotransmitter metabolism or with membrane structure. This dual emphasis derives from the need to understand both the initiating steps and the terminal functional roles associated with any given neurochemical response.

The growing emphasis on multidisciplinary research makes it likely that biochemists and electrophysiologists will soon find it worthwhile to collaborate in selecting standardized model systems of animal species and stimulation conditions, so that information from the two approaches can be more easily correlated. In mammalian studies, important species are the mouse, useful for biochemical studies because of the large number of neurological mutants available, and the rat, useful for physiological studies because of the ease with which electrodes can be attached to its skull.

The rapidly accumulating electrophysiological data on relatively simple non-mammalian nervous systems (e.g., *Aplysia*, cockroach) offers the possibility of detecting identifiable neurons that are responsible for long-term neuronal changes associated with well-defined behaviors. The biochemical analysis of such critical cells can be expected to furnish important clues to ways in which the metabolism of neurons in general are altered after stimulation.

Animals with neurological mutations will be of great interest in biochemical and physiological studies on memory mechanisms. There is evidence (Bovet, Boret-Nitti, & Oliverio, 1969) that certain mouse mutants learn in unusual manners. This could mean that some of their memory mechanisms are abnormal. The biochemical analysis of such strains offers the prospect of insight into the necessity of certain neurochemical responses for certain types of learning. Of great interest may be the behavior genetics of Drosophila (Hotta & Benzer, 1972).

Biochemists who are interested in the physical basis of memory will find themselves increasingly involved in multidisciplinary study at two different levels of analysis. The first is the study of ways in which neurochemical responses within specific cells are related to the electrophysiological and anatomical organization of brain tissue networks. Biochemistry will serve as the bridging framework between stimulation-induced changes in electrophysiological events and the much slower changes detected in anatomical structure. The second is the study of ways in which neurochemical responses to environmental, physiological, and hormonal stimulation are intrinsic to the dynamic biology of neurons. The determination of biological functions for the already known neurochemical responses is of primary interest. Some specific questions at this level include: Does the given neurochemical response affect, or perhaps reflect: (1) overall changes in the flux in particular general metabolic pathways? (2) changes in the life span of certain brain cells? (3) changes in physical contacts between particular sets of neurons? (4) rates of axonal flow of metabolites?, etc. The fundamental building block of the living brain is the individual cell, and much important information about the physical basis of cerebral plasticity will come from analyses carried out at the cellular level.

ACKNOWLEDGMENTS

The preparation of this chapter was supported by grants from the Alfred P. Sloan Foundation and the U.S. Public Health Service (NS07457). Some of the experiments discussed here were supported by grants from the U.S. Public Health Service (MH18136, NS07457), the U.S. National Science Foundation (GB35634X), and

the University of North Carolina. The authors thank Dr. P. W. Landfield for critical
review of the manuscript.

REFERENCES

Adair, L. B., Wilson, J. E., Zemp, J. W., & Glassman, E. Brain function and macromolecules. III. Uridine incorporation into polysomes of mouse brain during short-
term avoidance conditioning. *Proceedings of the National Academy of Sciences,
U.S.A.*, 1968, **61**, 917–922.

Agranoff, B. W. Actinomycin-D blocks formation of memory of shock-avoidance in
goldfish. *Science*, 1968, **158**, 1600–1601.

Albert, D. J. The effects of polarizing currents on the consolidation of learning.
Neuropsychologia, 1966, **4**, 65–77.

Barondes, S. H. Multiple steps in the biology of memory. In F. O. Schmitt (Ed.),
The neurosciences: Second study program. New York: Rockefeller Univ. Press,
1970. Pp. 272–278.

Barondes, S. H., & Cohen, H. D. Delayed and sustained effect of acetoxycycloheximide on memory in mice. *Proceedings of the National Academy of Sciences,
U.S.A.*, 1967, **58**, 157–164.

Barondes, S. H., & Cohen, H. D. Arousal and the conversion of "short-term" to
"long-term" memory. *Proceedings of the National Academy of Sciences, U.S.A.*,
1968, **61**, 923–929. (a)

Barondes, S. H., & Cohen, H. D. Memory impairment after subcutaneous injection
of acetoxycycloheximide. *Science*, 1968, **160**, 556–557. (b)

Baskin, F., Masiarz, F. R., & Agranoff, B. W. Effect of various stresses on the incorporation of [^{3}H]orotic acid into goldfish brain RNA. *Brain Research*, 1972, **39**,
151–162.

Beach, G., Emmens, M., Kimble, D. P., & Lickey, M. Autoradiographic demonstration
of biochemical changes in the limbic system during avoidance training. *Proceedings of the National Academy of Sciences, U.S.A.*, 1969, **62**, 692–696.

Bennett, E. L., Diamond, M. C., Krech, D., & Rosenzweig, M. R. Chemical and
anatomical plasticity of brain. *Science*, 1964, **146**, 610–619.

Bogoch, S. *The biochemistry of memory with an inquiry into the function of the
brain mucoids*. New York: Oxford Univ. Press, 1968.

Booth, D. A. Vertebrate brain ribonucliec acids and memory retention. *Psychological
Bulletin*, 1967, **68**, 149–177.

Bovet, D., Bovet–Nitti, F., & Oliverio, A. Genetic aspects of learning and memory
in mice. *Science*, 1969, **163**, 139–149.

Bowman, R. E., & Harding, G. Protein synthesis during learning. *Science*, 1969, **164**,
199–201.

Carreres-Quevedo, J., Gómez Bosque, R., Rodriguez Andres, M. L., & Coca Garcia,
M. C. Cytoplasmic RNA in the rat hippocampus after a learning experience:
An autoradiographic study. *Acta Anatomica*, 1971, **79**, 360–366.

Casola, L., Lim, R., Davis, R. E., & Agranoff, B. W. Behavioral and biochemical
effects of intracranial injection of cytosine arabinoside in goldfish. *Proceedings
of the National Academy of Sciences, U.S.A.*, 1969, **60**, 1389–1395.

Damstra-Entingh, T., Entingh, D. J., Wilson, J. E., & Glassman, E. Environmental
stimulation and fucose incorporation into brain and liver glycoproteins. *Pharmacology, Biochemistry and Behavior*, 1974, **2**, 73–78.

Davis, R. E., & Agranoff, B. W. Stages of memory formation in goldfish: Evidence for an environmental trigger. *Proceedings of the National Academy of Sciences, U.S.A.,* 1966, **55**, 555–559.

Deutsch, J. A., Hamburg, M. D., & Dahl, H. Anticholinesterase-induced amnesia and its temporal aspects. *Science,* 1966, **151**, 221–223.

Dunn, A., Entingh, D., Damstra-Entingh, T., Gispen, W. H., Machlus, B., Perumal, R., Rees, H. D., & Brogan, L. Biochemical correlates of brief behavioral experiences. In F. O. Schmitt & F. G. Worden (Eds.), *The neurosciences: Third study program.* Cambridge, Massachusetts: M.I.T. Press, 1974. Pp. 679–684.

Elias, M. F., & Eleftheriou, B. E. Reversal learning and RNA labeling in neurological mutant mice and normal littermates. *Physiology and Behavior,* 1972, **9**, 27–34.

Entingh, D. J., Damstra-Entingh, T., Dunn, A., Wilson, J. E., & Glassman, E. Brain uridine monophosphate: Reduced incorporation of uridine during avoidance learning. *Brain Research,* 1974, **70**, 131–138.

Flexner, J. B., Flexner, L. B., & Stellar, E. Memory in mice as affected by intracerebral puromycin. *Science,* 1963, **141**, 57–59.

Glassman, E. The biochemistry of learning: An evaluation of the role of RNA and protein. *Annual Review of Biochemistry,* 1969, **38**, 605–646.

Glassman, E. Macromolecules and behavior: A commentary. In F. O. Schmitt & F. G. Worden (Eds.), *The neurosciences: Third study program.* Cambridge, Massachusetts: M.I.T. Press, 1974, Pp. 667–677.

Glassman, E., & Wilson, J. E. The incorporation of uridine into brain RNA during short experiences. *Brain Research,* 1970, **21**, 157–168.

Grossman, S. P. *A textbook of physiological psychology.* New York: Wiley, 1967.

Grastyán, E., Lissak, K., Madarasz, I., & Donhoffer, H. Hippocampal electrical activity during the development of conditioned reflexes. *Electroencephalography & Clinical Neurophysiology,* 1959, **11**, 409–430.

Haywood, J., Rose, S. P. R., & Bateson, P. P. G. Effects of an imprinting procedure on RNA polymerase activity in the chick brain. *Nature,* 1970, **228**, 373–374.

Horn, G., Rose, S. P. R., & Bateson, P. P. G. Experience and plasticity in the central nervous system. *Science,* 1973, **181**, 506–514.

Hotta, Y., & Benzer, S. Mapping of behavior in *Drosophila* mosaics. *Nature,* 1972, **240**, 527–555.

Hydén, H. The neuron. In J. Brachet & A. E. Mirsky (Eds.), *The cell.* Vol. 4. New York: Academic Press, 1960. Pp. 215–323.

Hydén, H. Biochemical changes accompanying learning. In G. C. Quarton, T. Melnechuk, & F. O. Schmitt (Eds.), *The neurosciences: A study program.* New York: Rockefeller Univ. Press, 1967. Pp. 765–771.

Hydén, H., & Lange, P. W. Protein synthesis in the hippocampal pyramidal cells of rats during a behavioral test. *Science,* 1968, **159**, 1370–1373.

Hydén, H., & Lange, P. W. Protein changes in nerve cells related to learning and conditioning. In F. O. Schmitt (Ed.), *The neurosciences: Second study program.* New York: Rockefeller Univ. Press, 1970. Pp. 278–289. (a)

Hydén, H., & Lange, P. W. Brain-cell protein synthesis specifically related to learning. *Proceedings of the National Academy of Sciences, U.S.A.,* 1970, **65**, 898–904. (b)

Hydén, H., & Lange, P. W. Protein changes in different brain areas as a function of intermittent training. *Proceedings of the National Academy of Sciences, U.S.A.,* 1972, **69**, 1980–1984.

Irwin, L. N., & Samson, F. E. Content and turnover of gangliosides in rat brain following behavioral stimulation. *Journal of Neurochemistry,* 1971, **18**, 203–211.

Jarlstedt, J. Functional localization in the cerebellar cortex studied by quantitative determinations of Purkinje cell RNA. I. RNA changes in rat cerebellar Purkinje cells after proprio- and exteroreceptive and vestibular stimulation. *Acta Physiologica Scandinavica*, 1966, **67**, 243–252.

Jarvik, M. E. Effects of chemical and physical treatments on learning and memory. *Annual Review of Psychology*, 1972, **23**, 457–486.

John, E. R. Switchboard versus statistical theories of learning and memory. *Science*, 1972, **177**, 850–864.

Kahan, B. E., Krigman, M. R., Wilson, J. E., & Glassman, E. Brain function and macromolecules. VI. Autoradiographic analysis of the effect of a brief training experience on the incorporation of uridine into mouse brain. *Proceedings of the National Academy of Sciences, U.S.A.*, 1970, **65**, 300–303.

Kimble, D. P., & Kimble, R. J. Hippocampectomy and response perseveration in the rat. *Journal of Comparative & Physiological Psychology*, 1968, **60**, 474–476.

Kimble, G.. A. *Conditioning and learning*. 2nd ed. New York: Appleton, 1961.

Kottler, P. D., Bowman, R. E., & Haasch, W. D. RNA metabolism in the rat brain during learning following intravenous and intraventricular injections of ^{3}H-cytidine. *Physiology & Behavior*, 1972, **8**, 291–297.

Lajtha, A., & Marks, N. Protein turnover. In A. Lajtha (Ed.), *Handbook of neurochemistry*. Vol. 5. New York: Plenum, 1971. Pp. 551–629.

Landauer, T. K. Two hypotheses concerning the biochemical basis of memory. *Psychological Review*, 1964, **71**, 167–179.

Landfield, P. W., McGaugh, J. L., & Tusa, R. J. Theta rhythm: A temporal correlate of memory storage processes in the rat. *Science*, 1972, **175**, 87–89.

Lewy, A. J., & Seiden, L. S. Operant behavior changes norepinephrine metabolism in rat brain. *Science*, 1972, **175**, 454–455.

Lowry, O. H. The chemical study of single neurons. *Harvey Lectures*, 1962, **58**, 1–19.

Machlus, B., Entingh, D., Wilson, J. E., & Glassman, E. Brain phosphoproteins: The effect of various behaviors and reminding experiences on the incorporation of radioactive phosphate into nuclear proteins. *Behavioral Biology*, 1974, **10**, 63–73. (a)

Machlus, B., Wilson, J. E., & Glassman, E. Brain phosphoproteins: The effect of short experiences on the phosphorylation of nuclear proteins of rat brain. *Behavioral Biology*, 1974, **10**, 43–62. (b)

Marichich, E. S., & Izquierdo, I. The dependence of hippocampal RNA levels on the frequency of afferent stimulation. *Die Naturwissenschaften*, 1970, **57**, 254.

McGaugh, J. L., & Dawson, R. G. Modification of memory storage processes. *Behavioral Science*, 1971, **16**, 45–63.

Møllgaard, K., Diamond, M. C., Bennett, E. L., Rosenzweig, M. R., & Lindner, B. Quantitative synaptic changes with differential experience in rat brain. *International Journal of Neuroscience*, 1971, **2**, 113–128.

Neale, J. H., Klinger, P. D., & Agranoff, B. W. Camptothecin blocks memory of conditioned avoidance in the goldfish. *Science*, 1973, **179**, 1243–1246.

Olton, D., & Isaacson, R. L. Hippocampal lesions and active avoidance. *Physiology & Behavior*, 1968, **3**, 719–724.

Osborne, N. N., & Neuhoff, V. Neurochemical studies on characterized neurons. *Die Naturwissenschaften*, 1973, **60**, 78–87.

Penfield, W. & Milner, B. Memory deficit produced by bilateral lesions in the hippocampal zone. *A.M.A. Archives of Neurology & Psychiatry*, 1958, **79**, 475–497.

Perumal, R. Phosphorylation of synaptosomal proteins from mammalian brain during short-term behavioral experiences. (Doctoral dissertation, University of North

Carolina, Chapel Hill) Ann Arbor, Mich.: University Microfilms, 1973. No. 73–26, 227.

Pevzner, L. Z. Nucleic acid changes during behavioral events. In J. Gaito (Ed.), *Macromolecules and behavior*. New York: Appleton, 1966. Pp. 43–70.

Pohle, W., & Matthies, H. The incorporation of [³H]uridine monophosphate into the rat brain during the training period: A microautoradiographic study. *Brain Research*, 1971, **29**, 123–127.

Pribram, K. H. *Languages of the brain*. Englewood Cliffs, New Jersey: Prentice Hall, 1971.

Rees, H. D., Brogan, L. L., Entingh, D. J., Dunn, A. J., Shinkman, P. G., Damstra-Entingh, T., Wilson, J. E., & Glassman, E. Effect of sensory stimulation on the incorporation of radioactive lysine into protein of mouse brain and liver. *Brain Research*, 1974, **68**, 143–156.

Reinis, S. Autoradiographic study of ³H-thymidine incorporation into brain DNA during learning. *Physiological Chemistry and Physics*, 1972, **4**, 391–397.

Rensch, B., Rahmann, H., & Skrizipek, K. H. Autoradiographische Untersuchungen über visuelle "Engramm"-Bildung bei Fischen (II). *Pflügers Archiv*, 1968, **304**, 242–252.

Robustelli, F., Geller, A., & Jarvik, M. E. Detention, electroconvulsive shock, and amnesia. *Proceedings, 76th Annual Convention, American Psychological Association*, 1968, 331–332.

Rose, S. P. R. The biochemistry of neurones and glia. In A. N. Davison & J. Dobbing (Eds.), *Applied neurochemistry*. Philadelphia: Davis, 1968. Pp. 332–355.

Rose, S. P. R. Changes in amino acid pools in the rat brain following first exposure to light. *Brain Research*, 1972, **38**, 171–178.

Shashoua, V. E. RNA metabolism in goldfish brain during acquisition of new behavioral patterns. *Proceedings of the National Academy of Sciences, U.S.A.*, 1970, **65**, 160–167.

Soiero, R., Vaughan, M. H., Warner, J. R., & Darnell, J. E. The turnover of nuclear DNA-like RNA in HeLa cells. *Journal of Cell Biology*, 1968, **39**, 112–117.

Sokoloff, L. Local cerebral circulation at rest and during altered cerebral activity induced by anesthesia or visual stimulation. In S. S. Kety & J. Elkes (Eds.), *Regional neurochemistry*. London: Pergamon, 1961. Pp. 107–117.

Sokolov, E. N. Neuronal models and the orienting reflex. In M. A. B. Brazier (Ed.), *The central nervous system and behaviour*. New York: Josiah Macy, Jr., Foundation, 1960. Pp. 187–276.

Spevack, A. A., & Suboski, M. D. Retrograde effects of electroconvulsive shock on learned responses. *Psychological Bulletin*, 1969, **72**, 66–76.

Tomkins, G. M., Levinson, B. B., Baxter, J. D., & Dethlefsen, L. Further evidence for posttranscriptional control of inducible tyrosineaminotransferase synthesis in cultured hepatoma cells. *Nature, New Biology*, 1972, **239**, 9–14.

Uphouse, L. L., MacInnes, J. W., & Schlesinger, K. Effects of conditioned avoidance training on polyribosomes of mouse brain. *Physiology & Behavior*, 1972, **8**, 1013–1018. (a)

Uphouse, L. L., MacInnes, J. W., & Schlesinger, K. Effects of conditioned avoidance training on uridine incorporation into polyribosomes of parts of brain of mice. *Physiology & Behavior*, 1972, **9**, 315–318. (b)

Watson, W. E. An autoradiographic study of the incorporation of nucleic-acid precursors by neurones and glia during nerve stimulation. *Journal of Physiology*, 1965, **180**, 754–765.

Wegener, G. Autoradiographische Untersuchungen über gesteigerte proteinsynthese in tectum opticum von Fröschen nach optischer Reizung. *Experimental Brain Research*, 1970, **10**, 363–379.

Wilson, D. L., & Berry, R. W. The effect of synaptic stimulation on RNA and protein metabolism in the R2 soma of *Aplysia. Journal of Neurobiology*, 1972, **3**, 369–379.

Zemp, J. W., Wilson, J. E., & Glassman, E. Brain function and macromolecules. II. Site of increased labelling of RNA in brains of mice during a short-term training experience. *Proceedings of the National Academy of Sciences, U.S.A.*, 1967, **58**, 1120–1125.

Zemp, J. W., Wilson, J. E., Schlesinger, K., Boggan, W. O., & Glassman, E. Brain function and macromolecules. I. Incorporation of uridine into RNA of mouse brain during short-term training experience. *Proceedings of the National Academy of Sciences, U.S.A.*, 1966, **55**, 1423–1431.

Part III

VERTEBRATE SENSORY
AND MOTOR SYSTEMS

Chapter 8

Central Visual Processing

COLIN BLAKEMORE

University of Cambridge, England

The physiology of vision is a subject far too vast to review adequately in a single chapter: even a brief account of current trends is almost redundant in this fast-moving field. So I shall use my space in this book to try to pull together a few subjects of interest, past and present, which bear on the question of perceptual synthesis—the almost imponderable process by which data abstracted by sensory systems are incorporated into an animal's plan of action in response to the outside world.

To avoid too gross a bibliography I have, where possible, referred only to the most recent publications or review articles in each area. The interested reader can follow these references back to the original publications.

THE TRIGGER FEATURE HYPOTHESIS

Since the introduction of microelectrode techniques in the 1930s, it has become clear that neurons at each level in the visual pathway do not constitute homogeneous populations, signaling merely the presence or absence of light. Even in the retina itself there is active abstraction of information, apparently concerned with the retention of important spatial and temporal patterns of stimulation (Barlow, 1972). Retinal ganglion cells, whose axons comprise the optic nerve, fall into a limited number of distinct classes. Hartline's (1940) discovery that ganglion cells in the frog discharge briefly at the beginning or the end of a period of illumination, or at both (ON, OFF, and ON–OFF cells) was expanded by Barlow (1953), who was perhaps the first to speculate that visual neurons might each require a *trigger feature*—a particular pattern of stimulation—in order to produce a maximal response. He observed that ON–OFF ganglion cells discharge particularly strongly for a small stimulus *moving* across the *receptive field*, that area of the retina from which the cell receives input. He also discovered that light falling outside the actual receptive field of an ON–OFF cell can attentuate the response to a flashing light in the center of the field. This property of *lateral inhibition* makes the cell selectively sensitive to the size of an object appearing in the receptive field. Barlow speculated that these ganglion cells might be especially involved in the orienting and striking reactions of the frog in response to small insects.

The Invariance Principle

Lettvin, Maturana, Pitts, and McCulloch (1961) went even further in their analysis of frog retinal ganglion cells, using a good deal of inspired guesswork and a broad repertoire of visual stimuli. They classified the cells into five groups.

1. *Boundary or contrast detectors*—responding to local contrast between a light and a dark area (equivalent to Hartline's ON fibers).
2. *Moving convex edge detectors*—responding best to a dark convex boundary moving into the receptive field (perhaps equivalent to ON–OFF fibers).
3. *Moving edge or changing contrast detectors*—responding to a moving dark–light boundary (equivalent to ON–OFF fibers).
4. *Dimming detectors*—responding to darkening of the whole receptive field, no edge being necessary (equivalent to OFF fibers).
5. *Dark detectors*—responding tonically at a frequency inversely related to the light falling on a large, diffuse receptive field. (This group is very rare.)

Maturana and Frenk (1963) used the same approach in their work on pigeon optic nerve fibers, and the outcome was again a short list of different classes of ganglion cell with distinct trigger features.

1 and 2. *Verticality* and *horizontality detectors*. These *orientation detectors* respond to elongated straight edges, even if stationary, as long as they are at the correct orientation.

3. *General edge detectors*—similar to the frog's moving edge detectors.

4. *Directional moving edge detectors*—responding to edges of either contrast when they are moved in one direction across the receptive field.

5. *Moving convex edge detectors*—similar to those in frog retina.

6. *Luminosity detectors*—firing in relation to overall illumination.

Perhaps the most thorough analysis using this experimental strategy was that of Barlow, Levick, and their collaborators, on the rabbit's retinal ganglion cells (see Levick, 1967; Oyster, 1968). Again a short list suffices to describe every class of ganglion cell.

1. *Concentric* receptive field cells, or *contrast detectors*—having a roughly circular zone of ON responses surrounded by an annular region giving OFF responses, or the reverse arrangement.

2. *Large field* or *very fast movement detectors*—responsive to extraordinarily fast motion of an edge across the field.

3. *Direction selective cells*—like the pigeon's directional movement detectors.

4. *Orientation detectors*—some responding to vertical edges, others to horizontal.

5. *Local edge detectors*—responsive to moving edges but having a strong suppressive surround that makes the optimal stimulus a small moving spot or texture (similar to the convex edge detectors in frog and pigeon).

6. *Uniformity detectors*—inhibited by the presence of edges, responding tonically to uniform light or dark.

What is perhaps most remarkable about the results of these experiments on frogs, pigeons, and rabbits is that three species with such totally different habitats and behavioral demands on their visual worlds should have such similarities in their neural classification of visual features. Ignoring the subleties that genuinely distinguish the ganglion cells of these different species, it is possible to reduce the list of trigger features to a few key elements:

1. *Contrast* or *edge*
2. *Movement*
3. *Direction of movement*
4. *Convexity* or *size*

5. *Orientation of edge*
6. *Overall illumination*

To make the list really comprehensive, and perhaps adequate to describe visual coding in every vertebrate, one should add:

7. *Color*

To guide themselves and others in the search for the trigger features of visual neurons, Lettvin and his colleagues emphasized a characteristic that has been called *invariance*. The response of a cell to its trigger feature should persist unaltered, despite manipulations in other properties of the stimulus. This concept was an essential sequitur of the notion that each neuron is *encoding* a single attribute of the whole complex visual scene. Implicit is the idea that the absolute frequency of discharge is a relatively unimportant aspect of the message. A high impulse frequency might serve merely to increase the neural certainty that the trigger feature is actually present (Barlow, 1972). What matters is whether a cell is detectably active or not—a kind of all-or-none law of sensory coding (Figure 1).

The invariance principle is a compelling way of describing the responses of a cell with a concentric receptive field—a local contrast detector. The effect of a spot depends on how much lighter (for an ON center cell) or darker (for an OFF center cell) the spot is than its background. The response to contrast is invariant in the face of changes in overall illumination. This phenomenon is easily explained because the center and the surround of the receptive field are roughly equally influenced by overall changes in illumination, and thus the opposing effects cancel. Without knowledge of this simple piece of physiology the subjective counterparts of this invariance for general light level seem like perceptual miracles: when we see a patch which is lighter than a gray background it appears much brighter than the same patch on a black background (see Chapter 9 by Anstis in this volume). In fact many of the invariances that characterize human perception and pattern recognition can possibly be traced to the operation of invariance in the responses of feature-extracting neurons.

Gating and Tuning

In reality, invariance is an ideal form of coding that is never actually achieved. There is always more than one stimulus dimension that can modulate the discharge of a visual neuron.

Direction selectivity is possibly the most compelling example of a trigger feature: directional movement detectors signal the presence of

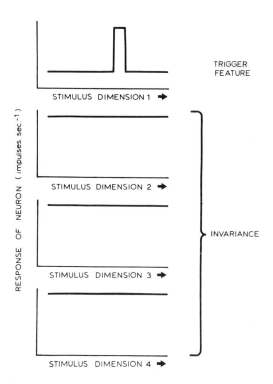

Figure 1. These graphs are hypothetical stimulus–response functions for a neuron with an extremely specific trigger feature. The neuron only responds to a stimulus of some particular value along stimulus dimension 1. As long as that single requirement is fulfilled the neuron's response remains invariant in the face of wholesale changes in other aspects of the stimulus, along dimensions 2, 3, and 4.

an object moving in the appropriate direction regardless of the sign of the contrast of the object against its background and whatever its shape and size. However, even direction selective cells disobey the invariance principle. For one thing, virtually all visual neurons have receptive fields smaller than the whole retina, so whatever the trigger feature, the position of the stimulus on the retina is a critical factor. Second, the luminance of the moving pattern, or more exactly its contrast against the background, must influence the response, if only to set a threshold value below which a stimulus cannot reliably excite the cell. But, even more damaging to the invariance principle, the velocity of movement as well as its direction are important variables. In the rabbit (Oyster, 1968), direction selective ganglion cells with ON–OFF receptive fields (plotted with flashing spots of light) prefer faster velocities, while those with ON fields prefer very slow movement (Figure 2). Thus, even classi-

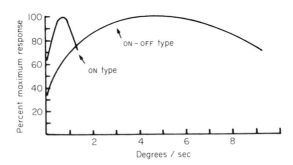

Figure 2. The responses of ON and ON–OFF rabbit direction selective ganglion cells (expressed as a percentage of the maximum response) are plotted as a function of the velocity of a stimulus moving in the preferred direction. The ON cells prefer very slow movement, while ON–OFF cells respond over a broad range of movement, from medium velocity to very fast. (Redrawn from Oyster, 1968).

cal direction selective ganglion cells carry information about the position, contrast, and velocity of the stimulus, as well as its direction of movement.

If the notion of total invariance must be abandoned, so too must the idea that the discharge frequency conveys no quantitative information about the stimulus. The *tuning* of a direction selective cell for stimuli moved in different directions is really very poor (Figure 3). So it may be more appropriate to think of sensory neurons having their response *gated* by certain characteristics of the stimulus, such as its position, contrast, and velocity, and then, as long as all these requirements are fulfilled, the cell could transmit by its frequency of firing an indication of the value of the stimulus along the dimension that lies beneath the cell's *tuning curve* (Figure 4).

While a scheme such as this, dependent on gating and tuning, could work very well if the tuning curve were a monotonic function, it has difficulties when the tuning curve has a maximum (such as directional tuning in Figure 3). Signals, other than the maximum one, are of course

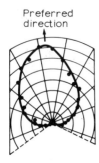

Preferred
direction

Figure 3. The directional "tuning" of an ON–OFF direction selective ganglion cell in the rabbit retina. The responses are plotted on polar coordinates indicating the direction of movement of the stimulus. (Redrawn from Oyster, 1968)

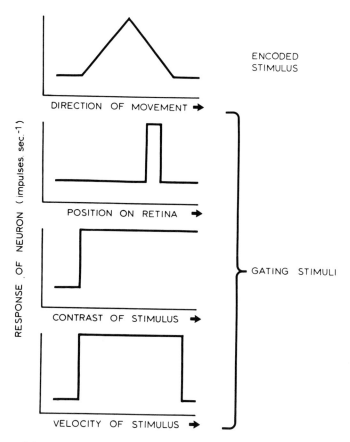

Figure 4. Stimulus specificities for a hypothetical direction selective cell, whose response is *gated* according to the position, contrast, and velocity of the stimulus and whose discharge frequency (as long as all the gating requirements are satisfied) signifies the direction of movement (top graph).

logically ambiguous, since they correspond to two unique values on the encoded stimulus dimension.

Object Detection

Another idea grew as a sibling of the trigger feature hypothesis: neurons might in fact respond optimally to real, behaviorally important objects or ensembles of features. The notion of object detection drew strength from the ethologists' reports that certain particular combinations of stimuli, called releasers, can provoke complex stereotyped patterns of behavior in animals. It was naturally exciting to speculate that the optimal stimuli for visual neurons might, in some cases, be the behaviorally important objects that release such patterns of behavior. Hence Barlow and Lettvin *et al.* spoke of some of the frog's ganglion cells as being "bug detectors" and thought that their signals might promote the orienting and snapping responses that frogs direct toward insects. The correspondence between neural optimal stimuli and the releasers of stereotyped behavior has been attacked in detail (Ingle, 1971), but in any case object detection is, by definition, very different from feature detection. What is implied is that a neuron can show multiple specificity, the response being gated along all the stimulus dimensions that contribute to the definition of the object in question (Figure 5).

The concept of object detection can be, and has been, carried to the point of absurdity (or simply incomprehensibility). It is all too easy to suggest that the whole visual system is designed simply for the construction (using input from lower-level feature detectors) of object-detecting neurons. There would be as many such cells as unique objects that can be recognized by the organism, each one exquisitely selective for a host of stimulus properties while remaining totally invariant in the face of a number of others.

With these simple notions of the trigger feature, invariance, gating, tuning, and object detection, I shall now turn to the cat's visual system (with occasional reference to the monkey) to see how these theories of coding might account for the properties of central visual neurons.

THE VISUAL PATHWAY

At first the visual systems of "higher" mammals seemed very different from those of rabbits, pigeons, and frogs. Certainly the receptive fields of cat retinal ganglion cells were found to have inhibitory surrounds (Kuffler, 1953) but they all seemed to have simple ON-center or OFF-center concentric organization. The trigger feature for a concentric receptive field is merely local contrast, a white or black spot of such a size that it just fills the center of the receptive field.

Even at the next synaptic level, in the dorsal lateral geniculate nucleus (LGN) the cells have very similar concentric receptive fields. Indeed simultaneous recording from LGN and retina has shown that only one or a small number of ganglion cells, usually all of the same type and close together in the retina, provide excitatory input to each LGN cell (Cleland, Dubin, & Levick, 1971). However, at the primary visual cortex (V1) a remarkable change in stimulus specificity occurs (Hubel & Wiesel, 1962, 1965). Although some cortical cells may also pick up from only a small number of afferent fibers they are virtually all *orientation*

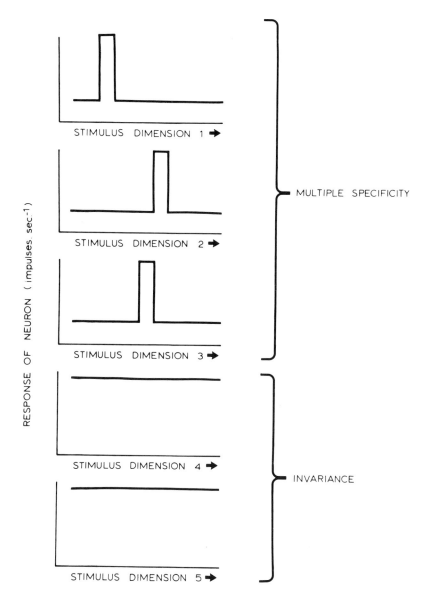

Figure 5. Stimulus requirements for a hypothetical object-detecting neuron. The cell shows precise specificity for particular values along a number of stimulus dimensions (1 to 3), and will only discharge if all of these requirements are met. Along other stimulus dimensions (4 and 5), the neuron shows complete invariance.

selective, each responding to a contour at a particular angle. Many of them are also direction selective, only firing if the contour moves in one direction, not the reverse. In terms of trigger features, there is a transition from local contrast detection at the retina to orientation and directional movement detection at the cortex (Figure 6).

In the cat's superior colliculus the superficial layers receive input directly from the retina and from the visual cortex (see Sterling, 1971). More than half these neurons are direction selective, and they usually respond optimally to movement away from the midline of the visual field toward the contralateral side.

X, Y, and W Systems

The Retina

A simple view of the cat's visual system, with shape information being processed in the cortex and directional movement in the colliculus, is now known to be too naive.

There is not one type of retinal ganglion cell, but three, and they have properties that are maintained, at least to some extent, at progressively higher levels in the pathway:

Y ganglion cells have large cell bodies and axons, which are fast-conducting (about 40 m sec^{-1}). They have concentric ON- or OFF-center receptive fields but give only transient responses to local illumination of center or surround. They have relatively large centers and much larger surrounds than X cells (see next paragraph) and often have lower maintained discharges. They are more frequent in the peripheral retina and they prefer faster velocities of stimulus movement. They will respond to movement of edges within the receptive field without overall change in flux and may always show only local changes in surround threshold after local light adaptation. The effects of center and surround stimulation summate nonlinearly and the response to a fine moving grating pattern consists of an unmodulated increase in firing rate (Cleland, Levick, & Sanderson, 1973; Enroth-Cugell & Robson, 1966).

X ganglion cells have smaller cell bodies and axons, which are slower conducting (about 20 m sec^{-1}). They also have concentric receptive fields but with smaller centers and surrounds and prolonged sustained responses to local illumination. They are much more common in the central retina and prefer slower stimulus velocities (up to about 20° sec^{-1}). They show simple linear summation of the effects of center and surround stimulation and hence produce a strictly modulated discharge in response to a moving grating. They will not respond vigorously to a moving pattern of edges unless it generates a change in the ratio

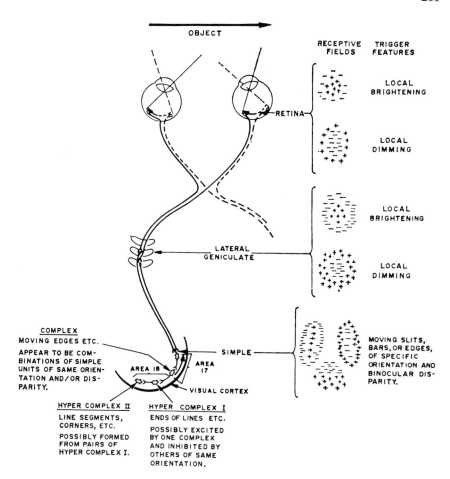

Figure 6. The geniculostriate pathway of the cat showing, at each level, the receptive fields of the neurons and their trigger features or optimal stimuli. $+$ = ON response; $-$ = OFF response. (From Barlow, Narasimhan, & Rosenfeld, 1972. Copyright 1972 by the American Association for the Advancement of Science.)

of flux falling on the center and the surround. They sometimes show nonlocalized changes in sensitivity after local adaptation of the surround (Cleland, Levick, & Sanderson, 1973; Enroth-Cugell & Robson, 1966).

W ganglion cells (see Hoffmann, 1973) have very slowly conducting axons (about 10 m sec^{-1}) and their cellular action potentials are extremely difficult to isolate, except with a fine micropipette; therefore they probably have tiny cell bodies. They fall into two groups:

1. "Excited by contrast." These cells give ON and OFF responses to flashing spots, they have no obvious surround region, and they are excited

by dark or light objects crossing the receptive field, rather like the edge detectors of rabbit, pigeon, and frog. They respond only to slow movement (up to 50° sec^{-1}) and many of them are direction selective.

2. "Inhibited by contrast." These cells are much like the rabbit's uniformity detectors: they have a maintained discharge in the presence of uniform illumination, which can only be *reduced* by any patterned visual stimulus.

In the last few years the central projections of X, Y, and W cells have been traced, mainly by the technique of measuring the latency for antidromic or orthodromic invasion of a neuron by electrical stimulation at points higher or lower in the visual pathway. The results of these experiments, some of them still very speculative, are summarized in Figure 7.

The Lateral Geniculate Nucleus

Certainly cells in the LGN can also mainly be classified as X or Y, on the basis of their responses to visual stimuli. In general LGN cells receive direct excitatory input from only one ganglion cell or from a very few ganglion cells of the same type (ON- or OFF-center, X or Y), and their own conduction velocities up to the visual cortex (to areas 17 and 18, V1 and V2) are also slow or fast, as appropriate. Only about 10% of geniculate cells have mixed X and Y properties, either because they have direct input from the very rare ganglion cells with mixed properties, or because they are driven by both X and Y ganglion cells (Cleland, Dubin, & Levick, 1971). There is also evidence that each type of LGN cell receives indirect inhibitory input from both types of ganglion cell and from the other type of LGN cell. W cells probably do not contribute to the geniculostriate pathway at all.

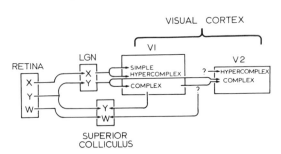

Figure 7. The visual pathway of the cat, showing the separate projections of W, X, and Y types of ganglion cell. The details of their different properties are described in the text.

The Visual Cortex

The fast axons of LGN Y cells and the slow axons of LGN X cells project up to the visual cortex. Recording from neurons in V1 and V2 while stimulating the optic chiasma and optic radiation showed that the slow axons terminate only in V1 while the fast axons end on cells in both V1 and V2, probably after bifurcating in the white matter. Some cells, in V2 particularly, have indirect fast afferent input, probably via two or more intracortical synapses (Stone & Dreher, 1973).

In V1, which has both fast and slow input, there is a clear correlation between the receptive field type of the cortical cell (as described by Hubel & Wiesel, 1962, 1965, 1968) and the type of afferent input (Hoffmann & Stone, 1971). The receptive field properties of cortical cells are as follows:

1. *Simple cells.* This type of neuron is particularly common in and around lamina IV, where the axons of geniculate cells terminate. In the monkey (Hubel & Wiesel, 1968), the physiological evidence for segregation of function in different laminae is even more distinct than in the cat. The defining characteristic of simple cells is that they have receptive fields that can be plotted, with stationary flashing stimuli (spots, or better, bars), into separate ON and OFF regions. There is relatively straightforward spatial summation within these separate zones and antagonism between them. The optimal type of stimulus (edge, black or white bar) and its optimal width (if it is a bar) can be predicted roughly from the map of ON and OFF areas.

However, the details of the responses to moving stimuli, in particular the contrast-independent direction selectivity demonstrated by many simple cells, cannot be accounted for by the receptive field plots using stationary stimuli (Bishop, Dreher, & Henry, 1972). The *response field* of a simple cell (the region within which moving stimuli generate a response) can in fact be described as a "complex" of excitatory and subliminal excitatory zones, each of which is selective for the nature of the moving stimulus (black or white edge), its orientation, and its direction of movement. The "excitatory complex" is swaddled in a roughly disk-shaped zone of "inhibitory side bands" that are not simply correlated with antagonistic flanks plotted with flashed stimuli (Figure 8). They can be revealed by sweeping stimuli across the receptive field of a simple cell whose normally low spontaneous discharge has been artificially increased with a jiggling bar shown to the "excitatory complex" of the same eye or the other one (Bishop, Coombs, & Henry, 1973). While the relative arrangement of "discharge centers" for edges of different contrast can roughly be accounted for on the basis of excitatory

Simple cell receptive field

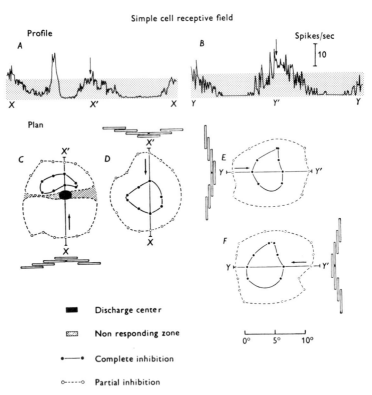

Figure 8. Maps of the receptive field of a simple cell in the cat cortex, showing the discharge center, or excitatory "complex," and the surrounding regions that produce partial or total inhibition depending on the orientation and direction of movement of a bright slit. (From Bishop, Coombs, & Henry, 1973) *A* and *B* are impulse density histograms generated by sweeping a short bar from X to X' and back (as shown in *C* and *D*), and from Y to Y' and back (shown in *E* and *F*). The spontaneous activity has been artificially raised to the stippled level (see text).

input from a group of lateral geniculate neurons with their concentric receptive fields in a row, the inhibitory side bands and such properties as direction selectivity seem to demand much more complicated interactions between neurons in the cortex.

Simple cells usually respond best to very slow movement, have low spontaneous activity, and tend to be very narrowly tuned for orientation: on average the response falls to half its maximum value if the stimulus orientation is changed by about 15°. In the cat, simple cells maximally sensitive to edges near horizontal or vertical are often more narrowly tuned for orientation than those optimally sensitive to diagonal lines (Rose & Blakemore, 1974).

There is evidence from the latency of simple cells to orthodromic stimulation from the optic radiation and optic chiasma that they have direct input from slow-conducting geniculate X cells and that they may be inhibited by cortical cells with Y input (Hoffmann & Stone, 1971; Stone & Dreher, 1973). A small number of simple cells have been definitely identified anatomically by a technique of intracellular dye injection after characterization of the receptive field, and they were usually *stellate cells*, short-axon Golgi type II neurons, which in some cases certainly have spiny dendrites with direct afferent input (Kelly & Van Essen, 1974).

2. *Complex cells.* These cells, common in both V1 and V2, are also orientation selective, though slightly less precisely tuned than simple cells (Rose & Blakemore, 1974). They have large response fields that cannot be subdivided into separate ON and OFF zones, and often do not respond at all to flashing stimuli. They respond to a bar whose optimal width (measured orthogonally to the preferred orientation) is always less than the size of the field, and the response is a prolonged discharge for the whole time that the bar is within the field. They prefer fast stimulus movement and have higher spontaneous activity. Most of them are pyramidal cells, large neurons found mainly in laminae III and V, whose axons leave the gray matter (Kelly & Van Essen, 1974).

An inspection of the latency of complex cells in V1 and V2, after stimulation of the optic chiasma or radiation, suggests that some complex cells receive direct synaptic input from LGN Y cells, whose fast-conducting axons probably often bifurcate, sending branches to both V1 and V2 (Stone & Dreher, 1973). Other cells with complex properties, more in V2 than V1, behave as if they have indirect fast input, that is, input through one or more synaptic relays from cortical neurons which themselves have fast input.

3. *Hypercomplex cells.* These neurons, while certainly present in V1, are more common in V2. They are also orientation selective, showing great variation in orientational tuning (Rose & Blakemore, 1974). They are generally found in the deepest and most superficial layers of the gray matter. Their defining characteristic is that they have a powerful inhibitory zone at one end or at both ends of the response field, so that if the moving bar or edge is extended into that zone the response is attenuated, or often abolished (Hubel & Wiesel, 1965). Thus edge-detecting hypercomplex cells with an inhibitory zone at only one end of the response field could be termed *corner detectors*, since they respond best to a 90° corner. In V2 some so-called "higher-order" hypercomplex cells have two orthogonal preferred orientations.

Some hypercomplex cells certainly have direct input from LGN X cells, since they produce longer latency, direct synaptically driven impulses in response to stimulation of the optic chiasma or radiation (Hoffmann & Stone, 1971). However, the sample was small and the question of input to hypercomplex cells is complicated by Dreher's (1972) suggestion that there are, in fact, two types of hypercomplex cells: those with complex properties (responding to faster movement) and those with simple properties (having distinct on and off zones and preferring slow movement). Indeed, many obvious simple and complex cells respond less strongly if the moving bar is extended beyond the receptive field. Hypercomplex cells seem to be merely extreme versions of simple and complex cells with particularly powerful "end inhibition," and are not a qualitatively-different class (Rose, 1974).

4. *Nonoriented cells.* Many cortical cells in lamina IV in the monkey have concentric receptive fields much like those of geniculate cells (Hubel & Wiesel, 1968). However, with notable exceptions (e.g., Baumgartner, Brown, & Schulz, 1965), cells without orientation selectivity are rarely reported in the cat's cortex, and they do not have simple center-surround organization. Some appear to have direct slow axon input (Hoffmann & Stone, 1971) while others have fast afferent input (Stone & Dreher, 1973). These nonoriented cells, despite their rarity, are required to provide intracortical inhibition in the model of simple cell receptive fields designed by Bishop and his colleagues (see Bishop, Coombs, & Henry, 1973).

The Superior Colliculus

In the cat's superior colliculus, cells in the superficial layers are, in general, sensitive to moving stimuli rather than stationary flashing spots and the majority of them are direction selective, usually responding best to movement into the contralateral hemifield, away from the area centralis. They have receptive fields with antagonistic surrounds and many of them show "hypercomplex" properties, preferring a stimulus much smaller than the whole receptive field (see Sterling, 1971).

About 90% of collicular cells, in the most superficial layers, have direct input from the axons of retinal W ganglion cells, as judged by their latency for stimulation of the optic chiasma or optic tract. These collicular cells respond to slow movement and are mainly direction selective.

Other collicular cells, in the deeper superficial layers have direct input from Y ganglion cells, whose axons branch, in the optic tract, to supply both the superior colliculus and the LGN. Collicular Y cells are not usually direction selective, which is rather unusual in view of the fact that many of them also receive indirect fast input from the axons of direction-selective complex cells in the visual cortex (Hoffmann, 1973).

The small proportion of direction-selective cells that persist in the superior colliculus after total ablation of the occipital cortex may depend for their properties on direct input from direction-selective retinal W cells (Hoffmann, 1973).

In the deeper layers of the cat's superior colliculus the neurons, which receive input from those in the superficial layers, are also often direction selective but the receptive fields are much larger. They respond better to fast movement, suggesting that their input is from the Y system, and many of them also respond to auditory or somatic stimulation. There is rough agreement between the optimal position in space and the preferred direction of movement for auditory and visual stimulation of the same neuron (Gordon, 1973).

Control of Eye Movements in the Monkey's Superior Colliculus

There is a great deal of evidence that the superior colliculus is involved in the control of the more automatic aspects of fixational eye movements (see Schiller, 1972) in both cats and monkeys. Indeed it has often been suggested that the tectal visual system is part of a general orienting system. In the cat, collicular neurons seem to be concerned with objects moving away from the fixation point, into the peripheral visual field, in other words objects that are tending to disappear from view. The cat's colliculus certainly has descending connections with the oculomotor system and the motoneurons responsible for head movement (Anderson, Yoshida, & Wilson, 1971). It is in studies of the monkey, however, that great strides have recently been made in the understanding of eye movement control.

In the monkey's superior colliculus, neurons in the superficial and intermediate layers are usually *nonspecific event detectors* responding best to small moving stimuli, *jerk detectors* responding to jerky movement, or *habituating novelty detectors* responding only to the initial presentation of a target. Surprisingly, very few cells are direction selective (Cynader & Berman, 1972). In the deeper layers the cells again have larger receptive fields, responding to fast movement. Some of these neurons have auditory and somatic input and most of them, in the conscious monkey, discharge before eye movements of a particular size and direction (Schiller & Stryker, 1972). In fact the "motor field"—that area of the visual field to which the fovea is carried when the cell is active—is roughly in the same position as the receptive field of the neuron before the eye movement takes place.

Electrical microstimulation of the superior colliculus produces a single saccade (or if it is prolonged, a staircase of identical eye movements), the direction and size of which match the position of the receptive fields and motor fields of cells in that part of the colliculus (Figure 9). This has led to the suggestion that the monkey's superior colliculus is a simple device for controlling *foveation*—the acquisition by the fovea of any interesting new object in the visual field (Schiller & Stryker, 1972). This idea has been somewhat weakened, however, by the finding that small lesions in the colliculus do no more than temporarily increase the latency of eye movements toward the represented part of the visual field (Wurtz & Goldberg, 1972). The fact that many collicular units show an enhanced response to a visual stimulus falling on the receptive field, if the monkey has been trained to look at it, suggests that the superior colliculus may play a role in facilitating foveation movements to those stimuli that the monkey has recognized as behaviorally important (Goldberg & Wurtz, 1972).

How well are collicular neurons fitted to perform the joint role of the detection of a novel stimulus and the identification of its position, in order to guide an eye movement? Both the Y system and the W system (which innervate the cat's superior colliculus) are concerned with the detection of moving rather than stationary stimuli—the W system for slow moving objects, the Y system for faster movement. Both systems seem rather ill-suited to the task of the identification of objects, since they respond to both light and dark patterns. Unfortunately they also seem badly fitted for the task of the exact identification of the position of a target. They respond to increments of movement anywhere within the receptive field, and the fields of collicular neurons (especially in the deeper layers) are very large compared with the accuracy of

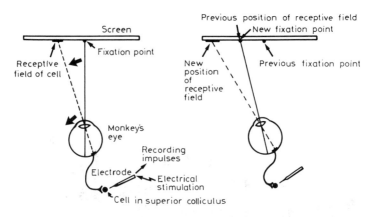

Figure 9. The results of microstimulation in the monkey's superior colliculus. Stimulation in the region of a cell with a receptive field to the left of the starting fixation point causes the eye to move in such a way that the fovea fixates the point previously occupied by the receptive field.

ballistic saccadic movements. The inevitable conclusion is that the message identifying the position of a target of interest (if indeed it inhabits the superior colliculus) consists of the relative activity of a large number of neurons with overlapping receptive fields.

This concept of collaboration between coding neurons to provide a signal more precise than any single coding element is familiar in the consideration of color recognition based on only three different absorption pigments. It is a concept that is appealed to again and again in discussions of neural coding, but it is inherently contradictory to the trigger feature hypothesis. We can no longer think of all neurons detecting a recognizable feature, even less an object, but must rather conceive of them providing only fragments of the message that identifies a stimulus.

Analysis of Disparity in the Visual Cortex

In both the visual cortex (Hubel & Wiesel, 1962, 1965, 1968) and the superior colliculus (see Sterling, 1971) most neurons, whatever their type, are binocularly driven. They have very similar receptive fields in the two eyes and they usually respond much more strongly during simultaneous stimulation of both retinas.

In the visual cortex of both cat and monkey there is now strong evidence that binocular neurons, or at least some proportion of them, play an important role in binocular depth discrimination (Barlow, Blakemore, & Pettigrew, 1967; Bishop & Henry, 1971; Hubel & Wiesel, 1970). Binocular cells, particularly the simple cells with small response fields, demonstrate remarkable selectivity for the retinal disparity of the images of an object of the appropriate orientation. As the disparity of the pattern is varied, equivalent to changing the distance of the object from the cat for a fixed eye position, each cell shows marked facilitation when the two images are in register on the two response fields and powerful occlusion of the discharge when the images are even slightly out of register in either direction.

Thus binocular cortical neurons can be termed *disparity detectors* (as well as being orientation selective). Moreover, there is some variation in the optimal disparity from one cell to another (although there is debate over the exact range of disparity variation: see Bishop & Henry, 1971). So different cells respond best to objects at different distances (Figure 10), and between them these binocular neurons could provide information for the interpretation of the relative distances of objects and for regulating vergence eye movements.

In the monkey, cells with obvious disparity selectivity and optimal

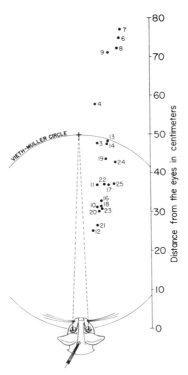

Figure 10. Results of an experiment to determine the optimal retinal disparity for a sample of binocular cortical neurons from a single cat. Each point indicates the projection onto the plane of fixation of the position in space at which an object of the correct orientation would have to be placed in order for its images to stimulate one particular neuron. The number of each neuron is shown next to the optimal position in space. (Redrawn from Barlow, Blakemore, & Pettigrew, 1967)

disparities very different from the zero disparity plane through the fixation point are found in a prestriate area of the visual cortex (Hubel & Wiesel, 1970).

Certain binocular cortical neurons in the cat are specialized not simply for the detection of *positional* disparities of the images of objects but for the interpretation of more subtle differences between the views obtained by the two eyes. For instance, cells are sometimes found that have opposite preferred directions of movement (though the same orientation preference) in the two eyes (Blakemore, Fiorentini, & Maffei, 1972; Pettigrew, 1973). Such neurons (if their optimal orientation is close to vertical) respond best to objects moving toward the cat or away from it. Likewise, binocular cells in V1 show a small ($\pm 15°$) range of *differences* in preferred orientation, and thus some binocular cells are sensitive to the tilt of a contour in three-dimensional space, toward the cat or away from it (Blakemore, Fiorentini, & Maffei, 1972). It is possible that binocular neurons also display small differences in preferred velocity of movement and bar width or bar length between the two receptive fields: such differences would also contain information about the arrangement of objects in three-dimensional space.

The Architecture of the Visual Cortex

In both cat and monkey there is a macroscopic organization of neurons according to certain aspects of their optimal stimuli. During a long diagonal penetration through the visual cortex, neurons with similar stimulus specificities are found in clusters, with sudden but relatively small changes in optimal stimulus between successive clusters. It seems, then, that the cortex is arranged into a matrix of radially organized *columns*, or probably more correctly *sheets*, of neurons, some 100 μm to 1 mm in width, each column containing a large number of neurons with some stimulus specificity in common, but with large variation in other stimulus requirements. Such clustering of neurons according to stimulus requirement has been described for the neurons' ocular dominance, preferred disparity, preferred position in oculocentric visual space, preferred direction of movement, and, in the monkey, the preferred color (Blakemore, 1970; Hubel & Wiesel, 1962, 1965, 1968, 1970). However the most substantial evidence for columnar "mapping" is for preferred orientation itself (Figure 11). Certainly, as a general rule, any cortical cell is surrounded most immediately by neurons with very similar optimal orientations (in the same column) and then at progressively greater distances by cells with progressively more different stimulus orientations (in the neighboring columns).

The cortex, viewed from the surface, is, then, functionally divided into a number of overlapping matrices of columns, each stimulus modality being mapped across the cortex and the whole array being mapped retinotopically with respect to the visual field. Each individual cell belongs to a number of columns and its stimulus specificities are defined by its columnar allegiances.

Blakemore and Tobin (1972) have suggested that this extraordinarily complicated arrangement of columns may be an essential feature of cortical organization, providing an opportunity for the sharpening of stimulus preferences by a process of local intracortical inhibition. Benevento, Creutzfeldt, and Kuhnt (1972) have demonstrated that the inhibitory input to cortical cells in the cat can itself be selective for certain aspects of the stimulus and therefore not simply derived from nonoriented cells, as suggested by Bishop, Coombs, and Henry (1973). Blakemore and Tobin (1972) have shown that cortical neurons can be inhibited even by contours falling outside the response field and that the

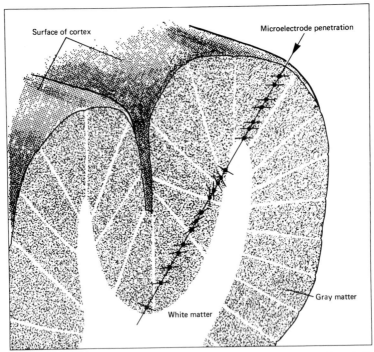

Figure 11. The typical result of a diagonal microelectrode penetration through the cat's visual cortex. Each point shows the position at which a cell is recorded and the short line indicates its preferred orientation. There are abrupt changes in optimal orientation between one columnar group and the next.

inhibition is itself orientation selective, being more broadly tuned than the cell's own excitatory orientational tuning curve but centered on the same optimal orientation (Figure 12). It could be that, given the columnar organization of preferred orientation, a simple process of lateral intracortical inhibition between neighboring cells, up to a distance of a small number of columns, could provide such an inhibitory input, which serves to sharpen each cell's orientation selectivity.

The same intracortical inhibitory process might therefore perform the same sharpening process for every stimulus modality that is mapped in a continuous or discontinuous columnar manner across the visual cortex.

Stimulus Analysis beyond the Visual Cortex

Beyond the primary visual cortex the anatomical arrangement of ascending projections becomes extremely complicated. In the cat there

Figure 12. (a) The tuning curve for a complex cortical cell, determined by measuring the mean number of impulses produced during eight sweeps of a bright bar at various orientations across the response field. The inset diagrams show the arrangement of the stimulus, the larger rectangle representing the response field and the small one the moving bar. The interrupted line shows the spontaneous discharge of the cell produced in the same time interval with no stimulus present.

(b) For the same cell the moving bar was *always* set at the optimal orientation: the abscissa now shows the orientation of a high-contrast grating moved randomly back and forth outside the response field itself. The solid line shows the level of response (about 53 impulses per presentation) elicited by the moving bar in the absence of any surrounding grating. Clearly the grating inhibited the cell over a very broad range of orientation centered on approximately the same orientation that was optimal for the neuron. Similar results are found for most complex cells and many simple cells. (From Blakemore & Tobin, 1972)

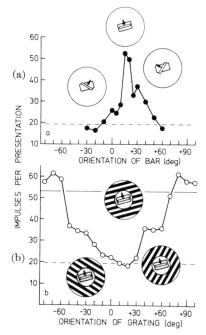

are projections from visual cortex to areas in the neighboring suprasylvian gyrus. Dow and Dubner (1969) have recorded in the anterior portion of the suprasylvian gyrus and found that about two-thirds of cells respond to visual stimuli. The cells are of three types: S *cells*, responding to stationary spots and having no inhibitory surrounds; M *cells*, responding to movement in any direction; and E *cells*, that are orientation selective but with broad orientational tuning.

In the lateral suprasylvian gyrus of the cat (Hubel & Wiesel, 1969) the cells are usually orientation selective, being either complex or hypercomplex in their properties. The distinctive property of cells in both of these areas is the very large size of their receptive fields, compared with those in V1, V2, and V3 (area 19).

In the monkey there is rather more physiological evidence about the fate of information leaving the primary visual cortex. Zeki has described a large number of localized visual cortical areas between V1 and the inferotemporal cortex (see Zeki, 1973). In some of these regions there seems to be a degree of specialization for certain stimulus attributes. Hubel and Wiesel's (1970) account of a region of disparity-detecting cells in the prestriate area has already been mentioned. Dubner and Zeki (1971) found that cells in the superior temporal sulcus (which receives from V1) respond particularly to moving stimuli and are often not selective for stimulus orientation. Similarly, neurons in another pre-striate area that Zeki calls V4 have strong color preferences but again are often nonspecific for stimulus orientation (Zeki, 1973).

In the inferotemporal cortex itself, Gross, Rocha-Miranda, and Bender (1972) have described neurons with extremely large receptive fields often covering both visual hemifields and always including the fovea. Although many of them are rather like V1 cells in their selectivity for movement, orientation, and color, a certain proportion seemed to have the most extraordinary degree of stimulus specificity. For instance, three cells responded best to the silhouette of an upward pointing monkey's hand, the very view that the monkey might habitually gain of its own hands.

PATTERNS OF SYNTHESIS

Now it might be appropriate to reexamine briefly the principles of coding discussed at the start of this chapter, in light of the present knowledge of the higher mammalian visual pathway. It is evident that even the concept of a neuron being gated in its responses along many stimulus dimensions and coding the stimulus value along one particular dimension (Figure 4) is too naive to account for all the properties of cortical and collicular cells.

The Multichannel Neuron

Each neuron in the visual cortex or superior colliculus can have spe-cificity for a very large number of stimulus attributes. While it may be reasonable to think of the receptive field simply gating the response according to position, it is clear that each cell has tuned characteristics for many stimulus dimensions. In the visual cortex an individual neuron can be selectively sensitive for the orientation and direction of movement, the width and length, the retinal disparity, the velocity, and the contrast of the stimulus.

The chances of all the stimulus requirements of such a cortical cell being exactly met (except in a neurophysiological experiment) are really very small.

A change in firing rate is multiply ambiguous, since it could imply an alteration in any one of these stimulus attributes. Each neuron in such a system as this must be considered as a *multichannel* coding device that contributes to the specification of information about many properties of the stimulus. So, neurons in the visual cortex and superior colliculus cannot really be said to possess a trigger feature, nor even to be object detectors. They detect assemblies of features with low probabilities of optimal stimulation. Each particular assembly is not uniquely related to any single object. The information about the individual features of the visual scene and the identity of the objects in it are present in the activity of those neurons, but no individual cell specifies a single feature or a single object.

One must come to the same conclusion for the visual system that Erickson and Schiffman (see Chapter 13 in this book) reach for the chemical senses. The identification of visual stimuli must rely on some comparison between the activities of these multichannel neurons.

Hierarchical Cascades, Generalization, and Parallel Processing

It is, of course, intriguing to consider the manner in which the information present in, say, neurons of V1 is interpreted. One very attractive possibility is that a serial, hierarchical arrangement of neurons might perform a process of *generalization*. The input to a higher-level neuron could consist of cells with identical stimulus requirements along one stimulus dimension but variation of specificity along other dimensions. This process would tend to produce true feature-detecting cells that demonstrate exact requirements for only one aspect of the stimulus.

The best-known example of such a hierarchical cascade is Hubel and Wiesel's (1962, 1965) model of the projection of simple cells onto complex and thence onto hypercomplex cells (see Figure 6). A complex cell would then have precise orientation specificity but would no longer require the stimulus to be exactly specified in retinal position. This theory was reinforced by the ideal arrangement of simple, complex, and hypercomplex cells within the same orientation column, and it is certainly true that the receptive fields of complex cells are on average larger than those of simple cells. Unfortunately the hierarchical theory has received a severe setback from the finding, already described, that simple, complex, and hypercomplex cells all can have independent direct input from the LGN, and that complex cells, in fact, may be the first to receive information, along the fast Y pathway.

Nevertheless, it is quite clear that sequentially arranged neurons almost always show an increase in receptive field size (for example from superficial to deep layers in the colliculus, and from V1 to inferotemporal or suprasylvian cortex). Generalization for position in the visual field seems to be a basic principle of visual neural coding.

There may be independent parallel hierarchies, each generalizing for most stimulus attributes while retaining precise information about a single quality (color in V4, orientation in the inferotemporal cortex, and movement in the superior temporal sulcus of the monkey). But in that case, what about the "hand-detecting" neurons also found in the inferotemporal cortex (Gross, Rocha-Miranda, & Bender, 1972)? Do they represent the summit of a different kind of hierarchy, one that truly detects objects by convergent input from large numbers of lower neurons each with very exact and different stimulus requirements? Perhaps that mode of object analysis, rejected earlier as untenable because it simply demands too many neurons, *is* a suitable form of coding for just those objects, like a hand, that are behaviorally very important because they are very frequently seen.

There may then be two modes of analysis within the visual system. One could utilize explicit object detectors for those things that figure most often and most crucially in the animal's visual environment. (The synthesis of such object detectors may in fact take place as a result of the animal's early visual experience: see Chapter 4 by Hirsch and Jacobson.) The second mode of analysis applicable for any type of object might employ a set of parallel hierarchies, each extracting a separate element of the stimulus.

REFERENCES

Anderson, M. E., Yoshida, M., & Wilson, V. J. Influence of superior colliculus on cat neck motoneurons. *Journal of Neurophysiology*, 1971, **34**, 898–907.

Barlow, H. B. Summation and inhibition in the frog's retina. *Journal of Physiology*, 1953, **119**, 69–88.

Barlow, H. B. Single units and sensation: A neuron doctrine for perceptual psychology. *Perception*, 1972, **1**, 371–394.

Barlow, H. B., Blakemore, C., & Pettigrew, J. D. The neural mechanism of binocular depth discrimination. *Journal of Physiology*, 1967, **193**, 327–342.

Barlow, H. B., Narasimhan, R., & Rosenfeld, A. Visual pattern analysis in machines and animals. *Science*, 1972, **177**, 567–575.

Baumgartner, G., Brown, J. L., & Schulz, A. Responses of single units of the cat visual system to rectangular stimulus patterns. *Journal of Neurophysiology*, 1965, **28**, 1–18.

Benevento, L. A., Creutzfeldt, O. D., & Kuhnt, U. Significance of intracortical inhibition in the visual cortex. *Nature New Biology*, 1972, **238**, 124–126.

Bishop, P. O., Coombs, J. S., & Henry, G. H. Receptive fields of simple cells in the cat striate cortex. *Journal of Physiology*, 1973, **231**, 31–60.

Bishop, P. O., Dreher, B., & Henry, G. H. Simple striate cells: Comparison of responses to stationary and moving stimuli. *Journal of Physiology*, 1972, **227**, 15–17P.

Bishop, P. O., & Henry, G. H. Spatial vision. *Annual Review of Psychology*, 1971, **22**, 119–160.

Blakemore, C. The representation of three-dimensional visual space in the cat's striate cortex. *Journal of Physiology*, 1970, **209**, 155–178.

Blakemore, C., Fiorentini, A., & Maffei, L. A second neural mechanism of binocular depth discrimination. *Journal of Physiology*, 1972, **226**, 725–749.

Blakemore, C., & Tobin, E. A. Lateral inhibition between orientation detectors in the cat's visual cortex. *Experimental Brain Research*, 1972, **15**, 439–440.

Cleland, B. G., Dubin, M. W., & Levick, W. R. Sustained and transient neurones in the cat's retina and lateral geniculate nucleus. *Journal of Physiology*, 1971, **217**, 473–496.

Cleland, B. G., Levick, W. R., & Sanderson, K. J. Properties of sustained and transient ganglion cells in the cat retina. *Journal of Physiology*, 1973, **228**, 649–680.

Cynader, M., & Berman, N. Receptive-field organization of monkey superior colliculus. *Journal of Neurophysiology*, 1972, **35**, 187–201.

Dow, B. M., & Dubner, R. Visual receptive fields and responses to movement in an association area of cat cerebral cortex. *Journal of Neurophysiology*, 1969, **32**, 773–784.

Dreher, B. Hypercomplex cells in the cat's striate cortex. *Investigative Ophthalmology*, 1972, **11**, 355–356.

Dubner, R., & Zeki, S. M. Response properties and receptive fields of cells in an anatomically defined region of the superior temporal sulcus in the monkey. *Brain Research*, 1971, **35**, 528–532.

Enroth-Cugell, C., & Robson, J. G. The contrast sensitivity of retinal ganglion cells of the cat. *Journal of Physiology*, 1966, **187**, 517–552.

Goldberg, M. E., & Wurtz, R. H. Activity of superior colliculus in behaving monkey. II. Effect of attention on neuronal responses. *Journal of Neurophysiology*, 1972, **35**, 560–574.

Gordon, B. Receptive fields in deep layers of cat superior colliculus. *Journal of Neurophysiology*, 1973, **36**, 157–178.

Gross, C. G., Rocha-Miranda, C. E., & Bender, D. B. Visual properties of neurons in inferotemporal cortex of the macaque. *Journal of Neurophysiology*, 1972, **35**, 96–111.

Hartline, H. K. The receptive fields of optic nerve fibers. *American Journal of Physiology*, 1940, **130**, 690–699.

Hoffman, K. P. Conduction velocity in pathways from retina to superior colliculus in the cat: A correlation with receptive-field properties. *Journal of Neurophysiology*, 1973, **36**, 409–424.

Hoffman, K. P., & Stone, J. Conduction velocity of afferents to cat visual cortex: A correlation with cortical receptive field properties. *Brain Research*, 1971, **32**, 460–466.

Hubel, D. H., & Wiesel, T. N. Receptive fields, binocular interaction and functional architecture in the cat's visual cortex. *Journal of Physiology*, 1962, **160**, 106–154.

Hubel, D. H., & Wiesel, T. N. Receptive fields and functional architecture in two non-striate visual areas (18 and 19) of the cat. *Journal of Neurophysiology*, 1965, **28**, 229–289.

Hubel, D. H., & Wiesel, T. N. Receptive fields and functional architecture of monkey striate cortex. *Journal of Physiology*, 1968, **195**, 215–243.

Hubel, D. H., & Wiesel, T. N. Visual area of the lateral suprasylvian gyrus (Clare-Bishop area) of the cat. *Journal of Physiology*, 1969, **202**, 251–260.

Hubel, D. H., & Wiesel, T. N. Cells sensitive to binocular depth in area 18 of the macaque monkey cortex. *Nature*, 1970, **225**, 41–42.

Ingle, D. J. Prey-catching behavior of anurans towards moving and stationary objects. *Vision Research*, 1971, Supplement 3, 447–456.

Kelly, J. P., & Van Essen, D. C. Cell structure and function in the visual cortex of the cat. *Journal of Physiology*, 1974, **238**, 515–547.

Kuffler, S. W. Discharge patterns and functional organization of mammalian retina. *Journal of Neurophysiology*, 1953, **16**, 37–68.

Lettvin, J. Y., Maturana, H. R., Pitts, W. H., & McCulloch, W. S. Two remarks on the visual system of the frog. In W. A. Rosenblith (Ed.), *Sensory communication*. Cambridge, Massachusetts: M.I.T. Press, 1961.

Levick, W. R. Receptive fields and trigger features of ganglion cells in the visual streak of the rabbit's retina. *Journal of Physiology*, 1967, **188**, 285–307.

Maturana, H. R., & Frenk, S. Directional movement and horizontal edge detectors in the pigeon retina. *Science*, 1963, **142**, 977–979.

Oyster, C. W. The analysis of image motion by the rabbit retina. *Journal of Physiology*, 1968, **199**, 613–635.

Pettigrew, J. D. Binocular neurones which signal change of disparity in area 18 of cat visual cortex. *Nature New Biology*, 1973, **241**, 123–124.

Rose, D. The hypercomplex cell classification in the cat's striate cortex. *Journal of Physiology*. (In press.)

Rose, D., & Blakemore, C. An analysis of orientation selectivity in the cat's visual cortex. *Experimental Brain Research*, 1974, **20**, 1–17.

Schiller, P. H. The role of monkey superior colliculus in eye movement and vision. *Investigative Ophthalmology*, 1972, **11**, 451–460.

Schiller, P. H., & Stryker, M. Single-unit recording and stimulation in superior colliculus of the alert rhesus monkey. *Journal of Neurophysiology*, 1972, **35**, 915–924.

Sterling, P. Receptive fields and synaptic organization of the superficial gray layer of the cat superior colliculus. *Vision Research*, 1971, Supplement 3, 309–328.

Stone, J., & Dreher, B. Projection of X- and Y-cells of the cat's lateral geniculate nucleus to areas 17 and 18 of visual cortex. *Journal of Neurophysiology*, 1973, **36**, 551–567.

Wurtz, R. H., & Goldberg, M. E. Activity of superior colliculus in behaving monkey. IV. Effects of lesions on eye movements. *Journal of Neurophysiology*, 1972, **35**, 587–596.

Zeki, S. M. Colour coding in rhesus monkey prestriate cortex. *Brain Research*, 1973, **53**, 422–427.

Chapter 9

What Does Visual Perception Tell Us About Visual Coding?

STUART M. ANSTIS

University of Bristol, England

INTRODUCTION

The brain is peculiarly difficult to study because it is a closely coupled, rapidly changing system which will not keep still long enough for one to have a good look at it. For this reason, psychologists have traditionally studied those aspects or parts of the brain that are relatively fixed, namely its input and output peripherals (the special senses and motor performances) and its store (memory). The central aim in studying perception is to find out the relationship between retinal images as the input, and the perceived world of visual objects as the output. Deducing this relationship from psychophysics is like trying to infer the wiring diagram of a pinball machine from watching the ways in which the machine's lights flicker on and off in response to the stimuli it receives from a rolling steel ball. It is not easy. With the brain, which is more complicated than the most glorious pinball machine, it is almost impossible. On the other hand, a neurophysiologist is like an engineer trying to understand a pinball machine when he has parts of its wiring diagram but has never seen the machine in play. In the last decade, there has been a highly successful two-pronged attack on visual problems by psychophysics and physiology. Physiologists have recently made exciting advances in tracing out the mechanisms of visual perception, progress that has resulted largely from their presenting just the right stimuli to excite the neurons in their preparations. These stimuli have often been suggested by psychophysics.

This chapter presents an arbitrarily selective review of some recent research in psychophysics and electrophysiology. The psychophysical phenomena of contrast and adaptation and of sensitivity to gratings are discussed. These have been explained in terms of neural summation and inhibition, both lateral and temporal. The phenomena were originally discovered with patches or gratings of different luminances, but have recently been explored using patches or gratings composed of numerous other visual dimensions such as color, disparate points, wavy lines, and so on. Models are reviewed here which attempt to find a quantitative match between these effects and the neural inhibitory processes which have been found at many levels of the visual system among populations of selectively sensitive neural units.

SIMULTANEOUS AND SUCCESSIVE CONTRAST

Figure 1 shows some examples of simultaneous contrast (top row) and of successive contrast or visual adaptation (middle and bottom rows). Note that simultaneous and successive contrast are close analogues of each other, and that nearly all visual dimensions show both effects. There is no a priori reason to expect that the parameters of the two effects should always coincide, since there is no reason to suppose that the space constants for simultaneous effects should correspond closely to the space and time constants of successive effects, but that they often do so has been pointed out by many authors (e.g., Ganz, 1966).

Figure 1(a–i) shows simultaneous contrast effects. In Figure 1a, brightness contrast (a phenomenon known to Aristotle and studied by Leonardo; see Boring, 1942) makes a gray disk set against a white background surround appear darker than the same disk seen against a black surround. The cause of brightness contrast is probably lateral inhibition in the retina, as has been well-documented (Ratliff, 1965; Békésy, 1968).

Many other visual dimensions show simultaneous contrast. Figure 1b shows spatial-frequency contrast (after Mackay, 1973). The spatial fre-

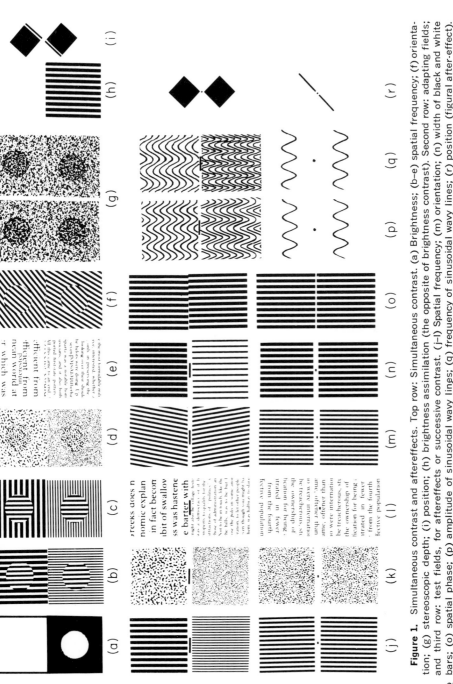

Figure 1. Simultaneous contrast and aftereffects. Top row: Simultaneous contrast. (a) Brightness; (b–e) spatial frequency; (f) orientation; (g) stereoscopic depth; (i) position; (h) brightness assimilation (the opposite of brightness contrast). Second row: adapting fields; and third row: test fields, for aftereffects or successive contrast. (j–l) Spatial frequency; (m) orientation; (n) width of black and white bars; (o) spatial phase; (p) amplitude of sinusoidal wavy lines; (q) frequency of sinusoidal wavy lines; (r) position (figural after-effect). For details, see text.

271

quency of a grating is defined as the number of cycles or pattern repeats per degree of visual angle. The gratings in the upper and lower disks are identical, but the upper disk in the coarse (low-frequency) surround looks finer (higher in spatial frequency) than the lower disk in the fine surround. The effect is orientation sensitive, being diminished when the inducing and test gratings are at right angles to each other as in Figure 1c. It should be mentioned that Georgeson (1974) suggests that this effect might be an artifact caused by eye movements. *Successive* contrast of spatial frequency is a well-known phenomenon (Blakemore & Sutton, 1969). After inspection of one grating, a subsequently viewed finer grating looks even finer, a coarser one even coarser (Figure 1j). It is important to make sure that Mackay's simultaneous contrast effect is not simply this aftereffect of adaptation. When viewing Figure 1b, if fixation drifted onto the background, even for a second or so, an aftereffect would be expected when the figure was refixated. The issue is still unresolved.

Figure 1d (after Mackay, 1973) shows analogous size contrast for sandpaper textures. Again, the two disks of sandpaper are identical, but the upper one looks finer. The same effect is shown for print in Figure 1e, with the word "interested" in the upper portion looking smaller than the same word in the lower portion. Both Figures 1d and 1e are instances of spatial-frequency contrast, with the size and spacing of the surround dots or letters exerting a contrast effect on the size and spacing of the test dots or letters. There is a suggestive similarity between Figure 1e and the geometrical size illusions, such as Titchener's circles. Orientation contrast is shown in Figure 1f (after Georgeson, 1973). The gratings in the upper and lower disks are exactly parallel, but their apparent orientations are displaced by their surrounds so that they appear markedly nonparallel. Georgeson (1973) found that the effect is frequency dependent, being strongest when the test and inducing gratings have the same frequency and disappearing if they differ in frequency by more than an octave (a factor of 2). So Figures 1b and 1f are a complementary pair, with Figure 1b showing orientation-dependent frequency contrast and Figure 1f showing frequency-dependent orientation contrast.

Figure 1i shows "position contrast" or "contour repulsion." The two diagonal lines are actually in exact alignment, but the inducing squares exert a repulsion effect on them, inducing an apparent vernier displacement. (The effect is so small in both this and the more familiar temporal analogue, the figural aftereffect, that a vernier judgment is necessary to make the effect show up clearly).

Depth contrast is shown in Figure 1g. The figure is a random-dot stereogram. Viewed stereoscopically, it shows two disks in the same depth plane. The upper disk is in a surround which lies nearer in depth, the lower disk in a surround which lies further away in depth. For some observers, the surrounds induce an apparent depth contrast on the two disks such that the upper disk appears to be further away than does the lower.

Figure 1h shows brightness *assimilation* (after Helson & Rohles, 1959). The gray test stripes appear darker against a dark surround than against a light surround. This effect only occurs when the test stripes are narrow enough to fall within spatial summatory zones (Westheimer, 1967). If the figures are made larger, the surround background extends into the inhibitory zones producing the brightness *contrast* effects, shown in Figure 1a. Assimilation effects for size have also been reported, as in the Delboeuf illusion. In this illusion, a circle appears larger when it is surrounded by a slightly larger concentric circle. Compare this with the size contrast effect of Titchener's circles, and note that—just as for brightness assimilation—size attraction or assimilation only occurs when the distance between inducing and test circles is rather small.

The temporal analogues of all these contrast effects are adaptations and negative aftereffects. Adaptation to any visual dimension can be demonstrated in four ways: (*1*) The "strength" of a sensation declines throughout adaptation; (*2*) the threshold for the adapting stimulus dimension rises; (*3*) imbalance in opponent channels causes a sensation of opposite value after adaptation; and (*4*) test stimuli of adjacent values along the adapted dimension appear displaced away from the stimulus value (Blakemore & Sutton, 1969). The fourth form of the effect, where the test and adapting fields are different, is shown in Figure 1j–t because it is the easiest way to demonstrate adaptation.

Figure 1j shows adaptation to spatial frequency (Blakemore & Sutton, 1969). The apparent spatial frequency of the test grating is raised after inspection of a low-frequency grating, but is lowered after inspection of a high-frequency grating. Note that the effect is only visible if the test grating has a spatial frequency within an octave of the adapting grating. Figures 1k and 1l show analogous adaptation to the fineness of grain of sandpapery textures (Walker, 1966) and of print. Cross-adaptation occurs: after adapting to sandpaper or gratings, aftereffects are visible on a test field of print and vice versa. This shows that these, too, are adaptations to spatial frequencies and not to contour positions or specific letters or other visual features (Anstis, 1974).

Figure 1m shows adaptation to orientation—the familiar tilt aftereffect. The direct effect decreases in strength if adapting and test gratings differ by more than 10°. This is consistent with Hubel and Wiesel's (1968) finding of units in the monkey visual cortex which were sensitive to contours within an orientation range of $\pm 10°$. Parker (1972) found no spatial-frequency tuning of the tilt aftereffect. However, all his gratings were of the same *physical* contrast, with no allowance for the fact that the human visual system is differentially sensitive to different spatial frequencies. Ware and Mitchell (1974b) used gratings of different spatial frequencies which had equal *subjective* contrast, i.e., relative to the subjects' contrast thresholds, and found that the tilt aftereffect was indeed greatest when the adapting and test gratings had the same spatial frequency. This resembled Georgeson's (1973) results for simultaneous tilt-contrast (Figure 1f). The "indirect" orientation aftereffect, in which both vertical and horizontal lines would be apparently rotated clockwise like a wheel after prior adaptation to a line tilted counterclockwise off vertical, might be attributed to adaptation of higher-order hypercomplex cells, which have two preferred orientations 90° apart (Coltheart, 1971).

Adaptation to the relative width of black and white bars in a grating is shown in Figure 1n (de Valois, 1973). Black test bars are shifted in apparent width away from the width of black (but not white) adapting bars, and white test bars are shifted in width away from the width of white (but not black) adapting bars. This suggests that there may be separate mechanisms for the detection of black and white objects, a hypothesis which correlates well with unit recordings showing the existence of on-center (white-excitatory) and off-center (black-excitatory) cells (de Valois, 1973).

Does the visual system adapt to relative spatial *phase* (vernier misalignment) between two gratings? Try it for yourself using Figure 1o. The upper adapting grating is displaced by one half-stripe (90° spatial phase angle) from the lower adapting grating. First, adapt with very steady fixation; next, adapt again with small scanning movements (as in adapting to spatial frequency). The results should be quite different in the two cases. If the eyes fixate the adapting pattern steadily, then the upper test grating does appear to be displaced to the left of the lower. But if the eyes scan the adapting pattern, then no such displacement is observed in the test pattern. This indicates that the absolute retinal position, not the relative phase, of the inspection gratings is adapted. Thus the aftereffect should be described as adaptation to position ("contour repulsion") rather than to phase. This is consistent with the receptive-field interpretation of visual responses to gratings, which is described later.

Figures 1p and 1q show adaptation to the amplitude and spatial frequency respectively of wavy sinusoidal lines. Again, try scanning or fixating during adaptation. It would be nice to think that one was adapting hypothetical tuned "wavyline" detectors. These might underlie the frequency response to wavy lines (Figure 5: Tyler, 1973); aftereffects from curved lines; or even the analysis of curved contours (see Figure 13). But this is very doubtful, since Figure 1p and 1q may be merely scrambled gratings of different spatial frequencies and orientations—a combination of Figure 1j and 1m.

Figure 1t shows an aftereffect of contour repulsion, an example of the "figural aftereffect" (after Köhler & Wallach, 1944). Other aftereffects (not shown) are adaptation to disparity (Blakemore & Julesz, 1971); adaptation to movement—the well-known movement aftereffect; and adaptation to gradual change of intensity (Anstis, 1967). These aftereffects can plausibly be attributed to adaptation of disparity detectors, movement detectors, and on- or off-units.

ADAPTATION DURING VERY STEADY FIXATION

Gazing at sandpaper textures can give two complementary forms of adaptation (in addition to size adaptation). If the eyes scan continuously over a sandpaper texture such as Figure 2a, this texture gradually appears more irregular and "lumpy." Conversely, Mackay (1964) reported that with very steady fixation the texture gradually appears to become more uniform and regular, like a tufted carpet. However, if the eyes are trans-

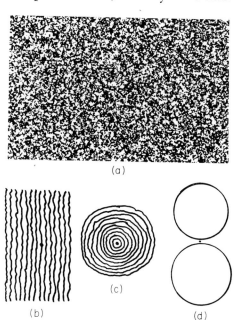

Figure 2. Adaptation with steady fixation. Adapt to each pattern in turn, with very steady fixation for about a minute.

(a) Random dots appear to become more evenly spaced (Mackay, 1964).

(b) Wiggly lines appear to become straighter and more parallel.

(c) Irregular circles become more circular.

(d) The two circles become more alike in size.

Each pattern normalizes toward "average values" as small irregularities adapt out. Any eye movement brings the irregularities back.

(a)

(b)

(c)

(d)

ferred to a fresh part of the texture it reverts to a less uniform, "lumpy" appearance. To see these effects, cover half the texture with gray paper while adapting, then remove it to provide an unadapted comparison test field. Thus when the test and adapting fields are identical, the dot spacings become apparently more uniform, as if the dots gradually repel each others' positions: when they are different, they become less uniform. I have noticed that if for "dot spacing" we substitute "orientation," "curvature," or "size," we find a new set of patterns which show analogous increases in uniformity during steady fixation and decreases in uniformity when the eyes move to a fresh part of the pattern (Figure 2b–d). Lines which are only roughly straight or parallel become markedly more straight or parallel, and the kinks disappear, during prolonged fixation, but markedly less so when the eyes move: roughly drawn concentric circles appear to become more circular and possibly more evenly spaced. Shapes which differ slightly in size get to look more alike. These effects may be related to the well-known normalization and figural aftereffect phenomena which occur during fixation of curved and tilted lines. Perhaps, in all these cases, steady fixation acts like the degraded vision of a stabilized retinal image and gradually loses finely detailed information about local departures from uniformity, so the percept regresses toward a low-information "Gestalt." Only the highly redundant, multiply represented uniformities—the average orientation of a line, or the average dot spacing—are transmitted. In normal, scanning vision, on the other hand, the high-information local variations are preferentially transmitted and it is the redundant average values that tend to be edited out of the neural message. Barlow (1959, 1961) has pointed out that redundancy reduction is a major principle of peripheral sensory coding.

In conclusion, we can distinguish between two kinds of adaptation (Stromeyer, 1969): (1) local adaptation to sharply localized visual properties which requires very steady fixation, e.g., the conventional afterimage, Mackay's (1964) adaptation, and adaptation to contour position in figural aftereffects; (2) global adaptation to visual properties which are spatially distributed, where the adaptation is not affected by scanning eye movements, e.g., spatial frequency and orientation.

CONTINGENT AFTEREFFECTS

Adaptation to spatial frequencies is *dependent* on orientation (Blakemore & Nachmias, 1971), on disparity (Felton, Richards, & Smith, 1972) and on wavelength (Lovegrove & Over, 1973: but Maudarbocus & Ruddock, 1973, disagree). This means that the aftereffect is strongest when the adapting and test fields are matched for orientation, disparity, and wavelength. This psychophysical linking between visual dimensions may reflect selective adaptation of subpopulations of units sensitive to frequency *and* orientation, or frequency *and* color, etc. *Contingent* aftereffects (CAEs) are more selective than dependent aftereffects. They single out adaptation to frequency-*and*-color, etc., but cancel out the adaptation to pure frequency or color. McCollough (1965) found that after adaptation to a vertical grating in orange light alternating with a horizontal grating in blue light, a vertical test grating seen in white light appeared tinged with blue-green, while a horizontal test grating appeared tinged with orange. McCollough attributed this CAE to color adaptation of orientation-specific edge detectors. Fidell (1970) found that the two gratings needed to differ in orientation by at least 11° in order to give a CAE. Held and Shattuck (1971) found a complementary CAE: after adapting to red grating tilted clockwise off vertical, alternating with a green grating tilted counterclockwise off vertical, a vertical test grating would appear to be tilted counterclockwise when red but clockwise when green. As the angle between the adapting and test gratings was increased from 0° to 75°, the magnitude of the color-contingent tilt aftereffect rapidly increased to a peak between 10° and 15°, and then dropped close to zero around 40°. So the angular specificities of the tilt-contingent colored aftereffect and of the color-contingent tilt aftereffects were consistent with estimates of the breadth of tuning for orientation of edge-sensitive channels in the human visual system.

It seems that any visual dimensions that give simple aftereffects when presented singly can give CAEs when presented in opposed pairs. Colored aftereffects contingent on spatial frequency were obtained by Harris (1970); Breitmeyer and Cooper (1972); Stromeyer (1972); and Leppman (1972). After adaptation to a coarse red horizontal grating alternating with a fine green horizontal grating, a coarse black/white test grating appeared tinged with green, and a fine black/white test grating appeared tinged with red. Leppman (1972) found that the spatial frequencies of the two test gratings needed to differ by at least an octave. He also found that the CAE was dependent only on frequency, not on orientation, which he took as evidence against McCollough's adaptation theory and in favor of an associative conditioning theory. Hepler (1968) and Stromeyer and Mansfield (1970) found colored aftereffects contingent on direction of movement, and conversely, Favreau, Emerson, and Corballis (1972) and Mayhew and Anstis (1972) independently reported color-contingent motion aftereffects. Anstis & Harris (in press) have found motion aftereffects contingent on binocular disparity. Colored aftereffects contingent on direction of gaze were reported by Kohler (1964) and by Leppman and Wieland (1966), though McCollough (1965) and Harrington (1965) were unable to replicate these findings. Mayhew (1973) found movement aftereffects contingent on direction of gaze. Wyatt (1974) found two CAEs involving three dimensions. One was a colored aftereffect contingent on both the orientation and spatial frequency of the test pattern. The other was an orientation-contingent colored aftereffect, produced by first generating a frequency-contingent color effect and then generating an orientation-contingent frequency effect. This implies visual channels tuned in three dimensions.

It appears that most pairs of visual dimensions can generate CAEs, and doubtless CAEs along other dimensions await discovery. The existence of CAEs raises a number of theoretical questions: (1) Are CAEs a form of neural adaptation, or of learning? (2) Where are CAEs localized within the nervous system? (3) What, if any, is their function or purpose? Each of these questions will be considered in turn.

1. Are CAEs caused by adaptation of double-duty neural units (McCollough, 1965; Stromeyer, 1969; Held & Shattuck, 1971), or by some kind of associative conditioning (Leppman, 1972; Murch, 1972)? McCollough suggested that if there are edge detectors in the human visual system, some might be wavelength specific. If some cells respond maximally to red vertical lines and others to green vertical lines, then steady viewing of red vertical lines might adapt out or fatigue the red vertical units. White vertical test bars, which normally excite red and green cells equally, would now produce a stronger response from the green vertical units, leading to a green aftereffect. Notice that the adaptation of the "red-and-vertical" units ("red-or-vertical" would not do), and correspondingly of "green-and-horizontal" units, would suffice to give a complementary pair of CAEs—McCollough's colored aftereffect contingent on orientation, and also Held and Shattuck's tilt aftereffect contingent on color, assuming here an orientation difference of less than 40° instead of 90°.

Murch (1972) and Leppman (1972) regard CAEs as being caused by learning or associative conditioning rather than by adaptation. Various arguments have been raised in support of a learning model: CAEs are weaker than simple aftereffects; they build up much more slowly than simple aftereffects and last much longer; and they show a pronounced reminiscence effect, being weaker immediately after the adapting session than half an hour later. Also, color vision remains normal, and CAE colors are only seen when an appropriate test pattern is presented. But the strongest support for a learning model comes from a new form of sensory learning discovered by Ramachandran and Braddick (1973). Inexperienced subjects show a delay of up to a minute before experiencing the stereoscopic percept from a random-dot stereogram. This perception time was progressively reduced with repeated exposures to a series of stereograms composed of tiny needles or line segments tilted 45° to the right. This perceptual learning showed complete transfer to an uncorrelated pattern of needles tilted to the right, but it showed no transfer to a pattern whose needles were tilted 45° to the left. Thus, the stereoscopic skill acquired seemed to be specific to, or contingent on, those orientation analyzers that were stimulated during the training period. It seems likely that some related form of perceptual learning may underlie CAEs.

2. Site of CAEs. Most simple aftereffects have a central location, originating in whole or in part after the point of binocular fusion. Most CAEs may have a more peripheral, prefusion location. Simple aftereffects of movement and of spatial frequency are at least partly postretinal in origin. This has been demonstrated by pressure-blinding the adapting eye, which effectively decouples it temporarily from the brain. Pressure-blinding one eye after adapting it to movement (Barlow & Brindley, 1963) or to a spatial frequency (Blake & Fox, 1972) did not prevent transfer of the aftereffect to the other eye. But pressure-blinding one eye before adapting it prevented the aftereffect from transferring to the other eye. Equal and opposite simple movement aftereffects could be built up in the two eyes simultaneously, and elicited separately by exposing each eye in turn to a stationary test field. This evidence indicates a prefusion component in simple aftereffects. But stroboscopic techniques of sharing the movement out between the two eyes have indicated the existence of a postfusion component also (Papert 1964; Anstis & Moulden 1970). Tyler (in press) demonstrated a post-fusion component in aftereffects of tilt and spatial frequency. He adapted his subjects to a random-dot stereograting similar to Figure 7, with no monocularly visible contours, and found that a briefly viewed test stereograting appeared subjectively shifted in orientation and spatial frequency. On the other hand, Blake & Fox (1974) showed that the spatial frequency aftereffect was peripheral to the neural site of binocular rivalry, since a monocularly inspected adapting grating was still capable of building up an aftereffect, even while the grating was subjectively invisible owing to suppression by a rivaling stimulus presented to the other eye. Mitchell and Ware (1973), Ware and Mitchell (1974a), and Movshon, Blakemore, and Chambers (1972) found that the amount of interocular transfer of a tilt aftereffect varied with the extent of stereopsis, subjects having no stereopsis showing no transfer. These subjects presumably lack any binocularly driven neurons at (or before) the level responsible for the aftereffects.

Whereas the simple spatial-frequency shift does show interocular transfer at about two-thirds of full strength, the McCollough aftereffect shows no transfer (Murch, 1972). It is possible, as with the simple movement aftereffect, to build up equal and opposite McCollough effects or color-contingent motion aftereffects in the two eyes simultaneously.

The nontransfer of CAEs may be due to color-selective, orientation-selective monocular units in the cortex (Hubel & Wiesel, 1968), but not to the separate populations of cells sensitive respectively to color and to orientation which are also found in the cortex (Dow & Gouras, 1973). Coltheart (1973) argues from the fact that colored CAEs show no interocular transfer, that no color-sensitive cortical cells are binocularly driven. He suggests that as one progresses further up the visual system, color-specificity gradually disappears, whilst orientation-specificity appears, and monocularity is converted to binocularity. So McCollough CAEs are almost certainly cortical. It goes without saying that motion aftereffects contingent on binocular disparity (Anstis & Harris, in press) must be post-fusional and hence cortical.

Mackay and Mackay (1973) exposed black–white oblique-left and oblique-right gratings to one eye and colors to the other, the colors alternating with the orientation. The color-adapted eye then showed the usual negative McCollough aftereffect. The pattern-adapted eye unexpectedly showed a positive McCollough aftereffect, with the test gratings tinged with the hue originally associated with the same orientation. These results are very puzzling. Mackay suggests that some transfer of information may take place between left and right eye channels before the stages at which color and form are associated, but whereas orientational information is transferred correctly, color transfer may be antagonistic, as if red light in one eye gives rise to "minus-red" in the other channel.

Contingent aftereffects are contingent on the *relative*, not the absolute, values of the paired dimensions. The McCollough effect has been demonstrated with transpositions of color (Stromeyer, 1969), orientation (McCollough, 1965), and color, intensity, and spatial frequency (Mayhew & Anstis, 1972). For instance, after adaptation to a red pattern rotating clockwise, alternating with a yellow pattern rotating anticlockwise, a yellow stationary test pattern gives an *anticlockwise* movement aftereffect when paired with a stationary green test field which gives a clockwise aftereffect (Mayhew & Anstis, 1972). So the locus of CAEs might come after some data processing which extracts ratios or relative values of color, orientation, and spatial frequency, such as might play a part in size and color constancy. The earliest plausible site for some of these processes is the lateral geniculate nucleus (LGN), but since many are orientation selective, they must occur at or after the visual cortex.

3. The function of CAEs. Simple aftereffects are a response to persistent stimuli, i.e., first-order redundancies. If the continued presence of a stimulus is indicated with progressively fewer nerve impulses, considerable economy of neural transmission is achieved (Barlow, 1961). CAEs are a response to pairwise correlations between stimuli, editing out second-order redundancies or invariances (such as correlations between color and orientation). First-order redundancies occur more frequently than do second-order redundancies, and it is possible that CAEs have rather longer time constants than do simple aftereffects in order to take longer time samples of the stimulus and thereby ensure that the correlations are reliably present.

The adaptation in both simple and contingent aftereffects can be likened to automatic gain control. In simple aftereffects adaptation in each of the many neural channels signaling a particular stimulus dimension, such as spatial frequency, will tend to line up the gains of each channel giving a locally flat overall frequency response (for stimuli with a reasonably flat frequency spectrum). CAEs probably arise in double-feature cortical units which form a transitional stage between scenes on the retina and the arrays of single-feature units in the cortex. It is not known what kind of cortical data processing segregates paired features into single features. Whatever the system, it is likely to be imperfect and to produce unwanted intermodulation or cross talk, in the form of spuriously colored edges, etc.

Such correlation errors could be removed by CAE adaptation, which would thus play a valuable internal housekeeping role in automatically editing out cross talk. Further cross talk is added optically in the form of colored fringes produced on the retina by chromatic aberration in the lens. These, like neural cross talk, could be removed by CAE adaptation. The color of these fringes depends on the polarity (black–white versus white–black) of edges, and Stromeyer (1973) has found a small colored CAE contingent on this edge polarity.

The persistent correlations imposed artificially in experiments on CAEs are presumably indistinguishable from spurious correlations caused by neural transmission errors, and they lead to the same processes of adaptation.

NEUROPHYSIOLOGICAL MODELS OF SIMPLE AFTEREFFECTS

A number of models have been put forward in the past few years to explain psychophysical effects by extrapolation from neurophysiological data (Ganz, 1966; Weisstein, 1969; Uttal, 1971; Coltheart, 1971; Over, 1971). The analogies on which these models are based can be classified under three headings:

1. The known neural interactive *processes* of (a) temporal adaptation, (b) lateral summation, and (c) lateral inhibition are sufficient (but perhaps not necessary) to account respectively for: perceptual adaptation and negative aftereffect, assimilation illusions and the falloff in responsiveness to high spatial frequencies, and simultaneous contrast and the falloff in responsiveness to low spatial frequencies. These neural processes have been observed in physiological responses to brightness and have been assumed, by extrapolation and analogy from psychophysical experiments, to occur in physiological responses to higher-order visual dimensions.

2. The *selective sensitivities* or trigger features of neural units (color, disparity, orientation, etc.) are along the same visual dimensions as the psychophysical effects of adaptation, contrast, etc.

3. Perhaps most impressively, the *ranges of sensitivity* found in neural units are quantitatively very similar to those found in psychophysical studies of masking, critical bands, adaptation, and contingent aftereffects.

Ganz (1966) explained both optical illusions of contour repulsion and figural aftereffects in terms of lateral inhibition between the neural correlates of brightness distributions. Any contour produces a distribution or ridge of neural activity, whose peak determines the seen position. Lateral inhibition from an inducing contour subtracts asymmetrically from the distribution caused by a test line, shifting the peak of the test contour away from the position of the inducing contour. This would explain optical illusions of contour repulsion, in which the test and inducing figures are present simultaneously. For figural aftereffects, the shift in peak location is said to be produced by the afterimage of the inducing contour, which remains after the line itself is removed. Ganz proved that the actual shape of the test distribution is not a critical variable.

The displacement aftereffect is typically maximal not when the inducing and test lines abut, but when they are separated by a few minutes of arc. This is the *distance paradox*. Ganz assumed that lateral inhibition of brightness decreased linearly with distance from the inducing contour, and thus attributed the distance paradox to Gaussian errors in fixation with a standard deviation of 5–6 minutes of arc, which would alter the relative positions of test and inducing contours, and hence the mean and variance of the settings of judged displacements. However, Westheimer (1967) found brightness facilitation within 5 minutes of arc of the point of stimulation and inhibition only at greater distances. Robinson (1972) has remarked that it is a pity that Westheimer's observations were not available to Ganz, since by incorporating them into his theory he would not have needed to invoke eye movements at all. Inhibition would cause a shift of the test contour away from the inducing contour, facilitation a move toward it. Such an explanation accommodates both the distance paradox and the attraction effects which occur at very small interfigural distances.

Coltheart (1971) summarized the physiological evidence for the existence of neural channels selectively sensitive to different overlapping ranges of edge orientation and explained the tilt aftereffect on the basis of selective adaptation of these tuned neural units. For any given stimulus edge, the neural channel whose preferred orientation is the same as that of the stimulus will respond maximally, and other channels progressively less so as their preferred orientation differs from that of the stimulus. The perceived orientation of the stimulus is assumed to correspond to the central tendency of the population of orientation-sensitive channels. Prolonged exposure to the edge(s) or bar(s) of an adapting stimulus at one particular orientation reduces the sensitivity of excited channels. When a test stimulus is then viewed, its central tendency will be shifted away from that of the adapting pattern, and thus its orientation will appear tilted away in the same manner.

Figure 3 shows how this might happen. For simplicity, I assume a triangular shape for the "tuning curve" of each orientation-sensitive channel. It is also assumed that the more a channel responds to an adapting edge of a particular orientation, the more it becomes adapted by it. Thus the adapting stimulus maximally depresses the channel which is most sensitive to it and depresses other channels proportionately less. Figures 3a and 3b show the response of *individual* channels to a range of stimulus orientations, before and after adaptation. Figure 3c is a replot of Figure 3b to show the response of adapted ranges or *populations* of channels to individual orientations, i.e., the patterns of activity that would be set up among the population of channels by particular stimulus orientations. Adaptation makes each population curve asymmetrical, shifting its median, or central tendency, away from the adapting stimulus. The shifts in central tendency, which I calculated graphically, are plotted as a function of test stimulus orientation in Figure 3d. Note the "distance paradox": the maximum shift in orientation occurs not at the adapting orientation, but about a quarter of a tuning-curve bandwidth away. This S-shaped shift curve resembles those found experimentally for the tilt aftereffect (Ware & Mitchell, 1973, Movshon, Chambers, & Blakemore, 1972); color-contingent tilt aftereffect (Held & Shattuck, 1971); figural aftereffect (see Ganz's review, 1966); and spatial frequency shift (Blakemore & Sutton, 1969).

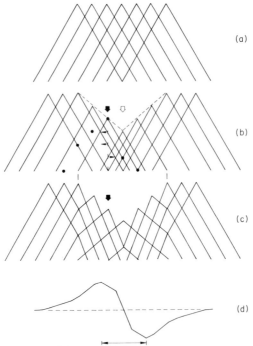

Figure 3. Tuning curves model.

(a) Schematic diagram of hypothetical channels tuned to orientations (or spatial frequencies, etc.). Tuning curves are arbitrarily assumed to be ∧-shaped.

(b) Adapting orientation (white arrow) produces a V-shaped depression in the population of tuning curves. The depth of depression is determined by exposure time, its width and shape by the width and shape of the individual tuning curves. Subsequent response of adapted population to a test orientation (black arrow) is shown by spots. The horizontal position of the spot indicates the channel to which it refers; the height of the spot indicates adapted sensitivity of that channel.

(c) Adapted *population* curves, as shown by spots in b, are plotted for various test orientations. (Note that these are not tuning curves of individual channels.) Adaptation has skewed the curves, shifting their central tendency, hence (it is argued) the apparent orientation of the test stimuli.

(d) Central tendency shifts ("tilt aftereffect") plotted as a function of test orientation. Notice the "distance paradox."

Orientation detectors may have special characteristics near the horizontal and vertical, since these orientations can be judged more accurately than oblique orientations, and oblique lines are judged as being nearer to the vertical or horizontal (whichever is nearer) than they really are (Bouma & Andriessen, 1968). The deviation is typically 1°–3°, but it can reach values as high as 10°. To explain these findings and others like them it has been suggested that the tuning curves for vertical and horizontal orientations are narrower (Andrews, 1965), more closely spaced (Bouma & Andriessen, 1968), or more sensitive (Bouma & Andriessen, 1970), than for oblique orientations.

In some respects, the theories of Ganz and of Coltheart are isomorphic. Both theories assume that some gradient of laterally spreading suppression shifts the peak of a distribution of excitation caused by the test figure, and that the position of this peak or central tendency determines the perceived position of the test contour. The gradient of desensitization is caused in Ganz's theory by lateral inhibition, and in Coltheart's theory by prior adaptation at the site of the inducing contour. For Ganz the distribution is of the neural correlate of brightness, point-by-point; for Coltheart it is the level of excitation within separate, parallel tuned channels or neural units, each of which has its own tuning curve. Tuned-unit or feature-detector theories such as that of Coltheart have commanded more support than Ganz's afterimage theory. This type of theory can be applied rather generally to most perceptual aftereffects, by substituting the words "spatial frequency," "movement," and so on in place of "orientation." Over (1971) has in effect done this.

Erickson (1968) distinguishes between two sensory coding principles. He argues that some stimulus properties, such as retinal location, are coded topographically in the nervous system, while others are coded nontopographically. In the former case, each neuron is responsive to a restricted range of the stimulus dimension, and the excitatory ranges of adjacent neurons overlap to only a limited degree. With other dimensions, such as wavelength (de Valois, 1965) or orientation (Hubel & Wiesel, 1968), coding is nontopographic with neurons broadly tuned such that each unit can signal most values along the dimension by variations in its discharge frequency. Topographic coding implies that the particular value of the stimulus dimension is signaled by *which* of a spatially arranged set of neurons are excited; when coding is nontopographic, different stimulus values are signaled by different *patterns* of neural activity within the total set of neurons sensitive to the relevant dimension.

Over (1971) explains all aftereffects in terms of adaptation and interactions between tuned neural units, but he attributes negative aftereffects to the adaptation of nontopographically coding units and figural aftereffects to the adaptation of topographically coding units. He raises (but does not answer) the difficulty that figural aftereffects *can* occur only if some neural units are involved in signaling both test and adapting contours (i.e., are more or less broadly tuned.) However, if this condition is met, as would seem essential for the appearance of figural aftereffects, the coding can hardly be considered to be topographic since information as to *which* neurons are responding would be insufficient to allow discrimination of test and inducing contours. The tuned-unit theory could be applied equally well to figural aftereffects if in fact position were coded nontopographically by broadly tuned units. There is evidence that this may be so. Consider the anatomy of the eye: for about 100 million rods and 7 million cones there are only 1 million ganglion cells with fibers going up the optic nerve. So there is at least a 7:1 ratio of convergence from cones to ganglion cells. Now, visual acuity as a function of retinal position is very roughly what one would expect from the number and spacing of the cones (Polyak, 1941). It follows that it is much higher than would be predicted from the number of ganglion cells. How is the high resolution retinal information transmitted up the optic nerve? The most likely answer is that retinal position is recoded nontopographically into the relative firing rates of ganglion cells reporting events in their receptive fields. The overlapping ganglion cell receptive fields are like the intersecting Venn circles in set theory, and, as n intersecting zones can specify up to 2^n different locations, position information can be transmitted both economically and precisely. If this speculation proves to be true, then figural aftereffects could be interpreted as resulting from adaptation of channels tuned to overlapping ranges of retinal position.

Uttal (1971) sounds a note of caution. He predicts that psychological theories based on current physiological technology will go the way of earlier psychological models based on the technology of hydraulic pumps, telephone switchboards, or even computers. He considers a number of visual effects which cannot be explained on the basis of simple neurological nets. For instance, the operations of a simple feature-filtering network can only be reflected in molar behavior if the feature is actually present in the stimulus. This is a "syntactical" axiom, since it deals with the specific geometrical (grammatical) position and structure of the parts of the pattern. The meaning or significance of a pattern (semantic content) should not be an effective variable. However, Uttal describes experiments, in which a part of the stimulus is actually missing, which suggest that some kinds of form perception may be influenced more by "closure" and "filling in." Examples are Leeper's (1935) study of fractured figures, where the arrangement of the parts of the figure suddenly becomes clear on the basis of clues which must be operating at a higher, semantic level, not at the syntactic level of feature detectors. Uttal also describes his own work on the perception of dotted letters in masking random-dot noise. The selection, from isolated dot stimuli, of a statistically connected set which form a letter, suggests higher-order decision processes of which no simple feature-filters would be capable.

But Kabrisky, Tallman, Day, and Radoy (1970) found that dotted letters form connected, blurred letters when they are put through a low-pass spatial filter. Perhaps feature-filters could operate on these blurred letters.

Uttal's comments do not invalidate perceptual theories based on selectively tuned neural units. But they do call attention to many perceptual phenomena which cannot be well explained by known neural mechanisms. Uttal suggests that perceptual theories should be based upon the statistical behavior of large populations of neural units, not of individual units, rather as a physicist might predict the behavior of a gas by using the external metrics of pressure, volume, and temperature—statistical estimates of the central tendency of individual particles.

SPATIAL FREQUENCY RESPONSE:
LUMINANCE GRATINGS

The transmission of spatial information through the visual system has been studied for about 25 years by measuring contrast thresholds to sinusoidal grating patterns. Excellent reviews are available by Campbell (1969, and in press) and Cornsweet (1970). The method was borrowed from electro-optical engineering, where it was developed for describing the performance of passive optical systems (Schade, 1956). A sinusoidal grating is a field of stripes whose intensity profile is a sine wave. Sensitivity is defined as the reciprocal of threshold contrast, i.e., the depth of modulation (max — min/max + min) at which the grating bars are just detectable. It is plotted as a function of spatial frequency in cycles per degree and a typical curve is shown in Figure 4 (from Campbell & Robson, 1968). Sensitivity to this pattern was greatest in the spatial frequency band from 2 to 5 cycles per degree falling off rapidly at the higher and lower frequencies on either side of this broad maximum. Similar curves have been published by Kelly (1973) and by Van Nes and Bouman (1967). These effects can be seen in Figure 5, which is a photograph of a sinusoidal grating with the spatial frequency swept from low at the left to high on the right, and the contrast swept from high at the bottom (high visibility) to low at the top (low visibility). The curve that separates the lower region, where the grating is visible, from the upper region, where the grating is below threshold, is in effect the same as Campbell and Robson's threshold visibility curve in Figure 4. The reader can test for himself the effects of viewing distance, illumination, and so on.

Thus the visual system acts as a band-pass filter for spatial frequencies. It is the spatial analogue of a hi-fi amplifier, whose sensitivity is limited

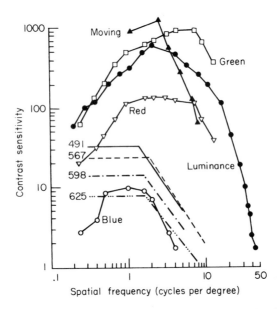

Figure 4. Graph of various MTFs (modulation transfer functions). Spatial frequency responses to gratings: ● = luminance grating (Campbell & Robson, 1968); ▲ = moving luminance grating (calculated from Van Nes's data, 1965): enhanced visibility at low frequencies suggests that movement reduces lateral inhibition.

Contribution of ▽ red-, ☐ green-, and ○ blue-sensitive spatial responses to seeing monochromatic luminance grating (Kelly, 1973, Figure 12). Results suggest lateral inhibition within and between red and green mechanisms in seeing brightness.

Sensitivity to equiluminous, hue-modulated gratings of mean wavelength: —491, —567, —598, and ···—···—··· 625 nm (Van der Horst, de Weert, and Bouman, 1967). Low sensitivity shows poor spatial acuity for color, and low-pass characteristic shows no lateral inhibition between color mechanisms, in seeing color.

at high frequencies by temporal integration from stray capacitance in parallel with the signal—an unavoidable weakness in design—and at low frequencies by temporal differentiation from coupling capacitors—a deliberate design feature to avoid gradual dc drift. In the same way the visual spatial sensitivity is limited at high frequencies by unavoidable optical integration, and at low frequencies by neural differentiation of lateral inhibition, which serves useful purposes.

Visual sensitivity to high frequencies is limited by area summation which is probably optical rather than neural in origin (Campbell & Gubisch, 1966). All passive optical systems, such as a lens or a photographic emulsion, show high-frequency falloff. But, the decrease in sensitivity at frequencies below 2 cycles per degree cannot be produced by any kind of passive incoherent optical process such as refraction or diffusion (Schade, 1956). Two explanations have been put forward for

Figure 5. Frequency-swept luminance grating (Campbell, 1968). The spatial frequency increases from left to right, and the amplitude or contrast increases from top to bottom. The stripes of about 5 cycles per degree are visible along more of their length than finer or coarser stripes, which are more difficult to see and look progressively shorter. The inverted-V-shaped border between the zone of visible stripes at the bottom and of invisible at the top is the same shape as the sensitivity curve in Figure 4 from Campbell and Robson (1968).

this low-frequency attenuation. Kelly (1973) has attributed it to lateral inhibition in the nervous system: this hypothesis is discussed later. But Hoekstra, van Goot, van Brink, and Bilsen (1974) believe that it is simply a methodological artifact. They note that most experimenters have used displays of limited size, which give very few cycles of gratings with a low spatial frequency. When they used a large display which gave enough cycles, they found that the frequency response curves of their subjects remained flat, with no sign of low-frequency fall-off, certainly down to .5 Hz, which was the limit of their apparatus. So the visual system behaves like a passive optical system. This paper poses a sharp challenge to existing theories.

The visual system's poor sensitivity for low spatial frequencies, however caused, accounts for the fact that blurred edges are hard to see and often disappear during steady fixation. This is the O'Brien–Craik–Cornsweet effect (see O'Brien, 1958; Cornsweet, 1970). Wundt's ring may be an example of the O'Brien–Craik–Cornsweet effect (Rogers, personal communication). A neutral gray ring, seen against a half-white, half-black background, shows little brightness contrast. The reason may be that the gradient of apparent brightness between the two halves of the ring, induced by simultaneous contrast, is too shallow to be seen, just as an O'Brien physical luminance gradient is hard to see. I have found that if the physical intensity difference of the black and white backgrounds is increased, by making the white background of brilliantly rear-illuminated ground glass, visibly different apparent brightnesses are induced in the two ring halves, joined by a *blurred* subjective contour where the ring crosses the junction between black and white grounds. (The difference between the two halves is still visible in an afterimage of the ring, which shows that eye movements are not important.) So there is no need to invoke what one textbook writer has called "Gestalt principles of cohesive forces exerted on all the part-processes of a good figure by the whole" (Osgood, 1953) in order to explain Wundt's ring.

Lateral-Inhibition and Spatial Frequency Response

Several patterns of neural connectivity would yield frequency response curves. The most likely is the concentric receptive field with on-center and off-surround, or vice versa (Patel, 1966; Thomas, 1970). According to this hypothesis, the geometry of the receptive field determines the spatial frequency response: its center determines the high-frequency response while both the center and its antagonistic surround control the low-frequency response (Kelly, 1973). The sensitivity to high frequencies is limited by spatial integration over small areas in relatively independent pathways, whereas sensitivity to low frequencies is limited by spatial differentiation caused by lateral inhibition between adjacent areas. Mathematically, integrating a sine wave halves its value each time its frequency is doubled, so the amplitude of an integrated sine wave falls off with a slope of -1 (-6 dB per octave) at high frequencies. Differentiating a sine wave halves its value for every halving of its frequency, so its amplitude falls off with a slope of 1 at low frequencies.

A receptive field can be simulated with a small disk-shaped photocell mounted concentrically in front of a larger photocell, and the two wired up so that the "surround" output voltage is subtracted from the "center" voltage. The edge response of this model was obtained by holding a stationary vertical edge at different distances from the field's center (Figure 6a). Its frequency response was obtained by scanning the photocells with slides of different gratings projected onto them (Figure 6b). The response was greatest when the stripes in the grating were the same width as the field center, and it fell off both for wider and for narrower stripes, giving a band-pass response about one octave wide which was symmetrical on log coordinates. However, if the surround "inhibition" was not equal to the center "excitation" sensitivity, the low-frequency response was enhanced asymmetrically, and the model gave a nonzero response to uniform diffuse light, which has zero spatial frequency. This asymmetrical curve (dotted line in Figure 6b) looked rather like the frequency response curves for retinal ganglion cells in the cat (Enroth-Cugell & Robson, 1966: Maffei & Fiorentini, 1973). The edge response can be predicted from the frequency response, both for the model and for ganglion cells (Enroth-Cugell & Robson, 1966). The symmetrical

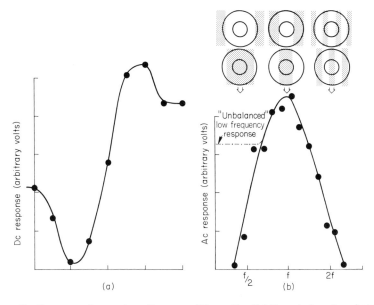

Figure 6. Response of on-center, off-surround "receptive field" made from two photocells. (a) Response to stationary edge held at different distance from receptive-field center. (b) Spatial frequency response. Half-amplitude half-bandwidth was one octave. Insets show that the response was greatest when one stripe of granting stimulus just covered center of receptive field. Response fell off for broader or narrower stripes. Dotted line indicates low-pass characteristic found if the surround inhibition did not exactly balance the center excitation. The unbalanced unit gave nonzero response to zero spatial frequency (uniform diffused light).

curve (solid line in Figure 6b) resembles the "tuning curves" for spatial frequency selective channels established psychophysically (Blakemore & Campbell, 1969).

Receptive fields can also be simulated on man's earliest computer—the abacus or bead frame (Figure 7). The firing rate of a ganglion cell is simulated by the number of beads in a column. Excitation adds beads, and inhibition removes them, and the receptive field profile is expressed as a set of rules for adding and subtracting beads to the columns next to the stimulated columns. Receptive fields are multiplied point-by-point (convoluted) into stimulus profiles to give the overall response. Receptive fields with lateral inhibition (Figures 7a and b) sharpen up stimulus contours; lateral summation (Figure 7c) would blur contours.

In summary, receptive fields with lateral summation and inhibition can act as spatial frequency filters. Some authors attribute human spatial frequency responses to populations of filters with overlapping tuning curves, probably about an octave wide. Other authors (e.g., Kelly, 1973) believe there are a few broadly tuned filters, or perhaps only one.

About 15 years ago, some radio astronomers measured the frequency response of the moon's surface (Smith, 1960). They bounced radar waves of different frequencies off the moon, measured the reflections, and deduced from this that the moon's surface must be covered with a very thick layer of fine dust. No fault was ever found with either their evidence or their reasoning. A few years later, the first astronauts landed on the moon and examined its physiognomy directly. Contrary to prediction, they were not swallowed up without trace by an ocean of dust when they landed. This was bad news for the theorists, though not for the astronauts. This true story enjoins caution in accepting deductions from frequency response curves.

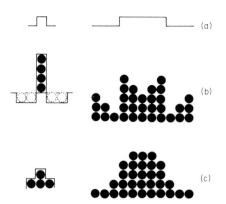

Figure 7. Abacus simulation of lateral inhibition. Each bead column represents a ganglion cell: number of beads gives its firing rate.

Point-spread function, here equal to receptive field, is shown on left. For example, in 7b, the rule is, increase the firing rate to the stimulated column by adding four beads and inhibit each next-door and next-but-one neighbor by subtracting one bead from each. Response to stimulus spot, five columns wide, is shown on the right.

(b) On-center, off-surround gives contour sharpening and Mach bands.

(c) Spatial summation gives blurring of stimulus.

We now turn to studies of frequency responses to gratings composed of colored stripes or of stripes of disparate or moving dots. These gratings, like luminance gratings, reveal band-pass frequency responses. This suggests the existence of lateral summation and inhibition between channels tuned to color, movement, and binocular disparity.

Colored Gratings

The existence of individual cones maximally sensitive to red, green, and blue was convincingly shown by Marks, Dobelle, and MacNichol (1964). The responses of the individual cones are probably transformed by various forms of data processing in the retina into one brightness signal $(R + G + B)$ and two chromatic signals $(|R - G|$ and $|B - Y|)$. De Valois (1965) has demonstrated this by recording from the LGN of the macaque. He found opponent color units which were excited by red light and inhibited by green $(+R - G)$ or vice versa $(-R + G)$, and similar opponent units for blue and yellow $(+B - Y$ and $-B + Y)$. He also found brightness-sensitive units with a broad-band response extending across the visible spectrum $(+R + G + B)$. De Valois found that these units showed almost exactly the same responses to color as does human color vision measured psychophysically: human color-naming data, discrimination of hue and saturation, and the apparent variation in hue produced by variations in intensity (the Bezold-Brücke phenomenon), all correspond to changes in firing rate in macaque units. De Valois concluded that the physiological mechanisms underlying color vision must be very similar in man and macaque.

The spatial interactions in the human color channels are quite different from those in the brightness channels, showing a much lower spatial acuity. Van der Horst, de Weert, and Bouman (1967) measured just-noticeable color differences in equiluminous gratings which were modulated only in hue. The frequency response curves they obtained for four different references wavelengths are shown in Figure 4. Unlike results from luminance gratings, they show no attenuation at all at low spatial frequencies. Since low-frequency attenuation is presumably caused by lateral inhibition, Van der Horst et al. conclude that no such inhibition is found within each color channel. This inference is confirmed by the reported absence of perceived Mach-color bands in equiluminous gradients of varying color, as reviewed by Green and Fast (1971).

Unlike luminance gratings, low-frequency colored gratings never become more visible when they are moving (Van der Horst & Bouman, 1969). This is not unexpected: if movement improves the visibility of low-frequency luminance gratings by interfering with lateral inhibitory processes, there will be no such improvement for the color channels which lack lateral inhibition.

Cavonius and Schumacher (1966) measured visual acuity for equiluminous colored stripe patterns. They found color acuity up to 1.3 minutes^{-1}. This value is the same as for visual acuity with differences in brightness. However, Van der Horst *et al.* (1967) point out that these results at high spatial frequencies are probably contaminated by luminance gratings induced by diffraction by the pupil.

These findings imply that the spatial organization of the color channels is quite different from that of the brightness channel. The color channels resemble low-pass filters, falling off above 1 cycle per degree, and therefore showing poor acuity for fine detail. Van der Horst (1969) estimated integration widths of 10 minutes for the R — G and of 25 minutes for the B — Y channels.

Kelly (1973) obtained very different frequency response curves for colored luminance gratings viewed under conditions of intense chromatic adaptation. With adapting colors that tended to isolate the red-, green-, and blue-sensitive spatial responses, all three results differed markedly from each other and from the neutral grating sensitivity. The green mechanism was much more sensitive than the others, the blue much less. The curves are shown in Figure 4; they look nothing like Van der Horst's results. Also in contrast to Van der Horst's findings, Kelly found that contrast sensitivity was reduced at low frequencies, which is typically what lateral inhibition causes. He concluded that there was spatial inhibition both within and between the red and green mechanisms.

Why the differences in results? The answer probably is that while Kelly measured luminance discrimination with no scope for color-opponent responses, since his gratings were the same color all over, Van der Horst measured hue discrimination with no scope for luminance discrimination, since his gratings were the same luminance all over.

Kelly's colored luminance gratings were sinusoidal light–dark gratings on an oscilloscope viewed through red, green, or blue Wratten filters. An intense uniform adapting field of complementary color (e.g., bluish green for a red filter) was superimposed so that ideally only the red receptors saw the grating. Subjects adjusted the *luminance* contrast of the grating to threshold. Van der Horst *et al.*'s equiluminous colored gratings were alternate stripes, of say red and orange, matched for luminance. Subjects adjusted the *hue* of one set of stripes to be just noticeably different from the other.

Presumably, Kelly's results show the contributions of the red, green, and blue receptors to the brightness (R + G + B) channel. The brightness is some kind of weighted sum of the R, G, and B responses with no *color opposition* process. Kelly's low-frequency attenuation indicates lateral inhibition, i.e., *spatial opposition,* between the output of the R, G, and B receptors both before and after they are pooled to give an overall brightness response. Van der Horst's results show the contribution of the red, green, and blue receptors to the color channels ($|R - G|$ and $|B - Y|$), which are based on color opposition. The lack of low-frequency attentuation in these results indicates the absence of spatial opposition in the color-opponent channels.

The lack of low-frequency attenuation in these results led Van der Horst to conclude that there is no spatial opposition in the color-opponent channels.

So the colored-grating experiments suggest that brightness channels are spatially opponent but not color opponent. At first it would seem logical to add that color channels must be color opponent but not spatially opponent. However, Ingling and Drum (1973) dispute this conclusion. Their calculations lead to the surprising conclusion that cells with an R+ center and a G— surround (or with a G+ center and an R— surround) would *not* account for Van der Horst's grating results and would *not* generate colored Mach bands (for details, see Ingling and Drum's paper). Thus the results from colored gratings are consistent with the existence of R+ center, G— surround cells, and vice versa, which I shall call "single-color opponent cells," but not with R + G— centers and R — G+ surrounds, and vice versa (double-opponent cells). Physiologists, unfortunately, have found all kinds of color cells—single-opponent, double-opponent, and nonopponent. This does not help to clarify matters.

In the monkey LGN, Wiesel and Hubel (1966) found that more than 75% of cells showed both color and spatial specificity (single-opponent cells). An R+ center, G— surround transmitted information about both color and size or position of a spot. The information was thus ambiguous, carrying the message "either long wavelength or small spot, or both." These units are neither spatially opponent nor color-opponent, but a bit of both. Daw (1973) reviewed many studies reporting double-opponent cells in various species, and it is hard to reconcile these with the colored-grating results or with the reported absence of colored Mach bands.

Dow and Gouras (1973) found that in the monkey's foveal striate cortex, spatially opponent cells specifying orientation and color-opponent cells specifying color involve two quite different populations carrying out parallel processes. The spatially opponent cells had small receptive fields, often less than $\frac{1}{16}^\circ$ across, whereas the color-opponent cells had fairly large receptive fields ($\frac{1}{3}$ to 1°). The red and green receptive fields of these units were coextensive, without inhibitory surrounds to contribute to spatial sharpening.

Walls (1954) found that colors could spread out or "fill in" over short distances, and Dow and Gouras (1973) suggested that color-opponent cells may be involved, since a small stimulus anywhere within their large receptive fields could activate such cells and create the impression of color over a larger area. The fact that human acuity for equiluminous colored patterns is much lower than acuity for patterns made up of differences in intensity has been used by television engineers. In color TV, the color information is presented with only half the bandwidth devoted to brightness, without the observer noticing. The chrominance and luminance signals are transmitted in parallel over separate frequency bands and are combined in the receiver. With the separate control of the black/white and color components of the picture that this independent transmission allows, the contrast of the color image as well as its saturation and hue can be controlled at every stage of the duplication process (Patterson, 1972). So a videotape image can go through as many as 10 generations before the loss of quality becomes appreciable, whereas a film would undergo quite unacceptable color distortion after only a few copies. The reason is that parallel processing prevents undesirable cross talk between the brightness and color information. The primate visual system separates out color information from brightness information (De Valois, 1965) very much as a TV system does, but it also separates out other visual dimensions, such as binocular disparity and movement. Such isolation, as demonstrated by the existence of specific neural detectors tuned to different visual features, may also serve to reduce intermodulation noise as the neural message undergoes multiple synaptic transmission.

Rushton (1961) proposed that all-or-none spike train coding in individual neurons is also a solution to the problems of distortion created by multiple transmissions. He suggested that the membrane time constant, determined by the high membrane capacitance and the high resistance of neural tissue (1 megohm per millimeter) would make the signal decay rapidly in a long length of nerve and that it must therefore be amplified repeatedly. In myelinated tissue, the nodes of Ranvier could act as booster amplifiers. Because there are hundreds of nodes between a receptor and the brain, even a small error in amplification would lead to huge errors, ruling out the possibility of using amplitude modulation to code information. Pulse code modulation is immune to this form of amplitude distortion, since the information is carried by the timing or frequency of the spikes, not by their amplitude. So, at the level of the single neuron, pulse code modulation is an elegant solution to the problems of amplitude distortion posed by transmission along the neuron. At a higher level of organization, the parallel processing of separate visual properties by specialized property-detectors may be a solution to the analogous problem of intermodulation distortion posed by multiple synaptic transmissions. If this suggestion is correct, it is also rather boring: it would mean that feature detectors are merely sophisticated forms of antinoise circuits.

The advantages of color vision may seem rather slight, given the way it must complicate neural circuits. (In fact color blindness hampers an individual so little that it was not even discovered until the eighteenth century.) But it may speed up visual processing because color is more invariant than brightness. Brightness varies not only with the reflectivity of objects, but also with the distance from, angle to, and brightness of, illumination sources (Yachida & Tsuji, 1971). Pattern recognition based on brightness would probably require the extraction of other invariants, such as brightness ratios between different parts of the visual field. Normalized color, however, is constant so long as the color temperature of the source remains the same. This allows for fast processing of visual information. Yachida and Tsuji (1971) have coupled a computer to a color TV camera and used it for automatic recognition of colored objects.

Disparity Gratings

Spatial responses to disparity can be measured with the aid of specially made gratings. Figure 8 shows a frequency-swept disparity grating (supplied by C. W. Tyler). It is a random-dot stereogram, which viewed stereoscopically looks like a piece of iron corrugated in depth. This disparity grating is the analogue of Campbell and Robson's frequency-swept luminance grating of Figure 5, with amplitude of depth replacing luminance contrast. (The corrugations run horizontally to prevent the valleys from being hidden by the crests.) The spatial frequency of the depth is swept from coarse (low spatial frequency) corrugations at the top to fine (high spatial frequency) corrugations at the bottom. The amplitude of depth is swept from deep corrugations on the left to shallow ones on the right.

Inspection of the fused disparity grating will show that the sinusoidal modulations of depth become invisible in the right half of the figure where the amplitude is low. The general shape of the threshold envelope curve, that separates the portions of the grating where depth is visible from the portions where it is invisible, is plotted in Figure 9. It has the same kind of band-pass characteristics as does the luminance grating,

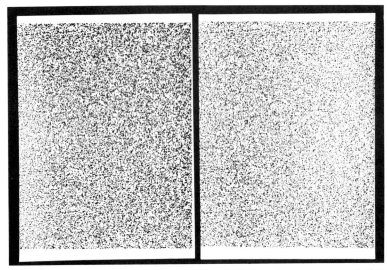

Figure 8. Frequency-swept disparity grating. This is the stereo analogue of Figure 5. Viewed stereoscopically, it looks like a surface which is corrugated in depth. The corrugations run horizontally (in Figure 5, the stripes run vertically). Frequency is swept from top to bottom, with coarse, low-frequency corrugations at the top, and narrow, high-frequency corrugations at the bottom. Amplitude or depth runs from right to left, with deep corrugations at the left tapering to an almost flat surface at the right. The V-shaped border between visible corrugations at the left and invisible at the right is the same shape as the disparity-grating curve in Figure 9. (Tyler, 1974)

but it is shifted about half a log unit down the spatial frequency scale. Whereas maximum visibility of a luminance grating is 2–5 cycles per degree, for the disparity grating it is about 1 cycle per degree. The shape of the curve is fairly insensitive to the ratio of black to white dots (Tyler, personal communication).

Spatial frequency response curves to luminance gratings tell one nothing about absolute thresholds for luminance detection, but rather tell one about just noticeable differences as a function of stimulus geometry, based on lateral interactions between luminance detectors after luminance has been detected. In the same way, data on disparity gratings, in which disparity is the analogue of luminance, tell one little about absolute sensitivity to disparity. This is rather poor in the grating task, reaching a maximum of only 1 minute of arc—an order of magnitude worse than conventional stereo acuity. Perhaps conventional stereo targets are more optimal stimuli than random-dot stereograms. Responses to disparity gratings do provide information as to lateral interactions between disparity detectors. Spatial frequency responses to luminance have been attributed to lateral summation and inhibition in receptive

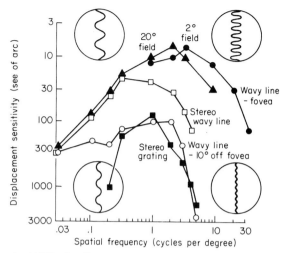

Figure 9. Graph of MTF's for disparity, etc. Spatial frequency response to sinusoidal wavy lines ▲● in fovea, ○10° off the fovea. Inset circles show the stimulus at the corresponding point in the graphical space. ▢, Stereo acuity for vertical lines waved in depth. (One eye saw wavy vertical line, the other a straight vertical line.) ■, Stereo acuity for the random-dot disparity grating shown in Figure 9. (Tyler, 1973 a, b, and unpublished)

fields whose inputs are the luminance of local points. The spatial frequency response to disparity gratings can be analogously explained by positing units with summatory receptive fields whose inputs are the outputs from disparity units. These lateral interactions would run across the frontoparallel plane. The low spatial frequency of 1 cycle per degree at which the disparity sensitivity reaches its peak indicates a rather poor ability to localize whereabouts in the frontoparallel plane a threshold depth stimulus lies, and this suggests that the receptive fields involved in disparity interactions may be rather large.

Tyler's stereo gratings lie in the frontoparallel plane. Since stereo-grating sensitivity shows falloff at high frequencies but not at low frequencies, there is probably lateral summation but not lateral inhibition between disparity units sensitive to different visual directions. On the other hand, physiologists report mutual inhibition between units sensitive to slightly different amounts of disparity (depth) in the same visual direction.

Cortical units sensitive to disparity have been found in the cat by Barlow, Blakemore, and Pettigrew (1966) and by Nikara, Bishop, and Pettigrew (1968). The receptive fields of these units can be plotted in conventional ways by stimulating each eye in turn. For instance, Bishop (1970) found simple units with a small excitatory zone, say .6° in diameter, partly enclosed in much larger inhibitory zones of about ten times the diameter—say 6°. No transient response took place in the surround either when a light was turned on or when it was turned off: the surround was purely inhibitory, whatever the stimulus. This inhibition was *within* a disparity unit, whose receptive field was the same shape for stimulation of either eye, but did not lie on corresponding points on the two retinas. The difference in position of the receptive field in each eye was the receptive-field disparity. Bishop (1970) also found inhibition *between* disparity units, of a kind consistent with the psychophysical grating responses. He found that a simple binocular unit that responded to a vertical moving edge presented monocularly gave a binocular response which was much greater than the sum of the monocular responses when the two edges were exactly at the unit's preferred receptive-field disparity. For positions on either side of this exact retinal correspondence the responses from the two eyes were virtually abolished by inhibition. A characteristic feature was the very sharp transition from complete facilitation to complete inhibition for very small shifts (10–30 minutes of arc) in binocular receptive-field alignment.

Moving Gratings

The visibility of low spatial frequencies is enhanced when they are moving (Figure 4) (Van Nes, 1968; Arend, 1973) or when they are exposed briefly (Kelly, 1973). Both movement and brief exposure probably interfere with the lateral inhibition that normally suppresses the low-frequency responses, so the enhancement of visibility in these conditions has nothing to do with movement perception per se. But Tyler (personal communication) exposed the random-dot patterns of Figure 8 alternately in a tachistoscope, which produced phi movement. An oscillating grating pattern was seen which was most visible at about 1 cycle per degree, as with the disparity grating which is seen when the same random-dot patterns are viewed stereoscopically. The oscillating grating does not exist in either random-dot pattern on its own, it exists only as a correlation between the two patterns. So here the perception of movement precedes that of form (as argued by Anstis, 1970), and there is no question of a stationary grating which becomes more visible when it moves, since there is no stationary grating.

The spatial frequency response to this movement-created grating is analogous to the response to a luminance grating. It is probably determined by the size of the receptive fields of movement detectors and by lateral interactions between them. Richards (1971) estimated the size of the receptive fields of movement detectors in the human visual system, using a psychophysical technique modified from Westheimer

(1967). Instead of measuring the detection threshold of a small spot as a function of the diameter of a veiling disk, he measured the detectability of a movement aftereffect on a small ($\frac{1}{3}°$) stationary test field as a function of the diameter of a rotating spiral adapting field. The strength of the movement aftereffect was assessed by finding the minimal contrast of the adapting field that would just induce an aftereffect. He found that the most effective adapting spiral had a diameter of 3°–6° in the fovea, ranging up to 20° in the periphery. The strength of the movement aftereffects increased up to these diameters, and then fell for still larger diameters. This gives an estimate of the size of the zones over which the adapting effect of movement summates; it is about twenty times the diameter of receptive fields underlying the response to sine wave gratings and to incremental thresholds. An eye that contained only movement detectors would therefore have an acuity of only about $20/400$, and this is the acuity of some amblyopes. Perhaps amblyopic vision is based on the responses of movement detectors (Richards, 1971). These diameters seem very large, but they are comparable to the size of receptive fields found for movement detectors in the rabbit retina (Barlow & Levick, 1965) and in many other species (see Grüsser & Grüsser-Cornehls, 1973, for a review).

Wavy Lines

Tyler (1973) measured the acuity for sinusoidal curvature of wavy lines as a function of spatial frequency of the waves (Figure 9). The smallest detectable peak-to-peak displacement of the wavy line was about 5 seconds of arc, which is of the same order of magnitude as vernier acuity. Sensitivity was maximal at about 3 cycles per degree and fell off for both higher and lower frequencies of curvature. The maximal orientation difference, or departure from straightness within a just-detectable wave was 20–30 minutes. Sensitivity measured 10° from the fovea was about a log unit lower. It was flat from .03 to 3 cycles per degree, then fell off very sharply at higher frequencies. The low-frequency falloff might be explained by lateral inhibition between orientation detectors, which has been observed physiologically (Blakemore & Tobin, 1972). A vernier target can be thought of as a vertical line which has been modulated with a low-amplitude spatial square wave, and it would be interesting to see whether vernier acuity, or even visual responses to more complex curves (Figure 13) could be predicted from responses to sinusoidal wavy lines.

Tyler (1973b) also measured the frequency response to lines which were waved in depth. One eye saw a vertical wavy line, as before, while the other eye saw a vertical straight line. Viewed stereoscopically, the sinusoidal disparity changes made different parts of the line appear at different depths. Stereo acuity for detecting depth was the same as monocular acuity for detecting curvature at low spatial frequencies of curvature, but it fell off much more rapidly at high frequencies, and stimuli with a finer grain than about 4 cycles per degree did not elicit depth perception, even though the sinusoidal curvature was clearly visible monocularly (Figure 9).

SPATIAL FREQUENCY ANALYSIS

The modulation transfer function (MTF) gives an admirably concise and complete description of the visual system's response to spatial stimuli. If the visual system is linear in its response to contrast—admittedly a highly questionable assumption—it allows us to predict the visual response to any spatial stimulus whatever, simply by calculating the response to each Fourier component of the stimulus and then summing these responses. The previously chaotic field of visual acuity can now be understood by considering the high-frequency response, and Mach bands, and other allied illusions by considering the low-frequency response. By means of such analysis a number of disparate and confusing phenomena have been brought together under some simplifying generalizations of great power.

But can the MTF do more than this? Is it merely a shorthand description of the visual-spatial response, or is it a clue to the mechanism by which we may analyze the visual world? Does the visual system perform a Fourier analysis on stimuli, transforming every picture into its set of Fourier components as the first stage of its perceptual analysis? There are tantalizing traces of evidence that this may be true, but no more than that. The mere fact that a system is frequency-sensitive is no proof that it performs a frequency analysis. After all, an MTF can be established for almost any passive optical, electrical, or mechanical system. A camera lens or a film emulsion has an MTF for spatial frequencies—in fact it was these MTFs that inspired the analogous work on the visual system—but these devices do not perform a Fourier analysis. A car suspension, a loudspeaker, or an amplifier has MTFs for temporal frequencies, but this does not make them into harmonic analyzers. When different parts of a system respond differentially or resonate to different frequencies, this may or may not be a sign of Fourier analysis. When a man sits on a vibrating table, the larger parts of his body will resonate to low frequencies of vibration and the small parts to high frequencies: his arms and legs can be set into involuntary motion by a few hertz (Hz) and his eyeballs by about 50 Hz (making reading very difficult). We would not normally call his body a frequency-to-place recoder, or a frequency analyzer, because the nervous system is not wired up to take advantage of these resonances. In the cochlea it is so wired, and it is correct to call the cochlea a Fourier analyzer. (In addition to this labeled-line or place coding, the auditory nerves show frequency coding for low-frequency stimuli. But this is an added complication.)

Different parts of the visual system have been shown to resonate to different spatial frequencies. A few different parts, or visual channels, would suffice for some form of frequency analysis, but many channels would be necessary for a true Fourier analysis. The psychophysical evidence for many tuned spatial filters, whose envelopes would form the visual frequency response, comes from critical-band effects—the fact that spatial frequencies interact in experiments on detection, adaptation, and masking only if they lie within about an octave of each other (a factor of 2). For instance, adaptation to a grating of one spatial frequency raises the threshold for test gratings of frequencies close to the adapting frequency (Blakemore & Campbell, 1969), and also displaces the apparent spatial frequency of such test gratings (Blakemore & Sutton, 1969). Test gratings within one octave of the adapting frequency were shown to be affected, and this gives an estimate of the width of the tuning curves for channels selectively tuned to spatial frequencies. Figure 3 and the accompanying discussion suggest possible mechanisms for this frequency shift.

Critical-band masking experiments suggest similar tuning-curve bandwidths. Julesz and Stromeyer (1972) masked a test grating with a band-limited "noise grating" of random vertical stripes, uncorrelated frame by frame at 60 Hz, on a television screen. The masking "noise grating" reduced the visibility of the test grating only if its spatial frequency band lay within ±1.5 octaves of that of the testing grating. Many other experiments suporting the multichannel model are reviewed by Campbell (1973).

Sachs, Nachmias, and Robson (1971) asked their subjects to detect the presence of a simple grating whose luminance was sinusoidally modulated with spatial frequency f, and of a complex grating whose luminance was modulated with the sum of two sinusoids with frequencies f and f'. They found an interaction between the detection of f and f' if f and f' differed by less than 25%, i.e., if f/f' lay between $\frac{4}{5}$ and $\frac{5}{4}$. If f and f' differed by more than this, they were detected independently. This gave a very much narrower estimate of the tuning curves of spatial frequency channels, with sensitivity down by about $\frac{1}{3}$, only 20% away from the peak frequency. These discrepant estimates of tuning curve bandwidths have not yet been reconciled.

Richards and Spitzberg (1972) accept these findings but reject the many-channel interpretation. They report narrow critical bands for adaptation to color, which is known to be mediated by only three broad-band photopigment response curves (red, green, and blue). They found that the chromatic thresholds for color naming of test hues were raised when they lay within $\frac{1}{6}$ octave of an adapting wavelength, suggesting very narrow-band channels each tuned to separate wavelengths (luminance threshold for *detecting* test hues without naming them were raised over a broad band of wavelengths). They conclude that spatial frequency responses may also be mediated by only a few broad-band filters, which together with appropriate neural interaction could produce many narrow-band channels.

So Campbell and others compare the spatial frequency response to hearing, while Richards and Spitzberg compare it to color vision. When we are discriminating a single pitch, the many channels in hearing are not of much more use than the few channels of color vision. But the many channels come to the fore when we are confronted with two stimuli simultaneously. Two or more pure tones in a chord can be told apart; this is Ohm's acoustic law. But a mixture of wavelengths on the retina cannot be analyzed into its components: we see only a weighted average of the wavelengths, and all other wavelength information is lost. This is the phenomenon of metamerism. It might be interesting to look for metameric effects in spatial frequency and orientation, as a test of the many-channels hypothesis.

Much of the evidence from detection, masking, and adaptation studies supports the notion of spatially tuned channels with a bandwidth of about one octave. Such channels would not look sharply tuned to an electronic engineer. The tuning factor, Q—defined as the reciprocal of the bandwidth—would have a value of only 1 or 2, or even for Sachs *et al.*'s more sharply tuned estimates, of 5 or 10. Electronic filters usually have a Q of hundreds or even thousands. It is worth noting that a bank of filters where $Q < 10$ will transform a signal, but *not* into its Fourier components. A Fourier transform preserves all the frequency, amplitude, and phase information in a signal and is fully invertible; no information is lost or discarded. Filters with $Q < 10$ would lose much of this information. Cascading n visual filters in series would sharpen up the tuning curves to the nth power (Sachs *et al.*, 1971). Maffei & Fiorentini (1973) found such sharpening in successive stages of the cat's visual system. The tuning curve bandwidths in the retina,

LGN, and cortex were, respectively: mostly low-pass, with some band-pass curves 3–4 octaves wide (retina); some low-pass, with most band-pass curves 2 octaves wide (LGN); and 1 octave wide (cortex). But none of these is nearly sharp enough to produce a Fourier transform of the visual input.

Other authors have looked for evidence that the visual system processes "real-life" stimuli in terms of their spatial frequency components. Weisstein (1972) found that gratings can mask bars, and bars can mask gratings, if they have the same fundamental frequency; Sullivan, Georgeson, and Oatley (1972) found cross-adaptation between bars and grating. It would be nice if the adaptation to print size and cross-adaptation between texture and print size (Anstis, 1973) could be taken as evidence that reading print involves a Fourier analysis. Unfortunately, print does contain visible periodicities with strong spatial frequency components clustering around the thickness of ink strokes and the letter spacings. This can be seen in Figure 10, where the superimposed grating has a maximal masking effect where its spatial frequency matches the major components of the print's ink strokes, at about 10–20 cycles per degree—a spatial-frequency range low enough for the eye to resolve comfortably but high enough to give a high packing density of words per page. The technology of movable type has evolved to match the eye's characteristics, but it has led to typeface being so like a slightly scrambled grating that we cannot conclude much about normal vision from experiments with typeface. The demonstration by Vienet, Duvernoy, Tribillon, and Tribillon (1973) that print can be transmitted over a very narrow bandwidth channel with little distortion supports the idea that print contains only a few spatial frequencies.

Kabrisky *et al.* (1970) have also suggested that the brain may perform a Fourier analysis on visual inputs. They prepared letter stimuli from

Figure 10. Spatial frequency masking. The grating masks the print most effectively, making it unreadable, where the predominant frequencies of grating and print coincide. Moving the grating over the print creates moiré fringes, which are stationary where the spatial frequencies coincide but moving everywhere else.

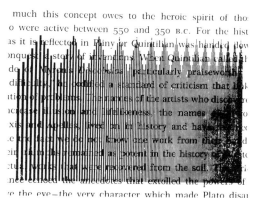

dots on an 11×11 matrix, or on a 21×21 matrix, and performed a low-pass spatial filtering operation on these with a computer. The stimuli were transformed into their Fourier components, the high spatial frequencies were removed, and the low spatial frequencies transformed back again to produce patterns looking like very blurred letters. Their subjects were able to recognize the letters without difficulty, but that they did so by performing a similar Fourier transformation is far from proven. Kabrisky et al.'s dot-pattern letters contained high-frequency noise or distortions, which may have been reduced by the computer blurring. This would make the letters easier to recognize. The visual system uses whatever information is available to it, and if the only information available is carried by the low-frequency components this will be used. It is possible that Kabrisky et al. would have obtained similar results if the low spatial frequencies were removed and the high spatial frequencies retained. For many visual stimuli the low- and high-frequency components are redundant, carrying much the same message.

Occasionally one can find visual stimuli in which the different spatial frequencies carry different information, so that the appearance of the object can be changed by selectively emphasizing high or low frequencies. Figure 11 is an example of this. The sharp edges of the squares convey the message of a radial checkerboard. But peripheral viewing, squinting, or optical defocussing loses this high-frequency edge information, leaving intact the low-frequency information which is carried by diagonal connectivity between adjacent squares. So, when blurred, Figure 11 changes into a set of overlapping logarithmic spirals. In these cases, the two different messages—checkerboard and spiral—are carried differentially by high and low frequencies. When both frequency bands are present, as in the sharp picture, the high-frequency information dominates. In Figure 12, the low frequencies carry picture information but the high frequencies carry masking noise. The figure was prepared from a high-quality photograph by quantizing its gray levels into concentric circles of graduated thickness. The circles transmit, but also mask, the low-frequency picture information, and at normal viewing distances the fovea is more sensitive to the spatial frequencies of the circle than of the picture. The picture can be unmasked by viewing it either from a distance or in peripheral vision. Harmon (1970) has published coarsely quantized pictures of a face, made by averaging local brightnesses within each square of an array of 14×18 squares. The information about the face is contained in the low spatial frequencies, but it is masked by the high-frequency noise caused by the quantization. It is only by suppressing the high-frequency components, by blurring the picture or viewing it at a distance,

Figure 11. Checkerboard spiral. Here, high and low spatial frequencies convey different messages. The high frequencies give information about the edges of the checkerboard squares, which one sees arranged in a radial pattern. Blurring the figure, or squinting at it, removes the high-frequency information. This enhances the low-frequency spirals formed by the diagonal connectivity between the centers of the squares.

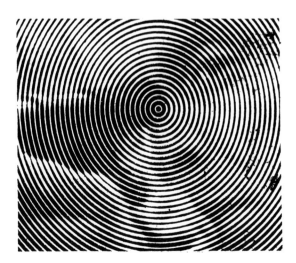

Figure 12. A picture quantized into circles. The low-frequency picture information is masked by the high-frequency components of the circles. It can be unmasked by defocusing, or by viewing from a distance or in peripheral vision.

307

that the identity of the face becomes apparent (at least to American readers) as Abraham Lincoln. Harmon and Julesz (1973) showed that critical-band masking accounts for difficulties in recognizing the face. They selectively removed various frequency bands of noise, using spatial filtering techniques, and found that most of the masking which suppressed recognition was caused by noise whose spatial frequency lay within 2 octaves of the picture's spectrum, and not by the very high frequency noise associated with the edges of the quantizing squares.

FEATURE DETECTION

The findings described in the preceding sections certainly support the notion that parallel processing applied to the retinal image generates a number of separate, independent maps specifying different stimulus features. The independence of each feature map from all the others is confirmed by their different spatial characteristics, as measured by the MTF studies. Hubel and Wiesel (1968) provide physiological evidence that confirms the existence of separate maps specifying different visual features and their independent representation in the cortex, at least for the rhesus monkey.

What is the significance of these exciting discoveries? There are two possibilities. The first, and one that has already been mentioned, is that we have found some anti-cross-talk circuits. This may be true, but the prevention of intermodulation distortion may be a fortuitous result, rather than the main aim, of the neural circuitry. The second, and more interesting, possibility is that we have found in the visual cortex the first stages of a pattern-recognition system similar to Selfridge's Pandemonium (Selfridge, 1959; Selfridge & Neisser, 1960).

This artificial recognition network consists of several layers of elements known as data, computational, cognitive, and decision *demons*. The data demons, or sensors, convert the input pattern into digital form and pass the information on to the computational demons which extract features or look for certain characteristics in the pattern by means of specific property-filters. These features might be of any kind: the presence of a line in a certain position and orientation, the total length of contour per unit area, the presence of certain spatial frequencies, and so on. Cognitive demons watch the responses of computational demons, and each cognitive demon is responsible for recognizing one pattern. Each cognitive demon looks for evidence for the presence of features associated with its particular pattern. The more features it finds, the closer the similarity between its own favorite pattern and the input pattern, and so the "louder it shouts." A decision demon listens to the pandemonium of shouting and selects the cognitive demon who is making the most noise as representing the pattern most probably present at the input. If such a theory goes any way toward describing the mechanisms of human pattern recognition, it suggests that the visual system collects n measurements (features) from the patterns to be classified, and that each pattern can then be represented as a point lying in an n-dimensional hyperspace, defined by a vector. Each measurement or feature defines the position of a pattern along one of the n visual dimensions. Pattern recognition is then the act of finding hyperplanes or curved hypersurfaces which will partition the vectors (patterns) into groups or clusters. Many computer programs have been written to do this (Nagy, 1968). Adaptation studies can be seen as exploring the effects of prolonged inspection of an adapting pattern upon test patterns which lie in a surrounding volume of the n-dimensional hyperspace. The larger this affected test volume proves to be, the broader is the tuning of the underlying feature detectors.

Features are extracted at successive stages in the visual system, with each stage requiring greater specificity in its trigger feature, while allowing less specificity, or greater generality, for the retinal position of the feature (Barlow, Narasimhan, & Rosenfeld, 1972). For instance, a complex cell in the cortex is triggered only by a line of its preferred direction, as defined by the spatial relations between the dots which compose a line or edge, but independent of the line's exact retinal position. It is tempting to extrapolate this increase in specificity and to speculate that there might be "canonical cells" or "grandmother detectors" in the visual system. Such cells would be tuned to very high-order visual invariants, based on the spatial relations between the different parts of a grandmother, but independent of retinal position or perhaps even which way grandma was facing or what she was doing. The weighted sum of these invariants might trigger the detector cell. If a grandmother is a cluster or n-dimensional zone in the visual hyperspace, as computer engineers tell us she is, than a grandmother detector must receive many ($\gg n$) inputs from detectors sensitive to the various dimensions. It must also receive nonvisual inputs supplying stored information of internal models or hypotheses about the world of the kind postulated by Gregory (1970, 1971) and Sutherland (1973).

Grandmother detectors, if they existed, would exemplify two different principles. One would be preferentially fast programs for frequently seen objects—a kind of perceptual Zipf's law. The other would be single-unit representations of complicated visual patterns. There is no direct evidence on this. Complex visual patterns presumably generate complex neural patterns initially; but these neural patterns could perhaps be handled more flexibly if they first converged onto "grandmother detectors"—rather as mathematical functions or computer subroutines can be manipulated more flexibly by first attaching names to them and then operating on the names.

It is not known whether grandmothers and other objects are represented in the visual system topographically by the firing of single units or nontopographically by patterns of neural firing, either within restricted clusters of neurons or in the form of diffuse patterns of activity resembling holograms. Nontopographic representation by neural firing patterns has two advantages. First, the number of possible patternings or combinations of n neurons vastly exceeds n, so that the brain could have enormously more possible states than it has neural components. Second, pattern coding has been found to be far more flexible than topographic coding for representing numbers, both on paper and in computers. Representing every visual object by a separate visual neuron or detector would be like representing every number by a separate special symbol on paper, or in a separate computer address. One would need to learn as many symbols as one had numbers. (The ancient Greeks had a clumsy numbering system something like this, using alphabetical letters as names for numbers. Perhaps this may be one reason why their arithmetic was so inferior to their brilliant achievements in geometry.)

Representing numbers topographically in a computer, by attaching number labels to specific addresses, would make it impossible to move numbers around to process them. Our arabic number system codes numbers in the form of an ordered *pattern* of digits. Ten basic symbols (two in a computer) plus the order in which they are arranged suffice to represent any number. An n-digit number is stored in a computer as a pattern spread out over n address positions. It can be moved around from one register to another, retaining its identity by retaining its pattern independent of its location. If the brain operates like this, it would mean that feature detectors are a special case, since they are a form of cortical address coding. Pattern coding might be the general rule for seeing grandmothers.

HIGHER PERCEPTUAL PROCESSES

Vision, that is the steps from the retinal image to maps of visual features in the cortex, can partly be explained mechanistically in terms of physiological data processing. But visual perception—the steps from visual feature extraction to the visual world of perceived objects—can still be explained only in rather descriptive terms.

Computer engineers distinguish between optical data processing and automatic pattern recognition—the computer equivalents of vision and perception respectively. Image processing includes noise suppression, edge extraction, spatial filtering, and so on (see Barlow *et al.*, 1972; Roetling, 1971; and Andrews, 1972, for introductory articles; Rosenfeld, 1969, and Andrews, 1970, for a longer treatment). Early and modern examples of image processing are shown in Figure 13. Image processing alone can never bring about pattern recognition. Early pattern-recognition programs, such as "Pandemonium" (Selfridge & Neisser, 1960) and others like it, extracted features on an empirical basis from the input, e.g., letters of the alphabet, and classified them (decided what letter was present at the input) by a decision process based on a weighted list of features present. More recent programs (Guzman, 1969; Clowes, 1971; Falk, 1972; Mackworth, 1973) had pictures of flat-sided bodies (polyhedra) as their inputs and inferred from these the three-dimensional structures of the objects they represented. The account that follows is based on Sutherland's valuable review (1973).

These more recent programs are in two stages. The input to the first stage is an array of points in the *picture* domain. From these, lines, vertices, and closed regions are recovered. Lines can have relationships between them which points cannot have; for instance, they can intersect each other or be parallel. (Falk's program recovers lines from a TV pic-

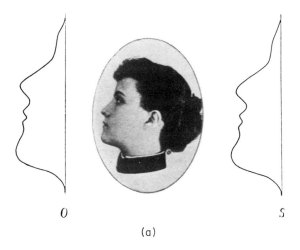

0   *S*

(a)

Figure 13. Fourier analysis, old and new.
(a) Miller (1916) analyzed a girl's profile into its first 18 Fourier components and then resynthesized it, using a synthesizer made of string and pulleys.

Stuart M. Anstis

(*1*) Original (Surveyor Spacecraft).

(*2*) Logarithm of the magnitude of the Fourier transform.

(*3*) Low-pass filtered original.

(*4*) High-pass filtered original.

(b)

Figure 13 (Continued).

(b) A digital computer is used to analyze a photograph (1) into its two-dimensional Fourier transform (2), which can be filtered before being transformed back into a picture. Low-pass filtering emphasized areas (3), while high-pass filtering emphasizes edges (4). (Moon scene supplied to Andrews by Fred Billingsley, Jet Propulsion Laboratory, Pasadena.) Does the visual system do Fourier transform on contours or pictures, using tuned "wavyline" detectors or channels tuned to spatial frequencies and orientations? (Andrews, 1970)

ture, while the programs of Guzman, Clowes, and Mackworth assume the recovery has already been carried out and accept inputs which have already been specified in terms of lines, vertices, and regions.) In the second stage, these lines and vertices are interpreted as representing three-dimensional objects in the *scene* domain. The first stage might be accomplished by some process of brightness differentiation leading to the detection of bars, edges, and corners, such as Hubel and Wiesel (1968) found in animals. The second stage begins where all known physiological processing leaves off. Guzman's program classified all vertices in the input picture into certain types (L, T, arrow, psi, fork) according to their appearance, then placed "links" between regions meeting at certain types of vertex. All regions joined up with enough links were grouped together as belonging to a single object. Such a program deals with the *syntax* of connectivity using vertices as heuristics (cues). Clowes (1971) also used vertices, but added a *semantic* element: the program not only determined which regions belonged together in the *picture* domain, but also specified all possible three-dimensional interpretations of the picture in the *scene* domain. Clowes realized that the types of vertex in a drawing not only give information about which regions belong to the same body, they also impose constraints on the relationships between three-dimensional surfaces corresponding to the regions. In addition, the type of vertex imposes constraints on the three-dimensional arrangement of edges meeting at the corresponding three-dimensional corner. By considering the way in which different vertices are successively arranged around regions, it is possible to eliminate systematically inconsistent edge-interpretations of lines and to reject "impossible" figures.

Falk's (1972) program went further: it could tell objects apart and tell when one object was standing on top of another one. Only nine different types of object were allowed to appear in the picture. Unlike the Guzman and Clowes programs, Falk's program did not always proceed in one direction, from the *picture* domain to the *scene* domain. It could backtrack, using its knowledge of structure in the *scene* domain to guide its processing in the *picture* domain.

Sutherland (1973) reviewed the relevance of these perceptual programs to human and animal perception, and concluded that "successful picture interpretation cannot normally be achieved by successive mapping from one domain to the next: it is necessary to set up hypotheses to guide further processing in lower domains." Gregory (1970, 1971) has also stressed that what we perceive at any instant cannot be explained by what we see at that instant: a very small input of information can lead to a great deal of behavior, so perception must rely heavily on information stored in the form of internal hypotheses or models of the world.

Any picture which changes in appearance without any change in the stimulus, such as an ambiguous figure, indicates that active perceptual computing must be going on. A retinal image of a scene could stand for an infinity of possible three-dimensional scenes, but normally we see only one of these possibilities, and it is quite stable. Figure 14 (O'Beirne, 1965) is "impossible," it is not consistent with any three-dimensional interpretation so it fluctuates in appearance. Clowes' program

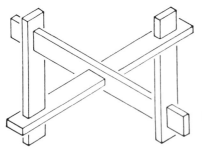

Figure 14. An impossible figure. (O'Beirne, 1965)

Figure 15. Figure-ground reversal. The two windmill patterns appear to alternate, and the circles lying on the figure appear to be in front of those lying on the ground. Sometimes the concentric circles themselves become the figure. (Horemis, 1970)

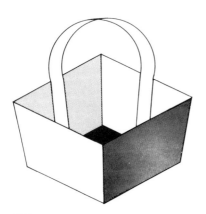

Figure 16. Perspective reversal. A reversible basket. Where are the handles fastened on the basket?

Figure 17. Who is this pyschologist? Which way is he facing? The parts of the figure that are irrelevant to one's current perspective interpretation are suppressed or ignored. (From the cover of S. Freud: *Therapy and Technique*, ed. by P. Rieff)

would reject such a figure. Ambiguous perspective pictures of the figure-ground type (Figure 15) or of the Necker cube type (Figures 16 and 17) show continuous reversals because the sensory information calls up two or three possible three-dimensional organizations with equal probability. As each organization is equally stable, and has equal sensory evidence to support it, the rival models each dominate in turn (Gregory, 1970). Wilson (1972) has programmed a computer to generate series of ambiguous figures: Figure 18 shows one of his transformations. Such series can be used to demonstrate perceptual hysteresis or inertia: a

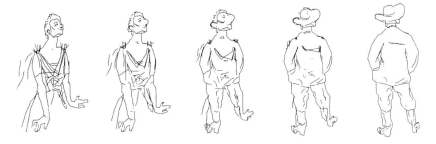

Figure 18. A sex change by computer. A series of ambiguous figures like these can be computer-generated between any desired pair of pictures. (Wilson, 1972)

subject who sees the pictures successively will often persist with his
initial perceptual hypothesis of Yvette Guilbert past the halfway point
where the current evidence from his senses ought to favor Aristide Briand.
The ambiguous portrait from ¡OHO! by Rex Whistler (Figure 19) shows
that perceptual identity is not always independent of orientation.

It is clear that the higher perceptual processes exemplified here cannot
be explained by any kind of edge extraction or spatial filtering, or any
known type of physiological mechanisms. They are of a different logical
type. Computer simulations are probably the most promising line of current research for understanding perception.

Finally, we can restore our humility as scientists by admiring our
own abilities as seeing-machines. Examine the multiple ambiguities in
the puzzle picture of Figure 20. Nothing in contemporary physiology
or computer expertise can give us an inkling of how we analyze a picture
like this with such extraordinary power and speed. Problems like this
will keep visual scientists occupied for many years to come.

Figure 19. Reversible face. (From ¡*OHO!* by Rex Whistler)

Figure 20. An ambiguous puzzle picture. How many heads are better than one? Keep looking for a while. Line-detectors alone cannot explain what happens when you look at this picture! (From cover of Columbia record CS 9781)

ACKNOWLEDGMENTS

This work was supported by Grant B/SR/4836–8 from the Science Research Council, and by a minor research grant from York University. I thank Tania Morley for her valuable help in reviewing the manuscript.

REFERENCES

Andrews, D. P. Perception of contours in the central fovea. *Nature,* 1965, **205,** 1218–1220.

Andrews, H. C. *Computer techniques in image processing.* London: Academic Press, 1970.

Andrews, H. C. Digital computers and image processing. *Endeavour,* 1972, **31,** 88–94.

Anstis, S. M. Visual adaptation to gradual change of intensity. *Science,* 1967, **155,** 710–712.

Anstis, S. M. Phi movement as a subtractive process. *Vision Research,* 1970, **10,** 1411–1430.

Anstis, S. M. Size adaptation to visual texture and print: Evidence for spatial frequency analysis. *American Journal of Psychology,* in press.

Anstis, S. M., & Harris, J. P. Movement aftereffects contingent on binocular disparity. *Perception,* in press.

Anstis, S. M., & Moulden, B. P. After effect of seen movement: Evident for peripheral and central components. *Quarterly Journal of Experimental Psychology,* 1970, **22**, 222–229.

Arend, L. E. Temporal determinants of the form of the spatial contrast-threshold MTF. Paper presented at the Association for Research in Vision and Ophthalmology, Sarsota, Florida, May 1973.

Barlow, H. B. Sensory mechanisms, the reduction of redundancy and intelligence. In *Mechanisation of thought processes.* London: National Physical Laboratory, H. M. Stationery Office, 1959.

Barlow, H. B. Possible principles underlying the transformations of sensory messages. In W. A. Rosenblith (Ed.), *Sensory communication.* Cambridge, Massachusetts: MIT Press, 1961. Pp. 217–234.

Barlow, H. B., Blakemore, C. B., & Pettigrew, J. The neural mechanism of binocular depth discrimination. *Journal of Physiology,* 1966, **193**, 327–362.

Barlow, H. B., & Brindley, G. S. Interocular transfer of movement after-effects during pressure blinding of the stimulated eye. *Nature* (London), 1963, **200**, 1347.

Barlow, H. B., & Levick, W. R. The mechanism of directionally sensitive units in the rabbit's retina. *Journal of Physiology,* 1965, **178**, 477–504.

Barlow, H. B., Narasimhan, R., & Rosenfeld, A. Visual pattern analysis in machines and animals. *Science,* 1972, **177**, 567–575.

Békésy, G. von. *Sensory inhibition.* Princeton, N.J.: Princeton Univ. Press, 1968.

Bishop, P. O. Beginning of form vision and binocular depth discrimination in cortex. In F. O. Schmitt (Ed.), *The neurosciences: A second study program.* New York: Rockefeller Univ. Press, 1970. Pp. 471–485.

Blake, R., & Fox, R. Interocular transfer of adaptation to spatial frequency during retinal ischaemia. *Nature New Biology,* 1972, **240**, 98, 76–77.

Blake, R., & Fox, R. Adaptation to invisible gratings and the site of binocular rivalry suppression. *Nature,* 1974, **249**, 488–490.

Blakemore, C., & Campbell, F. W. On the existence in the human visual system of neurons selectively sensitive to the orientation and size of retinal images. *Journal of Physiology,* London, 1969, **203**, 237–260.

Blakemore, C., & Julesz, B. Stereoscopic depth aftereffect produced without monocular cues. *Science,* 1971, **171**, 286–288.

Blakemore, C., & Nachmias, J. Orientation specificity on two visual aftereffects. *Journal of Physiology,* 1971, **213**, 157–174.

Blakemore, C., & Sutton, P. Size adaptation: A new aftereffect. *Science,* 1969, **166**, 245–247.

Blakemore, C., & Tobin, E. A. Lateral inhibition between orientation detectors in the cat's visual cortex. *Experimental Brain Research,* 1972, **15**, 439–440.

Boring, E. G. *Sensation and perception in the history of psychology.* New York: Appleton, 1942.

Bouma, H., & Andriessen, J. J. Perceived orientation of isolated line segments. *Vision Research,* 1968, **8**, 493–507.

Bouma, H., & Andriessen, J. J. Induced changes in the perceived orientation of line segments. *Vision Research,* 1970, **10**, 333–349.

Breitmeyer, B., & Cooper, L. A. Frequency specific colour adaptation in the human visual system. *Perception & Psychophysics,* 1972, **11**, 95–96.

Campbell, F. W. The human eye as an optical filter. *Proceedings of the Institute of Electrical & Electronic Engineers,* 1968, **56**, 1009.

Campbell, F. W. Trends in physiological optics. In W. Reichardt (Ed.), *Proceedings* of the *International School of Physics Enrico Fermi Course XLIII*. New York: Academic Press, 1969. Pp. 137–143.

Campbell, F. W. The transmission of spatial information through the visual system. In F.O. Schmitt & F. Worden (Eds.), *The neurosciences: Third study program*. Cambridge, Mass: MIT Press, 1973.

Campbell, F. W., & Gubisch, R. W. Optical quality of the human eye. *Journal of Physiology*, 1966, **186**, 558–578.

Campbell, F. W., & Robson, J. G. Application of Fourier analysis to the visibility of gratings. *Journal of Physiology*, 1968, **197**, 551–566.

Cavonius, C. R., & Schumacher, A. W. Human visual acuity measured with colored test objects. *Science*, 1966, **152**, 1276–1277.

Clowes, M. B. On seeing things. *Artificial Intelligence*, 1971, **2**, 79–116.

Coltheart, M. Visual feature-analysers and aftereffects of tilt and curvature. *Psychological Review*, 1971, **78**, 2, 114–121.

Coltheart, M. Colour-specificity and monocularity in the visual cortex. *Vision Research*, 1973, **13**, 2595–2598.

Cornsweet, T. *Visual perception*. New York: Academic Press, 1970.

Daw, N. W. Neurophysiology of color vision. *Physiological Reviews*, 1973, **53**, 571–611.

De Valois, K. K. Black and White: Equal but separate. Paper presented at the Association for Research in Vision and Ophthalmology, Sarasota, Florida, 1973.

De Valois, R. L. Analysis and coding of colour vision in the primate visual system. *Cold Spring Harbor Symposia*, 1965, **30**, 567–579.

Dow, B. M., & Gouras, P. Colour and spatial specificity of single units in rhesus monkey foveal striate cortex. *Journal of Neurophysiology*, 1973, **36**, 1, 79–100.

Enroth-Cugell, C., & Robson, J. G. The contrast sensitivity of retinal ganglion cells of the cat. *Journal of Physiology*, 1966, **187**, 517–552.

Erickson, R. P. Stimulus coding in topographic and nontopographic afferent modalities: On the significance of the activity of individual sensory neurons. *Psychological Review*, 1968, **75**, 447–475.

Falk, G. Interpretation of imperfect line data as a 3-dimensional scene. *Artificial Intelligence*, 1972, **3**, 101–144.

Favreau, O. E., Emerson, V. F., & Corballis, M. C. Motion perception: A color-contingent aftereffect. *Science*, 1972 ,**176**, 78–79.

Felton, T. B., Richards, W., & Smith, R. A. Disparity processing of spatial frequencies in man. *Journal of Physiology* (London), 1972, **225**, 349.

Fidell, L. S. Orientation specificity in chromatic adaptation of human edge-detectors. *Perception & Psychophysics*, 1970, **8**, 235–237.

Ganz, L. Is the figural aftereffect an *after*effect? A review of its intensity, onset, decay, and transfer characteristics. *Psychological Bulletin*, 1966, **73**, 2, 128–150.

Georgeson, M. A. Spatial frequency selectivity of a visual tilt illusion. *Nature*, 1973, **245**, 43–45.

Georgeson, M. A. Is texture-density contrast an inhibition or an adaptation? *Nature*, 1974, **249**, 85–86.

Gerrits, H. J. M., & Hendrik, A. J. H. Simultaneous contrast, filling-in process and information processing in man's visual system. *Experimental Brain Research*, 1970, **11**, 411–430.

Graham Smith, F. *Radio astronomy*. London: Penguin Books, 1960.

Green, D. G., & Fast, M. B. On the appearance of Mach bands in gradients of varying colour. *Vision Research*, 1971, **11**, 1147–1156.

Gregory, R. L. On how little information controls so much behaviour. In A. T. Welford & L. Houssiadas (Eds.), *Contemporary problems in perception*. London: NATO Advanced Study Institute, Taylor & Francis, 1970. Pp. 25–35.

Gregory, R. L. *The intelligent eye*. London: Weidenfeld and Nicholson, 1971.

Grüsser, O-J., & Grüsser-Cornehls, U. Neuronal mechanisms of visual movement perception and some psychological and behavioral correlations. In R. Jung (Ed.), *Handbook of sensory physiology*. Vol. VII/3: *Central processing of visual information*, A. New York: Springer-Verlag, 1973.

Guzman, A. Decomposition of a visual scene into three-dimensional bodies. In A. Grasselli (Ed.), *Automatic interpretation and classification of images*. New York: Academic Press, 1969.

Harmon, L. D. Some aspects of recognition of human faces. In J. O. Grüsser (Ed.), *Fourth Kybernetik Congress, Berlin, 1970*. Heidelberg: Springer, 1970.

Harmon, L. D., & Julesz, B. Masking in visual recognition: Effects of two-dimensional filtered noise. *Science*, 1973, 180, 1194–1196.

Harrington, T. L. Adaptation of humans to colored split-field glasses. *Psychonomic Science*, 1965, 3, 71–72.

Harris, C. S. Effect of viewing distance on a color aftereffect specific to spatial frequency. Paper presented to the Psychonomic Society, San Antonio, Texas, November 1970.

Held, R., & Shattuck, S. Color- and edge-sensitive channels in the human visual system: Tuning for orientation. *Science*, 1971, 174, 314–116.

Helson, H., & Rohles, F. H. A quantitative study of reversal of classical lightness-contrast. *American Journal of Psychology*, 1959, 72, 530–538.

Hepler, N. Color: A motion-contingent aftereffect. *Science*, 1968, 162, 376–377.

Hoekstra, J., van der Goot, D. P. A., van den Brink, G., & Bilsen, F. A. The influence of the number of cycles upon the visual contrast threshold for spatial sine wave patterns. *Vision Research*, 1974, 14, 365–368.

Horemis, S. *Optical and geometrical patterns and designs*. New York: Dover, 1970.

Hubel, D. H., & Wiesel T. N. Receptive fields and functional architecture of monkey striate cortex. *Journal of Physiology*, 1968, 195, 215–243.

Ingling, C. R., & Drum, B. A. Retinal receptive fields: Correlations between psychophysics and electrophysiology. *Vision Research*, 1973, 13, 6, 1151–1164.

Julesz, B., & Stromeyer, C. F. Spatial-frequency masking in vision: Critical bands and spread of masking. *Journal of the Optical Society of America*, 1972, 62, 10, 1221–1232.

Kabrisky, M., Tallman, T., Day, C. M., & Radoy, C. M. A theory of pattern perception based on human physiology. In A. T. Welford & L. Houssiadas (Eds.), *Contemporary problems in perception*. London: NATO Advanced Study Institute, Taylor & Francis, 1970.

Kelly, D. H. Lateral inhibition in human colour mechanisms. *Journal of Physiology*, 1973, 228, 55–72.

Kohler, I. The formation and transformation of the perceptual world. *Psychological Issues*, 1964, 3.

Köhler, W., & Wallach, H. Figural after-effects: An investigation of visual processes. *Proceedings of the American Philosophical Society*, 1944, 88, 269–357.

Leeper, R. A study of a neglected area of the field learning—the development of sensory organisation. *Journal of Genetic Psychology*, 1935, 46, 41–75.

Leppman, P. Spatial frequency dependent chromatic aftereffects. *Nature*, 1972, 242, 411.

Leppman, P. K., & Wieland, B. A. Visual distortion with two-colored spectacles. *Perceptual & Motor Skills*, 1966, 23, 1043–1048.

Lovegrove, W. J., & Over, R. Colour selectivity in orientation masking and aftereffect. *Vision Research*, 1973, 13, 895–902.

Mackay, D. M. Central adaptation in mechanism of form vision. *Nature*, 1964, 203, 993–994.

Mackay, D. M. Lateral inhibition between neural channels sensitive to texture density? *Nature*, 1973, 245, 159–161.

Mackworth, A. K. Interpreting pictures of polyhedral scenes. Paper submitted to the Third International Joint Conference on Artificial Intelligence, 1973.

Maffei, L., & Fiorentini, A. The visual cortex as a spatial frequency analyser. *Vision Research*, 1973, 13, 1255–1268.

Marks, W. B., Dobelle W. H., & MacNichol, J. R. Visual pigments of single primate cones. *Science*, 1964, 143, 1181–1183.

Maudarbocus, A. K., & Ruddock, K. H. The influence of wavelength on visual adaptation to spatially periodic stimuli. *Vision Research*, 1973, 13, 993–998.

Mackay, D. M., & Mackay, V. Orientation-sensitive after-effects of dichoptically presented colour and form. *Nature*, 1973, 242, 477–479.

Mayhew, J. E. W. Aftereffects of movement contingent on direction of gaze. *Vision Research*, 1973, 13, 877–880.

Mayhew, J. E. W., & Anstis, S. M. Movement aftereffects contingent on color, intensity and pattern. *Perception and Psychophysics*, 1972, 12, 77–85.

McCollough, C. Color adaptation of edge-detectors in the human visual system. *Science*, 1965, 149, 1115–1116.

McCollough, C. Conditioning of color perception. *American Journal of Psychology*, 1965, 78, 362–378.

Miller, D. C. The science of musical sounds. London: Macmillan, 1916.

Mitchell, D. E., & Ware, C. Interocular transfer of a visual aftereffect in normal and stereoblind humans. *Journal of Physiology*, 1973, 236, 707–721.

Movshon, J. A., Chambers, B. E. I., & Blakemore, C. B. Interocular transfer in normal humans, and those who lack stereopsis. *Perception*, 1972, 1, 483–490.

Murch, G. Binocular relationships in a size and color orientation specific aftereffect. *Journal of Experimental Psychology*, 1972, 93, 1, 30–34.

Nagy, G. State of the art in pattern recognition. *Proceedings of the Institute of Electrical & Electronic Engineers*, 1968, 56, 5.

Nikara, T., Bishop, P. O., & Pettigrew, J. Analysis of retinal correspondence by studying receptive fields of binocular units in cat striate cortex. *Experimental Brain Research*, 1968, 6, 353–372.

O'Beirne, T. H. Puzzles and paradoxes. London: Oxford Univ. Press, 1965.

O'Brien, V. Contour perception, illusion and reality. *Journal of the Optical Society of America*, 1958, 48, 112–119.

Osgood, C. E. Method and theory in experimental psychology. London: Oxford Univ. Press, 1953.

Over, R. Comparison of normalisation theory and neural enhancement explanation of negative aftereffects. *Psychological Bulletin*, 1971, 75, 4, 225–243.

Papert, S. MIT *Quarterly Progress Report*, April 1964.

Parker, D. M. Contrast and size variables and the tilt after-effect. *Quarterly Journal of Experimental Psychology*, 1972, 24, 1–7.

Patel, A. S. Spatial resolution by the human visual system: The effect of mean retinal illuminance. *Journal of the Optical Society of America*, 1966, 56, 689–694.

Patterson, R. Electronic special effects. *American Cinematographer*, 1972, **53**, 10, 1160–1183.

Polyak, S. L. *The retina*. Chicago: Univ. of Chicago Press, 1941.

Ramachandran, V. S., & Braddick, O. L. Orientation-specific learning in stereopsis. *Perception*, 1973, **2**, 371–376.

Ratliff, F. *Mach Bands: Quantitative studies on neural networks in the retina*. San Francisco: Holden-Day, 1965.

Richards, W. Motion detection in man and other animals. *Brain, Behavior and Evolution*, 1971, **4**, 1–16.

Richards, W., & Spitzberg, C. Spatial-frequency channels: Many or few? Paper delivered at meeting of the Optical Society of America, October 20, 1972.

Rieff, P. (Ed). *S. Freud: Therapy and technique*. New York: Collier, 1963.

Robinson, J. O. *The psychology of visual illusion*. London: Hutchinson, 1972.

Roetling, P. G. Image enhancement techniques. *Xerox Technical Review*, 1971, **1**, 41–49.

Rosenfeld, A. *Picture processing by computer*. New York: Academic Press, 1969.

Rushton, W. A. H. Peripheral coding in the nervous system. In W. Rosenblith (Ed.), *Sensory communication*. Cambridge, Mass.: MIT Press, 1961. Pp. 169–181.

Sachs, M. B., Nachmias, J., & Robson, J. G. Spatial-frequency channels in human vision. *Journal of the Optical Society of America*, 1971, **61**, 9, 1176–1186.

Schade, O. H. Optical and photoelectric analog of the eye. *Journal of the Optical Society of America*, 1956, **46**, 721–739.

Selfridge, O. Pandemonium: A paradigm for learning. In *Symposium on the mechanisation of thought processes*. London: H. M. Stationary Office, 1959.

Selfridge, O., & Neisser, U. Pattern recognition by machine. *Scientific American*, 1960, **203**, 60–68.

Stromeyer, C. F. Further studies of the McCollough effect. *Perception and Psychophysics*, 1969, **6**, 105–110.

Stromeyer, C. F. Edge-contingent color aftereffects: Spatial frequency specificity. *Vision Research*, 1972, **12**, 717–733.

Stromeyer, C. F., Lange, A. F., & Ganz, L. Spatial frequency phase effects in human vision. *Vision Research*, 1973, **13**, 2345–2360.

Stromeyer, C. F., & Mansfield, R. J. Colored aftereffects produced with moving edges. *Perception and Psychophysics*, 1970, **7**, 108–114.

Sullivan, G. D., Georgeson, M., & Oatley, K. Channels for spatial frequency selection and the detection of single bars by the human visual system. *Vision Research*, 1972, **12**, 383–394.

Sutherland, N. S. Intelligent picture processing. Paper delivered at the Conference on the Evolution of the Nervous System and Behaviour, Florida State University, Tallahassee, 1973.

Thomas, J. P. Model of the function of receptive fields in human vision. *Psychological Review*, 1970, **77**, 121–131.

Tyler, C. W. Periodic vernier acuity. *Journal of Physiology* (London), 1973, **22**, 8, 637–647. (a)

Tyler, C. W. Stereoscopic vision: Cortical limitations and a disparity scaling effect. *Science*, 1973, **181**, 276–278. (b)

Tyler, C. W. Depth perception in disparity gratings. *Nature*, 1974, **251**, 140–142.

Tyler, C. W. Stereoscopic tilt and size aftereffects. *Perception*, in press.

Uttal, W. R. The psychological silly season—or—What happens when neurophysiological data become psychological theories. *Journal of General Psychology*, 1971, **84**, 151–166.

Van der Horst, G. J. C., de Weert, C. M. M., & Bouman, M. A. Transfer of spatial chromaticity-contrast at threshold in the human eye. *Journal of the Optical Society of America*, 1967, **57**, 10, 1260–1266.

Van der Horst, G. J. C. Fourier analysis and colour vision. *Journal of the Optical Society of America*, 1969, **59**, 1670–1676.

Van der Horst, G. J. C., & Bouman, M. A. Spatiotemporal chromaticity discrimination. *Journal of the Optical Society of America*, 1969, **59**, 11, 1482–1488.

Van Nes, F. L., & Bouman, M. A. Spatial modulation transfer in the human eye. *Journal of the Optical Society of America*, 1967, **57**, 401–406.

Van Nes, F. L. Enhanced visibility by regular motion of retinal images. *American Journal of Psychology*, 1968, 366–374.

Vienet C., Duvernoy, J., Tribillon, G., & Tribillon, J. L. Three methods of information assessment for optical data processing. *Applied Optics*, 1973, **12**, 5, 950–960.

Walker, J. T. Textural aftereffects: Tactual and visual. Unpublished doctoral thesis, Univ. of Colorado, Boulder, Colorado, 1966.

Walls, G. L. The filling-in process. *American Journal of Optometry*, 1954, **31**, 329–340.

Ware, C., & Mitchell, D. E. On interocular transfer of various visual aftereffects in normal and stereoblind observers. *Vision Research*, 1974, **14**, 731–734. (a)

Ware, C., & Mitchell, D. E. The spatial selectivity of the tilt aftereffect. *Vision Research*, 1974, **14**, 735–738. (b)

Weisstein, N. What the frog's eye tells the human brain: Single-cell analysers in the human visual system. *Psychological Review*, 1969, **72**, 157–176.

Weisstein, N. Gratings mask bars and bars mask gratings: Visual frequency response to aperiodic stimuli. *Science*, 1972, 1047–1049.

Westheimer, G. Spatial interactions in human cone vision. *Journal of Physiology*, 1967, **190**, 139–154.

Westheimer, G. Visual acuity and spatial modulation thresholds. In *Handbook of sensory physiology*. Berlin: Springer-Verlag, 1972. Pp. 170–187.

Whistler, R. ¡OHO! London: John Lane The Bodley Head, 1946.

Wiesel T. N., & Hubel, D. H. Spatial and chromatic interactions in the lateral geniculate body of the rhesus monkey. *Journal of Neurophysiology*, 1966, **29**, 1115–1116.

Wilson, J. A. In M. J. Apter & G. Westby (Eds.), *The computer in psychology*. New York: Wiley, 1972.

Wyatt, H. J. Singly and doubly contingent after-effects involving color, orientation and spatial frequency. *Vision Research*, 1974, **14**, 1185–1194.

Yachida, M. & Tsuji, S. Application of color information to visual perception. *Pattern Recognition*, 1971, 3, 307–323.

Chapter 10

Central Auditory Processing

WILLIAM R. WEBSTER
LINDSAY M. AITKIN

Monash University, Victoria, Australia

INTRODUCTION

In this chapter an attempt will be made to summarize some of the recent single unit and anatomical studies of the central auditory pathway. It is not intended to provide a comprehensive review of central auditory processing, but to be selective and examine those areas of the classical lemniscal pathway in which single units studies have been carried out in the awake animal. If justification is required for such selectivity, then we believe that it is provided by the important information already produced using the awake preparation. For this reason we will not deal with the superior olivary complex. However, the reader is referred to the excellent and detailed review by Erulkar (1972) of the involvement of this region in sound localization.

In recent years, there have been many attempts to draw analogies between auditory processing and feature detection organizations of the visual system (Whitfield & Evans, 1965; Oonishi & Katsuki, 1965; Goldstein, 1968; Evans, 1968, 1970; Suga, 1972). The concept of "trigger feature" or "feature detector" has proved most heuristic for the study of the visual system (Blakemore & Campbell, 1969; Weisstein, 1969; Barlow, Narasimhan, & Rosenfeld, 1972). The most elegant theoretical model stems from the work of Hubel and Wiesel (1962, 1965, 1968) in cat and monkey. Although their hierarchical model, with its concepts of simple, complex, and hypercomplex receptive fields, has been very influential in general modeling in sensory processing it should be noted that later studies (Cleland & Levick, 1972; Henry & Bishop, 1972; Stone, 1972) suggest that it may not be the only organizational principle of the central visual system.

At the time of writing, no body of data exists which would support a model equivalent to that of Hubel and Wiesel for the auditory system, although a number of schemes have been suggested. Among the earliest was that of Katsuki, Watanabe, and Maruyama (1959). Their work, which preceded the seminal publications by Hubel and Wiesel (1962, 1965) suggested that plots of frequency versus threshold of auditory neurons became progressively sharper at succeeding auditory nuclei and that medial geniculate neurons were tuned to exceedingly limited ranges. Unfortunately, studies carried out by others (Kiang, 1965; Aitkin & Webster, 1972) lend little support to this model. Oonishi and Katsuki (1965) have tried to develop a hierarchical model within the auditory cortex also based on sharpness of threshold tuning, but other work (Abeles & Goldstein, 1970) has not supported their data.

It seems unlikely that tuning curves alone would define the receptive fields of higher auditory neurons (Aitkin, Anderson, & Brugge, 1970) although they may be adequate for characterizing auditory nerve fibers (Kiang, 1965). Furthermore, it seems unlikely that any one response characteristic could alone define the organization of the central auditory pathway or even the receptive field of an auditory neuron. What is required for this problem is a full documentation of the responses of auditory neurons to both static (steady tones) and dynamic (time varying and complex) stimuli in terms of tuning, spike counts, and discharge patterns. In view of the complexity of the connections and structure of the auditory pathway, it is important that each neuron studied should be carefully related to the cytoarchitecture of its parent auditory nucleus.

We have therefore carried out a selective review with three questions in mind. First: What structural details are available which may assist our understanding of central auditory processing? Second: Is there a body of consistent physiological data available for each auditory nucleus, and do correlations exist with structural features? Third: What new findings have emerged following the use of awake or unanesthetized animals?

We hope that the material which follows may help to clarify the differences in the behavior of neurons at each level so that some integrating principles may be formulated concerning central auditory processing.

COCHLEAR NUCLEUS

Anatomy

Classically the cochlear nucleus has been divided into a dorsal (DCN) and a ventral division (VCN). The VCN has been further subdivided by most workers into the anteroventral cochlear nucleus (AVCN), the posteroventral cochlear nucleus (PVCN), and the interstitial nucleus (IN). The auditory nerve enters the cochlear nucleus through the ventral nucleus and each fiber bifurcates into an ascending and a descending branch. Low frequency fibers bifurcate ventrally and high frequency fibers more dorsally. The AVCN is innervated by the ascending branch and the PVCN by the descending branch which then runs dorsally to innervate DCN. The interstitial nucleus is located laterally and is pierced by the incoming auditory nerve fibers.

Using Nissl stains, Osen (1969a) identified nine cell types in the cochlear nucleus of cat. As this method stains primarily the cell body, she combined this with the Glees method to demonstrate the terminals and dendrites. Osen (1969a) has divided AVCN into three areas on the basis of cell types (Figure 1). A rostral cap contains only large spherical cells each innervated with two or three large bulbs of Held. Bordering this is a region of small spherical cells innervated with two or three smaller bulbs of Held. The remainder of AVCN contains a mixture of globular, multipolar, and small cells (Figure 1). Similar cells are seen in IN with the globular cells being predominant. The anterior part of PVCN contains multipolar and small cells but the posterior part contains mainly octopus cells, which Osen defines by their large ventro-dorsally directed dendrites.

The DCN can be readily distinguished from the bulk of the cochlear nucleus by its laminations. Three layers have been described and Osen (1969a) reported four cell types in these layers: pyramidal or fusiform cells, giant cells, granular cells, and small cells (Figure 1). The fusiform layer is the most prominent and the fusiform cells are about 250 μm below the surface of DCN. In this layer, there are also granular cells and some small cells. The molecular layer which contains small cells, granular cells, and the superficial dendrites of the fusiform cells is superficial to the fusiform layer. The deepest layer, deep DCN, contains giant cells, granular cells, and small cells.

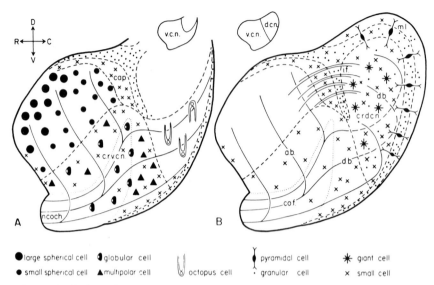

Figure 1. Sagittal diagrams of the cat cochlear nuclei showing the location of nine cell types in the ventral nucleus (A) and the ventral nucleus partly covered by the dorsal nucleus (B). (From Osen, 1969a)

Key to abbreviations:

a.b.	ascending cochlear branch
cap	peripheral cap of small cells
co.f.	cochlear nerve fiber
c.r.d.c.n.	central region of dorsal cochlear nucleus
c.r.v.c.n.	central region of ventral cochlear nucleus
d.b.	descending cochlear branch
m.l.	molecular layer
n. coch.	cochlear nerve

It should be kept in mind that a classification of cell types such as this will be an oversimplification, since there are always likely to be variants of cell shape and size which fall into no one category exactly but are intermediate between several classes.* Nonetheless, the rather precise geographical distribution of some types—e.g., spherical cells in AVCN, fusiform cells in DCN, octopus cells in PVCN—suggests that a clsssification scheme of this type may reveal more than it obscures.

The primary afferents enter DCN and arborize within all three layers. The incoming fibers end in both large and small terminals on each of the cell types. The electron microscope has shown that both endings have a similar synaptic complex but differ in the size and density of synaptic vesicles (Cohen, 1972). It is important to stress that there

---

* This is supported by a recent article by Morest *et al.* (1973).

are also numerous nonauditory sources of input to DCN in contrast to VCN (Rasmussen, 1964; Cohen, 1972).

As might be expected, the outputs of cochlear nucleus are complex and only the briefest of outlines can be attempted. There are three main tracts composed of projections from cochlear nucleus: the dorsal acoustic stria, the intermediate acoustic stria, and the trapezoid body. The fibers of the dorsal stria come from DCN and the bulk of them are axons of the fusiform and giant cells. The fusiform cells project to the contralateral lemniscal nuclei and inferior colliculus (Osen, 1970) but the destination of the giant cell fibers is not known. Most of the intermediate stria is made up of axons from the octopus cell area and these project to the ipsilateral dorsolateral periolivary nucleus and the contralateral ventral nucleus of the lateral lemniscus (van Noort, 1969; Warr, 1969). The trapezoid body contains projections from large and small spherical cells to the ipsilateral lateral superior olivary nucleus, to both medial superior olivary nuclei, and to the contralateral ventral nucleus of the lateral lemniscus (Stotler, 1953; van Noort, 1969). Axons of globular cells project to the contralateral medial nucleus of the trapezoid body and provide the large calycoid endings to the principal neurons (Morest, 1968; Osen, 1969b, Warr, 1972).

The above brief outline indicates that there is a great diversity of cell types and connections in the cochlear nucleus. These recent anatomical findings emphasize that the cochlear nucleus is a complex structure whose components are probably capable of a variety of transformations of the homogeneous input from the cochlear nerve.

Responses in Anesthetized Animals

The response patterns of auditory nerve fibers entering the cochlear nucleus are homogeneous with respect to simple tonal stimuli (Kiang, 1965; Rose, Brugge, Anderson, & Hind, 1967; Goldstein, 1968). These fibers respond with strong excitation followed by declining sustained discharge to tone and noise bursts. Their responses to clicks are related to their characteristic frequency in that knowledge of this frequency can allow predictions about poststimulus time histograms (PSTH) to tones, noise, and clicks. When continuous tones are presented, units respond with sustained excitation.

It has been well established that the responses of cochlear nucleus cells differ from those of auditory nerve fibers. Many will not fire to continuous tones, and responses to clicks cannot be predicted by responses to other stimuli (Kiang, 1965). It has also been shown that the divisions of cochlear nucleus are tonotopically organized (Rose,

Galambos, & Hughes, 1959). The responses of single units have been classified in a number of ways (Raab, 1971). We shall be highly selective and concentrate largely on a classification based on response patterns recorded at the best frequency of a unit (Pfeiffer, 1966a; Godfrey, 1971) (Figure 2). These workers have introduced some rather unappealing names for their response types. We hesitate to continue this tradition but in the interest of brevity we will employ them. Pfeiffer (1966a,b) has made an interesting correlation between response patterns and the anatomy of AVCN. In the region containing large and small spherical cells, all units fired with a "primarylike" response to short tone bursts. The PST histogram consisted of an initial peak of activity followed by a gradual decline to a lower level which was maintained for the duration of the tone burst (Figure 2, top left). The spike discharges of these neurons had a characteristic waveform which could be analyzed into pre- and postsynaptic components which were probably related to the large endings that contact these neurons. Similar waveforms were not found in recordings made from any other region of the cochlear nucleus. These units also had a characteristic pattern of spontaneous activity (Molnar & Pfeiffer, 1968).

Another clear correlation has been found in the octopus cell area, in the center of which Godfrey (1971) found that most units responded

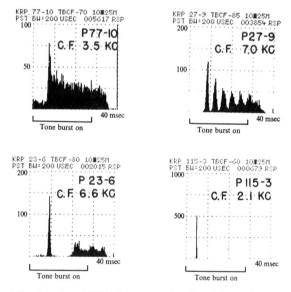

Figure 2. Poststimulus time (PST) histograms of unit activity to short tone burst stimulation, in which averaged firing rate is displayed as a function of time after the onset of the stimulus: "primarylike" (upper left), "chopper" (upper right), "pauser" (lower left), "onset" (lower right). (From Pfeiffer, 1966a)

with an onset response (Figure 2, bottom right) and had little or no spontaneous activity. In the periphery of this region Godfrey found three other response patterns previously described by Pfeiffer (1966a): "chopper" (Figure 2, top right), "pauser" (Figure 2, bottom left), and "primarylike." It is of interest that the part of the octopus cell area in which Godfrey found units responding with only onset spikes is one in which Cohen (1972) found only octopus cells. In the anterior PVCN, IN, and posterior AVCN, Godfrey (1971) found mainly "choppers" and "primarylike" response patterns. Unfortunately the small sample size does not permit any useful correlations. Given the variety of cell types in DCN, it is not surprising that a greater variety of response patterns was found. In the fusiform layer, two response patterns, "pauser" and "buildup," were most common, and Godfrey (1971) suggested that they correlate with the fusiform cells.

The importance of Godfrey's work lies in the strict histological controls he employed and the fact that complementary neuroanatomy was carried out simultaneously by Cohen (1972). Thus, two significant correlations have been established between cell type and firing pattern. However, this correlation does not prove that the spike discharges originate from these cells, and final proof awaits intracellular marking. The correlation also does not prove that these cells are incapable of other response patterns since Godfrey was not able to classify 60 out of 165 cells studied in posterior PVCN. Of 925 cochlear nucleus cells examined, 385 could not be adequately classified.

One important issue is: How stable are the unit response patterns with variation of stimulus frequency and intensity? Pfeiffer (1966a) and Godfrey (1971) gave no information about the response patterns of their cells other than at best frequency. In DCN, Greenwood and Maruyama (1965) have shown that response patterns can change as a function of changes in both frequency and intensity of the stimulus. They have shown that a nonmonotonic unit can change its pattern from a "primarylike" or sustained pattern to an onset response with an increase in intensity. Response patterns that appeared to be "pausers" could change to onset patterns when stimulus parameters were manipulated. Inhibitory side bands were common, and response patterns were the product of an interplay between excitation and inhibition.

Complex or dynamic sounds have not often been used to study cochlear nucleus response patterns (Erulkar, Butler, & Gerstein, 1968; Glattke, 1968; Møller, 1972). Using intracellular and extracellular recordings, Erulkar et al. (1968) found that the responses to complex stimuli of most cells could be predicted from the responses to simple stimuli. Møller (1972) reported that the tuning curves of some rat cochlear nucleus units became sharper when the sound frequency was rapidly varied. Unfortunately, the locations of the units were not reported in either of these studies. It would seem to us that a large number of physiological studies of the auditory pathway appear to lack accurate histological control. As a result, a correlation between cell morphology and physiological response is rendered difficult or impossible.

Responses in Unanesthetized Animals

Moushegian, Rupert, and Galambos (1962) recorded from VCN of the unanesthetized cat. Although no data were presented, they reported no differences in response patterns between anesthetized and unanesthetized preparations. Using a paralyzed preparation, Evans (1970) also found no differences in VCN, but reported dramatic changes in DCN response patterns in that the complex interactions between excitation and inhibition reported by Greenwood and Maruyama (1965) were even more apparent in the absence of anesthetic.

Webster (1973) found tonotopic organization present in the cochlear nucleus of the awake cat. Some onset units studied in the octopus cell area had little or no spontaneous activity, indicating that the onset response pattern may not always be dependent on the effects of anesthetic. When the responses of VCN cells were studied they appeared to vary very little as a function of tone frequency and intensity. "Primarylike" patterns remained "primarylike" over the response area and onset responses remained onset in type (Figures 3 and 4). "Chopper" patterns also appeared invariant over frequency. A few cells showed slightly more diverse patterns, and there was one significant deviation from a stable response pattern as a function of frequency: no matter what response pattern was obtained at the best frequency, if the unit could be driven by low frequency tones, then its spikes would also be phase locked (Figure 3).

Moushegian et al. (1962) have also reported that the ongoing activity of VCN could be suppressed during a tone. This result has not been confirmed in the awake cat (Webster, 1973) or in the anesthetized cat (Godfrey, 1971). However, similar effects have been seen in the DCN of the awake and anesthetized preparation (Greenwood & Maruyama, 1965; Evans, 1970; Godfrey, 1971; Webster, 1973).

Summary and a Speculation

The cochlear nucleus is clearly concerned with a variety of different transformations of information received from the auditory nerve. Both

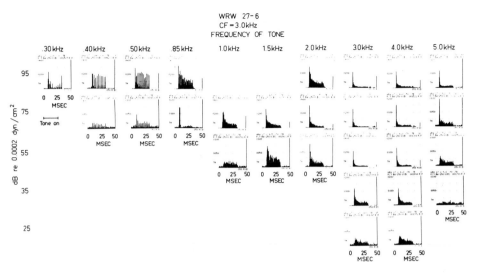

Figure 3. PST histograms as a function of frequency and intensity of ventral cochlear nucleus unit recorded in the awake cat showing "primarylike" PST histograms at high frequencies and phase locking at low frequencies. Number of stimuli = 300. (From Webster, 1973)

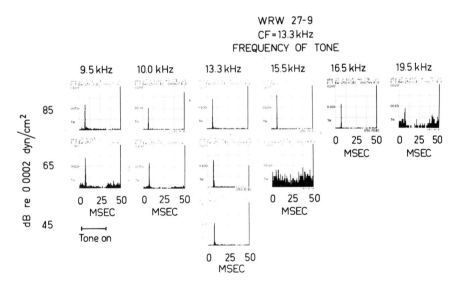

Figure 4. PST histograms as a function of frequency and intensity of ventral cochlear nucleus unit recorded in the awake cat showing "onset" pattern over all frequencies. Number of stimuli = 300. (From Webster, 1973)

VCN and DCN produce modifications of auditory messages, but more complex transformations appear to take place in the DCN, as is most clearly shown in experiments with unanesthetized animals. There is a lack of data on receptive fields of cochlear nucleus cells especially in the ventral nucleus. Attempts to study responses to complex stimuli appear premature without considerably more information on the properties of receptive fields in relation to simple stimuli.

These data raise the question: Just what are the functional differences between VCN and DCN? Although the present body of physiological information does not permit a theory about such differences, an analogy can be made between the visual and auditory systems.

In both systems the primary nerve bundle, optic or auditory, bifurcates to supply two nuclear masses. In the case of the visual system these two regions are geographically separate—the lateral geniculate body and the superior colliculus. In the auditory system they are not separate—the anteroventral, posteroventral, and dorsal cochlear nucleus form one mass, the cochlear nucleus. However, the laminar structure of the dorsal cochlear nucleus resembles that of the superior colliculus and both regions additionally receive a descending innervation and diverse input.

We would like to speculate that the ventral cochlear nucleus and the core of the auditory pathway to which it is eventually connected— superior olivary nuclei, central nucleus of inferior colliculus, ventral nucleus of medial geniculate body, and area AI (first auditory field)— form a pathway primarily connected with auditory perception. A second pathway—involving DCN, the nuclei of the lateral lemniscus, parts of the inferior colliculus (possibly including the external and pericentral nuclei), the medial division of the medial geniculate body, and the secondary auditory cortices—may be concerned with nonspecific acoustic features and may be the origin of acousticomotor connections. A similar speculation has been made by Graybiel (1973) for all sensory systems in her distinction between "lemniscal line systems" and "lemniscal adjunct systems." It would seem to us that the role of this latter system needs to be investigated by auditory physiologists.

INFERIOR COLLICULUS

Anatomy

It is generally accepted by most anatomists that the inferior colliculus may be subdivided into three gross regions (Ramón y Cajal, 1955;

Morest, 1964a; Berman, 1968; van Noort, 1969; Rockel & Jones, 1973a,b,c; Jones & Rockel, 1973). However, there is disagreement as to the limits of the three areas and their relative importance as a proportion of the whole inferior colliculus. All groups accept that a central (ICC) and an external nucleus (ICX) may be defined, but the dorsal cap of the nucleus, referred to by Berman (1968) as the pericentral nucleus (ICP) is believed by Morest to have a cortical structure and to form a dominant part of the whole nucleus (Morest, 1964a, 1966; Geniec & Morest, 1971). There is further disagreement between workers as to whether the pericentral nucleus is a layered structure, and it is possible that the more ventral parts of Morest's dorsal cortex are components of the central nucleus (Rockel & Jones, 1973a,c).

Morest (1964a) and Rockel and Jones (1973a) found that ICC is laminated and they suggested that these laminations might be the basis of tonotopic organization. Rockel and Jones (1973a) reported that laminations were only present in the ventral two-thirds of ICC and this agrees with the extent of the laminations described by Morest. In contrast to an earlier report from their laboratory (Rockel, 1971), Rockel and Jones (1973a) found that the laminae curve at their edges and those closest to the center of curvature form complete spheres. This observation could suggest that some microelectrode penetrations of ICC could show small reversals of best frequency as the same lamina is encountered a second time.

The laminations are produced by disk-shaped cells in man (Geniec and Morest, 1971) and bitufted and fusiform cells in cat (Rockel and Jones, 1973a), whose dendrites line up to form the laminae. These dendritic layers are paralleled by afferent fibers entering from the lateral lemniscus. Geniec and Morest (1971) have argued that the interdigitation of dendrites with incoming fibers would provide the optimal basis for preserving tonotopic organization. Geniec and Morest (1971) also report three types of stellate cells in ICC, and Rockel and Jones (1973a) report three types of multipolar cells. The descriptions of these two groups of cells are almost identical and the largest cell of each group is reported to have dendrites which cross many laminae.

Osen (1972) observed that lesions in the spherical cell area of AVCN produced degeneration in the contralateral ICC. If the lesions are small and discrete, then the pattern of degeneration indicates a tonotopic organization with low frequencies located dorsally and high frequencies ventrally placed. Osen also found that DCN projects to the contralateral ICC with a similar tonotopic organization. Van Noort (1969) obtained comparable results except that he predicted that tonotopic organization would be from lateral to medial in ICC. Both van Noort and Osen found that no part of the cochlear nucleus projected directly to the ipsilateral ICC. Van Noort (1969) further found that the medial superior olivary nucleus projects to both ICCs and that the major projection is to the ipsilateral ICC. Furthermore he observed that the lateral superior olivary nucleus projects equally to the contralateral and the ipsilateral ICC. These results indicate that a given cell in ICC could receive a very complex input from lower centers and in general agree with the earlier findings of Stotler (1953).

The ascending projections from the inferior colliculus consist of a large ipsilateral projection and a smaller contralateral one (Moore & Goldberg, 1963; Powell & Hatton, 1969). Most of the ipsilateral projection terminates in the rostral part of the ventral division and in the medial division of the medial geniculate body. The dorsal division of the medial geniculate body receives only a sparse projection but the lateral posterior nucleus of the thalamus also receives inferior colliculus fibers. The contralateral projection travels via the commissure to the contralateral ICC and to the contralateral brachium. These fibers terminate in much the same way as ipsilateral fibers except none appear to reach the caudal tip of the ventral medial geniculate body. The inferior colliculus also has strong projections to the superior colliculus, the cerebellum, the pretectum, and the midbrain reticular formation. Detailed studies of the projections of ICP and ICX are not yet available (Geniec & Morest, 1971).

Diamond, Jones, and Powell (1969) found that the auditory cortex projects to ICP or the dorsal cortex and to ICC with a possible projection to ICX. The corticofugal fibers enter the laminations in ICC from the opposite end to that of the ascending fibers from the lemniscus. The two groups of fibers run in opposite directions and overlap (Diamond et al., 1969).

Responses in Anesthetized Animals

The central nucleus of the inferior colliculus in cat and rabbit has been shown to be tonotopically organized with respect to best frequency (Rose, Greenwood, Goldberg, & Hind, 1963; Aitkin et al., 1970; Aitkin, Fryman, Blake, & Webster, 1972). When a microelectrode penetrates ICC from either a dorsolateral to medioventral or a dorsal to ventral direction, then cells are encountered with progressively higher best frequencies. This pattern of tonotopic organization is consistent with the laminar structure of ICC and with Osen's (1972) findings. It is not consistent with van Noort's account of how tonotopic organization could be structured in ICC. Rose et al. (1963) reported a reverse sequence in

ICX, with high best frequencies encountered dorsally and low ventrally. Webster and Veale (Figure 5) found a short reverse sequence in ICP or cortex and the extent of the sequence was consistent with the extent of the cortex described by Morest (1964a). The demonstration of tonotopic organization in this region would not directly follow from the projection patterns to the area, since both lemniscal and ICC projections terminate only in the lower regions. However, recent investigations by Rockel and Jones (1973c) suggest that one source of input to ICP may be the dorsal nucleus of the lateral lemniscus.

Most studies of the central nucleus have concluded that it is not possible to form a response pattern classification to simple tonal stimuli presented at best frequency (Rose *et al.*, 1963; Geisler, Rhode, & Hazelton, 1969; Stillman, 1971). Not only is there variation across units, but response patterns of a given unit can change markedly with variations in stimulus parameters such as intensity and tone frequency. For example, a response pattern may change from a "primarylike" to a "pauser" with increasing intensity at best frequency. Some of these changes are related to the fact that a large number of cells have nonmonotonic functions relating spike count to stimulus intensity (Rose *et al.*, 1963), which could indicate that complex interactions between excitation and inhibition occur more often in the ICC than in the VCN. Given this

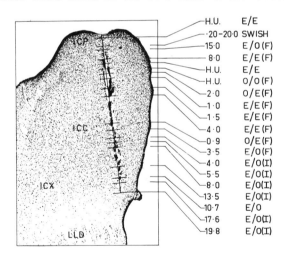

Figure 5. Sagittal section of inferior colliculus showing electrode track and best frequencies of units recorded along the track. A short tonotopic sequence from 15.0 kHz to 2.0 kHz is seen in the pericentral nucleus and a tonotopic sequence from 1.0 to 19.8 kHz is seen in the central nucleus (ICC). The column of letters summarizes the binaural responses of a cell, e.g., E/E (F) indicates that unit is excited by contra and ipsilateral ear stimuli and that the binaural response is one of facilitation. H.U. indicates an habituating unit. (From Webster and Veale, 1971)

response pattern variation and given that there are only two types of principal cells or cells forming the laminations, it might not be possible to find simple correlations between cell type and response pattern. It should be pointed out, however, that some of the claims about diversity of response patterns are based on studies with small sample sizes (Rose *et al.*, 1963; Geisler *et al.*, 1969) in which a small number of cells are described in great detail. However, at least some inferior colliculus units respond to tonal stimuli in a fashion little influenced by frequency or intensity. The unit illustrated in Figure 6 has a powerful onset burst and sustained firing near the best frequency (8 kHz) and fires in an onset fashion at the edges of the response area.

In both rabbit ICX (Aitkin *et al.*, 1972) and cat ICP (Webster & Veale, 1971), the response patterns of cells are quite different from those recorded in ICC. The tuning about the best frequency is often broad and in many cases it is so broad that the concept of best frequency becomes almost inappropriate (Figure 7). It is also difficult to obtain best frequencies for many cells since they habituate very quickly to repeated stimuli. Some units cease firing after two or three stimulus presentations. These observations suggest that both the tonotopic orga-

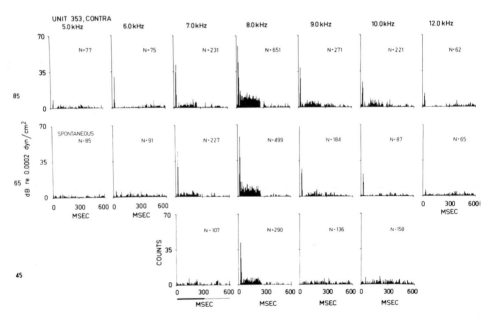

Figure 6. Response area of unit recorded in the inferior colliculus of the anesthetized cat to tones of 300 msec duration. Best frequency = 8.0 kHz; number of stimuli = 20. (From Webster and Veale, 1971)

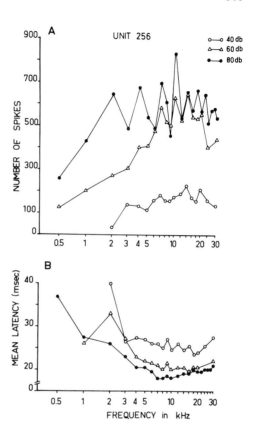

Figure 7. (A) Number of spikes from unit in ICP of anesthetized cat plotted as a function of frequency for three levels of intensity for contra-lateral stimulation. Each point is based on 20 stimulus presentations. (B) Mean latency for same unit plotted as a function of frequency for three levels of intensity. Both figures show the broad tuning of this unit. (From Webster and Veale, 1971)

nization and function of these neurons are quite different from the cells of ICC. Geniec and Morest (1971) have suggested that layer IV of the ICP cortex is capable of integrating the input and output of ICC with projections to other layers from the auditory cortex. Jane, Masterton, and Diamond (1965) on the basis of behavioral studies have suggested that the apex of inferior colliculus (including the cortical regions) is involved in the ability of cats to attend to auditory stimuli. Such data are quite compatible with physiological results and suggest that very complex forms of processing occur in ICP. Further studies of this area are needed to determine more precisely the relationship between the structure and the firing pattern of the cells.

If a hierarchical organization exists in the auditory system, as it appears to in the visual system, it would seem important to determine the relationship between responses to complex stimuli and the response area. It has been argued that relatively simple stimuli such as pure tones and clicks cannot reveal the complexity of higher levels of the auditory pathway, and that complex, dynamic stimuli are required to strongly activate neurons (Nelson, Erulkar, & Bryan, 1966). Nelson *et al.* found that the responses of many inferior colliculus cells to complex stimuli could not be predicted from knowledge of their response areas to simple tones since these neurons responded selectively to aspects of amplitude and frequency modulated (FM) tones. These results appear to contrast with published findings for the cochlear nucleus (Erulkar *et al.*, 1968) and may represent an example of the integration and refining of information as it ascends the auditory pathway.

Responses in Unanesthetized Animals

The responses recorded in the awake animal were considerably different from those recorded under anesthetic (Bock, Webster, & Aitkin, 1972). The response areas of cells were much broader and discharge rate was usually a monotonic function of stimulus intensity. A great variety of response patterns was observed over a response area, but sustained responses at the best frequency were a common observation. In the anesthetized cat, a very common response to a tone consisted of an onset burst followed by a pause and then a sustained response throughout the tone. This response pattern was infrequently seen in the awake animal. Many more cells responded with a purely inhibitory pattern although this was probably because the identification of this pattern was aided by the high level of spontaneous activity in the awake preparation.

The locations of many of the above cells in the inferior colliculus were not determined as the histological controls were inadequate, but later work (Bock, 1973) showed all these effects in neurons located in ICC. Tonotopic organization was also observed in ICC without anesthetic although determination of threshold curves was difficult because of the high levels of spontaneous activity. Since free-field stimulation was employed in these experiments, it is difficult to make detailed comparisons with earlier work under anesthetic. It is possible that this form of stimulus presentation eliminated nonmonotonic intensity functions, but to us this appears an unlikely explanation. If, as Stillman (1971) suggests, the response pattern carries a neural code, then these data indicate that one should be careful in drawing conclusions about sensory coding and auditory feature detectors in ICC from data obtained with the anesthetized preparation.

In view of the preponderance of sustained effects to tone stimuli in the unanesthetized cat inferior colliculus, it may well be that sustained firing or depression is the normal response to a sustained stimulus throughout the primary auditory pathway and that more complex patterns either indicate the presence of a pharmacological agent such as an anesthetic or indicate that the stimulus is inappropriate for influencing the unit under study.

Concluding Remarks

Within the central nucleus of the inferior colliculus we have been unable to correlate unit discharge pattern with cell morphology, although the laminar structure of this region seems well correlated with tonotopic organization of best frequency. A gross correlation between regions in the inferior colliculus and response pattern is possible since many cells in ICP and ICX have response patterns quite different from those of the central nucleus. It would seem that more single unit studies of the inferior colliculus need to be carried out with both simple and complex stimuli before any conclusions can be reached about the way inferior colliculus units modify information ascending from the cochlear nucleus. These experiments should be carried out with unanesthetized animals, since anesthesia considerably modifies response patterns in the inferior colliculus.

MEDIAL GENICULATE BODY

Anatomy

Considerable progress in understanding the functional properties of neurons in the medial geniculate body (MGB) has been made possible by a series of important neuroanatomic studies of this region carried out in the last decade. The MGB is composed of three major subdivisions—the dorsal, medial, and ventral divisions—of which only one, the ventral division, can be considered as a specifically auditory thalamic relay nucleus (Morest, 1964b, 1965a,b).

The projection patterns of fibers related to each subdivision are distinctly different. Thus, fibers from the ventral division terminate primarily in the first auditory field (AI) (Rose & Woolsey, 1949; Woolsey, 1964), and also distribute to areas AII and Ep (Wilson & Cragg, 1969; Graybiel, 1973). Neurons in the ventral division receive ascending fibers from the central nucleus of the inferior colliculus (Moore & Goldberg, 1963). Neurons in the medial division project fibers to the auditory cortex but while these terminate in and around the ectosylvian cortex they are not concentrated in any specific cortical subdivision (Rose & Woolsey,

1958; Woolsey, 1964; Graybiel, 1973). Furthermore, neurons in the medial division, although supplied to some extent by fibers from the inferior colliculus, receive a multimodal innervation (Nauta & Kuypers, 1958; Altman & Carpenter, 1961; Graybiel, 1972). Finally, input to the dorsal nucleus of MGB originates from structures outside the central nucleus of the inferior colliculus (Morest, 1965b) and fibers from the dorsal division terminate mainly in the insular and temporal cortical fields (Rose & Woolsey, 1958; Diamond, Chow, & Neff, 1958; Graybiel, 1973). The separate nature of these subdivisions is made further manifest by the differences in the corticofugal projections to each region of the medial geniculate body (Diamond et al., 1969).

The use of the Golgi technique has enabled the classification of medial geniculate neurons in terms of their dendritic branching patterns (Morest, 1964b). The dendrites of neurons in the ventral division are tufted, i.e., are segregated at each end of the cell body. The long axes of these neurons are oriented parallel to the incoming fibers of the brachium of the inferior colliculus, and together dendrites and fibers form laminae roughly parallel to the lateral surface as seen in Figure 8 (LV). Neurons in the dorsal division are characterized by radially

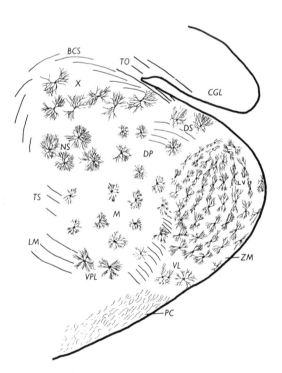

Figure 8. Golgi–Cox stained transverse section of 15-day-old cat. Typical distribution of principal neuron types at the junction of the anterior and middle thirds of the medial geniculate body. Subdivisions of MGB abbreviated as follows: LV = pars lateralis; VL = nucleus ventro-lateralis; ZM = zona marginalis; M = divisio medialis; DS = nucleus dorsalis superficialis; DP = nucleus dorsalis profondis. (From Morest, 1964b)

oriented dendrites (Figure 8: DS, DP) while neurons in the medial division (Figure 8: M) may be of the radiate type (especially posteriorly) or tufted (especially anteriorly). Dendrites of radiate neurons of the medial division may be very long, often exceeding 300 μm, and are pierced along their length by longitudinal fibers (Morest, 1964b).

The preceding comments refer to the principal neuron whose axon projects beyond the nucleus of origin. However, neurons with short axons (Golgi type II) are common in each region and in some cases their axons appear to contact the principal cells (Morest, 1971) (Figure 9). The origin of afferent input to the Golgi type II cell of the medial geniculate body is not clear—it may be from the incoming fibers of the brachium of the inferior colliculus, from the dendrites of the principal cells, or from corticofugal fibers. There do not appear to be many recurrent collaterals from principal cell axons of the ventral division compared

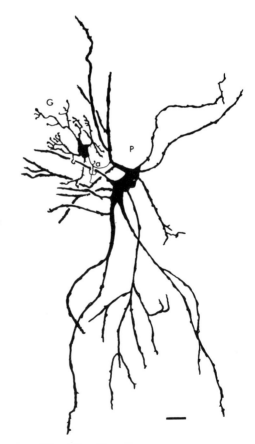

Figure 9. Mature Golgi type II neuron (G). Its axon (a) and dendrites end in association with intermediate dendrites of the principal neuron (P) in the ventral nucleus. (From Morest, 1971 [Berlin–Heidelberg–New York: Springer])

with the number seen in the dorsal division of the medial geniculate body or the lateral geniculate body (Morest & Webster, unpublished observations).

The electron microscope has revealed added complexities at the synaptic level (Majorossy & Réthelyi, 1968; Morest, 1971; Jones & Rockel, 1971). Terminals of fibers from the brachium of the inferior colliculus and from Golgi type II cells enter into a complex relationship with dendrites of principal and Golgi type II cells in synaptic "nests" (Morest, 1971), "clusters" (Majorossy & Réthelyi, 1968), or "glomeruli" (Jones & Rockel, 1971). Additionally, corticofugal fibers terminate on principal cell dendrites distal to the glomeruli. It is interesting that a similar type of synaptic specialization occurs in three of the principal thalamic sensory nuclei—the medial geniculate body, lateral geniculate body (Szentagothai, 1963; Jones & Powell, 1969), and ventroposterior nucleus (Jones & Powell, 1969).

Responses in Anesthetized Animals

Responses to tonal stimuli of units in the three regions of the medial geniculate body are distinctively different in the pentobarbital-anesthetized cat. Neurons in the dorsal division do not appear to respond to tonal stimuli in any clearly defined manner (Aitkin, 1973), although units in the chloralose-anesthetized preparation are frequently excited by click stimuli (Altman, Syka, & Shmigidina, 1970). On the other hand, the majority of units in both the medial and ventral divisions are influenced by tonal stimuli and mostly discharge only at the onset of the tone pip (Adrian, Lifschitz, Tavitas, & Galli, 1966; Aitkin & Webster, 1972). However, with more intense tonal stimuli (80–100 dB SPL) a variety of late discharges may be evoked although these often occur after the stimulus-on period (Dunlop, Itzkowic, & Aitkin, 1969). When onset discharges are evoked in a given unit they are usually invariant in pattern with regard to stimulus frequency or intensity.

Units in the ventral division discharge regularly to stimuli presented at rates of 1 per second, but this feature is only true for one population of units in the medial division. Units in a second group in the medial division show considerable irregularity in their firing and habituate to stimulus rates as low as 1 per 4 seconds (Aitkin, 1973). Many of the discharge properties of these units are strikingly similar to neurons in the pericentral nucleus of the inferior colliculus (p. 338) and in the suprasylvian association areas (Irvine, Wester, & Thompson, 1973).

A marked difference between units in the medial and ventral divisions lies in the shape of the threshold tuning curves which may be measured for them. Ventral division neurons are usually sensitive to a restricted range of tonal frequency at threshold (Figure 10 A), while those in the medial division are often extremely broadly tuned (Figure 10 B). However, not all elements in the medial division are broadly tuned—unit 72-15-3 (marked with asterisks in Figure 10 B) is indistinguishable from curves in Figure 10 A.

It is possible that a correlation exists between morphology and discharge characteristics. Thus tufted neurons of the ventral division receive a dense innervation from a relatively restricted number of preterminal fibers ascending from the inferior colliculus (Morest, 1964b), which correlates with the relatively sharp tuning curves of neurons in the ventral division. The larger dendritic fields of medial radiate neurons receive few terminals from any one fiber but a dendrite contacts many fibers (Morest, 1964b). This correlates with the broadly tuned, irregular firing of units in this region if it be assumed that fibers supplying them arise from the inferior colliculus. Furthermore, the laminae of tufted neurons

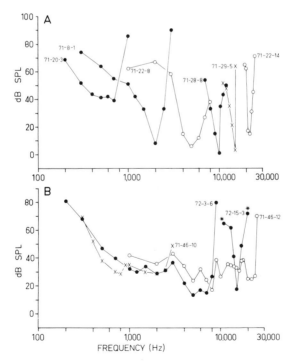

Figure 10. Threshold tuning curves from units in the ventral (A) and medial (B) divisions of MGB. Numbers appended to each curve indicate the year, cat number, unit number. (From Aitkin, unpublished observations)

and fibers of the ventral division correspond to a tonotopic organization of this structure (Aitkin & Webster, 1971, 1972) in which low frequency elements are located laterally and high frequency units medially, as illustrated in Figure 11. No similar data have emerged for the medial division and it would seem that, because of the broad tuning of the majority of units in this region, such an organization would at best be very blurred.

The function of the Golgi type II cells and the synaptic "nests" with which they participate is uncertain. The perikarya of these neurons are smaller than those of principal neurons (Morest, 1971) and, if in fact they produce action potentials, it is possible that few recordings have been made of their spike discharges with the relatively large metal microelectrodes that most workers have used. However, they may subserve an inhibitory function since strong and widespread inhibitory effects are produced by auditory stimuli in medial geniculate neurons (Aitkin, Dunlop, & Webster, 1966; Webster & Aitkin, 1971), and these effects appear to be generated within the medial geniculate (Aitkin & Dunlop, 1969).

Little information is available relating to the specific ways that medial geniculate body neurons respond to FM or AM stimuli, although Whitfield and Purser (1972) and Watanabe (1972) report that the modes of response of these neurons are very similar to those of cortical neurons. Thus at this stage it is not clear what further elaboration of feature extraction with regard to changing tonal frequency occurs at this level but it has been clearly shown that certain other auditory trigger fea-

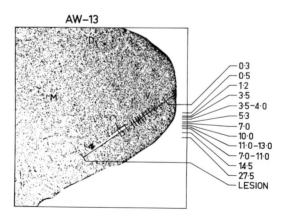

Figure 11. Microelectrode penetration through the MGB of experiment AW-13. Numbers to the right of the illustration are the best frequencies of units or unit clusters located along the electrode track at the points marked. V = MGB, ventral division; M = MGB, medial division; D = MGB, dorsal division. (From Aitkin & Webster, unpublished observations.)

tures—for example, the direction of sound movement with clicks (Altman et al., 1970) and interaural time and intensity differences (Aitkin & Webster, 1972)—are recoded at the medial geniculate body. As with other areas, more parametric data are needed which relate receptive fields to responses to complex stimuli.

Responses in Unanesthetized Animals

Most of the conclusions reported in the preceding section should be considered in the light of the profound effects anesthetic agents may have upon the activity of medial geniculate units. The most obvious feature of the unanesthetized preparation is that the spontaneous activity of these units is greatly increased compared with the anesthetized preparation (Aitkin et al., 1966; Imig, Weinberger, & Westenberg, 1972; Whitfield & Purser, 1972). Such a high background firing rate renders the estimations of threshold, best frequency, and threshold tuning curves very difficult, as has been mentioned for the inferior colliculus (p. 340), but it also reveals that a high proportion of units in the medial geniculate are inhibited by tonal stimuli (Whitfield & Purser, 1972). This latter finding confirms the observations made in anesthetized cats (Dunlop et al., 1969). Sustained discharges are also very common in the unanesthetized animal and may show substantial modifications as a function of tone frequency and intensity for a given unit (Whitfield & Purser, 1972; Aitkin & Prain, 1974). However, onset responses to tone stimuli do occur in the medial geniculate of the awake cat, suggesting that the onset discharges of the anesthetized cat are not necessarily the result of the depressive effects of anesthetic agents.

One of the most interesting observations made in the medial geniculate body of the unanesthetized preparation is that of the "W-shaped" response area in which a given unit exhibits a central region of excitation (or lessened inhibition) at one frequency and inhibition for neighboring frequencies above and below this "best" frequency (Whitfield & Purser, 1972) (Figure 12). In the latter study these response areas formed more than 50% of the sample examined. It seems reasonable to believe that these response areas can also occur in the anesthetized preparation but are not so readily apparent due to the lack of background firing with which to contrast the inhibiting effects of the stimuli. Some supporting data may be seen in the studies of Dunlop et al. (1969, p. 154) and Aitkin and Webster (1972, p. 372).

Examples of the discharge patterns of a unit showing a "W-shaped" response area are illustrated in Figure 13 (Aitkin & Prain, unpublished). A striking variation in response pattern occurs as a function

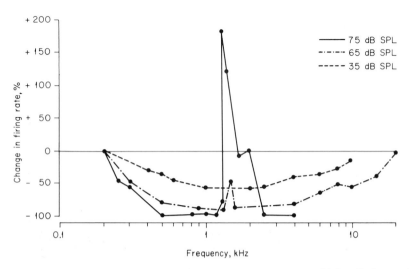

Figure 12. Relative spike counts as a function of tone frequency and intensity for a unit demonstrating a "W-shaped" response curve. (From Whitfield & Purser, 1972 [Basel: S. Karger])

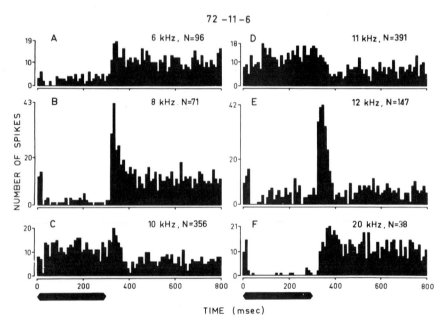

Figure 13. Unit 72–11–6, medial division of MGB. Response histograms at 60 dB SPL, indicating a "W-shape" in this unit's response curve. N = total number of spikes during the tone presentation, where duration is indicated by the bar beneath both abscissas. (From Aitkin & Prain, unpublished observations)

of frequency—not only do spike counts during the tone vary from inhibition at higher and lower frequencies to excitation in the center of the frequency range, but strong offset responses also occur at intermediate frequencies.

Concluding Remarks

The laminar structure of the ventral division of this nucleus may be correlated with a tonotopic organization. Although this finding has been questioned in the literature (Purser & Whitfield, 1972) as being a property more of the ascending fibers supplying the medial geniculate body than of the cells within it, the highly organized projections from AI to ventral MGB (Diamond *et al.*, 1969) are consistent with tonotopic arrangement being a key feature of the organization of ventral MGB. The different properties of tufted and radiate neurons of the ventral and medial division remain to be examined in the unanesthetized preparation. If differences prevail, it seems likely that the broadly tuned radiate neurons of the medial division may be intimately concerned with multimodal interactions as both anatomical and physiological findings (Poggio & Mountcastle, 1960; Wepsic, 1966) suggest, while tufted neurons of the ventral division may be related to the relaying of auditory information to the cortex. The role of inhibition in coding at this thalamic nucleus is enlarged in view of recent experiments with unanesthetized animals. It is possible that integration in the ventral MGB may be mediated via Golgi type II cells.

AUDITORY CORTEX

Anatomy

The majority of single unit studies of auditory cortical neurons have been carried out in area AI and, in view of this fact, we will concentrate on the anatomy and physiology of this region. We feel, however, that the lack of comparative unit data from primary and secondary auditory cortex is one of the major deficits in the current knowledge of this area.

Area AI receives projections from the anterior principal division of MGB (Rose & Woolsey, 1958) or, as defined by Morest (1964b), the ventral division of MGB. Probably AI also receives projections from medial MGB (Rose & Woolsey, 1958; Woolsey, 1964; Graybiel, 1973) and from contralateral AI (Diamond, Jones, & Powell, 1968a). Reciprocal corticocortical connections also exist between AI and AII of the same side (Downman, Woolsey, & Lende, 1960), and probably between all auditory cortical areas except the suprasylvian fringe region and Tunturi's area AIII (Diamond, Jones, & Powell, 1968b).

A point-to-point projection appears to exist between the principal division of MGB and AI (Rose & Woolsey, 1949). The projection of the anterior part of the principal division (PP) to AI ". . . are of a concentrated character with a relatively sharp projection from part of PP to a limited area of AI" (Woolsey, 1964). However, small lesions of the ventral division of MGB lead to a rather widespread zone of termination of degenerating fibers in the cortex (Wilson & Cragg 1969).*

As for most other cortical regions, the auditory cortex of the cat is a six-layered structure, although the degree of lamination and the cell populations differ in the various auditory fields (Rose 1949). The columnar structuring of cells perpendicular to the cortical surface is well marked in AI. With regard to the zones of termination of thalamocortical axons in a given layer of cortex, little precise information exists for the auditory cortex. It seems likely that, as for other sensory cortices, most thalamocortical fibers end in or below layer IV and rarely penetrate to more superficial levels (Colonnier 1966; Lorente de No, 1951; Wong, 1967). This pattern of termination is generally similar for AI (Cragg, personal communication) with a difference in that it includes a considerable component in layer I (Wilson & Cragg, 1969).

The dendritic patterns of auditory cortical neurons are diverse, and neurons may be classified as pyramidal, stellate (star), double bouquet, fusiform, and "special" (including inverted pyramids) (Tunturi, 1971). Small pyramids tend to be more superficially located than large pyramidal cells, and star cells are concentrated in layers III and IV. Pyramidal neurons tend to have circular basal dendritic fields while stellate neurons may exhibit oval or circular fields (Wong, 1967).† In Holmes preparations the consistent grid pattern of vertical and horizontal fibers in layers I–IV, which is apparent in the visual cortex, appears to be lacking in the auditory cortex (Wong, 1967). With the exception of the report of de Lorenzo (1960) little information exists about the synaptic morphology of the auditory cortex in relation either to depth within the cortex or to the specific types of cell found therein.

---

* Recent publications of Sousa–Pinto (1973b) and Niimi and Naito (1974) support a point-to-point projection from MGB to AI.

† These findings are in general consistent with those of a recent study by Sousa–Pinto (1973a).

Physiology

In view of the probable complexity and diversity of the connections of cortical neurons it is not surprising that no clear picture has arisen of the response characteristics of units in AI. Goldstein, Hall, and Butterfield (1968) have concluded that ". . . coding of acoustic stimuli by units of the primary auditory cortex of the cat appears to be a highly individualistic and variegated matter, which almost defies quantification" (p. 454). Although this statement is largely applicable at the time of writing, certain experimental studies are worthy of detailed note since they have made clear some important directions for future investigation.

Tonotopic Organization and Organization in Depth

It has previously been indicated that the ventral nucleus of the medial geniculate body of the cat is tonotopically organized and that it projects upon AI in an orderly fashion. A number of unit studies have been carried out in that cat in which the tonotopic organization of units in AI has been examined (Hind, 1960; Evans, Ross, & Whitfield, 1965; Goldstein, Abeles, Daly, & McIntosh, 1970). The results of these experiments in general reveal only an irregular relationship between best frequency and location across AI. A number of reasons may account for these results. First, there may in fact be a "scrambling" of the orderly connections from ventral MGB at an intracortical level. Both the data of Wilson and Cragg (1969) on the projection patterns from MGB and the observations by Wong (1967) that there is a lack of grid patterning of fibers parallel and perpendicular to the cortical surface lend support to this argument. Second, and more probable in the view of the present authors, tonotopic organization may be present but it may be more difficult to reveal than in subcortical auditory structures.

In addition to the inherent problems involved in superimposing units isolated in many experiments, at different cortical depths, on a single "typical" surface map of AI, two of the major recent studies have been carried out with unanesthetized animals. From what we have described in the present chapter, the contention of J. E. Rose (1968) that "deep anesthesia tends to simplify the picture" (p. 290) may indeed be true in the case of tonotopic organization of the auditory cortex. In the unanesthetized state, ascending thalamic-originating impulses may represent only a small part of the ongoing cortical activity and it may therefore be difficult to extract accurate information about best frequency from the response of a given neuron. This notion is supported by the recent observations by Brugge and Merzenich (1971) and Merzenich and

EXPT. 70-200

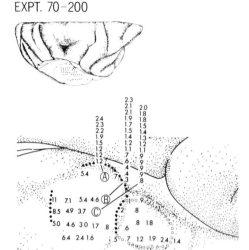

Figure 14. Surface map of the auditory cortex of the rhesus monkey, *Macaca mulatta*, showing the distribution of best frequencies of 42 penetrations into primary auditory cortex and fields rostral and lateral to it. Approximate boundaries of A1, except laterally, are indicated by the line of closed circles (••). Line of small open circles (∘∘) encloses the small rostrodorsal field. Penetrations labeled A, B, C pass down the bank of an elevation on the superior temporal plane. Columns of numbers with these points are best frequencies of neurons encountered at regular intervals within these penetrations. (From Merzenich and Brugge, 1973)

Brugge (1973) that units in the auditory cortex of the anesthetized monkey are organized tonotopically in a systematic fashion. The clarity of such organization is illustrated in Figure 14: Not only is best frequency organized according to a surface map but it is also systematically related to depth in a sulcus.°

Several studies have focused attention on the relationship between depth along a given electrode track and the functional properties of neurons encountered (Hind, 1960; Oonishi & Katsuki, 1965; Abeles & Goldstein, 1970), but it is unfortunate that in all cases only extracellular spike discharges were measured. Since extracellular spikes may be recorded from a unit over substantial microdrive movements in the auditory cortex (Erulkar, Rose, & Davies, 1956), it is difficult to assign a given unit response to a unique cortical layer. Furthermore, intracellular studies combined with the injection of intracellular stains have to date been unsuccessful, so no correlations between cell morphology, layering, and response characteristics are yet available (Goldstein, 1971, p. 336, and personal communication).

Nonetheless, certain consistent findings have emerged from these studies. Units in a given radial penetration may have a similar best frequency

° Merzenich, Knight, and Roth (1973) have now clearly shown that area AI of the cat is organized tonotopically.

(Hind, 1960; Oonishi & Katsuki, 1965; Abeles & Goldstein, 1970) particularly if they are narrowly tuned. Units with broad tuning curves or multipeaked curves are frequently encountered in the same penetration. Oonishi and Katsuki (1965) have suggested that narrowly tuned units are located in the deeper layers of AI while multipeaked and broad tuning curves are derived from units more superficially located. Abeles and Goldstein (1970) have been unable to confirm these findings. It is possible that broad tuning curves, rather than being the result of intracortical integration, may be related to neurons in medial MGB which project upon AI. Finally, Abeles and Goldstein (1970) have been unable to find any distinctive feature of the responses of units in a given radial penetration to tonal stimuli other than their best frequency. At present, therefore, we must conclude that organization-in-depth in the auditory cortex remains to be resolved.

Responses in Unanesthetized Animals

The majority of recent auditory cortex unit studies have been carried out in unanesthetized animals, either paralyzed (Goldstein, 1968; Goldstein et al., 1968; Hall & Goldstein, 1968; Abeles & Goldstein, 1970; Goldstein et al., 1970; Goldstein, de Ribaupierre, & Yeni–Komshian, 1971; Abeles & Goldstein, 1972) or unparalyzed preparations (Hubel, Henson, Rupert, & Galambos, 1959; Katsuki, Murata, Suga, & Takenaka, 1959; 1960; Bogdanski, 1960; Katsuki, Suga, & Kanno, 1962; Gerstein & Kiang, 1964; Evans & Whitfield, 1964; Evans et al., 1965; Whitfield & Evans, 1965; Funkenstein, Nelson, Winter, Wollberg, & Newman, 1971; Wollberg & Newman, 1972; Miller, 1971; Miller, Sutton, Pfingst, Ryan, Beaton, & Gourevitch, 1972; Swarbrick & Whitfield, 1972; Brugge & Merzenich, 1973). In contrast to this, only three recent studies employing accurately calibrated tonal stimuli have been carried out in the anesthetized preparation (Hind, 1960; Oonishi & Katsuki, 1965; Brugge, Dubrovsky, Aitkin, & Anderson, 1969). As a consequence, it is very difficult to make comparisons between responses in each state although, as for lower centers, it is apparent that background and evoked firing is severely depressed in the anesthetized state (Goldstein, 1968). In the anesthetized cat many units do not appear to respond to tones, and, for those that do, onset responses are prevalent (Erulkar et al., 1956; Brugge et al., 1969). In the absence of anesthetic, onset responses may still be observed but sustained responses—excitation or inhibition—are frequent (Evans & Whitfield, 1964; Goldstein et al., 1968).

In a study of single units in the macaque auditory cortex, Brugge and Merzenich (1973) employed molded ear inserts to enable them to use calibrated binaural stimuli in the awake animal. As for the medial geniculate body, nonmonotonic functions were reported for many units, and discharge patterns could vary in shape across the response area of a given unit. They have also shown that response properties relevant to sound localization—sensitivity to interaural time and intensity differences—are as accurately registered by auditory cortical neurons as they are in the anesthetized animal.

Feature Extraction

Perhaps the most interesting studies of units in the unanesthetized auditory cortex are those in which auditory stimuli are employed in a more complex situation. A number of workers have suggested that, at higher levels of the auditory nervous system, the significance of the auditory stimulus to the awake animal is of paramount importance, and that the use of simple stimuli such as pure tones may not enhance our understanding of auditory processing at these levels (Swarbrick & Whitfield, 1972; Funkenstein et al., 1971; Miller, 1971). Whitfield and Evans (1965) have shown that some units relatively unresponsive to steady tones may be vigorously excited by frequency-modulated tones. Continuing work from this laboratory (Swarbrick & Whitfield, 1972) suggests that the concept of stimulus "shape" is relevant to the discharge of some cortical neurons. A very small population of units in the awake squirrel monkey auditory cortex is selectively responsive to squirrel monkey vocalizations and, for some cells, tonal stimuli are ineffective when vocalizations produce strong responses (Funkenstein et al., 1971). Furthermore, units responding to vocalizations can range from those sensitive to parts of a call to those discharging to many components of different vocalizations. Such studies as this require a thorough understanding of the responses of each cell to simple tonal stimuli before deductions as to the neuronal mechanisms responsible for vocalization selectivity can be made. One result of such careful attention to accurate stimulus conditions has been the observation by Brugge and Merzenich (1973) that cortical neurons which are nonmonotonic in their response to changes in stimulus intensity have many properties which suggest that they may act as feature detectors for sound intensity.

The use of "natural" stimuli is one way of circumventing the problem of acoustic "significance" with awake animals; another method is to endow a tonal stimulus with significance by using it as a signal for a task to be performed, in a conditioning procedure. This technique has been used by Miller and his group (1971, 1972), who have recently shown that the rates of discharge of certain units in the awake monkey auditory cortex to a given range of tonal stimuli can be radically altered when tonal stimuli signal the onset of a behavioral task, compared with when the same stimuli are presented in a situation of little significance to the animal.

Concluding Remarks

While some degree of understanding exists about the relationship between neuronal morphology and unit response characteristics in the medial geniculate body, no definitive statements are possible for area AI. Furthermore, detailed comparisons between responses in primary and secondary fields in the unanesthetized animal have not been carried out, tonotopic organization in the cat auditory cortex remains disputed, and the physiological significance of auditory cortical layering remains obscure. Although recent experiments utilizing complex stimuli have been very fruitful and indicate that cortical units show greater variability and selectivity to acoustic stimuli than those in lower auditory regions, such results have not been accompanied by substantial parametric data using simpler acoustic stimuli such as tone pips. In view of the large gaps in our knowledge of the auditory cortex it seems fruitless to us to attempt to fit such data as there are into organizational frameworks equivalent to those described for visual cortex.

There are many problems in experiments with awake animals which are not present with anesthetized preparations and which are brought to a head in cortical studies. These include such factors as the recording stability of the preparation, the effects of states of arousal and degree of attention, the "significance" of complex auditory stimuli and "nonsignificance" of tonal stimuli, and the problems of acoustic measurement with free field stimulation. It seems likely that these problems will be largely overcome in the near future and that our understanding of central auditory processing will impove considerably.

CENTRAL AUDITORY PROCESSING: CONCLUSION

We have attempted to concentrate on two major issues—the relationships between unit discharge characteristics in the auditory pathway and the precise anatomical location and morphology of the responding element, and the influence of anesthetic agents. We feel that these comparisons are required before any serious attempt can be made to evaluate the functions of auditory neurons, including their role in feature detection. At this stage the correlations between morphology and response patterns are few in number and are inadequate to form the foundation of a coherent scheme on which to base the organization of the ascending auditory pathway. Furthermore, the experiments in which such correlations have been attempted have utilized anesthetized animals.

In view of the paucity of information relating to systematic relationships between physiology and anatomy in the auditory system, we would like to make the following suggestions as to the directions which future research could take. First, at all levels of the central auditory system, there is a need for parametric data on the response areas of cells to simple tonal stimuli. The specifications of the receptive fields of such cells, at and above the brainstem level, should include an analysis of their binaural capability. These studies should be based on large sample sizes and they might allow us to establish classes of auditory receptive fields. Second, in conjunction with the establishment of receptive fields, the responses of these cells should be studied with complex or dynamic stimuli and, where appropriate, "natural" stimuli. Third, since anesthesia has profound effects on central auditory neurons, the above data should be obtained where possible in unanesthetized preparations. Fourth, there is a need for studies which are concerned with establishing the continuity of auditory connections, using anatomical methods in conjuction with sensory and electrical stimulation. One major example would be the elucidation of the system which may arise in relation to DCN—the nuclei of the lateral lemniscus, parts of the inferior colliculus and medial geniculate body, and the secondary auditory cortices. Such studies will need careful histological controls and might be directed toward acousticomotor integration. Finally, there is a pressing need for intracellular studies at all levels of the central auditory pathways. In particular, identification of cell types with intracellular marking would allow confirmation of correlations between cell morphology and receptive field type. The recent study of Caspary (1972) on the cochlear nucleus of the kangaroo rat suggests that intracellular marking may be quite feasible in the lower auditory system.

We feel that when we have data from some of the suggested experiments we might be in a better position to make statements about more complex aspects of auditory physiology, such as the existence of neurons which act as feature detectors. At the moment, we feel that only the vaguest characterization can be given of what constitutes an auditory feature detector. If we were able to specify systematically the receptive field organization at each level of the auditory pathway, then we might be able to construct a useful model of central auditory organization. However, all our preceding proposals have been confined to physiological experiments. In the visual system, psychophysical approaches to feature detection have been very fruitful in suggesting the sort of features for which the neurophysiologist might search (Julesz, 1971, pp. 66–74). In particular, the technique of selective adaptation (McCullough, 1965; Lovegrove & Over, 1972) has been most powerful in this regard. An auditory study (Eimas & Corbit, 1973) using this technique has indicated possible complex auditory feature detectors in the human.

We feel that an increasing use of this general approach might allow auditory neurophysiologists to ask more specific questions about auditory receptive fields.

REFERENCES

Abeles, M., & Goldstein, M. H., Jr. Functional architecture in cat primary auditory cortex: Columnar organization and organization according to depth. *Journal of Neurophysiology*, 1970, **33**, 172–187.

Abeles, M., & Goldstein, M. H., Jr., Responses of single units in the primary auditory cortex to tones and tone pairs. *Brain Research*, 1972, **42**, 337–352.

Adrian, H. O., Lifschitz, W. M., Tavitas, R. J., & Galli, F. P. Activity of neural units in medial geniculate body of the cat and rabbit. *Journal of Neurophysiology*, 1966, **29**, 1046–1060.

Aitkin, L. M. The medial geniculate body of the cat: Responses to tonal stimuli of neurons in the medial division. *Journal of Neurophysiology*, 1973, **36**, 275–283.

Aitkin, L. M., Anderson, D. J., & Brugge, J. F. Tonotopic organization and discharge characteristics of single neurons in nuclei of the lateral lemniscus of the cat. *Journal of Neurophysiology*, 1970, **33**, 421–458.

Aitkin, L. M., & Dunlop, C. W. Inhibition in the medial geniculate body of the cat. *Experimental Brain Research*, 1969, **7**, 68–83.

Aitkin, L. M., Dunlop, C. W., & Webster, W. R. Click-evoked response patterns of single units in the medial geniculate body of the cat. *Journal of Neurophysiology*, 1966, **29**, 109–123.

Aitkin, L. M., Fryman, S., Blake, D. W., & Webster, W. R. Responses of neurons in the rabbit inferior colliculus. I. Frequency-specificity and topographic arrangement. *Brain Research*, 1972, **47**, 77–90.

Aitkin, L. M., & Prain, S. Medial geniculate body: Unit responses in the awake cat. *Journal of Neurophysiology*, 1974, **37**, 512–521.

Aitkin, L. M., & Webster, W. R. Tonotopic organization in the medial geniculate body of the cat. *Brain Research*, 1971, **26**, 402–405.

Aitkin, L. M., & Webster, W. R. Medial geniculate body of the cat: Organization and responses to tonal stimuli of neurons in the ventral division. *Journal of Neurophysiology*, 1972, **35**, 365–380.

Altman, J., & Carpenter, M. B. Fiber projections of the superior colliculus in the cat. *Journal of Comparative Neurology*, 1961, **116**, 157–178.

Altman, J. A., Syka, J., & Shmigidina, G. N. Neuronal activity in the medial geniculate body of the cat during monaural and binaural stimulation. *Experimental Brain Research*, 1970, **10**, 81–93.

Barlow, H. B., Narasimhan, R., & Rosenfeld, A. Visual pattern analysis on machines and animals. *Science*, 1972, **177**, 567–575.

Berman, A. L. *The Brain Stem of the Cat. A cytoarchitectonic atlas with stereotaxic coordinates*. Madison: The Univ. of Wisconsin Press, 1968.

Blakemore, C., & Campbell, F. W. On the existence of neurones in the human visual system selectivity sensitive to the orientation and size of retinal images. *Journal of Physiology*, 1969, **203**, 237–260.

Bock, G. R. Discharge patterns of single units in the inferior colliculus of the alert cat. Unpublished doctoral dissertation, Monash University, 1973.

Bock, G. R., Webster, W. R., & Aitkin, L. M. Discharge patterns of single units in inferior colliculus of the alert cat. *Journal of Neurophysiology*, 1972, **35**, 265–277.

Bogdanski, D. F. Chronic microelectrode studies. In G. L. Rasmussen & W. F. Windle (Eds.), *Neural mechanisms of the auditory and vestibular systems*. Springfield, Ill.: Thomas, 1960.

Brugge, J. F., Dubrovsky, N. A., Aitkin, L. M., & Anderson, D. J. Sensitivity of single neurons in auditory cortex of cat to binaural tonal stimulation; effects of varying interaural time and intensity. *Journal of Neurophysiology*, 1969, **32**, 1005–1024.

Brugge, J. F. & Merzenich, M. M. Representation of frequency in auditory cortex in the macaque monkey. In M. B. Sachs (Ed.), *Physiology of the auditory system*. Baltimore: National Educational Consultants, Inc., 1971.

Brugge, J. F., & Merzenich, M. M. Patterns of activity of single neurons of the auditory cortex in monkey. In A. Møller (Ed.), *Basic mechanisms in hearing*. New York: Academic Press, 1973.

Caspary, D. Classification of subpopulations of neurons in the cochlear nuclei of the kangaroo rat. *Experimental Neurology*, 1972, **37**, 131–151.

Cleland, B., & Levick, W. R. Physiology of cat retinal ganglion cells. *Investigative Ophthalmology*, 1972, **11**, 285–290.

Cohen, E. S. Synaptic organization of the caudal cochlear nucleus of the cat: A light and electron microscopical study. Unpublished doctoral dissertation, Harvard University, 1972.

Colonnier, M. L. The structural design of the neocortex. In J. C. Eccles (Ed.), *Brain and conscious experience*. Berlin: Springer-Verlag, 1966.

de Lorenzo, A. J. Electron microscopic observations of the auditory cortex with particular reference to synaptic junctions. In G. L. Rasmussen and W. F. Windle (Eds.), *Neural mechanisms of the auditory and vestibular systems*. Springfield, Ill.: Thomas, 1960.

Diamond, I. T., Chow, K. L., & Neff, W. D. Degeneration of caudal medial geniculate body following cortical lesions ventral to auditory area II in the cat. *Journal of Comparative Neurology*, 1958, **109**, 349–362.

Diamond, I. T., Jones, E. G., & Powell, T. P. S. Interhemispheric fiber connections of the auditory cortex in the cat. *Brain Research*, 1968, **11**, 177–193. (a)

Diamond, I. T., Jones, E. G., & Powell, T. P. S. The association connections of the auditory cortex of the cat. *Brain Research*, 1968, **11**, 560–579. (b)

Diamond, I. T., Jones, E. G., & Powell, T. P. S. The projection of the auditory cortex upon the diencephalon and brain stem in the cat. *Brain Research*, 1969, **15**, 305–340.

Downman, C. B. B., Woolsey, C. N., & Lende, R. A. Auditory areas, I, II and Ep: Cochlear representation, afferent paths and interconnections. *Bulletin of the Johns Hopkins Hospital*, 1960, **106**, 127–142.

Dunlop, C. W., Itzkowic, D. J., & Aitkin, L. M. Tone-burst response patterns of single units in the cat medial geniculate body. *Brain Research*, 1969, **16**, 149–164.

Eimas, P. D., & Corbit, J. D. Selective adaptation of linguistic feature detectors. *Cognitive Psychology*, 1973, **4**, 99–109.

Erulkar, S. D. Comparative aspects of spatial localization of sound. *Physiological Reviews*, 1972, **52**, 237–360.

Erulkar, S. D., Butler, R. A., & Gerstein, G. L. Excitation and inhibition in cochlear nucleus. II. Frequency modulated tones. *Journal of Neurophysiology*, 1968, **31**, 537–548.

Erulkar, S. D., Rose, J. E., & Davies, P. W. Single unit activity in the auditory cortex of the cat. *Bulletin of the Johns Hopkins Hospital*, 1956, **99**, 55–86.

Evans, E. F. Upper and lower levels of the auditory system: A contrast of structure and function. In E. R. Caraneillo (Ed.), *Neural networks*. New York: Springer-Verlag, 1968.

Evans, E. F. Central mechanisms relevant to the neural analysis of simple and complex sounds. In O. J. Gruesser and R. Klinke (Eds.), *Pattern recognition in biological and technical systems*. New York: Springer-Verlag, 1970.

Evans, E. F., Ross, H. P., & Whitfield, I. C. The spatial distribution of unit characteristic frequency in the primary auditory cortex of the cat. *Journal of Physiology*, 1965, **179**, 238–247.

Evans, E. F., & Whitfield, I. C. Classification of unit responses in the auditory cortex of the unanaesthetized and unrestrained cat. *Journal of Physiology*, 1964, **171**, 476–493.

Funkenstein, H. H., Nelson, P. G., Winter, P., Wollberg, Z., & Newman, J. D. Unit responses in auditory cortex of awake squirrel monkeys to vocal stimulation. In M. B. Sachs (Ed.), *Physiology of the auditory system*. Baltimore: National Educational Consultants, Inc., 1971.

Geisler, C. D., Rhode, W. S., & Hazleton, D. W. Responses of inferior colliculus neurons in the cat to binaural acoustic stimuli having wide-band spectra. *Journal of Neurophysiology*, 1969, **32**, 960–974.

Geniec, P., & Morest, D. K. The neuronal architecture of the human posterior colliculus. *Acta Oto-Laryngologica*, 1971, Suppl. **295**.

Gerstein, G. L., & Kiang, N. Y.-S. Responses of single units in the auditory cortex. *Experimental Neurology*, 1964, **10**, 1–18.

Glattke, T. J. Unit responses of the cat cochlear nucleus to amplitude-modulated stimuli. *Journal of the Acoustical Society of America*, 1968, **45**, 419–425.

Godfrey, D. A. Localization of single units in the cochlear nucleus of the cat: An attempt to correlate neuronal structure and function. Unpublished doctoral dissertation, Harvard University, 1971.

Goldstein, M. H., Jr., Single unit studies of cortical coding of simple acoustic stimuli. In F. D. Carlson (Ed.), *Physiological and biochemical aspects of nervous integration*. Englewood Cliffs: Prentice-Hall, 1968.

Goldstein, M. H., Jr. Comment. In M. B. Sachs (Ed.), *Physiology of the auditory system*. Baltimore: National Educational Consultants, Inc., 1971. P. 336.

Goldstein, M. H., Jr., Abeles, M., Daly, R. L., & McIntosh, J. Functional architecture in cat primary auditory cortex: Tonotopic organization. *Journal of Neurophysiology*, 1970, **33**, 188–197.

Goldstein, M. H., Jr., de Ribaupierre, F., & Yeni-Komshian, G. H. Cortical coding of periodicity pitch. In M. B. Sachs (Ed.), *Physiology of the auditory system*. Baltimore: National Educational Consultants, Inc., 1971.

Goldstein, M. H., Jr., Hall, J. L. II, & Butterfield, B. O. Single-unit activity in the primary auditory cortex of unanaesthetized cats. *Journal of the Acoustical Society of America*, 1968, **43**, 444–455.

Graybiel, A. M. Some extrageniculate visual pathways in the cat. *Investigative Ophthalmology*, 1972, **11**, 322–332.

Graybiel, A. M. The thalamocortical projection of the so-called posterior nuclear group: A study with anterograde degeneration methods in the cat. *Brain Research*, 1973, **49**, 229–244.

Greenwood, D. D., & Maruyama, N. Excitatory and inhibitory response areas of auditory neurons in the cochlear nucleus. *Journal of Neurophysiology*, 1965, **28**, 863–892.

Hall, J. L. II, & Goldstein, M. H., Jr. Representation of binaural stimuli by single units in primary auditory cortex of unanesthetized cats. *Journal of the Acoustical Society of America*, 1968, 43, 456–461.

Henry, G. H., & Bishop, P. O. Striate neurons: Receptive field organization. *Investigative Ophthalmology*, 1972, **11**, 357–368.

Hind, J. E. Unit activity in the auditory cortex. In G. L. Rasmussen and W. F. Windle (Eds.), *Neural mechanisms of the auditory and vestibular systems.* Springfield, Ill.: Thomas, 1960.

Hubel, D. H., Henson, C. O., Rupert, A., & Galambos, R. Attention units in the auditory cortex. *Science*, 1959, **129**, 1279–1280.

Hubel, D. H., & Wiesel, T. N. Receptive fields, binocular interaction and functional architecture in the cat's visual cortex. *Journal of Physiology*, 1962, **160**, 106–154.

Hubel, D. H., & Wiesel, T. N. Receptive fields and functional architecture in the two non-striate visual areas (18 & 19) of the cat. *Journal of Neurophysiology*, 1965, **28**, 229–289.

Hubel, D. H., & Wiesel, T. N. Receptive fields and functional architecture of monkey striate cortex. *Journal of Physiology*, 1968, **195**, 215–243.

Imig, T. J., Weinberger, N. M., & Westenberg, I. S. Relationships among unit discharge rate, pattern and phasic arousal in the medial geniculate nucleus of the waking cat. *Experimental Neurology*, 1972, **35**, 337–357.

Irvine, D. R. F., Wester, K. G., & Thompson, R. F. Acoustic tuning of cells in medial suprasylvian "association" cortex of cat. *Proceedings of the Australian Physiological and Pharmacological Society*, 1973, **4**, 61–62.

Jane, J. A., Masterton, R. B., Diamond, I. T. The function of the tectum for attention to auditory stimuli in the cat. *Journal of Comparative Neurology*, 1965, **125**, 165–192.

Jones, E. G., & Powell, T. P. S. Electron microscopy of synaptic glomeruli in the thalamic relay nuclei of the cat. *Proceedings of the Royal Society of London, Series B*, 1969, **172**, 153–171.

Jones, E. G., & Rockel, A. J. The synaptic organization in the medial geniculate body of afferent fibers ascending from the inferior colliculus. *Zeitschrift fur Zellforschung und mikroskopische Anatomie*, 1971, **113**, 44–66.

Jones, E. G., & Rockel, A. J. Observations on complex vesicles, neurofilamentous hyperplasia and increased electron density during terminal degeneration in the inferior colliculus. *Journal of Comparative Neurology*, 1973, **147**, 93–118.

Julesz, B. *Foundations of cyclopean perception.* Chicago: Univ. of Chicago Press, 1971.

Katsuki, Y., Murata, K., Suga, N., & Takenaka, T. Electrical activity of cortical auditory neurons of unanaesthetized and unrestrained cat. *Proceedings of the Japanese Academy*, 1959, **35**, 571–574.

Katsuki, Y., Murata, K., Suga, N., & Takenaka, T. Single unit activity in the auditory cortex of an unanaesthetized monkey. *Proceedings of the Japanese Academy*, 1960, **36**, 435–438.

Katsuki, Y., Suga, N., & Kanno, Y. Neural mechanism of the peripheral and central auditory system in monkeys. *Journal of the Acoustical Society of America*, 1962, **34**, 1396–1410.

Katsuki, Y., Watanabe, T., & Maruyama, N. Activity of auditory neurons in upper levels of brain of cat. *Journal of Neurophysiology*, 1959, **22**, 343–359.

Kiang, N. Y.-S. Stimulus coding in the auditory nerve and cochlear nucleus. *Acta Oto-Laryngologica*, 1965, **59**, 186–200.

Lorente de No, R. Cerebral cortex: Architecture, intracortical connections, motor projections. In J. F. Fulton (Ed.), *Physiology of the nervous system*. New York: Oxford Univ. Press, 1951.

Lovegrove, W. J. & Over, R. Color adaptation in spatial frequency detectors in the human visual system. *Science*, 1972, **176**, 541–543.

Majorossy, K., & Rethelyi, M. Synaptic architecture in the medial geniculate body (ventral division). *Experimental Brain Research*, 1968, **6**, 306–323.

McCullough, C. Color adaptation of edge-detectors in the human visual system. *Science*, 1965, **149**, 1115–1116.

Merzenich, M. M., & Brugge, J. F. Representation of the cochlear partition on the superior temporal plane of the macaque monkey. *Brain Research*, 1973, **50**, 275–296.

Merzenich, M. M., Knight, P. L., & Roth, G. L. Cochleotopic organization of primary auditory cortex in the cat. *Brain Research*, 1973, **63**, 343–346.

Miller, J. M. Single unit discharges in behaving monkeys. In M. B. Sachs (Ed.), *Physiology of the auditory system*. Baltimore: National Educational Consultants, Inc., 1971.

Miller, J. M., Sutton, D., Pfingst, B., Ryan, A., Beaton, R., & Gourevitch, G. Single cell activity in the auditory cortex of rhesus monkeys: Behavioral dependency. *Science*, 1972, **177**, 449–451.

Møller, A. R. Coding of sounds in lower levels of the auditory system. *Quarterly Reviews of Biophysics*, 1972, **5**, 59–155.

Molnar, C. E., & Pfeiffer, R. R. Interpretation of spontaneous spike discharge patterns of neurons in the cochlear nucleus. *Proceedings of Institute of Electrical and Electronics Engineers*, 1968, **56**, 993–1004.

Moore, R. Y., & Goldberg, J. M. Ascending projections of the inferior colliculus in the cat. *Journal of Comparative Neurology*, 1963, **121**, 109–136.

Morest, D. K. The laminar structure of the inferior colliculus of the cat. *Anatomical Record*, 1964, **148**, 314. (a)

Morest, D. K. The neuronal architecture of the medial geniculate body of the cat. *Journal of Anatomy*, 1964, **98**, 611–630. (b)

Morest, D. K. The laminar structure of the medial geniculate body of the cat. *Journal of Anatomy*, 1965, **99**, 143–160. (a)

Morest, D. K. The lateral tegmental system of the midbrain and the medial geniculate body: Study with Golgi and Nauta methods in cat. *Journal of Anatomy*, 1965, **99**, 611–634. (b)

Morest, D. K. The cortical structure of the inferior quadrigeminal lamina of the cat. *Anatomical Record*, 1966, **154**, 389–390.

Morest, D. K. The collateral system of the medial nucleus of the trapezoid body of the cat, its neuronal structure and relation to the olivo-cochlear bundle. *Brain Research*, 1968, **9**, 288–311.

Morest, D. K. Dendrodendritic synapses of cells that have axons: The fine structure of the Golgi type II cell in the medial geniculate body of the cat. *Zeitschrift für Anatomie und Entwicklungsgeschichte*, 1971, **133**, 216–246.

Morest, D. K., Kiang, N. Y.-S., Kane, E. C., Guinan, J. J., Jr., & Godfrey, D. A. Stimulus coding at caudal levels of the cat's auditory nervous system. II. Patterns of synaptic organization. In A. Møller (Ed.), *Basic mechanisms in hearing*. New York and London: Academic Press, 1973.

Moushegian, G., Rupert, A., & Galambos, R. Microelectrode study of ventral cochlear nucleus of the cat. *Journal of Neurophysiology*, 1962, **25**, 515–529.

Nauta, W. J. H., & Kuypers, H. G. J. M. Some ascending pathways in the brain stem reticular formation. In H. H. Jasper (Ed.), *Reticular formation of the brain*. Boston: Little, Brown, 1958.

Nelson, P. G., Erulkar, S. D., & Bryan, J. S. Responses of units of the inferior colliculus to time-varying acoustic stimuli. *Journal of Neurophysiology*, 1966, **29**, 834–860.

Niimi, K., & Naito, F. Cortical projections of the medial geniculate body in the cat. *Experimental Brain Research*, 1974, **19**, 326–342.

Oonishi, S., & Katsuki, Y. Functional organization and integrative mechanism of the auditory cortex of the cat. *Japanese Journal of Physiology*, 1965, **15**, 342–365.

Osen, K. K. The intrinsic organization of the cochlear nuclei in the cat. *Acta Oto-Laryngologica*, 1969, **67**, 352–359. (a)

Osen, K. K. Cytoarchitecture of the cochlear nuclei in the cat. *Journal of Comparative Neurology*, 1969, **136**, 453–472. (b)

Osen, K. K. Course and termination of the primary afferents in the cochlear nuclei of the cat. *Archives of Italian Biology*, 1970, **108**, 21–51.

Osen, K. K. Projection of the cochlear nuclei on the inferior colliculus in the cat. *Journal of Comparative Neurology*, 1972, **144**, 355–369.

Pfeiffer, R. R. Classification of response patterns of spike discharges for units in the cochlear nucleus: Tone burst stimulation. *Experimental Brain Research*, 1966, **1**, 220–235. (a)

Pfeiffer, R. R. Anteroventral cochlear nucleus: Wave forms of extracellularly recorded spike potentials. *Science*, 1966, **154**, 667–668. (b)

Poggio, G. F., & Mountcastle, V. B. A study of the functional contributions of the lemniscal and spinothalamic systems to somatic sensibility. *Bulletin of the Johns Hopkins Hospital*, 1960, **106**, 266–316.

Powell, E. W., & Hatton, J. B. Projections of the inferior colliculus in cat. *Journal of Comparative Neurology*, 1969, **136**, 183–192.

Purser, D., & Whitfield, I. C. Thalamo-cortical connexions and tonotopicity in the cat medial geniculate body. *Journal of Physiology*, 1972, **222**, 161P.

Raab, D. H. Audition. *Annual Review of Psychology*, 1971, **22**, 95–118.

Ramón y Cajal, S. *Histologie du Système Nerveux de l'Homme et des Vertébrés*. Vol. 2. Paris: Maloine, 1911.

Rasmussen, G. L. Anatomic relationships of the ascending and descending auditory systems. In W. S. Fields and R. R. Alford (Eds.), *Neurological aspects of auditory and vestibular disorders*. Springfield, Ill.: Thomas, 1964.

Rockel, A. J. Observations on the inferior colliculus of the adult cat stained by the Golgi technique. *Brain Research*, 1971, **30**, 407–410.

Rockel, A. J., & Jones, E. G. The neuronal organization of the inferior colliculus of the adult cat. I. The central nucleus. *Journal of Comparative Neurology*, 1973, **147**, 11–60. (a)

Rockel, A. J., & Jones, E. G. Observations on the fine structure of the central nucleus of the inferior colliculus of the cat. *Journal of Comparative Neurology*, 1973, **147**, 61–92. (b)

Rockel, A. J., & Jones, E. G. The neuronal organization of the inferior colliculus of the adult cat. II. The pericentral nucleus. *Journal of Comparative Neurology,* 1973, 149, 301–333. (c)

Rose, J. E. The cellular structure of the auditory region of the cat. *Journal of Comparative Neurology,* 1949, 91, 409–440.

Rose, J. E. Comment. In A. V. S. de Reuck and J. Knight (Eds.), *Hearing mechanisms in vertebrates.* London: Churchill, 1968.

Rose, J. E., Brugge, J. F., Anderson, D. J., & Hind, J. E. Phase-locked response to low-frequency tones in single auditory nerve fibers of the squirrel monkey. *Journal of Neurophysiology,* 1967, 30, 769–793.

Rose, J. E., Galambos, R., & Hughes, J. F. Microelectrode studies of the cochlear nuclei of the cat. *Bulletin of the Johns Hopkins Hospital,* 1959, 104, 211–251.

Rose, J. E., Greenwood, D. D., Goldberg, J. M., & Hind, J. E. Some discharge characteristics of single neurons in the inferior colliculus of the cat. I. Tonotopical organization, relation of spike count to tone intensity and firing patterns of single elements. *Journal of Neurophysiology,* 1963, 26, 294–320.

Rose, J. E., & Woolsey, C. N. The relations of thalamic connections, cellular structure and evocable electrical activity in the auditory region of the cat. *Journal of Comparative Neurology,* 1949, 91, 441–466.

Rose, J. E., & Woolsey, C. N. Cortical connections and functional organization of the thalamic auditory system of the cat. In H. F. Harlow and C. N. Woolsey (Eds.), *Biological and biochemical bases of behavior.* Madison: Univ. of Wisconsin Press, 1958.

Sousa–Pinto, A. The structure of the first auditory cortex (AI) in the cat. I. Light microscopic observations on its organization. *Archives of Italian Biology,* 1973, 111, 112–137. (a)

Sousa–Pinto, A. Cortical projections of the medial geniculate body in the cat. *Advances in Anatomy Embryology and Cell Biology,* 1973, 48, 1–42. (b)

Stillman, R. D. Pattern responses of low-frequency inferior colliculus neurons. *Experimental Neurology,* 1971, 33, 432–440.

Stone, J. Morphology and physiology of the geniculocortical synapse in the cat: The question of parallel input to the striate cortex. *Investigative Ophthalmology,* 1972, 11, 338–345.

Stotler, W. A. An experimental study of the cells and connections of the superior olivary complex of the cat. *Journal of Comparative Neurology,* 1953, 98, 401–432.

Suga, N. Analysis of information-bearing elements in complex sounds by auditory neurons of bats. *Audiology,* 1972, 11, 58–72.

Swarbrick, L., & Whitfield, I. C. Auditory cortical units selectively responsive to stimulus "shape." *Journal of Physiology,* 1972, 224, 68–69.

Szentagothai, J. The structure of the synapse in the lateral geniculate body. *Acta Anatomica,* 1963, 55, 166–185.

Tunturi, A. Classification of neurons in the ectosylvian auditory cortex of the dog. *Journal of Comparative Neurology,* 1971, 142, 153–166.

van Noort, J. *The structure and connections of the inferior colliculus: An investigation of the lower auditory system.* Assen: Van Gorcum, 1969.

Warr, W. B. Fiber degeneration following lesions in the posteroventral cochlear nucleus of the cat. *Experimental Neurology,* 1969, 23, 140–155.

Warr, W. B. Fiber degeneration following lesions in the multipolar and globular cell areas in the ventral cochlear nucleus of the cat. *Brain Research,* 1972, 40, 247–270.

Watanabe, T. Fundamental study of the neural mechanism in cats subserving the feature extraction process of complex sounds. *Japanese Journal of Physiology,* 1972, **22**, 569–583.

Webster, W. R. Single unit studies of the cochlear nucleus in the awake cat. *Proceedings of the Australian Physiological and Pharmacological Society,* 1973, **4**, 89–90.

Webster, W. R., & Aitkin, L. M. Evoked potential and single unit studies of neural mechanisms underlying the effects of repetitive stimulation in the auditory pathway. *Electroencephalography and Clinical Neurophysiology,* 1971, **31**, 581–592.

Webster, W. R., & Veale, J. L. Patterns of binaural discharge of cat inferior colliculus units. *Proceedings of the Australian Physiological and Pharmacological Society,* 1971, **2**, 1.

Weisstein, N. What the frog's eye tells the human brain: Single cell analyzers in the human visual system. *Psychological Bulletin,* 1969, **72**, 157–176.

Wepsic, J. G. Multimodal sensory activation of cells in the magnocellular medial geniculate nucleus. *Experimental Neurology,* 1966, **15**, 299–318.

Whitfield, I. C., & Evans, E. F. Responses of auditory cortical neurons to stimuli of changing frequency. *Journal of Neurophysiology,* 1965, **28**, 655–672.

Whitfield, I. C., & Purser, D. Microelectrode study of the medial geniculate body in unanaesthetized free-moving cats. *Brain, Behaviour and Evolution,* 1972, **6**, 311–322.

Wilson, M. E., & Cragg, B. G. Projections from the medial geniculate body to the cerebral cortex in the cat. *Brain Research,* 1969, **13**, 462–475.

Wollberg, Z., & Newman, J. D. Auditory cortex of squirrel monkey: Response patterns of single cells to species-specific vocalizations. *Science,* 1972, **175**, 212–214.

Woolsey, C. N. Electrophysiological studies on thalamocortical relations in the auditory system. In A. Abrams, H. H. Garner, & J. E. P. Toman (Eds.), *Unfinished tasks in the behavioral sciences.* Baltimore: Williams and Wilkins, 1964.

Wong, W. C. The tangential organization of dendrites and axons in the auditory areas of the cat's cerebral cortex. *Journal of Anatomy,* 1967, **101**, 419–433.

Chapter 11

Auditory Localization

G. BRUCE HENNING

University of Oxford

INTRODUCTION

This chapter on auditory perception will deal with a limited range of binaural phenomena. The central issue will be sound localization, and sound localization is hardly the central issue in auditory perception. That dubious distinction belongs to the phenomena of "pitch" perception, however difficult and varied may be the uses of that word. Clearly, the most important auditory function in man is speech perception, but in view of Békésy's apt comment (1960), "Theories of hearing are little more than theories of pitch," it is equally clear that theories of hearing have yet to make any really significant contribution to the study of speech. The residue pitch, or the problem of the "missing fundamental" as it is sometimes called, might be a suitable "auditory percept" for this chapter, but the whole topic is very closely linked to the too often mooted central problem of pitch for adequate separate treatment. Moreover, the experiments of Houtsma & Goldstein (1972) apart, the topic has been subject of a large symposium (Plomp & Smoorenenburg, 1970).

In spite of its suitability by default, sound localization offers attractive possibilities as a topic of some interest in itself. More important perhaps is the fact that insight into the binaural system gained by studying sound localization may allow us to make some inferences about auditory mechanisms not possible from experiments involving monaural listeners. After all, however amusing the phenomena may be to explore, and aesthetic judgments apart, it is the inferences we can make about the underlying mechanism or class of mechanisms that give experiments any lasting value they may have.

CLASSIC EXPERIMENTS

The ability to locate sound sources precisely is an obvious advantage possessed by two-eared observers, yet our understanding of the mechanisms involved is rudimentary to say the least. The general lines of current representations of the way in which we locate sound sources (Mills, 1972) were essentially formulated in Lord Rayleigh's Sidgwick Lecture of 1906: A tone generated by a source directly ahead of an observer facing the source arrives at both ears simultaneously and with equal amplitude. As the source moves to one side, sound from the source reaches the ear on that side first and, by and large, with greater amplitude. Thus an observer with a stationary head has differences in both interaural time of arrival and interaural intensity on which to base judgments of source direction.

Interaural time differences may of convenience be of two sorts. The difference in the time at which a transient signal starts (or stops) at each ear is termed the onset (offset) or transient delay. In contrast, the relative time of arrival of particular features of an ongoing or continous sound has been called the ongoing or continous delay (Tobias & Schubert, 1959). The relative effectiveness of interaural intensity and timing cues in determining the apparent direction of a sound source and in determining the precision with which that direction is indicated depends on several parameters of the sound generated. The effects of one of these, signal frequency, have been thoroughly explored.

In Rayleigh's experiments involving mistuned tuning forks, broken Winchester bottles, and Lady Rayleigh it was established, for example, that interaural time differences could affect the apparent localization of pure tones of frequencies below about 1500 Hz and that interaural amplitude differences, the remaining cue for fixed headed observers listening to pure tones, was the crucial cue at higher frequencies. Rayleigh further established for pure tones of relatively long duration that continuous or ongoing delay was more important than transient delay, a point emphasized by Tobias and Schubert (1959).

Subsequent work, much of it done nearly fifty years after Rayleigh, provided precise measures of the smallest detectable changes in interaural time and intensity differences, the minimum audible change in the direction of a sound source, and the relative effectiveness of interaural time and intensity differences both in opposition and as determinants of the apparent location of sound sources (Zwislocki & Feldman, 1956; Stevens & Newman, 1936; Mills, 1958). All these things were determined as a function of signal frequency, sometimes as a function of signal intensity, rarely as a function of signal duration and almost always with pure tones.

A notable synthesis of the results of these psychophysical experiments and the physical measurements of interaural intensity and time differences produced by real sources (Fedderson, Sandel, Teas, & Jeffress, 1957; Weiner, 1947) was achieved by Mills (1958) whose researches provide strong support for Rayleigh's contention that "when a pure tone of low pitch is recognized as being on the right or the left, the only alternative . . . is to suppose that the judgment is founded upon the difference of phases at the two ears." An interaural phase difference in pure tones is indistinguishable from an interaural delay if we restrict delays to be less than one-half the period of the tone, but comparison of locations produced by phase differences at a number of frequencies indicates that interaural delay, rather than interaural phase, is the determining factor.

ROLE OF ORIENTATION

One particularly important relation determined in the more recent experiments (but anticipated by Rayleigh in his advice to fogbound mariners attempting to estimate the bearing of a fog signal) was that the precision of sound localization was strongly dependent upon the orientation of the observer's head relative to the source (Mills, 1972). Sounds coming from directly before the observer's head might be located with great precision; changes of 1 or 2 degrees of arc might be detected. But sounds arriving from greater than 60 degrees to the left or right were poorly localized; that is, movements of 7 degrees or more might be required to detect changes in the source location even for tones of low frequency (Fedderson et al., 1957). Such observations indicated the importance of head movements to source-locating ability—an observer moving his head to bring a source directly in front of him could achieve nearly an order of magnitude greater precision of localization than if constrained to listen to a sound incident from one side or the other. Moreover, an observer free to move his head might use the features of a sound field taken at a number of pairs of locations to make inferences about the source location—to become a many-eared rather than a two-eared detector. If the signal lasted long enough and if the field depended more on the source location and less on the acoustic characteristics of the observer's environment we might try to model the sound localization process in terms of such a multiple detector array.

There is, however, a characteristic of sound localization that suggests that by moving our heads we do not become multiple-ear arrays. The phenomenon has been called the precedence effect and was reported in one of its forms, more than half a century before Rayleigh's 1906 lecture (Henry, 1851; Gardner, 1968). The classic demonstration, however, is that of Wallach, Newman, and Rosenzweig (1949). In the experiments of Wallach *et al.*, two pairs of clicks were presented binaurally to the observer. Provided the time between the clicks is not too long a single sound is heard as incident from a particular direction. The second pair of clicks is usually called the echo pair, and provided the echo-pair clicks occur at least 1 msec later than the initial pair they produce remarkably little effect on the apparent location of the apparent single souce. That the echo pair have relatively little influence on the apparent location of the source could be determined by arranging the interaural difference of the second pair to give information about the source location that conflicted with that provided by the first pair; that is, by arranging the interaural delays to be of opposite sign in the initial click and its echo and by varying the magnitude of the interaural difference in the second pair needed to offset a given interaural difference in the first pair. To establish trading ratios of this sort seems to have a great appeal for sensory psychologists; we have made the observer into a null sensing device, thereby acquiring all the advantages associated with such an instrument. In experiments in which cues are opposed in order to center an image, for example, exchanging the waveforms leading to each earphone should produce no change in the apparent location of the source. But in creating an observing instrument convenient to study, we run the risk of discovering not the mechanism underlying the phenomena we hope to explore but the mechanisms underlying the observing instrument we have created. This point is forcefully made by Hafter and Jeffress (1968) and again by Hafter and Carrier (1971) in their discussions of experiments in which interaural time and intensity cues are opposed—experiments designed to determine the relative effectiveness of those two cues. In the case of the latter type of "trading" experiment, it is not even the general principle that makes the inadequacy of the procedure obvious but rather the order of magnitude variability in the results (Green & Henning, 1969).

That amount of variability does not yet exist in the few measures of the relative effectiveness of echoes in determining the apparent location of a sound source, and it is in any case clear that interaural information in delayed "copies" of a transient signal is suppressed—or at least given less weight in determining the sound source location than it would have had as the initial signal. This finding suggests that to explain the effects of head movements in sound localization as resulting from the increase in information produced by multiple sampling would be difficult, if not misleading.

The question of what other information in echoes might be suppressed and the mechanism responsible for that suppression has been raised by McFadden (1973) who pointed out that suppression of echoes is not restricted to conditions of binaural stimulation. The question of the requisite conditions leading to a precedence effect—the answer to the question of what constitutes a copy of the initial transient—remains virtually unexplored.

Equally unexplored are the more likely head-movement-dependent interaural cues suggested by Wallach (1940) only just over a quarter of a century ago. Head movements produce dynamic changes in interaural cues—changes in interaural time and intensity differences that depend both on the source location and on the head movement. The generation of time-varying interaural differences was a significant technical problem overcome in an engaging but crude fashion in Wallach's experiments; the generation of the appropriate signals now poses no technical difficulty and we may hope soon to see serious attacks on the properties of, and the mechanisms underlying, dynamic cues in auditory localization.

Let us return to the fixed-headed observer and some experiments that extend Rayleigh's formulation. David, Guttman, and van Bergeijk (1962) published experiments that may be interpreted to confirm the report of Leakey, Sayers, and Cherry (1958) that observers can detect interaural delays in the envelopes of ongoing waveforms even if the components of the waveform are restricted to the frequency region above 1500 Hz where interaural delays in any component heard in isolation would produce no measureable change in the apparent location of the source. The fusion of the signals at each ear and movement of the apparent source occurred even if the sinusoids carrying the delayed envelopes were of different frequency. The experiments of David *et al.* (1962) used clicks or bursts of noise filtered so that they contained negligible energy below 1500 Hz. The experiment, inevitably, was one in which interaural delays were balanced against interaural intensity differences with conflicting information to produce an image of the sound centered in the observer's head. The trading ratio extracted from this experiment was greater than that obtained with unfiltered transients but well within the range of trading ratios found for low-frequency sinusoidal signals of a similar mean intensity. Moreover, the observers produced similar ratios when the bursts of noise carrying the delayed envelopes came from different oscillators. The important inferences to be drawn from this experiment and from that of Leakey *et al.* are that: (1) timing information can be carried in high-frequency complex wave-forms and used by observers in locating sound sources, (2) the timing information may be carried in the envelope of the waveform, and (3) the carriers on which the timing information is imposed may have differ-ent frequencies at the observer's ears and may indeed be incoherent.

These points have subsequently been confirmed by Ebata, Sone, and Nimura (1968) and Henning (1974), and merit emphasis because they are inconsistent with views of binaural processing based on Rayleigh's insight. Models of binaural interaction that have a cross-correlator (or the equivalent described in neurophysiological terms) as their basic mechanism for extracting information about interaural delays will fail to predict the result that observers can detect delays in the envelopes of carriers (Jeffress, 1948; Licklider, 1956; Sayers & Cherry, 1957).

No model of binaural interaction that has been elaborated in sufficient detail to make its predictions clear can account for even the limited sets of observations outlined in this brief chapter. Physiological evidence sufficient to account for the same narrow range of results will not be available for decades. Can we find psychophysical techniques of sufficient precision and reliability to reduce the enormous error in our measure-ments? And can we find the insight to match Rayleigh, or indeed David, Guttman, and van Bergeijk, delimiting on even this narrow front the class of mechanisms that might lead to our ability to locate sources of sound?

REFERENCES

Békésy, G. von. *Experiments in hearing.* New York: McGraw-Hill, 1960.

David, E. E., Guttman, N., & van Bergeijk, W. A. Binaural interaction of high-fre-quency complex stimuli. *Journal of the Acoustical Society of America,* 1962, 31, 774–782.

Ebata, M., Sone, T., & Numura, T. Binaural fusion of tone bursts different in fre-quency. Sixth International Congress on Acoustics, Tokyo, Japan, 1968, A33–A36.

Feddersen, W. E., Sandel, T. T., Teas, D. C., & Jeffress, L. A. Localization of high-frequency tones. *Journal of the Acoustical Society of America,* 1957, 29, 988–991.

Gardner, M. B. Historical background of the Haas and/or Precedence Effect. *Journal of the Acoustical Society of America,* 1968, 43, 1243–1248.

Green, D. M., & Henning, G. B. Audition. *Annual review of Psychology*, 1969, **20**, 105–128.

Hafter, E. R., & Carrier, S. C. Binaural interaction of low-frequency stimuli: The inability to trade time and intensity completely. *Journal of the Acoustical Society of America*, 1971, **51**, 1852–1862.

Hafter, E. R., & Jeffress, L. A. Two-image lateralization of tones and clicks. *Journal of the Acoustical Society of America*, 1968, **44**, 563–569.

Henning, G. B. The detectability of interaural delay in high-frequency complex waveforms. *Journal of the Acoustical Society of America*, 1974, **55**, 84–90.

Henry, J. *Scientific writings of Joseph Henry.* Washington: Smithsonian Institution, 1851.

Houtsma, A. J. M., & Goldstein, J. L. The central origin of the pitch of complex tones: Evidence from musical interval recognition. *Journal of the Acoustical Society of America*, 1972, **51**, 520–530.

Jeffress, L. A. A place theory of sound localization. *Journal of Comparative and Physiological Psychology*, 1948, **41**, 35–39.

Leakey, D. M., Sayers, B. McA., & Cherry, C. Binaural fusion of low- and high-frequency sounds. *Journal of the Acoustical Society of America*, 1958, **30**, 222–223(L).

Licklider, J. C. R. Auditory frequency analysis. In C. Cherry (Ed.), *Information theory.* New York: Academic Press, 1956.

McFadden, D. Precedence effects and auditory cells with long characteristic delays. *Journal of the Acoustical Society of America*, 1973, **54**, 528–530 (L).

Mills, A. W. On the minimum audible angle. *Journal of the Acoustical Society of America*, 1958, **30**, 237–246.

Mills, A. W. Auditory localization. In J. V. Tobias (Ed.), *Foundations of modern auditory theory.* New York: Academic Press, 1972.

Plomp, R., & Smoorenenburg, G. F. (Eds.) *Frequency analysis and periodicity detection in hearing.* Leiden, The Netherlands: A. W. Sijthoff, 1970.

Rayleigh, Lord (J. W. Strutt). On our perception of sound direction. *Philosophical Magazine*, 1907, (Ser. 6) **13**, 214–232.

Sayers, B. McA., and Cherry, E. C. Mechanisms of binaural fusion in the hearing of speech. *Journal of the Acoustical Society of America*, 1957, **29**, 973–987.

Stevens, S. S., & Newman, E. B. The localization of actual sources of sound. *American Journal of Psychology*, 1936, **48**, 297–306.

Tobias, J. V., & Schubert, E. D. Effective onset duration of auditory stimuli. *Journal of the Acoustical Society of America*, 1959, **31**, 1595–1605.

Wallach, H. The role of head movements and vestibular and visual cues in sound localization. *Journal of Experimental Psychology*, 1940, **27**, 339–368.

Wallach, H., Newman, E. B., & Rosenzweig, M. R. The precedence effect in sound localization. *American Journal of Psychology*, 1949, **62**, 315–336.

Wiener, F. M. On the diffraction of a progressive sound wave by the human head. *Journal of the Acoustical Society of America*, 1947, **19**, 143–146.

Zwislocki, J., & Feldman, R. S. Just noticeable differences in dichotic phase. *Journal of the Acoustical Society of America*, 1956, **28**, 860–864.

Chapter 12

The Somatosensory System

PATRICK D. WALL

University College, London, England

MODELS OF THE BRAIN

One of the most fundamental of the problems of the somatosensory system is that there seems to be not one but several systems. This is not special to the somatosensory system: all sensory systems in vertebrates show the feature of multiple pathways. Implicit in all physiological writing about sensory mechanisms is an attitude toward the reason for this multiplicity. In the choice of which part of the system to investigate and in the design of experiments, the physiologist inevitably has some model of the brain in mind which assigns a significance to the existence of multiple pathways. Three types of model have been proposed or hinted at.

Specialized Model

Here it is assumed that each pathway has its own specialized function. A subdivision of the specialization approach proposes that the information transmitted over the system is split into specialized categories. From this we get the use of terms such as the "light touch pathway" and "the pain and temperature pathway" referring to the dorsal columns and the ventrolateral white matter. Another version of this specialization function is that the target organs of the different pathways have specialized functions. We see an example of this in the separation of spino-cerebellar pathways and spinothalamic pathways which carry somewhat similar information from spinal cord to structures which are believed to have completely different and specialized functions.

Hierarchical Model

Here it is believed that parallel pathways represent some stage of evolutionary development and that more recently evolved pathways over-lay and suppress more ancient pathways. Some of the names overtly signify this difference as in the neospinothalamic system versus the paleo-spinothalamic system. Sir Henry Head's clinical studies led him to de-scribe two types of sensation, epicritic and protopathic, subserved by two anatomical systems with differing evolutionary histories. There are many versions of this subdivision of new and old with additional correla-tions, fast and slow, myelinated and unmyelinated, paucisynaptic and polysynaptic. Many of these words have a well-known semantic aura about them in which new, fast, direct are associated with better and the words old, slow, indirect are associated with inferior functional qualities.

Redundant Model

Here it is proposed that pathways are simply duplicated to provide protection in case of damage. This seems a refuge in intellectual bank-ruptcy where too simple a frame of reference has been used for analysis. A Martian who paid a brief visit to earth and sampled only a small area in France in 1916 might send back a report "The function of the human finger is to pull a trigger. Ten redundant fingers have evolved to allow the function to continue in spite of the loss of up to nine digits."

Each of these explanations contains an implicit model of the brain and each seems to me to fail to fit the experimental facts and also to be inherently unreasonable. Therefore it is necessary to suggest new theories and experiments and to define the nature of the trap into which the classical approach has led us. Information about events on the surface of the body and from its depths is carried over pathways to three major destinations in the brain: (1) to cerebellum; (2) to medulla, pons, and midbrain; (3) to thalamus and cortex. The cerebellum has been excluded from playing a role in sensation because cerebellar lesions are not associ-ated with sensory disturbances. As we shall see, exactly the same type of evidence has been used to support the redundancy theory in other structures where lesions fail to produce sensory loss. Since destruction of parts of the cerebellum leads to incoordination of movement, it is assigned a motor function. This way of thinking accepts the debatable assumption that motor and sensory functions are separable. The system projecting to the hindbrain and midbrain has been given the function of directing various automatic movements and body-righting and orientation movements, since these are the main forms of somatosensory controlled behavior which remain in a mammal from whom the forebrain has been removed. Finally all detailed sensation and perception is classically thought to be fed by information transmitted directly or indirectly to thalamus and cerebral cortex. The evidence that somatosensory information does flow over these pathways is quite certain. Anatomical investigations show that the pathways exist, and physiological investigation, in many cases with remarkable detail, shows the type of messages which are being sent over the various tracts. The assignment of function to the pathways and to their end stations depends largely on conclusions from behavioral and clinical studies following lesions. The conclusion that a function may be assigned to part of the brain if it disappears when that part of the brain is destroyed is quite reasonable if the brain is a set of independent parts. However it has been fashionable since the days of Sherrington to pay lip service to the idea of an integrative nervous system. If the parts of the brain should be viewed as interconnected and interdependent, then the removal of one part may have many and curious consequences on the function of what remains. The assignment of the lost property strictly to the lost part then has no more logic than the statement "The weight carrying property of this bridge is the property of this girder because if I take it out the bridge falls down" or "The strength of this piece of string is located at a point 6.2 cm from the top and because if I cut it there the ends part."

THE DORSAL COLUMN—
MEDIAL LEMNISCAL SYSTEM

Let us follow the attempts which have been made to assign a function to the dorsal columns which are the most obvious of the various somatosensory input systems. This tract of axons makes up the dorsal fasciculus of the spinal cord. It is present in all mammals and becomes progressively larger pari-passu with the development of the cerebral cortex until in man it occupies about one-third of the cross-sectional area of the spinal cord. A major component in it are afferent axons which run from the periphery and terminate on the dorsal column nuclei. Axons from the cells of these nuclei cross to the opposite side and run to the thalamus by way of the medial lemniscus. Posterior thalamus projects to somatosensory cerebral cortex. Physiological studies show that a major stream of information about events on the surface of the body arrives at the cerebral cortex having passed only two synaptic regions, one in the dorsal column nuclei and the other in the thalamus. All textbooks of neurology teach that a group of signs and symptoms occur in a patient suffering a progressive destruction of dorsal columns. They are in order of appearance as the condition progresses: (1) decrease of vibration sense, (2) decrease of position sense, (3) decrease of ability to state if the skin is touched in one or two places, (4) increase of light-touch threshold, (5) increased difficulty in identifying a number drawn on the skin or a common object placed in the hand. The signs and symptoms exist but their specific association with destruction of dorsal columns is a myth. The origin of the myth and its near immortality is a moral story. Tabes dorsalis, a syphilitic infection of the central nervous system, was clearly described in the nineteenth century. Patients with this disease show a progressive development of the five signs. Examination of the spinal cords showed degeneration of fibers in the dorsal columns. Hence the loss of sensory discriminative ability was attributed to loss of dorsal column fibers and this conclusion fitted a number of other developing attitudes of the time. Hughlings Jackson under the influence of Darwin saw the brain as exhibiting a hierarchy in which evolution had developed a series of higher and higher centers with the cortex as the highest center containing the mechanisms of elaborate discrimination. Darwinism had also stimulated a revived interest in comparative anatomy and the dorsal columns were known to have a parallel evolution with the cortex and to feed the cortex. Tabes provided a powerful link between the various facts and theories for the identification of the location of special functions in specialized structures and in their specialized inputs. Unfortunately by the turn of the century, as microscopy improved, it was realized that the degeneration of the dorsal columns was secondary to the destruction of axons in the dorsal roots and that there was a loss of input not only to dorsal columns but also to the spinal cord gray matter and therefore to all other input systems. This new finding did little to shake the classical story which was so simple and neat. Once the function of dorsal columns had been totally accepted as understood, the loss of input to other systems was dismissed as irrelevant by circular arguments which claimed that it had already been shown and was well-known that these older more indirect systems were not responsible for higher functions.

There the matter rested for half a century, maturing from theory to doctrine to dogma. When in 1965 Cook and Browder cut dorsal columns in man and found that the patients failed to exhibit prolonged disturbances of sensation, the neurological community totally ignored the apparently heretical report. Even the carefully controlled studies in 1964 by de Vito, Ruch, and Patton, which showed that monkeys with histologically verified complete dorsal column sections could still discriminate weights and textures, failed to have any discernible impact on opinion. Since that time a large number of studies have been completed on the effect or rather lack of effect of dorsal column lesions (Wall, 1970), and we shall now consider in detail two papers.

A Consideration of Two Studies

The first paper (Eidelberg & Woodbury, 1972) contains its conclusion in its title "Apparent redundancy in the somatosensory system in monkey." The summary states:

> Monkeys were used for recording the cortical responses to electrical stimulation of the superficial radial nerve skin territory. High cervical section of fasciculi gracilis and cuneatus with or without concurrent lesions of the spinocervical tract did not modify the cortical evoked responses. Combined dorsal funiculus–lateral lemniscus either eliminated or markedly attenuated the early responses and only late residual components were present in primary and nonprimary cortex. We conclude that fast efficient conduction of information from skin to cortex occurs both through the dorsal funiculus–medial lemniscus and through the neospinothalamic system.

To emphasize the authors' point, evoked cortical responses were unaffected by section of either one of the two major paucisynaptic somatosensory pathways which lead to cortex. We must first ask what is meant by redundancy because this word has two separate meanings, one redundancy of components and the other redundancy of function. If a pessimist holds up his trousers with two pairs of suspenders, he is using component redundancy. If he achieves the same end by wearing both a belt and suspenders, he is using redundant but different components with the same function.

To ask which use of the word redundancy is used by Eidelberg and Woodbury, we must examine the anatomy and physiology of the two systems to see if they contain the same or different components. The number of subcortical synapses in the two systems may be the same but the physiology of the two is strikingly different. Microelectrode studies of the dorsal columns, the dorsal column nuclei, the ventral posterior lateral nucleus of the thalamus, and the sensory cortex have been carried out in considerable detail. For two reasons, somewhat less work has been done on systems which relay in the spinal cord. First, physiologists naturally selected the system that was traditionally labeled the important one and, second, the spinal cord relay system exhibits a complexity which makes description more difficult. There are four outstanding characteristics of the dorsal column–medial lemniscus system. (1) The cells within each relay nucleus and the fibers in the projecting tracts are arranged so that their receptive fields form an extraordinarily precise and repeatable somatotopic map. If a few cells are located and their receptive fields plotted, then the location of the receptive fields of most of the other cells in the nucleus can be predicted with considerable accuracy. (2) As a corollary of the detailed somatotopic mapping, the individual excitatory receptive fields are small and usually surrounded by inhibitory receptive fields which further enhance the spatial resolution capability of the system. (3) Most cells transmit information from a single variety of peripheral afferent fiber. For example, if a cell responds to movement of one type of hair, it is unlikely to respond to other types of hair or to the stimulation of touch corpuscles. In other words, the majority of cells in this system respond to a limited number of fibers which are restricted in their spatial origin and in the type of information they transmit. (4) Descending controls and intruding fibers from other systems exist but, as we shall discuss later, these systems seem to be inactive under most conditions of operation and only become active in certain special behavioral circumstances (Ghez & Pisa, 1972). These four characteristics add up to a transmission system which approaches that of a reliable relay system in its rigidity, predictability, and absence of cross talk.

By contrast, most of the interneurons in the spinal cord gray matter show strikingly different characteristics, and these differences are exaggerated as one moves from the most dorsal cells to more ventral ones (Wall, 1973). (1) A somatotopic map exists within the cord but it is only in lamina IV in the dorsal part of dorsal horn that this map approaches a poor imitation of the other systems accurate maps. (2) The receptive field size of individual cells in the most dorsal laminae is somewhat larger than those in dorsal column nuclei and the field size expands markedly as more ventral cells are sampled. Although the cord's cells frequently have inhibitory zones associated with their excitatory receptive fields, these zones do not completely surround and therefore do not act to restrict the region from which information is obtained. (3) Convergence is the general rule with few exceptions. Both inhibitory and excitatory convergences of more than one type of specific afferent converge on most cord cells. This rapidly leads to complex input–output functions depending on the nature and location of simultaneously active afferents. (4) Many powerful and varied descending control systems play on cord cells. These can not only control the general level of a cell's excitability but by affecting differentially one or another of the convergent afferent systems can, in effect, change the modality of a cell. For example, in lamina VI, a descending system from the brain stem holds these cells in the decerebrate cat in a mode whereby they respond preferentially to muscle afferents. If this descending system is blocked, the cells now respond preferentially to cutaneous stimuli. It is clear from what has been said that the great majority of cells feeding the two systems do not duplicate each other. As in the case of all biological generalizations there are exceptions. Some axons originating in dorsal horn project through the dorsal columns or converge on the dorsal column nuclei. These make a minority of the input and furthermore there is no evidence for the reciprocal cross-over from dorsal to lateral lemniscus. We can conclude that if the two systems are said to be redundant, the redundancy must be mainly of a functional variety and not a component duplication.

Let us turn then to examine the function tested by the experiment which produced a reaction of equal size and speed by activation of either pathway. Unfortunately we must examine all four key words, pathways, reaction, size, and speed to see if a meaningful demonstration of equal alternatives has in fact been demonstrated. First the word "pathway." This word may be used to mean no more than the fact that an anatomical connection exists and may imply nothing about its function, but usually the word is taken to mean a set of axons actually in use during normal working hours. When an electric shock is used as the stimulus, the volley produced differs from a naturally occurring volley in three ways. (1) The shock will generate impulses in a combination of afferents which would never be simultaneously active after any known natural stimulus. Since the cells along the transmission pathway are subject to both inhibitory and excitatory convergences, the cells may transmit impulses because of an unusual combination of excitatory impulses or fail to transmit because of inhibition. (2) An electrical shock generates a synchronous volley so that impulses arrive at synapses in such a way that there is an optimal chance for temporal summation. It has been known for a long time that electrical stimulation of skin produces the appearance of very large receptive fields in cells with relatively small fields when any known natural stimulus is applied. This was normally attributed to the recruitment of a subliminal fringe by temporal summation. While this may frequently be the case, it has been shown (Merrill & Wall, 1972) that the phenomenon occurs even where no signs of a subliminal fringe exists and where the naturally effective afferents have been blocked. This work raises the possibility of the existence of ineffective synapses which normally neither excite nor facilitate the cell and yet, under the artificial conditions of synchronization, excite the cell. There is some reason to think that these ineffective synapses may trigger only local dendritic spikes which do not propagate from distant parts of the dendrite unless large numbers of such spikes occur at the same time. (3) Electric shocks normally generate only a single impulse in the peripheral axons. While the synchrony optimizes the degree of temporal summation, there are some synaptic connections which require repetitive bombardment before the excitability of the cell can be raised to firing level. These three factors mean that electrical stimuli and other stimuli which the system was not evolved to handle may demonstrate the existence of pathways which are never normally active or may fail to show pathways which would have been active if stimuli had been used which generate volleys tuned to the synaptic properties of the transmitting cells.

The reaction which was chosen here for measurement was the evoked cortical response. The main reason for this choice is because it is there, like Mount Everest, and it is easy to record. What exactly is being recorded remains unknown after some 50 years of proposals: summed action potentials, summed excitatory postsynaptic potentials, prolonged terminal arborization depolarizations, potassium accumulation, glial depolarization, etc. If the actual cause of the potential change is unknown, it is not surprising that its functional significance is unknown beyond the fact that something is happening in the cortex. Whatever that something is, the evoked cortical potential seems a very poor indicator of the state of the entering and leaving impulse traffic from the cortex except at the threshold transition from nothing to something, when of course many measures are correlated but not causally related. A particular breakdown of the usefulness of the cortical evoked response as an index of cortical functioning occurs when its size is used. The generator of the potential change easily saturates long before the inputs and outputs reach their maximal discharge rates. Therefore shape and size while easy to measure are either poorly correlated or even not correlated at all with the impulse traffic.

Like Mr. Toad of Toad Hall, neurophysiologists are obsessed with speed. Speed is a legitimate measure in the nervous system for establishing the monosynaptic versus polysynaptic transference of impulses from one area to another. Even this must be handled with the greatest care since a cell may respond monosynaptically to a slow weak input long after it has responded to a fast large polysynaptic input. The bias in favor of speed comes from a number of sources. The use of electrical stimuli or sudden brief onset stimuli generates an initially synchronous volley and the central events then spread out in time depending on conduction velocities, synaptic delays, and number of synapses. There may be a legitimate reason for concentrating on the fastest event on the grounds that this is the simplest in the sense that it involves the smallest number of linked cells but this is by no means always the case because of different conduction velocities and synaptic efficiences. For example, the spinocervical tract system delivers impulses to cortex slightly ahead of the dorsal column–medial lemniscus system in spite of the fact that an additional synapse is involved. However the real reason is often unspoken; it is that recently evolved systems are fast and direct and that they deliver impulses "fastest with the mostest" and dominate subsequent events and are therefore the events on which analysis should concentrate. There is no clear evidence that this domination occurs. The physiologist is presumably looking in these cortical pathways for the basis of discriminative behavior. This type of behavior shows three temporal aspects which seem to me to rule out even the simplest version of this "horse race" of competing nerve impulses. If a sudden flash occurs, the subject shows a startle reaction which takes 25–50 msec from stimulus to response of the hand. This shows that a pathway exists from eye to hand muscles capable of producing contraction in that time. However, if a highly trained subject is asked to press a button when a light comes on, the reaction does not occur for about 150 msec. This means that the fastest simplest discriminated events take a very long time as compared with the minimal conduction times and synaptic delays. Furthermore all discrimination tasks show the phenomenon of "metacontrast." A brief stimulus is presented for discrimination. Up to 100–150 msec after the event to be discriminated, a second stimulus is added. The second similar stimulus can eliminate the ability to detect the first. This is a curious horse race where the horse who has a delayed start always wins. It seems more likely that the time taken for discrimination includes periods of integration and perhaps recurrent feedback processes in which information is reprocessed by mechanisms which do not give preferential treatment to the first impulses to arrive. This must particularly be the case in the usual form of discrimination where the subject is not presented passively with a brief stimulus but with an ongoing event which he may explore and question. It might be claimed that even in the case of discrimination based on multiple successive stimuli, each change in the situation will be signaled by new and old, fast and slow systems so that the fast would have a chance to cancel or override the slow. To counter this argument that relative times of arrival allow the fast to dominate, we should not forget the behavior of the two genetic mutant mice who have lost the ability to myelinate their axons in the central nervous system. The effect of this will be that the relative times of arrival of nerve impulses will be radically altered by a single gene mutation. These animals die from epilepsy but, in other respects, they are surprisingly normal. For example, their auditory thresholds have been tested by operant conditioning techniques and found to be normal. This means that not only is their auditory discrimination normal but that they have the ability to learn and manipulate a Skinner box environment apparently just as well as normal mice. This implies that at least for these tasks, the details of the relative times of arrival of information over various systems are not crucial for discrimination. Since we evidently may not give automatic precedence to early large components, it is important now to emphasize that in the paper under discussion, smaller delayed components were still recorded even after section of both major tracts. By delay they mean some 30 msec and by smaller they mean some 50%. These changes do not automatically relegate these responses to limbo as we shall see in the next paper to be discussed. We have gone over the paper under discussion here in some detail not to criticize it as such but to consider the relevance of a group of classical neurophysiological techniques, i.e., brief stimuli, lesions, measurement of evoked potentials, speed, size, shape, etc., to the implied interest of all who work on the nervous system, which is to understand the mechanisms of behavior. We must question whether this type of physiology may not be becoming a private game played like that of the masters in Hesse's *Magister Ludi*, as an intellectual parody without ever testing its relevance to the real world.

Single unit studies have been made in the somatosensory cortex in the normal cat and following dorsal column section. A study by Dobry and Casey (1972a) showed that dorsal column section reduces the number of responding cells, reduces the number with sharp edges to their field, increases the average receptive field size, lengthens the average latency, and increases the number of cells with a prolonged depression following excitation. These studies emphasized the relative insensitivity of evoked potential measurements versus single cell recording. They showed that the cortical responses produced by different pathways are not identical.

Let us move then to the second paper which we shall analyze (Schwartz, Eidelberg, Marchok, & Azulay, 1972). It examines the behavior of monkeys in a situation which one would reasonably expect a monkey to face in its natural environment, but with lesions of the central nervous system which may never occur naturally. The paper is entitled "Tactile discrimination in the monkey after section of the dorsal funiculus and lateral lemniscus." The summary states:

> The generally accepted view that tactile discrimination in primates is mediated by the dorsal funiculus or the spinothalamic tract (or both) was examined in this study. The absolute intensity detection threshold for electrical stimuli of one wrist was mildly but transiently elevated by ipsilateral dorsal funicular section. Lesions of the contralateral lateral lemniscus caused a significantly longer lasting increase of the threshold. Discrimination of subtle tactile differences in texture, form, pattern and hardness was unchanged or only slightly affected by single or combined lesions of both afferent systems, despite the fact that the latter markedly attenuated or eliminated cortical evoked responses. Either a remarkable intrapathway redundancy exists so that a very few residual fibres may mediate these complex discriminations or other pathways may function vicariously or equipotentially.

The reader will understand that classical theory is, to put it mildly, in serious trouble. A major pathway, the dorsal funiculus, has been sectioned and produces transient changes in the very functions which classical theory predict should be abolished. This substantiates and extends the results of a large number of independent studies of the past 20 years. In most of the previous studies the authors have explained away their lack of results by assuming that another alternative pathway took over. The favorite alternative pathway was the neospinothalamic system, seen as a likely substitute for the dorsal column funiculus since this system, which sends axons directly from spinal cord to thalamus, evolves in the higher mammals. In the study discussed here, the authors took the logical step of sectioning both pathways and yet the function continues like a pot of gold at the end of the rainbow. Obviously there is a limit, short of total spinal cord section. It is known that if the entire dorsal quadrant on the ipsilateral side and the entire ventral quadrant on the contralateral side is sectioned, then animals are incapable of tactile discrimination on one side. Even such extensive destruction does not produce total anesthesia. It has been known since the last century that complete hemisection of the cord on one side and complete hemisection of the cord on the other side, two segments rostral, still allows the animal to respond to relatively intense stimuli. This has recently been studied again and the results show that some information meanders along multiple cell pathways which zigzag from side to side and still deliver some useful information to the head. How are we to face up to this retention of functional capacity of the whole organism in the face of such apparently devastating lesions? The studies after 1950 are accompanied by histological pictures of the lesions and yet the excuse is still offered that perhaps the lesions were not complete. A frequently quoted study shows that cats can continue pattern discrimination when only 2% of the optic tract remains (Galambos, Norton, & Frommer, 1967). Is this the explanation of the negative results in the somatosensory system? I believe not for a number of reasons. As we have said, the dorsal column–medial lemniscus system is characterized by marked somatotopic mapping not only in the thalamic and dorsal column nuclei and cortex but also in the interconnecting tracts of white matter. By contrast, the optic nerve has a jumbled arrangement of fibers connecting the detailed map of retinal ganglion cells to the exactly mapped lateral geniculate and to the relatively exactly mapped superior colliculus. Therefore while optic tract partial sections whittle down the input at random, partial dorsal column or lemniscal lesions hit particular areas of a mapped projection and therefore should produce scotomata, "blind" spots, just as partial destruction of retina or visual cortex produces scotomata. Very detailed anatomical studies of afferent fiber projections have been made and it is clear that 2% of the fibers do not project outside the dorsal columns into the medial lemniscus system. Complete dorsal funiculus lesions have been made without disturbing function, not even threshold function. Therefore we can no longer maintain that the negative results can be explained by the presence of partial lesions of the system but it can be maintained that other systems are functionally redundant.

If the argument of functional redundancy is to be advanced, we have to propose some plausible argument as to why the pressures of survival advantage should have led to the evolution of exactly equivalent but alternative ways of carrying out the same task. Are there any known diseases or injuries from which any animal survives which destroy one or the other of the major afferent pathways? The answer seems to me unequivocably no. All the known diseases, injuries, or infections or metabolic diseases of animals which seriously affect these systems singly or multiply are lethal and there is no evidence that a monkey survives severe brain injury or severe central nervous system infection any better than a rat, frog, or duckbill platypus. One is free to speculate that there was a period of evolution in which the world was populated by mad neurosurgeons or viruses or mutant genes who went around knocking off one projection system while leaving others and that these neurosurgeons or viruses or genes have left no trace of their existence except that we have duplicated nervous systems left as hangovers of this historical period. This hypothesis seems a trifle implausible but is the implied defence of those who propose that we have a redundant nervous system evolved to handle a challenge which does not now exist. It is just conceivable that environmental factors including bacteria and viruses might have disappeared without trace but if redundant pathways were a genetic answer to genetic errors, then one would expect viable mutants to occur with missing redundant systems. In all the plethora of remarkable genetic or congenital disorders of the central nervous system (CNS), some of them apparently symptomless, such as aplasia of the corpus callosum, none are known in which major somatosensory pathways are absent.

Even if the evolution of true redundancy is implausible, let us turn to a test of redundancy. If a function continues after the destruction of one pathway and this is explained by saying that a second pathway can also fulfill the same task, then one should be allowed to cut the second pathway and show that the first can also continue the carrying out of the task. This has been done in rats for the dorsal funiculi in contrast to other systems (Wall, 1970). All parts of the thoracic cord were sectioned except for the dorsal funiculi which were left intact. The head and forelimbs of the animal reacted normally to stimuli applied to them. When stimuli were given to the hind legs, there were local reflex reactions but no behavioral reactions could be detected in head and forelimbs. The animal failed to orient or arouse or to show a startle reaction or to wake up from sleep when large electrical stimuli were delivered to the hind legs. More subtle tests were carried out. If a normal rat with its forelimbs and head in a tube was bar pressing to obtain food, very small stimuli such as the bending of a single hair on the tail would lead to an interruption of the feeding pattern. In an animal with only dorsal columns intact, no such interruption would occur. In a common Skinner box, small electrical stimuli delivered to the hind quarters are quickly associated as a signal that bar pressing produces reward in the normal rat but in the animal with only dorsal columns no such learning occurs. In summary, no behavior has been elicited in the forequarters or head by stimuli to the hind quarters when information is transmitted only by the dorsal columns. In spite of this absence of behavior, large cortical evoked responses are recorded following the hind leg stimuli, a rather extreme example of the lack of correlation between behavior and cortical responses. We must conclude from these experiments that the dorsal columns are in no sense an alternative pathway to other ascending systems. Furthermore the information delivered has no apparent "meaning," that is to say it fails to evoke simple or discriminative behavior if other pathways are absent. A reasonable conclusion would seem to be that the dorsal column system must function in relation to the functioning of other systems. What can that function be?

The Function of the Dorsal Columns

Two types of answer have been given recently on the function of the dorsal columns, both of which I believe must be dismissed as irrelevant to the functioning of the main component. The first type of answer concentrates on the initial transient effects of the lesion (Gilman & Denny Brown, 1966). It is implied here that the "true" function of a system is revealed by the immediate effects of destruction which are observed before other systems have a chance to take over. This answer is a hidden version of the redundancy theory and needs not only alternative systems but also some completely unknown mechanism for the slow turning on of inactive systems. It depends therefore on a combination of two ad hoc theories unsubstantiated by any evidence. An alternative and far simpler explanation for transient effects is that: (1) the system is an integrated one, (2) the removal of any component drops the excitability of the rest, (3) slow homeostatic metabolic mechanisms restore membrane potentials to their normal operating range. The papers under review here were concerned with the long-term effect of lesions after initial adjustments were complete.

The second type of positive answer recently reported must surely be attributed to section of muscle afferents rather than the main component which projects to the thalamus (Melzack & Bridges, 1971; Dubrovsky & Garcia-Rill, 1973). The most obvious components of this variety are the large proprioceptive afferents which run in the dorsal columns from the hind legs to terminate in Clarke's column in the lower thoracic cord and from the forelimbs to terminate in the external cuneate nucleus. Section of dorsal columns at C1 in the rat, cat, or monkey leads to a number of clear disturbances of movement and in the use of limbs. If at least 90% of the dorsal columns are cut, simple observation may show that the affected limbs are left in strange positions at rest and are used in a clumsy way during movement. More detailed experiments have shown that these animals have a severe deficit in their ability to reach for or to jump on a moving target. The authors attribute these deficits to the loss of the exclusive path for fibers from muscle spindles and low threshold joint receptors from the forelimbs which pass through the dorsal columns. These defects are mainly limited to forelimbs even though the lesion also cuts the dorsal columns arising from the hind limbs. The information from hind limbs about position is transmitted to the head by systems other than dorsal columns and therefore escapes. Mid–upper thoracic dorsal column lesions do not produce these gross disorders of directed movement nor do lesions of dorsal column nuclei. While these relatively gross disorders have been reported in the cat with high cervical lesions and in the monkey soon after section, in the chronic monkey fine control of the digits was impaired but the eye–hand coordination task of picking objects off a rotating disk was only slightly impaired. If a disk was rotated at 8 rpm preoperatively, animals retrieved 100% of the objects, and postoperatively, 80–100%. If the disk was speeded up to 20 rpm, retrieval in the controls dropped to 45% and to 28% for the operated animals. If this helps to explain the movement disorders we are left with the question of why no obvious deficit in skin sensation is detected.

One of the most telling positive clues we have is that the learning of a sensory task is affected by dorsal column (DC) lesions even though once learned the task does not appear to depend on there being intact dorsal columns (Kitai & Weinberg, 1968; Dobry & Casey, 1972b). In all but two of the animal studies, the animals were first trained on some discrimination task, then the lesion was made, and then the animals were retrained, and it was shown that the animals performed normally. However, in two studies, the animals were trained to a new task after the lesion. In the first paper Kitai and Weinberg were able to train cats to discriminate roughness after DC lesions but they were slower than normals. In the later work of Dobry and Casey, cats were trained after real or sham operations to discriminate four graded levels of roughness. Six cats with high cervical lesions were compared with four intact controls. Two cats with 97% and 100% destruction of their total dorsal column cross-sectional area failed to reach criterion on the second hardest roughness discrimination test. However, the other four cats, with between 38% and 86% of their dorsal columns destroyed, learned the highest discrimination grade as quickly as intact controls. These cats represent the only clearly positive results in the literature on the effect on somatosensory discrimination of DC destruction. All other positive findings are clues or suggestions rather than results if we eliminate the transient changes following many lesions and the motor disorders following high cervical lesions which we have already suggested are produced by interference with proprioceptive afferents. In a detailed examination of one patient who had a surgical section of unknown extent of one dorsal column (Wall, 1970), sensory thresholds and two-point discrimination were found to be normal. However, if the man was presented with conflicting stimuli so that his attention was forced to shift from one stimulus to another, he initially failed on the side of the lesion although he soon learned the skill. In rats, with high thoracic DC lesions or lesions of the dorsal column nucleus supplied by one hind limb, the affected limb would slip into an unusual position when the animal was resting on a difficult perch. In the monkeys tested on a number of tactile discriminations, all discriminations were relearned following dorsal column destruction. However, certain tasks appeared very much more difficult to relearn. Discriminations of sandpaper roughness, the differentiation between a hexagonal and a square rod, and the differentiation between rods with horizontal or vertical grooves were all learned more easily the second time in spite of having a dorsal column lesion between the first and second training sessions. However the differentiation between round and oval rods seemed to become more difficult on retraining after the section of dorsal columns.

This then is the extent of the known positive results of the section of a major pathway in the face of the large number of negative results. Even the positive results have the curious nature that for each affected discrimination, the animal can carry it out in the absence of dorsal columns if the task has been learned before the lesion. This points to the most likely general conclusion: some sensory information is used during the process of learning to discriminate and yet this information is not necessary for the final skilled discrimination itself. Obviously it is necessary to define the nature of this sensory information, how it is handled, and why it is necessary only during the learning stage. It is quite clear that we can no longer expect it to be anything so simple as a single modality or any of the classical sensory fractions since all particular abilities seem intact. The simplest suggestion is the possibility that quantity of information is an important factor in the early stages of learning. Specifically this could mean that multiplication of the input allowed spatial summation during the buildup of weak connections at early stages of learning. Once the input–output characteristics were established the stimulus–response threshold might drop and an adequate output could be achieved with a smaller input. This is merely a restatement of the redundancy hypothesis which we have already ruled out by showing that dorsal columns alone seem incapable of triggering any behavioral reaction no matter how well the reaction is established.

Let us turn then to the earlier suggestion that information in dorsal columns is relevant only when activity exists in other systems and as we now know only during the period of learning or under conditions of conflict between stimuli. For learning to occur, alerting arousal and orientation must be present. Although it is difficult to quantitate these phenomena they appear to be present and normal in an animal without dorsal columns and absent in an animal with only dorsal columns. Could it be that information flowing over dorsal columns plays a role in the transition from the stage where the animal knows that there is something to be discriminated to the stage where the animal has identified the essence of the stimulus? Much of learning theory but little of experimental psychology deals with this transition phase from general to selective attention, from error to success, from hypothesis to proof. Could it be that the special type of sensory information which is required in this phase is the information acquired by exploration and that certain learning tasks require this? Exploration involves the active generation by the animal of a sensory input in contrast to the passive reception of the stimulus which characterizes many psychological experiments especially those in psychophysics. In the standard neurological examination of a patient, he is reduced to a passive alerted object on which stimuli are applied to which he can do nothing and from which he has to extract only the binary decision present or absent. In more usual tasks, the animal enters an active exploration of his world and one correlate of cortical evolution is the evolution of complex exploration. Exploration is not a random search or overall scan, it is movement directed toward the testing of hypotheses. It is of interest as we develop this theme to remember that descending controls on the dorsal column nuclei act in only one known condition and that is during shifts of movement pattern in purposive behavior as distinct from appetitive or reflex responses (Ghez & Pisa, 1972). Could it be that the role of dorsal columns is to feed information to cortex on the basis of which cortex interprets information fed by older systems? Furthermore dorsal columns might fulfill the role of continuously feeding information with minimal censorship or mixing so that cortex could set filters across other pathways. The setting of filters would limit the total amount of information transmitted but would select particular types of information. Exploration by palpation asks a series of questions such as: Are there discrete objects? Are there edges? Movements are organized to detect such objects. It is not unreasonable to suggest that during movement, which is directed by descending control, there is a simultaneous descending control of the sensory relay cells in spinal cord and subsequent stations. If the cortex "knew" how to set the filters either because of other information or because of previous learning, it would not be necessary to go through the stage of detecting significant features by external search or by repetitive reexamination of afferent volleys. Under these circumstances, the dorsal columns would play no role since the situation would be familiar and the tactics for handling sensory information would have been established. If this were the case, we would predict that dorsal columns would play no role in the passive detection of stimuli, where no exploration was possible or necessary, nor would they play a role where skills of detection were fully developed. When the physician says to the patient "Tell me if I touch you with the sharp or blunt end of the pin" the patient has received auditory information on how to set the filters on his somatosensory pathways and the dorsal columns would play no role.

When an animal has been pretrained to a task, he recognizes the object of the test situation by visual and other cues and can set his sensory pathways into the detection mode relevant to the known problem to be discriminated. All of the detection tasks we have discussed are severely affected by destruction of the cortex. The cortex might be generating search tactics for examination of sensory signals arriving over the older pathways, or, if the information in the arriving signals is not sufficient to allow discrimination, the cortex might be responsible for generating the exploratory movements which would bring in the required information. Certainly one of the characteristics of cortical lesions is an absence of fine exploratory movement. Even the placing reaction may be a movement of exploration. Even in the absence of triggering clues delivered by the dorsal columns, the cortex might circulate through its existing repertoire of analysis strategies in order to solve a discriminative problem.

We would suggest therefore that the following hypothesis fits the facts known at present about the dorsal column–medial lemniscus system. When a novel situation requiring discrimination occurs on the skin, the dorsal column–medial lemniscus system delivers uncensored unanalyzed information about the event to the cortex. The physiology of the system is designed to deliver a discrete separated mosaic of detail. Cortex examines the detail and forms a hypothesis and tests the hypothesis by setting filters across the older pathways by descending controls. If the information arriving over the filtered controlled pathways confirms the hypothesis on the nature of the event, the relevant behavior follows. If the information is insufficient, exploratory movements are initiated by cortex to generate more information. This suggestion fits the fact that no one sensory behavior disappears when dorsal columns are cut and the fact that dorsal columns alone seem incapable of initiating behavior. It is of obvious interest to speculate whether the projection pathways to cortex which subserve the other senses follow a similar pattern. It is apparent that somatosensory cortex can still play a dominating role in directing discriminative behavior even when cut off from one or more of its major sources of afferent information. In other sensory systems, particularly the visual system, the cortex may be more dependent on its major input and unable to demonstrate its properties independent of the input.

It is clear from the experiments reported here that we are forced away from the classical view that each sensory modality has its own sensory projection system. Evidently we must search for much subtler properties. Submammalian species evolved excellent sensory detection mechanisms with levels of sensitivity and resolution comparable to the mammals. We did not need to evolve a cortex and its associated pathways to detect the presence of isolated events. Cortex is presumably associated with the complex and subtle behavior of mammals in contrast to the relatively rigid behavior of submammals. The testing of the subtle role of ascending and descending systems will require observations of these systems in behaving animals, not only during the process of discrimination but during the period of learning to discriminate—not an easy task.

Since this chapter was written, a most important paper has appeared by Vierck (*Experimental Brain Research*, **20**, 1974, 331–346) entitled "Tactile movement detection and discrimination following dorsal column lesions in monkeys." He shows that monkeys with dorsal columns cut can discriminate between a stationary and a moving stimulus. However, three monkeys with such a lesion had difficulties in discriminating the direction of the movement. The animals were not completely incapable of doing the task but made more errors. No one would claim that *the* function of dorsal column has been unmasked by these experiments. However, we have an indication of the type of complex sensory analysis in which they might be involved. Detection of direction of movement involves a short memory of where preceding stimuli were located. It involves an analysis of the relations between stimuli and, as such, would involve the type of internal examination of relations which I have called "internal exploration."

REFERENCES

Cook, A. W., & Browder, E. J. Functions of posterior columns in man. *Archives of Neurology*, 1965, **12**, 72–92.

de Vito, J. L., Ruch, T. C., & Patton, H. D. Analysis of residual weight discriminatory ability and evoked cortical potentials following section of dorsal columns in monkeys. *Indian Journal of Physiology and Pharmacology*, 1964, **8**, 117–126.

Dobry, P. J. K., & Casey, K. L. Coronal somatosensory unit responses in cats with dorsal column lesions. *Brain Research*, 1972, **44**, 399–416.(a)

Dobry, P. J. K., & Casey, K. L. Roughness discrimination in cats with dorsal column lesions. *Brain Research*, 1972, **44**, 385–397.(b)

Dubrovsky, B., & Garcia-Rill, E. Role of dorsal columns in sequential motor acts requiring precise forelimb projection. *Experimental Brain Research*, 1973, **18**, 165–177.

Eidelberg, E., & Woodbury, C. M. Apparent redundancy in somatosensory system in monkeys. *Experimental Neurology*, 1972, **37**, 573–581.

Galambos, R., Norton, T. T., & Frommer, G. P. Optic tract lesions sparing pattern vision in cats. *Experimental Neurology*, 1967, **18**, 8–25.

Ghez, C., & Pisa, M. Inhibition of afferent transmission in cuneate nucleus during voluntary movement in the cat. *Brain Research*, 1972, **40**, 145–151.

Gilman, S., & Denny Brown, D. Disorders of movement and behaviour following dorsal column lesions. *Brain*, 1966, **89**, 397–418.

Kitai, S. T., & Weinberg, J. Tactile discrimination study of the dorsal column lemniscus system in the cat. *Experimental Brain Research*, 1968, **6**, 234–246.

Melzack, R., & Bridges, J. A. Dorsal column contributions to motor behaviour. *Experimental Neurology*, 1971, **33**, 53–58.

Merrill, E. G., & Wall, P. D. Factors forming the edge of a receptive field: The presence of relatively ineffective afferent terminals. *Journal of Physiology*, 1972, **226**, 825–846.

Schwartz, A. S., Eidelberg, E., Marchok, P., & Azulay, A. Tactile discrimination in the monkey after section of the dorsal funiculus and lateral lemniscus. *Experimental Neurology*, 1972, **37**, 582–596.

Wall, P. D. The sensory and motor role of impulses travelling in the dorsal columns towards cerebral cortex. *Brain*, 1970, **93**, 505–524.

Wall, P. D. Dorsal horn electrophysiology. In A. Iggo (Ed.), *Handbook of sensory physiology*. Vol. II. Berlin: Springer-Verlag, 1973. Pp. 253–270.

Chapter 13

The Chemical Senses:
A Systematic Approach

ROBERT P. ERICKSON
SUSAN S. SCHIFFMAN

Duke University

Progress and understanding in all sciences have depended on the development of a few general concepts which encompass and provide systematic form for large and diverse areas of data. In sensory physiology and psychophysics, the development of such concepts as "stimulus parameters" or "dimensions" has provided basic support of this nature. Knowledge of the proper stimulus parameters, for example, has given useful guidelines in the problem areas of receptor stimulation and stimulus encoding by the nervous system, as well as of psychophysical phenomena.

In the chemical senses, "parameters" and other related concepts have not been well developed, and thus these sensory systems have remained the "poor cousins" of sensory physiology. Therefore, in this chapter we shall present an approach to a few such concepts which cover a large proportion of the research in the physiology and psychophysics of the chemical senses. Preliminary definitions and findings will be presented concerning a broad group of concepts which lend themselves to a general systematization ("parameters," "primaries," "specificity," and "fiber

393

types"). Tentative approaches to the related problem areas of neural coding and the stimulation process will also be discussed.

THE PARAMETERS: TWO APPROACHES

Just as the knowledge of the relevant parameters has provided in the nonchemical senses a powerful approach to the development of data, the data also contain useful clues to the disclosure of the parameters underlying them. First we shall demonstrate a development of stimulus parameters from psychophysical data, and then we shall demonstrate how similar dimensions derive from considerations of existing neural data. These techniques are applicable to both taste and olfaction.

Psychophysical Approach

At the present time, the stimulus-receptor mechanisms that are responsible for the sensations of taste and smell are not fully understood. Thus, it has not been possible to order stimuli along their relevant dimensions on the basis of their chemical or physiological mechanisms. However, mathematical methods have recently been developed for ordering stimuli on the basis of their psychological similarities (Guttman, 1968; Kruskal, 1964; Shepard, 1962); these multidimensional scaling procedures have been used in the nonchemical senses. The multidimensional spaces which result from application of these procedures accommodate the data such that stimuli close to each other in a space are perceptually similar to each other; less similar stimuli occur farther apart. At the very least, these spaces could demonstrate the "quality" or psychophysical relationships among chemical stimuli based on quantitative measurement. The resulting arrangements may also lead us to parameters of chemical stimulation which produce the organization of taste and olfaction sensations as discussed in the next two sections.

Taste

An ordering of gustatory stimuli in a stimulus space is shown in Figure 1 (Schiffman & Erickson, 1971). Nineteen chemical stimuli chosen for their range of gustatory quality and chemical composition were equated for intensity, and then paired with one another so that similarity judgments between all combinations of the stimuli could be made. Guttman's nonmetric multidimensional scaling procedure (Guttman, 1968) was applied to these ratings, and the results demonstrated that these stimuli can be ordered in the three-dimensional Euclidean space shown.

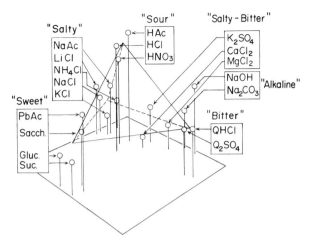

Figure 1. Taste space based on psychophysical (stimulus similarity) measures of simple stimuli (Schiffman & Erickson, 1971). The stimuli group roughly in accord with Henning's taste tetrahedron, with an additional "alkaline" group outside the tretahedron approximately opposite sweet.

It was also shown (Schiffman & Erickson, 1971) that an underlying set of three dimensions (hedonic quality, molecular weight and |pH-7|) accounts well for this space.

These dimensions were developed as follows. Each stimulus was rated by the subjects on a series of 45 semantic differential scales (for example, good——bad, salty——not salty, etc.). Also, a series of physical continua (molecular weight, pH, and a number of parameters that proved to be irrelevant) were examined for these 19 stimuli. Both the psychological semantic differential scales and the physical continua were used as trial dimensions with the result that three major dimensions emerged to describe the three-dimensional space. Figures 2(a), (b), and (c) represent planes through the similarity model defined by the three major dimensions which emerged: hedonic value, molecular weight, and |pH-7|. The plane in 2(a) is the floor of the model in Figure 1.

The model in Figure 1 demonstrates that the stimuli fall loosely into the five groups: sweet, sour, salty, bitter, and alkaline. In Figure 1 it is shown that parts of this arrangement (all groups except alkaline) are roughly approximated by Henning's taste tetrahedron; the alkaline group falls outside the tetrahedron, off the salty, sour, bitter surface. It should be emphasized that this is merely an ordering in a multidimensional space and does not necessarily implicate these groups as "primaries" (see pp. 409–410).

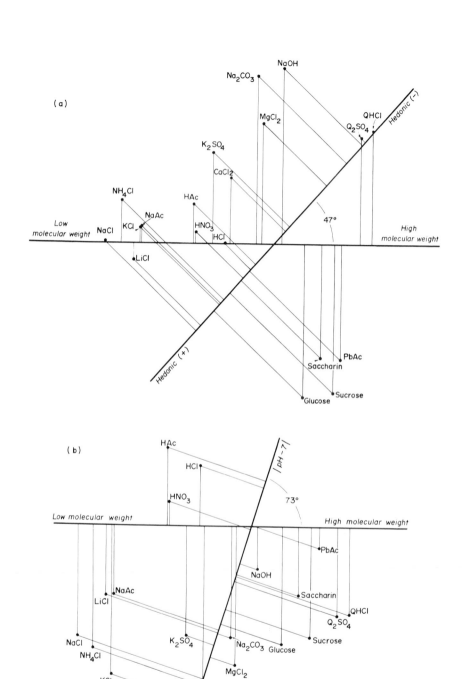

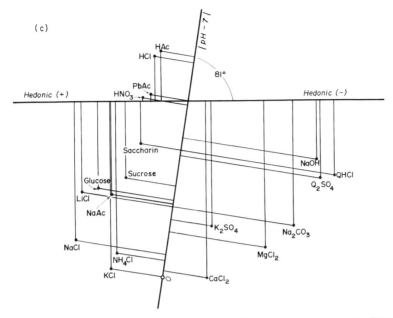

Figure 2. Three planes (Schiffman & Erickson, 1971) describing the space in Figure 1. (a) Molecular weight and hedonic dimensions (floor of Figure 1). (b) Molecular weight and |pH-7| dimension. (c) Hedonic and |pH-7| dimensions.

Only rather simple stimuli were arranged in Figure 1. Schiffman and Dackis (in press) have ordered more complicated stimuli, including such essential nutrients as amino acids, vitamins, and fatty acids (Figure 3). Similarity judgments were made between 21 stimuli including NaCl, glucose, quinine HCl, NaOH, and ascorbic acid (this was used as the sour standard rather than HCl). This space is similar to that given in Figure 1 in several respects. It is three-dimensional, and tetrahedral with respect to sweet, sour, salty, and bitter, suggesting that this solution has some generality. Sucrose, glycine, serine, and alanine form a sweet group. NaCl represents the salty corner of the tetrahedron while QHCl and ascorbic acid represent the bitter and sour corners respectively; these stimulus positions relate this space to that in Figure 1.

The placement of the other stimuli in Figure 3 can be understood with respect to this arrangement when the stimuli are related to the adjectives by the semantic differential method. Lysine HCl which has a strong salty component, is located near NaCl. Histidine HCl which has sour, salty, and bitter components, lies on the sour, salty, bitter face of the tetrahedron. Riboflavin, retinoic acid, tryptophan, leucine, and aspartic acid, which group together in the center of the space,

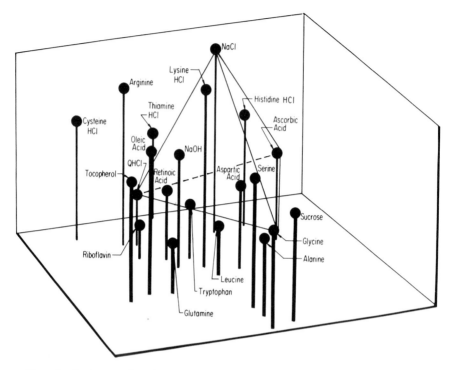

Figure 3. Taste space based on psychophysical (stimulus similarity) measures of complex stimuli (Schiffman & Dackis, unpublished). Solution comparable to Figure 1.

all have a strong dry component. However, tryptophan and riboflavin are nearer QHCl since they both have bitter components, and aspartic acid is placed toward ascorbic acid due to its sour components; leucine and retinoic acid are virtually tasteless. Thiamine HCl is located near QHCl but in the direction of ascorbic acid indicating its bitter-sour taste. Arginine is located near the bitters as well as NaCl indicating its bitter-salty taste. Arginine does fall slightly outside the tetrahedral structure, however, due to an alkaline component. Tocopherol, oleic acid, and glutamine tend to form a fatty group; the first two are definitely described as fatty while glutamine is considered meaty with a fatty component. Cysteine HCl falls off the side of the tetrahedral structure; this is probably due to its sulfur component. It is described as bitter but its sulfur was noted by most subjects ("contains sulfur," or "tastes like raw sewage"). All subjects wore nose plugs, but it is still possible that sulfurous odor reached the olfactory receptors via the nasal pharynx, thus affecting the placement of cysteine HCl in the space. NaOH fell in the center of the space in this experiment rather than off the side of the tetrahedral structure as it did in Figure 1; this probably was due to inclusion of a wider range of stimuli in Figure 3 as well as the fact that NaOH was the only stimulus used in liquid form. NaOH also proves to be a problem in the neural space (Figure 7).

Another means of ordering essential nutrients was employed by Schiffman and Dackis (in press). Thirty-four stimuli (18 amino acids, 4 fatty acids, 8 vitamins, NaCl, QHCl, NaOH, and sucrose) including the 21 stimuli in Figure 3 were ordered in a multidimensional space on the basis of their "semantic differential" measures. Incorporating the results from this semantic differential ordering (not shown here) with the results in Figure 3, several hypotheses can be made about the physicochemical properties of these essential nutrients.

In Figure 1, the sweet stimuli were lighter than the bitter ones; here the three amino acids with the smallest molecular weights (glycine, alanine, and serine) taste sweet, whereas the heavier amino acids tended to be bitter. Amino acids with aliphatic side chains and those with side chains containing hydroxylic (OH) groups tended to be either pleasant or tasteless; those containing sulfur or aromatic rings tended to be quite unpleasant. The amino acids with side chains containing basic groups (arginine, lysine HCl, and histidine HCl) have in common sharp, bitter, and salty components even though two are crystallized from acid solution. Amino acids containing acidic groups are sour. Linolenic, linoleic, and oleic acids, which are insoluble in water but soluble in ether, are considered to have a fatty taste; whether "fatty" is a taste, odor, or tactile quality is not clear. Substances which are not soluble or have low solubility in water (e.g., tyrosin, ergocalciferol, vitamin K, and retinoic acid) tend to be tasteless.

Other physicochemical aspects of the molecules were investigated to understand their relationship to gustatory quality, such as melting point, boiling point, density, crystal type, and molecular shape; none of these appear powerful compared with functional grouping in determining taste quality.

Olfaction

In addition to their use in taste, multidimensional scaling techniques have also been used to order olfactory stimuli. Schiffman and Erickson (1972) applied multidimensional scaling techniques to two sets of psychophysical olfactory data. Guttman's multidimensional scaling procedure was first applied to Wright and Michels' (1964) set of correlations among 50 odorants; the result is given in Figure 4. The space is two-dimensional, with the stimuli falling into two subsets. The large group

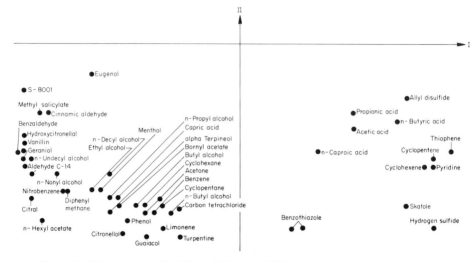

Figure 4. Olfactory space (Schiffman & Erickson, 1972) based on psychophysical (stimulus similarity) measures (Wright & Michels, 1964). This two-dimensional space does not resemble Henning's three-dimensional odor prism.

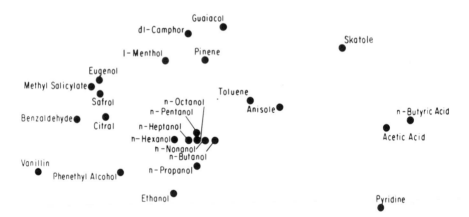

Figure 5. As Figure 4 (Schiffman & Erickson, 1972; based on Woskow, 1964). Note similar relationships between stimuli common to Figures 4 and 5 (see text).

to the left is affectively rather pleasant, and the group to the right rather unpleasant. Guttman's technique was also applied to Woskow's (1964) stimulus similarity data (Figure 5), in which he compared 25 olfactory stimuli, many of which also had been employed by Wright and Michels. The result is similar to that in Figure 4 in several respects: both solutions are two-dimensional, and in both vanillin, benzaldehyde, methyl salicylate, eugenol, and ethanol appear grouped to the left, while skatole, butyric acid, acetic acid, and pyridine are grouped to the right. From these analyses we conclude, tentatively, that: (1) an olfactory space can be derived from the psychophysical measures which accounts well for the data; (2) the space is of small dimensionality, probably two-dimensional; (3) there are no clear psychophysical "groups" of stimuli (see p. 409) to support the idea of "primaries."

An analysis of the data (Schiffman, 1974) showed that the following factors alone do not determine olfactory quality when considered *individually*: molecular size, molecular shape, structure, number of double bonds, molecular weight, boiling point, water solubility, or dipole moment. All of the stimuli in Figure 5 were ether (fat) soluble; fat solubility may be essential for stimulation to occur. The Raman spectra (discriminant function analysis) separated the two major groups (pleasant and unpleasant) but did not describe the space within each group. Functional groups were important in that the aldehydes, esters, phenols, and alcohols were in the pleasant group, with nitrogen and sulfur (when these were not associated with oxygen) and carboxylic acids in the unpleasant group. Also, cyclic compounds tended to group together.

Because no single physicochemical variable can order olfactory stimuli, Schiffman (1974) weighted a series of variables in an attempt to reproduce the space in Figure 4 showing that when considered together, several variables can account for the space. She used a technique which incorporated the Mahalanobis d^2 measure of distance to weight the mean Raman intensities, cyclic aspects, molecular weight, number of double bonds, and functional groups. She was successful in weighting the variables in such a way that the distances, and thus the space arrangements among the stimuli (see Figure 6), are quite similar to those found using Wright and Michels' psychophysical data. In comparing the "regenerated space" (Figure 6) with the "original" (Figure 4) the error in fitting interstimulus distances is 24%. The weights applied to scores for selected physicochemical variables are seen in Table 1. The variables used here for prediction are correlated only slightly with each other. When variables were highly correlated, such as molecular weight, boiling point, size, and number of atoms in the molecule, only one (here, molecular weight) was used for prediction. It can be seen from the size of these weights that molecular weight, some Raman ranges, as well as functional groups, are quite important in odor quality. Although the error in the present solution is small, with incorporation of a wider range of physicochemical variables, better prediction can undoubtedly be made.

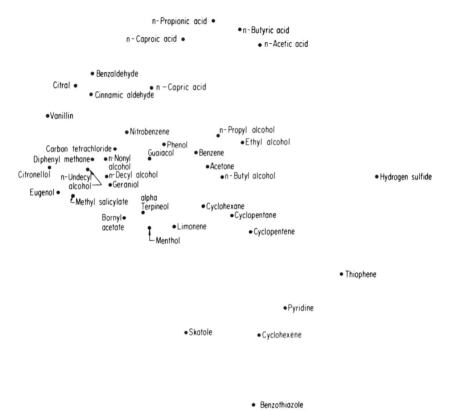

Figure 6. Reconstruction of stimulus arrangements in Figure 4 from ratings of the stimuli on several physicochemical scales (Table 1) (Schiffman, 1974).

Summary

In both taste and olfaction, it is clear that multidimensional scaling techniques can be used successfully to order stimuli. Further it is apparent that these derived orderings may aid in the search for the relevant physicochemical variables. An estimation of the full success or failure of this recently developed methodology awaits its systematic use with broad ranges of stimuli.

Neural Approach

As with psychophysical data, the neural responses in the chemical senses also may be treated with multidimensional scaling techniques to reveal the organization of the stimuli producing these responses. This organization for taste (Figure 7) resembles that obtained from the psy-

TABLE 1 Weightings[a] of Physicochemical Characteristics of Stimuli in Figure 4 to Form the Reconstruction of This Space Shown in Figure 6

Physicochemical characteristic	Weighting factor
Molecular weight	6.24
Number of double bonds	.51
Phenol	2.33
Aldehyde	3.21
Ester	.24
Alcohol	2.54
Carboxylic acid	5.50
Sulfur	3.44
Nitrogen	3.15
Benzene	−0.14
Halogen	−0.34
Ketone	−0.19
Cyclic	4.56
Mean Raman intensity below 175 cm^{-1}	0.01
Mean Raman intensity from 176 to 250 cm^{-1}	3.57
Mean Raman intensity from 251 to 325 cm^{-1}	−0.75
Mean Raman intensity from 326 to 400 cm^{-1}	3.81
Mean Raman intensity from 401 to 475 cm^{-1}	1.65
Mean Raman intensity from 476 to 550 cm^{-1}	−3.63
Mean Raman intensity from 551 to 625 cm^{-1}	−0.69
Mean Raman intensity from 626 to 700 cm^{-1}	−1.16
Mean Raman intensity from 701 to 775 cm^{-1}	0.07
Mean Raman intensity from 776 to 850 cm^{-1}	3.04
Mean Raman intensity from 851 to 925 cm^{-1}	0.24
Mean Raman intensity from 926 to 1000 cm^{-1}	0.36

[a] Weighting factors indicate importance of each characteristic in reconstructing this space (e.g., molecular weight is very important). Functional groups were coded by their number in a particular molecule (benzaldehyde has one aldehyde group; cyclic compounds were coded "1" and non-cyclic "0"). Large negative weights imply that these variables are inversely related to olfactory quality.

chophysical data as discussed previously (Figure 1) in that it is roughly approximated by Henning's (1916) taste tetrahedron. Although not shown here, it was also found that the neurons do not fall into "primary" groups (Doetsch & Erickson, 1970; Erickson, 1963, 1967; Schiffman & Falkenberg, 1968; Schiffman & Erickson, 1971). (This is discussed further on pp. 409–410).

When this analysis is applied to olfaction (Figure 8), a space similar to that generated from the psychophysical data (Figure 4) is generated. Both spaces are two-dimensional and the stimuli common to both analy-

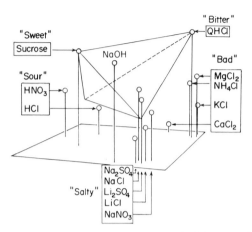

Figure 7. Taste space based on neural responses from individual rat chorda tympani neurons. As the psychophysically based space in Figure 1, the solution is three-dimensional, approximating Henning's taste tetrahedron plus a "bad" group opposite "sweet." The placement of NaOH—alkaline—falls off the salty, sour, bitter face of the tetrahedron as in Figure 1.

ses bear similar spatial relations to each other, i.e., in both solutions (Figures 8 and 4) propanol, butanol, and benzene are "grouped," as are benzaldehyde, nitrobenzene, and geraniol, with decanol and menthol falling between these groups. Neither the neural nor the psychophysical spaces resemble the three-dimensional olfactory prism of Henning (1915) (see Boring, 1942).

To demonstrate how such an approach may be valid for the chemical senses, it is instructive to see how well these methods succeed in another sensory system in which the organization of the stimulus dimensions, the "answer," is already known. For these purposes we will demonstrate how the responses of visual neurons may be used to reveal the underlying wavelength parameter

For the purposes of deriving the relevant dimensions, the responses of a number of visual receptors to several different wavelengths were analyzed; this is analogous to the situation in taste or olfaction where the parameters are unknown but the responses of a number of neurons to several stimuli are known. The data were analyzed by a method which placed both the stimuli and the neurons in a space (Schiffman & Falkenberg, 1968) such that the response of each neuron to the various stimuli was greatest to the stimuli closest to that neuron and least for the stimulus furthest away; the results are shown in Figure 9.

It is clear that with these methods the arrangements of stimuli and neurons is properly recovered. That is, the relation of each stimulus to all the others is correct ("orange" is between "yellow" and "red,"

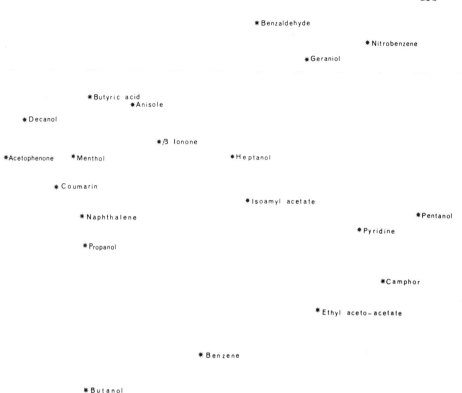

Figure 8. Olfactory space based on neural responses from individual frog mitral cells (MacLeod, 1971). As psychophysically based spaces in Figures 4 and 5, the space is two-dimensional, with similar groupings of the stimuli common to psychophysical (Figure 4) and neural solutions (Figure 8) (see text).

etc.), as are the points of maximum sensitivity for each neuron, "blue" neurons appearing in the "blue" region, etc. In addition it is clear that, except for the "double cone," the receptors fall within three groups. If it had not been known before, we might observe from this arrangement that "wavelength" would be a useful parameter.

With this analysis we conclude that the method is generating the correct stimulus and neuron relationships, and may be attempted with the chemical senses; Figures 7 and 8 are the results of the analysis.

RELATED PROBLEM AREAS

There are a number of important research areas which are all closely related to the issue of stimulus parameters. Typically, there is a lack

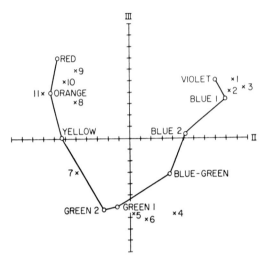

Figure 9. Validation of method used in Figures 7 and 8 (Schiffman & Falkenberg, 1968). Correct placements obtained for visual stimuli (and neurons). A double cone is represented by number 7.

of systematization in these areas. For example, there are no standard definitions for such commonly used and basic terms as "stimulus primaries," "specificity," etc., and the relations between the various definitions of each term are usually not made clear.

Primaries

The problem of "primaries" in taste and olfaction is a very ancient one going back at least to the time of Linnaeus (see discussion in Boring, 1942). There are three basic questions about primaries which have gone unanswered; indeed, they have usually gone unasked. First, what is meant by the term "primary"? Second, do any data in the chemical senses demand the existence of primaries for their explanation? And third, are primaries logically useful or essential in the chemical senses?

Definition of Primaries

There are at least four kinds of definitions of "primaries": stimulus, psychophysical, neural, and phenomenological. The first definition refers to the fact that there may be distinct *stimulus* types such as "salty" in taste, or "ethereal" in olfaction. *Psychophysical* primaries are based on laws of stimulus mixtures, or on descriptions of tastes and odors in controlled, experimental situations. The third refers to the matter of *neuron* types such as "salty" taste fibers. It is of course possible that one of these kinds of primaries exists without the others, or several types of primaries might exist ("salty" stimuli and "salt" neurons) with or without a close relation between the two.

The fourth type of "primary" definition expresses the idea that there are phenomenologically very "pure" or "primary" chemical sensations. Such impressions might or might not have any relation to the other kinds of primaries. However, the phenomenological "primaryness" of certain tastes has provided the useful thrust to search for the other primaries. The logical status of the phenomenological definition is not clear, and thus this type of "primary" will not be discussed other than to point out that the existence of such primaries has never been established; that is, it has not been demonstrated what clear, pure tastes or odors are. It would be difficult to design such an experiment. It should be noted in passing, however, that the phenomenological primacy of a few colors has played a strong role in the history of color vision. The implication for the chemical senses is that there might be classes of taste neurons or stimuli, such as "salty" and classes of olfactory neurons or stimuli such as "fruity." Data concerning the nonphenomenological types of primaries will be discussed.

Do the Data Demand the Existence of Primaries?

The existence of "primaries" has been argued on stimulus, psychophysical, and neural grounds.

Stimulus Primaries

In the chemical senses, our understanding of the issue of stimulus primaries is only rudimentary. Our state of ignorance about this topic may be highlighted by consideration of Moncrieff's discussion of this topic (Moncrieff, 1967). He requires more than sixty pages to present the assorted pieces of knowledge about the basic odor stimuli, and more than forty pages to discuss the nature of taste stimulation. No other discussion of these topics has ever reduced the nature of the stimulus primaries to a very simple order, and all discussions are more notable for the exceptions to the rules than for the rules themselves. The point here is that the chemistry of taste and olfactory stimulation does not resolve to a few neat and simple patterns.

While the chemistry of gustatory and olfactory stimulation has not given itself to notable simplification, some neural studies have lent support to the idea of stimulus types. The proportions of populations of

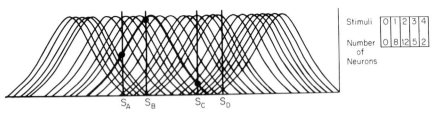

Figure 10. Hypothesized neural and stimulus organization for the chemical senses. This model is presented as one-dimensional for simplicity. It is to be generalized to two dimensions for the olfactory sense (Figures 4–8), and to three dimensions for taste (Figures 1–3). Stimuli (S_A, S_B, S_C, S_D, etc.) arranged to form baseline. This arrangement is seen as continuous (Figures 1–8) rather than stimuli falling in tight "primary" groups. Neural response functions for twenty-seven neurons shown as bell-shaped curves on baseline. Each neuron responds best to one stimulus, but also responds to broad ranges of stimuli (e.g., heavily drawn response is highest at stimulus B, but encompasses stimuli A, B, and C.) The neurons fall in a continuous fashion along this baseline, rather than appearing in primary groups (Doetsch & Erickson, 1970; Erickson, 1963, 1967; Schiffman & Falkenberg, 1968, for taste; Schiffman & Erickson, 1972, for olfaction).

Insert: The number of neurons responding to 0, 1, 2, 3, or 4 stimuli is in accord with nonparametric predictions based on "independence" of stimuli (see text). Independence suggests primacy, but may also indicate special relative positions on a dimension as is true for the various sets of color primaries.

neurons responding to each of certain stimuli are distributed in a manner that would suggest that the stimuli were *unrelated*. From this it is argued that the stimuli are independent of each other, and thus are distinct stimulus types. Some typical data (Frank & Pfaffmann, 1969) are given in Figure 10 (section in brackets). Each of the 4 stimuli activate about half of the 27 neurons shown (NaCl and HCl about 60% each, sucrose and quinine about 40% each). If these stimuli operated independently (i.e., stimulus A was not coupled with stimulus B, etc.), then 2 neurons should respond to all 4 stimuli, 5 should respond to 3, 12 to 2, and 8 to 1; this proves to be the case. That these four stimuli could also be related members of a continuous baseline parameter is illustrated in the figure. However, when the *amount* of response of each neuron is considered (rather than whether or not it responded), and when larger numbers of stimuli are employed, it appears that the stimuli do not group into independent types, but fall in a scattered fashion throughout the stimulus space (Doetsch & Erickson, 1970; Erickson, 1963, 1967; Schiffman & Falkenberg, 1968; Schiffman & Erickson, 1971; Scott & Erickson, 1971; Erickson, Doetsch, & Marshall, 1965).

Since taste and olfactory stimulation do not resolve easily into a few "primary" mechanisms, perhaps "primary" mechanisms do not provide the optimal approach to this matter. Thus, we will continue to examine the possibility of stimulus "continua" in the chemical senses, the circumstance typically obtained in the nonchemical senses (as wavelength for color, or frequency for auditory pitch).

Psychophysical Primaries

As primaries, certain stimuli should combine to produce a wide range of tastes. VonSkamlik (1926, see also Boring, 1942) showed that certain simple tastes could be duplicated by mixtures of his four primary tastes. For example, he showed that certain inorganic salts, which seem to have complex tastes, could be imitated by combinations of sucrose, sodium chloride, tartaric acid, and quinine. However, the breadth of his success does not correspond to the extent to which various colors may be matched by combinations of three primary colors.

An interesting issue at this point is the problem of whether taste stimuli are simple or complex. For example, should sodium chloride simply be considered a salty stimulus, or is it more complicated, e.g., with the sodium and chloride ions acting independently. If this is correct, then many chemical stimuli would have to be plotted at more than one point along taste dimensions. However, we have made the suggestion elsewhere (Erickson, 1968) that the taste system is a "synthetic" system in which several stimuli are "fused," as the visual system would fuse two stimulating wavelengths into a single perceived color. This is apparently similar to VonSkramlik's view. Thus, it would be expected if the chemical stimulus had two components, they would be "fused," appearing at one point in the stimulus dimensions, and they would be perceived as one unique taste.

Also, as is clear in visual research, various sets of stimuli could be used as primaries. This would suggest that VonSkramlik's stimuli may not necessarily be the only ones that could have been used.

In close agreement with VonSkramlik's conclusions, Henning felt that the *continua* of taste sensations could be described by a tetrahedron in which four stimulus primaries define the corners; he also felt that olfactory sensations could be encompassed by six primary stimuli describing a prism. The taste tetrahedron, but not the olfactory prism, was given partial support in the preceding discussions. In our present state of relative ignorance on the subject, these points can be used as reference points for ordering large numbers of stimuli, but it is not clear that the corners of these spaces have any special value as "primaries."

Neural Primaries

Neurally, the term "primary" states the position that there exist only a few classes of neurons, each of which is maximally sensitive, or singularly sensitive, to one taste quality. This issue is unresolved, but several

lines of evidence (Doetsch & Erickson, 1970; Erickson, 1963, 1967; Schiffman & Falkenberg, 1968; Schiffman & Erickson, 1971) suggest that
there is a continuum of fiber sensitivities (as in Figure 10) with no
grouping into neuron classes (see pp. 402–404).

The evidence which has been used to support the idea of primary
fiber types is also compatible with the point of view that there is a
continuum of fiber sensitivities. An argument for the existence of fiber
types in the chemical senses derives in part from the fact that the "four
primary classes" of stimuli are maximally efficient at different parts of
the tongue (Kiesow, 1894) and different portions of a central nervous
system (CNS) relay (Halpern, 1966), and different olfactory stimuli
are maximally efficient at particular portions of the olfactory mucosa
(Mozell, 1964, 1970) or bulb (Adrian, 1953; Mozell, 1964). However,
this argument is entirely equivocal (see Figure 10) since the same is
true for the point of maximum stimulation with several stimuli in a
continuous system (as could be exemplified by any four tones in the
cochlea, or in any auditory CNS relay). Presumably, *any* stimulus, primary or not, would have a point of maximum stimulation efficiency
on the tongue or olfactory bulb, but this would not indicate the existence
of primary neuron types.

That various drugs, or variations in temperature, may affect tastes
differentially (Pfaffmann, 1959) has also been used as evidence that
there are a few neural taste primaries. Again, this is not necessarily
evidence that primary types of neural mechanisms have been isolated,
but may simply indicate that certain portions of a neural continuum
are more sensitive to temperature or drug effects than another (see
Figure 10).

Summary

We conclude that no data in the chemical senses, stimulus, psychophysical or neural require the concept of "primaries" for their parsimonious description.

Are "Primaries" Necessary in the Chemical Sense?

The only sensory systems other than taste and olfaction in which
primaries have been postulated are color vision and perhaps the temperature sense, although the existence of "warm" and "cold" primaries is
debatable (Poulos, 1971). For the others, including audition, kinesthesis,
vestibular sensitivity, touch, pain, and even visual form, there seems
to have been no need to postulate the existence of sensory primaries
for adequate neural coding.

Why then, have "primaries" been postulated for the chemical senses? In color vision the idea of primaries seems to have derived from the laws of color mixture, according to which nearly all color sensations may be aroused by the proper combinations of a few primary colors. It is not at all clear that any such general laws of stimulus combinations are valid for taste or olfaction, i.e., the success with which one may reproduce all of the chemical sensations with a few primaries is indeed very slight.

Since sensory systems may operate adequately without the support of primaries, and since the laws of stimulus mixture, as found in color vision, do not clearly apply in the chemical senses, the existence of chemical primaries would not seem parsimonious unless demanded by the data. Since the data do not demand primaries (see pp. 407–410) this concept may be more of a liability than an asset.

Specificity

"Specificity," a term related to neural responses, is probably the most overused and underdefined term in sensory physiology. The term has at least three definitions (Erickson, 1968) referring to the narrowness of tuning of the individual neural elements, to the question of fiber types, and to the selectivity of the neurons within several domains of stimuli.

Narrowness of Tuning

The problem of the selectivity of sensory neurons to particular stimuli may be one of the oldest in the history of neurophysiology. This topic is a direct outgrowth of the theory of specific nerve energies, and thus goes back at least to Bell and Magendie in the early nineteenth century, through Johannes Mueller's statement of the law of specific nerve energies in the mid-nineteenth century. This doctrine has evolved into a theory of "specific sensory fibers" (Boring, 1950) or "labeled lines" (Perkel & Bullock, 1968). The general import of the specific nerve energies theory was that the activity in any sensory nerve, such as the auditory nerve, was the adequate signal for a particular kind of sensation, in this case auditory. In the modern "specific fiber" or "labeled line" point of view, a more explicit sensation is mediated by each individual neuron and thus each neuron should be specific for one elementary sensation. This obviously follows from the concept that the neuron is the functional unit of the nervous system (see p. 416).

In the realm of the chemical senses it seems clear that both gustatory and olfactory neurons are typically tuned to very wide ranges of diverse stimuli (Dethier, 1971, 1973; Doetsch & Erickson, 1970; Erickson, 1963, 1967, 1968, 1973; Erickson *et al.*, 1965; Frank & Pfaffmann, 1969; Ganchrow & Erickson, 1970; Gesteland, Lettvin, Pitts, & Rojas, 1963; MacLeod, 1971; Mathews, 1972; O'Connell & Mozell, 1969; Pfaffmann, 1959; Schiffman & Erickson, 1972; Scott & Erickson, 1971; Yamada, 1971) (see Figures 11–14). For insects it had appeared that the neurons might be tuned to rather limited ranges of stimuli, but in recent reviews of this matter it is concluded that even for insects each neuron is quite broadly tuned (Dethier, 1971, 1973). This general idea is characterized in Figure 10 by the breadth of tuning of the individual neurons along the stimulus dimension shown. Considering the response curve that is drawn with a heavy line, the fact is demonstrated that a neuron typically responds to many stimuli, as those indicated along the baseline. In this sense, taste and olfactory neurons are clearly "nonspecific."

A rather interesting view of specificity in taste has been offered by Békésy (1964). He suggests that individual neurons are very precisely tuned, but that the *stimuli* are not specific; that is, he feels that a salty

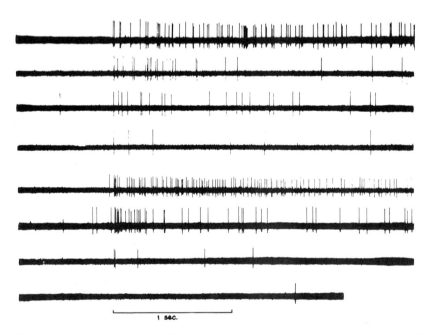

Figure 11. Responses of a rat peripheral taste (chorda tympani) neuron to various stimuli. With records unlabeled, it is clear that single neuron stimulus representation is confused. This typical neuron responds to diverse stimuli and with graded responses to variation in stimulus intensity. Stimuli are, from top to bottom: .03 *M* HCl, .1 *M* KCl, 1.0 *M* sucrose, .01 *M* QHCl, .1 *M* NaCl, .01 *M* NaCl, .001 *M* NaCl, water.

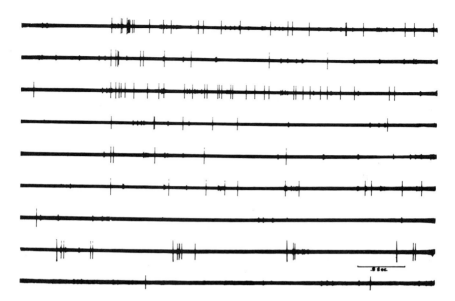

Figure 12. Responses of a rat second-order taste neuron (nucleus tractus solitarius) indicating responsiveness to warming, cooling, and mechanical stimulation, as well as taste stimuli. From top to bottom, the stimuli are: .1 *M* NaCl, .1 *M* KCl, .03 *M* HCl, .01 *M* QHCl, 1.0 *M* sucrose, cooling (28–16°C jump; threshold change is about 2°C), warming causes inhibition (27–33°C); mechanical stimuli (artifacts shown before each group of impulses), water.

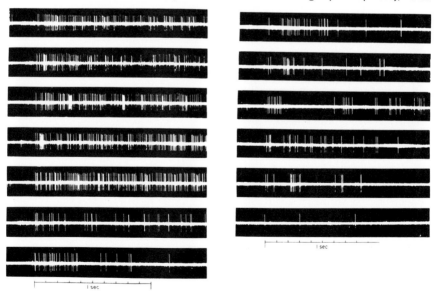

Figure 13. As Figure 12, showing very broad responsiveness of these neurons. Stimuli are, top to bottom, left column first: .1 *M* NaCl, .1 *M* NaNO₃, .1 *M* Na₂SO₄, .1 *M* LiCl, .1 *M* Li₂SO₄, .3 *M* KCl, .3 *M* CaCl₂, .1 *M* Mg Cl₂, .1 *M* NH₄Cl, .03 *M* HCl, .03 *M* HNO₃, .01 *M* QHCl, 1.0 *M* sucrose (Doetsch & Erickson, 1970).

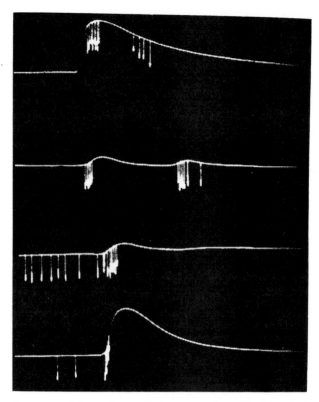

Figure 14. Responses of a frog primary olfactory neuron to a wide range of stimuli. Stimuli, from top to bottom: canphor, limonese, carbon disulfide, ethyl butyrate (Gesteland *et al.*, 1963).

stimulus may actually have in addition to its saltiness other tastes such as sweet or bitter. In his view the reason why each neuron responds to many stimuli is not that the neuron is broadly tuned, but that the stimulus is broadly tuned. That there is a logical difference between these two points of view is not entirely clear since the two realms (neuron and stimuli) are defined in terms of each other. For example, it could be logically argued that our view of monochromatic color stimuli in vision is not correct, and that each wavelength is actually multichromatic, a muddy stimulus which would stimulate a range of specific neurons. Thus each color-coded neuron could be considered to be entirely specific, with the stimuli being broadly tuned. This position would have no obvious advantage, and either position could be taken. Considering the broad sensitivities of sensory neurons in general (Erickson, 1968, 1973), it seems more parsimonious, or at least in line with a useful tradition, to consider the sensory neurons to be broadly tuned with stimuli remaining specific.

Fiber Types

The term "specificity" is sometimes used to describe the fact that there are only a few neuron types. For example, we may speak of specific color-coded neurons in the sense that there are only a few types, and a neuron which is maximally sensitive in the orange part of the spectrum may be termed a specific "orange" receptor.

In the chemical senses, various types of neurons have been postulated, such as "salt neurons." Typologies of this nature generally derive the type name according to the stimulus that produces the best response. Referring to Figure 10, it is seen that even broadly tuned neurons will respond best to one stimulus; such a typology, as with color-coded neurons, does not demand that a neuron is narrowly tuned.

It is clear that responding best to one stimulus does not necessarily indicate that there are fiber types, although this method is often used to define typologies. For example, if four stimuli were used, such as a sweet, a sour, a salty, and a bitter stimulus, then as indicated in this figure each neuron would respond better to one of these stimuli than to another. But as is clear in the figure, it does not follow from this that four fiber types have been isolated. If, instead of determining only the best stimulus for each neuron, the responses to a broad range of stimuli are closely examined, the question of fiber types may be examined. This approach has been used in taste (Doetsch & Erickson, 1970; Erickson, 1963, 1967; Schiffman & Falkenberg, 1968) and olfaction (Schiffman & Erickson, 1972), and no evidence of fiber typologies was seen. Instead, it appeared that there was a continuum of fiber sensitivities as in Figure 10, similar to the auditory sense where the sensitivities of the neurons are seen to be distributed in a rather even fashion over the frequency dimension.

Modality Specificity

A third type of specificity is that which describes the selectivity of the afferent neurons to a particular sensory domain. For example, some tactile neurons are nonspecific in the sense that they are also influenced by temperature stimuli. Taste neurons may be sensitive to temperature and mechanical stimuli (Figure 12) and the rate of response to any taste stimulus depends on the temperature of the stimulating solution (Sato, 1967). It is possible that this responsiveness is seen strictly as noise by the nervous system, but until proven otherwise, the possibility that these neurons may also signal mechanical and temperature stimulation must remain open.

Summary

There is evidence to indicate a "continuum" rather than a "typology" of neurons in the chemical senses. There appears to be no clear evidence for neural specificity for taste or olfaction in terms of "fiber types" or "narrowness of tuning" of individual neurons.

Neural Coding

The most obvious problem in coding in taste and olfaction may be described in reference to Figures 11–14. These figures show that individual gustatory and olfactory neurons respond to very broad ranges of stimuli. To make clear the problem that this poses for the nervous system, only the neural responses are given; the responses are not labeled to indicate what the stimulus was since the nervous system is not provided the luxury of such labels (see the figure legends). Even with the clue that the first record in Figure 11 was produced by .03 M HCl, it is not possible to deduce the stimuli producing the other responses. It would be very difficult for the nervous system to "decode" the responses of any of these neurons, since each responds to so many different stimuli, including mechanical and thermal stimuli (Figure 12); each is also sensitive to the intensity of the stimulus (Figure 11). Thus, the fact that the neuron responds to each stimulus with the greater or lesser rate of firing may indicate that the stimulus was changed in quality, *or* the same stimulus may have changed in intensity (Erickson, 1963, 1968, 1973).

Some investigators have suggested that the responsiveness of each neuron to so many stimuli may be treated by the nervous system as noise, and perhaps at higher levels of the nervous system this noise is weeded out such that each neuron then responds to only one stimulus. However, this is not the case; neurons at higher levels of the nervous system still respond to a broad array of stimuli (Figures 12 and 13) (see Scott & Erickson, 1971, for thalamic neurons; adequate records are not available for cortical neurons).

The problem then is that since the individual neurons are presumably the functional units of the nervous system, they should be the encoders of stimuli; but each responds to a bewildering array of stimuli. The solution to this as well as to other problems in neural coding (Erickson, 1968, 1973) may be found in the color-vision theory of Helmholz. Following Young, he conceived of color vision as being mediated by a

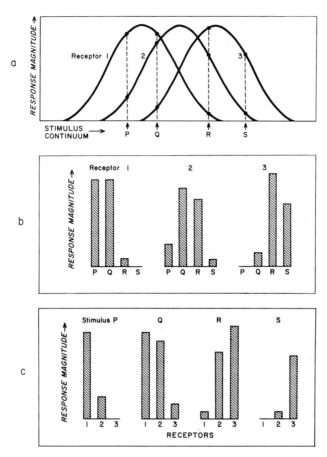

Figure 15. Hypothesized neural coding process in the chemical senses.

(a) Afferent neurons or receptors. Curves 1, 2, and 3 represent the responsiveness of three neurons along a hypothetical stimulus dimension. P, Q, R, and S represent four stimuli along this stimulus dimension. The responsiveness of a neuron to one of these stimuli is indicated by the intersection of the response curve and the ordinate erected at the stimulus.

(b) Responsiveness of the three neurons to the four stimuli in (a). In each of the bar graphs is shown the responsiveness of one of the neurons to each of the stimuli in (a). These show the breadth of responsiveness of each neuron to various stimuli. One neuron cannot adequately represent a stimulus; that is, it cannot differentiate between stimuli.

(c) Across-fiber patterns. In these bar graphs are shown the patterns of activity across the three neurons produced by the four stimuli in (a). Each stimulus produces a characteristic pattern across this small population of neurons. There would be as many across-fiber patterns as stimuli. The nature of the stimulus and its intensity are adequately given in these patterns (Erickson, 1963).

few groups of broadly tuned neurons as shown in Figure 15. The breadth of tuning of such neurons is illustrated in part (a) of this figure. In Figure 15(b), the responses of these neurons to four stimuli are detailed, making it clear that any individual neuron could not unequivocally encode color or intensity. In part (c) of this figure, which represents the view of Helmholtz, it is shown how in the *combined* activities these few neurons could unequivocally encode very many different colors, as well as their intensities. What appears to have been a *problem* in the chemical senses, i.e., the broad tuning of each neuron, Helmholtz used as a *solution.*

Beyond the area of color vision, this type of "across fiber pattern" coding of stimulus quality and intensity is of utmost value in the neural encoding of stimulus events in general; the details of this are discussed elsewhere (Erickson, 1968, 1973). In brief, in this approach the individual neuron is unburdened in its coding role, and an aggregation of neurons becomes the "functional unit" for the neural message.

Very briefly, the advantages of this method of neural encoding compared with the labeled line view are: (1) There may be many more messages encoded by the nervous system than there are neurons. For example, if each neuron could assume ten usably different levels of activity, then for each group of neurons the number of messages would be 10^n, where n is the number of neurons. In this example, eleven neurons could encode as many messages as there are neurons in the nervous system, i.e., 10^{11}. (2) Each message would be given much neural weight; this would not be true if each neuron were given the responsibility of carrying a message. (3) Each message would be given great stability in the face of variability of the neuron's response and in the face of neuron losses.

Summary

The breadth of tuning of the individual neurons in the chemical senses is probably functional in providing an economical coding of many stimuli by means of the "across-fiber pattern" of activity in a limited number of neurons. This has been detailed in both taste (Doetsch & Erickson, 1970; Erickson, 1963, 1967, 1968, 1973; Erickson *et al.*, 1965; Ganchrow & Erickson, 1970; Pfaffmann, 1959; Scott & Erickson, 1971) and olfaction (Cain & Bindra, 1972; O'Connell & Mozell, 1969; Schiffman & Erickson, 1971).

The Neural Code and the Parameters

The neural codes should relate very closely to the underlying stimulus dimensions. In the example for color vision in Figure 15 (c) the stimuli which are close to each other on the wavelength parameter are neurally similar, the dissimilarity increasing with stimuli that are further apart.

TABLE 2 Correlation Matrix for All Pairs of All Stimuli[a]

	Na2SO4	NaNO3	LiCl	Li2SO4	NaOH	KCl	NH4Cl	MgCl2	CaCl2	HCl	QHCl	Sucrose	
	.82 (28)	.86 (24)	.91 (44)	.71 (16)	.78 (21)	.02 (56)	.06 (57)	.36 (16)	.00 (39)	.37 (42)	.30 (48)	.19 (41)	NaCl
		.83 (25)	.82 (28)	.75 (16)	.77 (19)	-.27 (28)	-.22 (28)	-.07 (16)	-.18 (27)	-.09 (23)	.05 (26)	.21 (26)	Na2SO4
			.77 (24)	.75 (16)	.79 (19)	.04 (24)	.05 (24)	.39 (16)	.11 (23)	.12 (23)	.32 (24)	.40 (24)	NaNO3
				.69 (16)	.83 (21)	.03 (45)	.11 (46)	.44 (16)	.03 (39)	.37 (37)	.33 (43)	.28 (41)	LiCl
					.66 (13)	-.21 (16)	-.16 (16)	.12 (16)	-.09 (15)	.10 (16)	.09 (16)	.25 (16)	Li2SO4
						-.05 (21)	.15 (21)	.32 (13)	.11 (18)	.17 (21)	.21 (21)	.31 (21)	NaOH
							.85 (59)	.77 (16)	.87 (39)	.49 (43)	.46 (50)	.04 (43)	KCl
								.91 (16)	.84 (40)	.57 (43)	.45 (51)	-.05 (44)	NH4Cl
									.91 (15)	.48 (16)	.79 (16)	.09 (16)	MgCl2
										.41 (34)	.39 (38)	.07 (37)	CaCl2
											.56 (43)	.12 (37)	HCl
												.34 (42)	QHCl

(Inset legend: box labeled **r** above, box labeled **n** below)

[a] The number in the upper left corner of each cell is the correlation (Pearson's r). The number in the lower right of each cell is the number of pairs on which the correlation is based (Erickson *et al.*, 1965). According to the present hypothesis, the capacity of an organism to distinguish between these stimuli is inversely related to the magnitude of the correlation; e.g., LiCl and NaCl should taste very much more alike than, say, LiCl and QHCl. The correlations are inversely related to distances between stimuli (Figures 1 and 7).

For example, the neural patterns for two stimuli which are close together, such as P and Q, are more similar than those which are produced by stimuli which are further apart, such as P and S.

If the chemical senses are organized as color vision, the similarity of the across-fiber patterns set up across taste or olfactory neurons by two stimuli should indicate the distance apart that these stimuli are on the stimulus dimensions. The degree of difference between two across-fiber patterns may be discovered in several ways, one of the simplest being in the correlation between the points constituting the two patterns. Correlations between a number of taste stimuli are given in Table 2. These correlations should relate closely with the stimulus dimensions, as well as with psychophysical estimates of stimulus similarity. Such correlations do, in fact, relate very well with the dimensions derived from neural activity shown in Figure 7, and with the psychophysical dimensions (Figure 1) and psychophysical estimates of similarity in animals (Morrison, 1967, 1969).

These correlations have also been shown to relate to cross-adaptation studies in which the degree of adaptation by one stimulus on another is hypothesized to indicate the similarity of the two stimuli, i.e., of any two stimuli correlating very highly with each other, one should adapt the neurons to the responses of the other. The correspondence between the correlations and cross-adaptation has been shown to be very high neurally (Smith & Frank, 1971) and psychophysically (Mc-Burney, 1969).

Summary

It is seen that even in the absence of detailed parameters for taste and olfaction, some progress can be made in the understanding of the neural code for chemical stimuli. As progress in our understanding of the stimulus parameters will facilitate our understanding of the neural coding process, so also has our understanding of the neural coding process improved our understanding of the stimulus parameters.

The Stimulation Process

The stimulation process has attracted a great variety of approaches. Some theorists have been concerned with the general character of the process, whether it is an adsorptive, vibrational, enzymatic, or a membrane-puncturing process. Other theorists have dealt with the issue of how a stimulus is matched to its appropriate receptor site, and still others with how the neural response is related to stimulation.

In general, these topics have only been approached separately, in piecemeal fashion, i.e., what is the receptor process for "sweet"? But treated broadly and systematically, these pieces should form a coherent whole if there are, in fact, general principles underlying the chemical senses. It is beyond our capacity to attempt such a general synthesis; however, we will point to ways that the various mechanisms of stimulation may be systematically related to the stimulus parameters discussed above.

Character of the Process

It is obvious that contact of a chemical with the specialized membranes of the receptor cells of taste and smell is necessary to produce stimulation. Although it has been hypothesized that penetration of the chemical into membrane initiates stimulation (Davies, 1970), the stimulation process is generally believed to be initiated by the action of the chemical at the receptor membrane.

Beidler (1954) noticed in taste the hyperbolic relation between stimulus concentration and the amount of response, and thus hypothesized that the reaction was adsorptive. By contrast, the hypothesis that for olfaction the initial process involves enzymatic mediation has been advanced by Baradi and Bourne (1951), since this would also yield a hyperbolic relationship.

Kaissling (1969) pointed out that the hyperbolic relationship between stimulus concentration and amount of response is congruent with both adsorptive and enzymatic processes. Emphasizing the similarities of these two kinds of processes, he introduced a general scheme that covers the attachment mechanisms for both adsorptive and enzymatic processes for olfaction; this approach is also applicable to taste. His scheme involves the specification of discrete attachment sites on the receptor membrane to which molecules attach and subsequently either simply detach (in the case of pure adsorption), or, alternatively (by enzymatic action), become unavailable for reattachment to the original site due to either some chemical change or transfer to another site.

Emerging from these considerations is the concept that the receptor cell surface in both taste and smell is a mosaic of receptor sites (which are presumably bound into a lipid or protein matrix) to which molecules are attracted and bind in some manner, and subsequently disengage, or undergo transformation or transfer. The thrust that the concept of "sites" has provided is discussed in the next section.

Matching of Stimulus and Receptor Site

Some theorists have concentrated on the matching of stimuli with the appropriate receptor sites. Amoore, Palmieri, & Wanke (1967), have attempted a broad theory that the shape of the odorant molecule is matched by an appropriately shaped receptor site. However, the more limited approaches to this topic have usually met with more success; e.g., what is the stimulating mechanism for "sweet"? For example, Shallenberger and Acree (1971) have postulated specific types of molecular configurations for limited ranges of stimuli which are conducive to hydrogen bonding between stimulus and site. Evidence for specific taste receptor proteins (providing sites) has been advanced by Dastoli, Lopiekes, & Doig (1968) for bitter substances, and by Dastoli and Price (1966) for sweet substances. Dzendolet (1968, 1969) hypothesizes that sweet or sour stimuli accept protons from, or donate protons to, polar receptor sites. Wright and Burgess (1970) see olfactory stimulation as resulting from molecule-site vibrational resonances. Considered together, these separate approaches might provide a general holistic view congruent with the receptor site approach.

The concept of sites seems valuable, since stimulus attachment to receptor site types could show both the sensitivity and specificity evidenced in taste and smell. The attachments could be expected to be sensitive to a number of modulating effects such as ion-pair formation, steric hindrances, and double-layer charging effects to name but a few. Such modulating effects are important since, for example, the taste of salts with a common cation changes slightly with changing anions, and a high degree of discrimination is seen in both taste and smell between stereoisomers.

However, the broad and general concept of sites does not directly provide the sought-for simplification of the stimulating process, since the interaction between stimulus molecules and receptor sites is a complex of a number of processes. For example, at the physicochemical level the selectivity and sensitivity of attraction and binding of chemical species to receptor sites will be governed by field forces, and by Van der Waal forces between ions and molecules. Field forces arise from ions possessing net charges through deficiencies and excesses of electrons; Van der Waal forces on the other hand arise from differences in charge separation within molecules and higher order effects due to particular electron distributions within ions and molecules. Methods are available for calculating the net attractive force between ionic or molecular pairs (Ling, 1962), but these are complex, necessitating simultaneous consideration of a number of physicochemical parameters of both receptor site and stimulating species; ionic radius, charge, dipole moment, and electron distribution are a few of these parameters.

The complexity of these considerations leave little a priori grounds to expect a simple dimensionality in the chemical senses specifiable by a few physicochemical parameters. However, since the psychophysical and neural responses have been accounted for in spaces of small dimensionality, the stimulation parameters of main influence may be correspondingly few and may be ascertained by inspection of the spaces (given that the spaces contain a sufficient number and range of stimuli). The complex range of *possible* parameters may resolve to a few *functional* parameters.

Summary

The concept of "tuned" receptor sites seems most useful at this time to provide bases to which particular chemical species may adsorb. Many mechanisms have been proposed for the attachment of the correct stimulus to the correct site: the primary mechanisms of attachment may possibly be revealed by determination of the dimensions of the arrangement of these chemicals in a "stimulus space."

GENERAL SUMMARY

Research in the chemical senses is notable for its lack of broad organization. It appears that some systematization in the psychophysical and neural branches of these fields is possible. In view of the great catalyst to progress afforded by systematization in the other senses, and in science in general, attention is called to its possibilities in the chemical senses.

ACKNOWLEDGMENTS

This chapter was supported in part by grants from NSF-GB-33464X, (R.P.E.); and NICHD-HD NS 06853-01, (S.S.S.). We are grateful to Mr. Gary Heck for his contributions to the section on "The Stimulation Process."

REFERENCES

Adrian, E. D. The response of the olfactory organ to different smells. *Acta Physiologica Scandinavica*, 1953, **29**, 5–14.

Amoore, J. E., Palmieri, G., & Wanke, E. Molecular shape and odour: Pattern analysis by PAPA. *Nature*, 1967, **216**, 1084–1087.

Baradi, A. F., & Bourne, G. H. Localization of gustatory and olfactory enzymes in the rabbit, and the problems of taste and smell. *Nature*, 1951, **168**, 977–979.

Békésy, G. von. Sweetness produced electrically on the tongue and its relation to taste theories. *Journal of Applied Physiology*, 1964, **19**, 1105–1113.

Beidler, L. M. A theory of taste stimulation. *Journal of General Physiology*, 1954, **38**, 133–139.

Boring, E. G. *Sensation and perception in the history of experimental psychology.* New York: Appleton, 1942.

Boring, E. G. *A history of experimental psychology.* (2nd ed.) New York: Appleton, 1950.

Cain, D. P., & Bindra, D. Response of amygdala single units to odors in the rat. *Experimental Neurology*, 1972, **35**, 98–110.

Dastoli, F. R., Lopiekes, D. V., & Doig, A. R. Bitter-sensitive protein from porcine taste buds. *Nature*, 1968, **218**, 884–885.

Dastoli, F. R., & Price, S. Sweet-sensitive protein from bovine taste buds: Isolation and assay. *Science*, 1966, **154**, 905–907.

Davies, J. T. Recent developments in the "penetration and puncturing" theory of odour. In G. E. W. Wolstenholme and J. Knight (Eds.), *Taste and smell in vertebrates: A Ciba Foundation symposium.* London: Churchill, 1970. Pp. 265–281.

Dethier, V. G. A surfeit of stimuli: A paucity of receptors. *American Scientist*, 1971, **59**, 706–715.

Dethier, V. G. Electrophysiological studies of gustation in Lepidopterous Larvae. *Journal of Comparative Physiology*, 1973, **82**, 103–134.

Doetsch, G. S., & Erickson, R. P. Synaptic processing of taste-quality information in the nucleus tractus solitarius of the rat. *Journal of Neurophysiology*, 1970, **33**, 490–507.

Dzendolet, E. A structure common to sweet-evoking compounds. *Perception and Psychophysics*, 1968, **3**, 65–68.

Dzendolet, E. Theory for the mechanism of action of "miracle fruit." *Perception and Psychophysics*, 1969, **6**, 187–188.

Erickson, R. P. Sensory neural patterns and gustation. In Y. Zotterman (Ed.), *Olfaction and taste*. Oxford: Pergamon Press, 1963. Pp. 205–213.

Erickson, R. P. Neural coding of taste quality. In M. Kare and O. Maller (Eds.), *The chemical senses and nutrition*. Baltimore: Johns Hopkins Press, 1967. Pp. 313–327.

Erickson, R. P. Stimulus coding in topographic and nontopographic afferent modalities: On the significance of the activity of individual sensory neurons. *Psychological Review*, 1968, **75**, 447–465.

Erickson, R. P. Parallel "population" neural coding in feature extraction. In G. Quarton, T. Melnechuk, F. Schmitt (Eds.), *The neurosciences: A study program*. New York: Rockefeller Univ. Press, 1973.

Erickson, R. P., Doetsch, G. S., & Marshall, D. A. The gustatory neural response function. *Journal of General Physiology*, 1965, **49**, 247–263.

Frank, M., & Pfaffmann, C. Taste nerve fibers: A random distribution of sensitivities to four tastes. *Science*, 1969, **164**, 1183–1185.

Ganchrow, J. R., & Erickson, R. P. Neural correlates of gustatory intensity and quality. *Journal of Neurophysiology*, 1970, **33**, 768–783.

Gesteland, R. C., Lettvin, J. Y., Pitts, W. H., & Rojas, A. Odor specificities of the frog's olfactory receptors. In Y. Zotterman (Ed.), *Olfaction and taste*. Oxford: Pergamon Press, 1963. Pp. 19–34.

Guttman, L. A general nonmetric technique for finding the smallest coordinate space for a configuration of points. *Psychometrika*, 1968, **33**, 469–506.

Halpern, B. P. Chemotopic coding for sucrose and quinine hydrochloride in the nucleus of the fasciculus solitarius. In T. Hayashi (Ed.), *Olfaction and taste*. Vol. II. Oxford: Pergamon Press, 1967. Pp. 549–562.

Henning, H. Der Geruch. I. *Zeitschrift für Psychologie*, 1915, **73**, 161–257.

Henning, H. Die Qualitatenreihe des Geschmacks. *Zeitschrift Psychologie*, 1916, **74**, 203–219.

Kaissling, K. E. Kinetics of olfactory receptor potentials. In C. Pfaffmann (Ed.), *Olfaction and taste*. Vol. III. New York: Rockefeller Univ. Press, 1969. Pp. 52–70.

Kiesow, F. Beiträge zur physiologischen Psychologie des Geschmackssines. *Philos. Stud.*, 1894, **10**, 329–368.

Kruskal, J. B. Multidimensional scaling by optimizing goodness of fit to a nonmetric hypothesis. *Psychometrika*, 1964, **29**, 1–27.

Ling, G. N. *A physical theory of the living state*. New York: Blaisdell, 1962.

MacLeod, P. An experimental approach to the peripheral mechanisms of olfactory discrimination. In G. Ohloff and A. F. Thomas (Eds.), *Gustation and olfaction*. London: Academic Press, 1971. Pp. 28–44.

Mathews, D. F. Response patterns of single units in the olfactory bulb of the rat to odor. *Brain Research*, 1972, **47**, 389–400.

McBurney, D. H. Effects of adaptation on human taste function. In C. Pfaffmann (Ed.), *Olfaction and taste*. Vol. III. New York: Rockefeller Univ. Press, 1969. Pp. 407–419.

Moncrieff, R. *The chemical senses*. Cleveland: CRC Press, 1967.

Morrison, G. R. Behavioural response patterns to salt stimuli in the rat. *Canadian Journal of Psychology/Review of Canadian Psychology*, 1967, **21**, 141–152.

Morrison, G. R. Taste psychophysics in animals. In C. Pfaffmann (Ed.), *Olfaction and taste*. Vol. III. New York: Rockefeller Univ. Press, 1969. Pp. 512–516.

Mozell, M. M. Olfactory discrimination: Electrophysiological spatiotemporal basis. *Science*, 1964, **143**, 1336–1337.

Mozell, M. M. Evidence for a chromatographic model of olfaction. *Journal of General Physiology*, 1970, **56**, 46–63.

O'Connell, R. J. & Mozell, M. M. Quantitative stimulation of frog olfactory receptors. *Journal of Neurophysiology*, 1969, **32**, 51–63.

Perkel, D. H., & Bullock, T. H. Neural coding. *Neurosciences Research Program Bulletin*, 1968, **6**, 221–348.

Pfaffmann, C. The sense of taste. In J. Field, H. W. Majoun, & V. E. Hall (Eds.), *Handbook of physiology: Neurophysiology I*. Washington, D.C.: American Physiological Society, 1959. Pp. 507–533.

Pfaffmann, C. The afferent code for sensory quality. *American Psychologist*, 1959, **14**, 226–232.

Poulos, D. A. Trigeminal temperature mechanisms. In R. Dubner & Y. Kawamura (Eds.), *Oral-facial sensory and motor mechanisms*. New York: Appleton, 1971. Pp. 47–72.

Sato, M. Gustatory response as a temperature-dependent process. In W. Neff (Ed.), *Contributions to sensory physiology*. New York: Academic Press, 1967. Pp. 223–251.

Schiffman, S. S. Physico-chemical correlates of .olfactory quality. *Science*, 1974, **185**, 112–117.

Schiffman, S. S. & Dackis, C. Taste of nutrients: Amino acids, vitamins, and fatty acids. To be published in *Perception and Psychophysics*.

Schiffman, H., & Falkenberg, P. The organization of stimuli and sensory neurons. *Physiology and Behavior*, 1968, **3**, 197–201.

Schiffman, S. S., & Erickson, R. P. A theoretical review: A psychophysical model for gustatory quality. *Physiology and Behavior*, 1971, **7**, 617–633.

Schiffman, S. S., & Erickson, R. P. Organization of olfactory stimuli and neurons. Presented at Proceedings of the Symposium Mediterranean sur L'Odorat, Cannes, France, 1972.

Scott, T., & Erickson, R. P. Synaptic processing of taste quality information in thalamus of rat. *Journal of Neurophysiology*, 1971, **34**, 868–884.

Shallenberger, R. S., & Acree, T. E. Chemical structure of compounds and their sweet and bitter taste. In L. M. Beidler, (Ed.), *Handbook of sensory physiology*. Vol. IV. Part 2. New York: Springer-Verlag, 1971. Pp. 221–277.

Shepard, R. The analysis of proximities: Multidimensional scaling with an unknown distance function. II. *Psychometrika*, 1962, **27**, 219–246.

Skramlik, E. von. Die physiologie des geschmachssinne. In *Handbuch der Physiologie der niederen Sinne*. Leipzig: G. Thieme, 1926. Pp. 346–520.

Smith, D., & Frank, M. Cross adaptation between salts in chorda tympani nerve of the rat. *Physiology and Behavior*, 1972, **8**, 213–220.

Woskow, M. *Multidimensional scaling of odors*. Doctoral dissertation, Univ. of California, Los Angeles, 1964.

Wright, R. H., & Burgess, R. Specific Physicochemical Mechanisms of Olfactory Stimulation. In G. E. W. Wolstenholme and J. Knight (Eds.), *Taste and smell in vertebrates.* London: Churchill, 1970. Pp. 325–337.

Wright, R. H., & Michels, K. Evaluation of far infrared relations to odor by a standards similarity method. *Annals of the New York Academy of Sciences,* 1964, **116,** 535–551.

Yamada, M. A search for odour encoding in the olfactory lobe. *Journal of Physiology,* 1971, **214,** 127–143.

Chapter 14

Motor Coordination: Central and Peripheral Control during Eye–Head Movement

EMILIO BIZZI

Massachusetts Institute of Technology

INTRODUCTION

How the central nervous system produces coordinated motor output has long been one of the major problems in neurophysiology. In order that movement be coordinated, a set of appropriate muscles must be selected, each of the agonist muscles must be activated in the proper temporal relationship to others, and precise amount of inhibition must be delivered to each antagonist muscle.

From the time of Sherrington to the present, investigators have held different views as to how the spatiotemporal properties of a given motor coordination may be achieved by the nervous system. Those supporting the "peripheral control theory" have stressed the importance of sensory feedback in eliciting the motor output. At the other extremes, researchers held that the central nervous system contains all the information necessary for spatiotemporal patterning (Székely, Czeh, & Voros, 1969; Weiss,

427

1934; Wiersma, 1947; Willow & Hoyle, 1969; Wilson & Waldron, 1968).
Although there are numerous instances of centrally patterned motor
outputs that are not dependent on sensory input, experimental evidence
points to an interaction between central activity and peripheral feedback
in the large majority of cases.

In general, the relative contribution of central and peripheral feedback
to motor coordination is better understood in invertebrates. In fact, the
work of Wiersma (1947), Kennedy (1968), and Wilson and Waldron
(1968) has shown the importance of central, built-in mechanisms for
pattern generation and has also indicated that the role of feedback varies
from species to species. In vertebrates, Weiss (1934) and Székely et
al. (1969) found convincing evidence for central initiation and patterning
of locomotory movements. In mammals, studies on deafferentation in
monkeys by Taub and Berman (1968) and Bossom and Ommaya (1968)
have shown that monkeys were able to use their limbs in locomotion
although motor deficits were present. A study by Engberg and Lundberg
(1969) on the electromyographic activity in the hind limb musculature
of the unrestrained cat outlines a hypothetical scheme for locomotion
where central patterning accounts for the onset of extensor activity and
the peripheral feedback for hind limb flexion.

In the following, as a model for studying coordinated movement in
general, we shall examine the notion of central programming and the
role of peripheral feedback in the specific case of coordinated eye–head
movements. Two main ideas will be stressed: first, the concept of motor
coordination and, second, the notion of strategies of motor coordination.

STRUCTURE OF EYE–HEAD MOTOR COORDINATION

To begin with, we will describe the spatiotemporal characteristics
of the central programs underlying sequentially ordered eye and head
movements with the aim of distinguishing the role played by the central
command and its interaction with peripheral feedback from receptors
(vestibular, neck afferents) excited by centrally initiated eye and head
movements.

Before describing the structure of the motor output underlying
eye–head movement, we will first have to distinguish between the various
strategies of eye–head coordination. It is well-known that in primates
there are several modes of active eye–head coordination. To mention
just a few: coordinated movements of the head and eyes occur while
tracking a moving target, during exploration or scanning of the surround,
as a response to an unexpected stimulus, or in anticipation of a stimulus

Figure 1. Typical coordinated eye and head response to sudden appearance of a target. Five superimposed tracings showing consistency of the eye–head pattern. Horizontal bar = 100 msec; left vertical bar indicates eye movement amplitude in steps of 10 degrees starting from the center (EOG output is nonlinear); right vertical bar indicates head movement amplitude in steps of 10 degrees starting from the center.

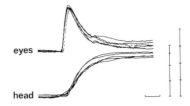

whose occurrence can be predicted. Here we will examine first the strategy of eye and head movements which follow the unexpected presentation of a visual target. We have named this strategy "triggered."

TRIGGERED EYE–HEAD MOVEMENTS

We know from work in man (Fleming, Vossius, Bowman, & Johnson, 1969) and monkeys (Bizzi, Kalil, & Tagliasco, 1971) that the appearance of a target in the visual field is usually followed by an orderly sequence of three movements. First, a saccadic eye movement directed toward the target which carries the fovea to the image of the target. Second, after a latency of 20–40 msec, a head movement follows in the same direction. Third, a compensatory eye movement which, by being counter to that of the head, allows the fovea of the eye to remain on the target it has just acquired (Figure 1). To achieve this orderly sequence of movements, i.e., to direct the eyes and the head toward the target and ultimately fixate it with the fovea, the subject must make a number of computations. To begin with, he must compute the angular distance between the initial line of sight and the position of the target which is to be acquired. The magnitude of this angular distance will determine, as a first approximation, the amplitude of the saccadic eye movement as well as that of the head movement. Although we do not have a clear understanding of the manner in which the retinal error is computed, relevant information on coding of retinal input by the visual system is presented in this volume. Once the retinal input has been elaborated by cortical and subcortical visual areas, this "signal" is then translated into the oculomotor system and used to initiate saccadic eye movements. There is, in fact, a linear relation between the magnitude of target displacement and the amplitude of the saccadic eye movement (Robinson, 1968). Furthermore, it has been shown that the motor control system for the head uses the same retinal error signal to produce a movement of the head (Bizzi, Kalil, Morasso, & Tagliasco, 1972). This interpretation is based upon the fact that electromyographic records show that impulsive motor commands are delivered in near synchrony to eye muscles and neck muscles (discussed later). From this it follows that both the oculomotor and head motor control systems must be making use of the same retinal error information at approximately the same time. As a result, amplitudes of eye and head movements are produced which are well-correlated with the magnitude of target displacement. It has been shown that this finding is valid in the special case when head movement begins from the straight ahead position and the eyes are centered in the head at the time the target is presented. Usually, however, the eyes will not be centered in the head at the time of target appearance. Then, in order to produce an appropriate head turning, the head movement control systems must have access to information concerning the position of the eyes in their orbits and must then combine this information with the retinal error signal if head movements are indeed to be coordinated with eye movements. Information about eye position can be supplied by eye muscle proprioceptors or by efferent motor collaterals. Indeed the existence of eye muscle afferents has already been described (Cooper & Daniel, 1963; Cooper & Fillenz, 1955). Furthermore, there is evidence to suggest that the position of the eyes in the head is controlled by a special population of cells in the cortical area 8 (frontal eye field) (Bizzi, 1968).

CENTRAL PROGRAMMING OF EYE–HEAD MOVEMENTS: ELECTROMYOGRAPHIC FINDINGS

After the onset of the target light to which an animal is trained to respond, the first event to take place is invariably a synchronous burst of electromyographic (EMG) activity in all agonist neck muscles. Electromyographic activity is suppressed in all the antagonists either *concurrently* or *preceding* the agonist activation. This finding indicates that in this mode, the suppression of the antagonists might be the result of descending supraspinal activities rather than being reflexly induced by way of the inhibitory connections of primary afferents.

The next point of importance is that the electromyographic burst appears simultaneously in all agonists regardless of the initial head position of the animal. However, the amplitude and duration of the bursts are dependent upon the starting position and the extent of the intended head movement.

The second electromyographic event to take place is an *eye muscle burst,* which has always been shown to occur approximately 20 msec *after* the beginning of neck muscle activity.

Taken together, these electromyographic results indicate the spatio-temporal characteristics underlying sequentially ordered eye and head

Figure 2. Comparison of eye saccades and gaze. a. Eye saccade to a suddenly appearing target wih head fixed. b. Coordinated eye saccade (E) and head movement (H) to the same target with head free. The gaze movement (G) represents the sum of E and H. Note the remarkable similarity of eye saccade in (a) and gaze trajectory in (b) as well as reduced saccade amplitude in (b). Time calibration 100 msec. [From Morasso et al., 1973.]

movement. The neural commands are, in fact, delivered simultaneously to all neck muscles and shortly thereafter also to the eye muscles. However, the overt sequence of movements that results from these commands does not reflect this order insofar as the head movement actually occurs *after* each saccade (10–30 msec) (see Figure 1). Peripheral factors such as the longer contraction time of the neck muscles, as well as the inertial properties of the head, are responsible for this delay. Thus, the central command initiates these movements but does not by itself specify their serial order.

INTERACTIONS BETWEEN THE CENTRAL
PROGRAM AND AFFERENT REFLEX ACTIVITY

The activation of eye and neck muscles will lead not only to movements of the eyes and head, but also to the activation of a number of sensory receptors. Neck muscle spindles and neck tendon organs, joint receptors from the cervical segments, and, of course, vestibular afferents, will all be activated by the muscle contraction that follows arrival of central impulses. The question then arises as to what extent all these afferents modify the ongoing motor program. This interaction between a peripheral feedback and the central programs is, of course, a very general problem in all cases of motor coordination. Although we will deal here only with the specific case of eye and head coordination, it should be clear that the same kind of analysis can be applied to more complex motor coordinations such as those involving eye–head and arm.

To begin with, we will consider the following questions: First, is there any modification, let's say, of eye movements due to the activation of

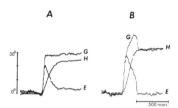

Figure 3. Eye–head coordination in intact (A) and in labyrinthectomized monkey 40 days after surgery (B). Eye movements (E); head movements (H); gaze movement (G) represents the sum of E plus H. Note gaze overshoot and corrective saccade. [From Dichgans *et al.*, 1973.]

neck or vestibular receptors? The answer is yes, and experimentally it can be shown that saccades made during head movement (see Figure 2) are smaller in amplitude than saccades made to the same target, but with the head restrained (Atkin, 1964; Morasso, Bizzi, & Dichgans, 1973).

Since modification of saccades made during head movement does exist, what type of afferent impulse is responsible for the modulation of saccadic amplitude (vestibular, neck, and visual afferents)? Again, the experimental evidence indicates that the vestibular afference alone is responsible for the observed phenomenon (Dichgans, Bizzi, Morasso, & Tagliasco, 1973). Visual afference has been discounted by the demonstration that saccade characteristics are not changed by turning off the target light just before the saccadic movement is initiated. Likewise, the effect of neck afferents was found to be negligible (Dichgans *et al.*, 1973). Positive evidence of the crucial role of vestibular afference was demonstrated by surgical interruption of the pathway linking the vestibular receptors to the brainstem. After the operation, the saccade amplitude during head turning was identical to saccade amplitude with head fixed; this resulted in a remarkable overshoot of the gaze (eye movement plus head movement) (Figure 3).

Finally, we would like to raise the question of the "advantage" to the animal of saccadic adjustments during head movements. An examination of Figure 2 will provide this answer. In comparing the saccade made with head restrained in 2a with the gaze (G) in 2b, it is clear that the dynamic properties of saccadic eye movements made with head restrained are indistinguishable from those resulting from the sum of the eye and head movement (gaze). It follows that if the animal's goal is to bring the image of the target to the fovea as quickly and as precisely as possible, then the use of combined eye–head movement (i.e., gaze) and the use of eye movements alone are equally effective behaviorally. This effectiveness rests, of course, upon a precise modulation of saccadic amplitude due to reflex vestibular activity initiated by the head movement.

This finding represents a contribution to the understanding of eye–head coordination insofar as it indicates that the output of parallel, independently active neural centers which initiate these movements (see p. 434) is "coordinated" in a reflex way. This organization has distinct advantages (Atkin, 1964). It simplifies the task of the programming centers which need not be informed about the overall picture of the movement (Bizzi, 1968). It makes it unnecessary to postulate a separate population of neurons with the exclusive responsibility for "coordination" of eye and head (Bizzi, Kalil, & Tagliasco, 1971). By relying on loops which monitor the *actual* movement of the head, the resulting adjustment in eye movements will include all the unpredictable peripheral loads and resistances which might change the course of the centrally initiated (intended) head movement.

It is important to stress that modulation of saccadic amplitude during head movement occurs as a result of reflex action initiated by the stimulation of the vestibular receptors, rather than by centrally preprogrammed modification of saccadic parameters. This fact can be viewed as a challenge to the traditional view that saccadic eye movements are "ballistic," i.e., with trajectories which are totally predetermined by the initial oculomotor command. It is intriguing at this point to speculate whether, in the course of other "ballistic" movements such as those of the arm, interactions between the central program and a reflex activity from muscle and tendon organs might also occur.

The modification of saccade characteristics is one aspect of the interaction between central programming and reflex activities. Although this interaction plays a decisive part in the process of target acquisition by a combined eye and head movement, the role of feedback from peripheral sensory organs (vestibular and neck afference) extends beyond saccadic modulation to control and generate the compensatory eye movement.

COMPENSATORY EYE MOVEMENTS

In the previous section, some aspects of the mechanism whereby the image of the target is brought to the fovea have been examined. The problem of maintaining fixation during the completion of the head movement now arises. In fact, since the eyes move first and with higher velocity than the head, their line of sight reaches and fixates a target while the head is still moving (see Figure 1). Then, for the duration of the head movement, the eyes maintain their fixation by performing a rotational movement which is counter to that of the head and compensates for it. Consequently, this is called a compensatory eye movement.

These movements may be easily observed by asking a friend to fixate a visual target and rotate his head from side to side while maintaining fixation. You will see that, as the head moves, the eyes rotate in an equal and opposite way so that they will remain on target. These compensatory eye movements may be extremely quick; try fixating your index finger while rapidly rotating our head, and then move your index finger with equal rapidity while holding your head motionless. In the first instance, the vestibular system quickly moves the eyes opposite to the head and the target is seen clearly, but in the second instance, the slower "visual corrective loop" to the eye muscles cannot compensate for movement of the target, which consequently is reduced to a blur, as the eyes try in vain to match its velocity.

Compensatory eye movements have been studied by several investigators and, although it is generally agreed that these eye movements are influenced critically by visual, vestibular, and proprioceptive reflexes, the suggestion has been made that they are initiated centrally (Fleming et al., 1969). Recently, it has been shown that the stabilization of the eye during head rotation is not centrally programmed, but results from the reflex action of the vestibular system (Bizzi et al., 1971, 1972). As a consequence of the head movement, vestibular receptors are stimulated and their activity induces a compensatory eye movement which allows the fovea to remain fixed in relation to a point in visual space. This fixation permits a second retinal error to be computed, and if the eye is off target, the same sequence can then be repeated. The closed loop structure of triggered eye–head coordination is schematized in the following diagram.

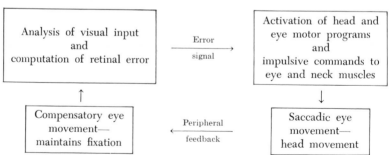

If we take the repetitive sequence of eye saccade–head movement–compensatory eye movement as one definition of eye–head coordination, then it is clear that a central nervous system motor program only initiates in an impulsive manner coordinated movements of the head and eyes. Since there is no central programming of the compensatory eye movement, it follows that the *behavioral coordination* of head and eyes is the joint result of a central initiation accompanied by the intervention of feedback from the periphery. In view of this, it is not necessary to postulate a population of "executive neurons" with exclusive responsibility for *coordinating* the eyes and head.

STRATEGIES OF MOTOR COORDINATION

So far we have described the timing and the characteristics of eye–head coordination which are elicited by the appearance of a visual target. We have already mentioned that in primates there are several modes of eye–head coordination such as tracking with eye and head, scanning, etc., and we will now proceed to describe another mode of coordination. We will call this mode "predictive" because it occurs only after the animal has memorized a set of reward contingencies and is able to make an appropriate predictive movement which anticipates the presentation of the stimulus (Bizzi *et al.*, 1972). Under these conditions, the timing between eye and head movements and the pattern of neck muscle activation are different from those found in the visually triggered mode (Figure 4). In fact, during predictive eye–head coordination the head begins to move well before the saccade of the eye is initiated. In addition,

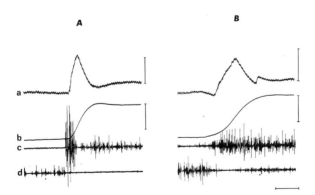

Figure 4. Strategies of eye–head coordination. A. Eye–head coordination triggered by the presentation of a visual stimulus (see dot at the bottom). B. Predictive eye–head coordination, target appeared after head turning was initiated (see dot at the bottom). a. Horizontal eye movement; b. horizontal head movement; right (c) and left splenii capitus—time calibration 200 msec. Eye calibration in A, 30° in B, 30°. Head calibration 40°. [From Bizzi *et al.*, 1972, pp. 45–48.]

the head movement is achieved by a gradual increase in activity of the agonist muscles which is mirrored by a decrease in the activity of the antagonists (Figure 4).

This pattern of reciprocal, gradual antagonist–agonist activity is in marked contrast with the EMG bursting invariably present in the agonist during the triggered mode of coordination (Figure 4a). The fact that motor output subserving eye–head coordination is not fixed, but instead is organized differentially with regard to specific behavioral situations, raises several interesting questions. For instance, are there separate cortical or subcortical regions for the programming of different modes of coordination? And if so, are the various modes of coordination subserved by a totally different and separate neuronal network? How is the switching from one mode to another accomplished? These are, of course, formidable questions for the neurophysiologist interested in outlining realistic models for these motor coordinations—yet the investigation of these problems is now not beyond reach. In fact, the newly developed technique of single cell recording in intact trained animals (Evarts, 1966), combined with the traditional neurophysiological and behavioral approaches, are beginning to unravel some aspects of the supraspinal patterns of motor activity (Delong, 1971; Evarts, 1967; Thatch, 1970a,b).

REFERENCES

Atkin, A. Effect of head movement on gaze movement velocity. *The Physiologist,* 1964, 7, 82.

Bizzi, E. Discharge of frontal eye field neurons during saccadic and following eye movements in unanesthetized monkeys. *Experimental Brain Research,* 1968, 6, 69–80.

Bizzi, E., Kalil, R. E., & Morasso, P. Two modes of active eye-head coordination in monkeys. *Brain Research,* 1972, 40, 45–48.

Bizzi, E., Kalil, R. E., Morasso, P., & Tagliasco, V. Central programming and peripheral feedback during eye-head coordination in monkeys. In J. Dichgans & E. Bizzi (Eds.), *Cerebral control of eye movements and motion perception.* Basel: Karger, 1972.

Bizzi, E., Kalil, R. E., & Tagliasco, V. Eye-head coordination in monkeys: Evidence for centrally patterned organization. *Science,* 1971, 173, 452–454.

Bossom, J., & Ommaya, A. K. Visuo-motor adaptation (to prismatic transformation of the retinal image) in monkeys with bilateral dorsal rhizotomy. *Brain,* 1968, 91, 161–172.

Cooper, S., & Daniel, P. M. Muscle spindles in man; their morphology in the lumbricals and the deep muscles of the neck. *Brain,* 1963, 86, 563–586.

Cooper, S., & Fillenz, M. Afferent discharges in response to stretch from the extraocular muscles of the cat and monkey and the innervation of these muscles. *Journal of Physiology,* London, 1955, 127, 400–413.

DeLong, M. R. Activity of pallidal neurones during movement. *Journal of Neurophysiology,* 1971, **34,** 414–427.

Dichgans, J., Bizzi, E., Morasso, P., & Tagliasco, V. Mechanisms underlying recovery of eye-head coordination following bilateral labyrinthectomy in monkeys. *Experimental Brain Research,* 1973, **18,** 548–562.

Engberg, I., & Lundberg, A. An electromyographic analysis of muscular activity in the hindlimb of the cat during unrestrained locomotion. *Acta Physiologica Scandinavica,* 1969, **75,** 614–630.

Evarts, E. V. Methods for recording activity of individual neurons in moving animals. In R. F. Rushmer (Ed.), *Methods in medical research.* Chicago: Year Book Medical Publisher, 1966.

Evarts, E. V. Representation of movements and muscles by pyramidal tract neurons of the precentral motor cortex. In M. D. Yahr & D. P. Pupura (Eds.), *Neurophysiological basis of normal and abnormal motor activities.* New York: Raven Press, 1967.

Fleming, D. G., Vossius, G. W., Bowman, G., & Johnson, E. L. Adaptive properties of the eye-tracking system as revealed by moving-head and open-loop studies. *Annals of the New York Academy of Sciences,* 1969, **156,** 825–850.

Kennedy, A. Input and output connections of single arthropod neurons. In F. D. Carlson (Ed.), *Physiological and biochemical aspects of nervous integration.* Englewood Cliffs, New Jersey: Prentice-Hall, 1968.

Morasso, P., Bizzi, E., & Dichgans, J. Adjustment of saccade characteristics during head movements. *Experimental Brain Research,* 1973, **16,** 492–500.

Robinson, D. A. Eye movement control in primates. *Science,* 1968, **161,** 1219–1224.

Székely, G., Czeh, G., & Voros, G. The activity pattern of limb muscles in freely moving normal and deafferented newts. *Experimental Brain Research,* 1969, **9,** 53–72.

Taub, E., & Berman, A. J. Movement and learning in the absence of sensory feedback. In S. J. Freedman (Ed.), *The neuropsychology of spatially oriented behavior.* Homewood, Illinois: Dorsey Press, 1968.

Thatch, W. T. Discharge of cerebellar neurons related to two maintained postures and two prompt movements. I. Nuclear cell output. *Journal of Neurophysiology,* 1970, **33,** 527–536. (a)

Thatch, W. T. Discharge of cerebellar neurons related to two maintained postures and two prompt movements. II. Purkinje cell output and input. *Journal of Neurophysiology,* 1970, **33,** 537–547. (b)

Weiss, P. Function of de-afferented amphibian limbs. *Proceedings of the Society for Experimental Biology and Medicine,* 1934, **32,** 436–438.

Wiersma, C. A. G. Giant nerve fiber system of the crayfish. A contribution to comparative physiology of synapse. *Journal of Neurophysiology,* 1947, **10,** 23–38.

Willows, A. O., & Hoyle, G. Neuronal network triggering a fixed action pattern. *Science,* 1969, **166,** 1549–1551.

Wilson, D. M., & Waldron, I. Models for the generation of the motor output pattern in flying locusts. *Proceedings of the IEEE,* 1968, **56,** 1058–1064.

Woodworth, R. S., & Schlosberg, H. *Proceedings of the Experimental psychology.* New York: Holt, 1954.

Part IV

INTEGRATION
AND REGULATION
IN THE BRAIN

Chapter 15

Psychobiology of Attention

MICHAEL I. POSNER

University of Oregon

INTRODUCTION

A few years ago Worden (1966) reviewed the literature relating attention and auditory electrophysiology. He found little in the way of an organized analysis of the field of attention at the psychological level. Accordingly, he found it difficult to relate the psychological concept of attention with neurophysiological studies. The last several years have seen an enormous outpouring of books, chapters, and articles devoted to the subject of attention (Broadbent, 1971; Evans & Mulholland, 1969; Kahneman, 1973; Keele, 1973; Mackworth, 1969; McGaugh, 1971; Moray, 1970; Mostofsky, 1970; Regan, 1972). These review the results of studies attacking the problem by one of two main approaches. First are studies of human information processing in simple tasks. Second are physiological studies involving recording of autonomic activity, measures derived from the electroencephalograph (EEG), or responses of one or a small group of cells at some specific level of the nervous system.

Despite Worden's effort to direct researchers toward a joint consideration of behavioral and physiological studies, most of the literature continues to be divided according to the technique used for the study of attention. Books in physiological psychology (Milner, 1970; Thompson,

1967) deal only in a cursory way with human performance studies although they are sometimes casually introduced at key points to provide an intuitive feel as to why some measure reflects upon attention. On the other hand, books stressing human performance rarely deal with other methods. There are some exceptions, such as Mackworth's effort to study habituation and vigilance by both evoked potential and performance methods and Kahneman's study of mental effort by both performance and autonomic techniques. A more thorough effort at integrating several techniques is found in Weinberger's review of attentive processes (McGaugh, 1971).

The consequence of this fragmentation by method has been damaging to an understanding of attention. This is especially true because the nature of behavioral studies of attention has changed radically, often without those using other methods being aware of this shift. In the period from the 1930s through the 1950s, experimental psychology drew its inspiration from investigations of multitrial learning tasks. The studies primarily viewed the organism as a black box. Attention was tacked on to account for the obvious fact that the output did not reflect all dimensions of the input. These studies were relatively unanalytic to questions of interest to psychobiologists because they did not distinguish between different levels of the nervous system or detail the functions which attention might perform within the brain.

As Worden (1966) suggests, problems of attention derive their basis from the behavior and experience of humans. While the questions can be referred to nonhuman organisms, their basic anthropomorphic character remains. It would seem desirable to view attentional questions within a general framework derived from studies of human information processing. In the last dozen years such studies have begun to develop methods and theories which seek to determine the internal operations underlying performance. A detailed understanding of autonomic and central nervous system activity related to these operations may allow a more meaningful reference of problems of attention to organisms suited for detailed physiological studies. This chapter attempts to bring together information within such a framework.

History of Performance Methods

The electrophysiology of color and form vision has been built upon a long tradition of human psychophysics. There is also precedent in psychobiology for the study of internal mental operations related to attention through observations of human performance. One technique has involved the measurement of the time required to perform simple tasks. Helmholtz's original effort to measure the speed of neural conduction employed this method, and systematic work on the time demands of mental operations began with studies of the Dutch physiologist Donders, reported in 1868. Mental chronometry has been the subject of intensive study (Posner, 1969; Sternberg, 1969).

A second technique for the study of internal mental operations was first suggested by the famous biologist Jaques Loeb (1898). It attempted to measure the degree of voluntary effort required by a task by assessing interference with the strength of hand grip. This technique is directly related to the study of attention because it stresses the inhibitory consequences of some aspects of the processing of one signal upon other activity. Loeb's method is not a very useful one but the idea that some aspects of information processing involve a special limited capacity mechanism which produces widespread interference with the processing of other signals has been a very important feature of attentional theories. By studying those aspects of processing which produce interference between signals, experimental psychologists have sought to understand the attention demands of mental activity.

Components of Attention

Although there is no generally agreed upon definition of attention, three senses of the term predominate in the literatures of psychology and biology. There is no pretense that these definitions are either mutually exclusive or exhaustive. Rather, the goal of this chapter will be to study model situations in which the mechanisms underlying each sense of the general term "attention" can be isolated and their functions observed.

One sense of the term attention is called alertness and concerns the study of an organismic state which affects general receptivity to input information.

The second sense of attention involves selection of some information from the available signals for special treatment. We may select a position in space, a physical characteristic, or a type of form. The selected item is more likely to affect our awareness, memory, or behavior than are other items presented simultaneously. When the basis of selection is a simple physical property of the stimulus such as its location, which does not depend on its prior identification, selection is said to involve "stimulus set." When the basis of selection depends upon prior identification by the nervous system (e.g., select consonants rather than vowels) the term "response set" is often used.

Finally, there is a sense of attention related to the degree of conscious effort which a person invests. This aspect of attention is related to what the subject consciously perceives at a given moment and what consequently will have an increased likelihood of being available for later recall. Studies suggest a specific brain mechanism which is correlated with this subjective experience under certain conditions.

Each of the sections of this chapter reviews evidence relating to attention in one of these three senses. This organization is used on the hypothesis that the three senses of attention outlined above involve separate although interrelated mechanisms. It is unlikely that each sense of attention has only a single mechanism. Rather, each sense of attention may itself consist of a number of different mechanisms. Where possible, specific hypotheses have been advanced as to the nature of these mechanisms. Because in many sections the evidence is fragmentary and even contradictory these hypotheses are frankly speculative. Moreover, I am a psychologist whose main work involves the study of human information processing, and thus data in that area may be used more accurately than evidence from other areas. I have been encouraged to risk error on the details of other methods because of the importance of bringing different techniques of analysis into closer contact.

ALERTNESS

Alertness and Arousal

The term arousal has been applied to a wide variety of behavioral and electrophysiological changes which accompany the transition from sleep to wakefulness. An organism fully awake, rapid in its response, and whose electroencephalogram (EEG) is desynchronized, is said to be at a high level of arousal. In recent years, the unity of the arousal concept has been questioned from a number of sources (Broadbent, 1971; Kahneman, 1973; Routtenberg, 1971).

The term arousal has its origin in the idea of a general drive state which potentiates all behavior. The idea is that all activity must be driven by some internal energy and the availability of this energy corresponds to the arousal level of the organism. That view is more closely related to a stimulus–response psychology than it is to an information processing account. When one shifts to an information processing framework the concept of arousal shifts as well. An organism may be in an optimal state for processing of external signals, for concentrating upon signals from its own long-term memory, for responding, or for any number of different kinds of information processing activities. It is not obvious that these states should be identical. Thus, there seems no a priori reason to expect a single state to be related to all aspects of optimal performance. Indeed, it is possible that states favoring sensory intake and motor outflow are antagonistic (Routtenberg, 1971).

Alertness as used here refers only to the problem of receptivity to external signals. Receptivity may vary either because some part of a particular pathway from sensory input to response has changed in its character or because the state of the organism has changed the reception to all or at least to a broad class of stimuli (Groves & Thompson, 1970). Alertness is concerned with mechanisms of the latter type. Changes of receptivity in specific neural pathways will be dealt with under the rubric of selection.

Most physiological discussion of alertness has centered around the observations that lesions of the ascending reticular activating system inhibit both desynchronization of the EEG and behavioral wakefulness (Moruzzi & Magoun, 1949; Thompson, 1967). A view that the reticular system operated as an undifferentiated whole has given way to more specific proposals about the role of different midbrain and thalamic nuclei. For example, Sharpless and Jasper (1956) identified slower tonic changes in alertness with midbrain levels and faster phasic changes with thalamic levels. A more detailed functional account of what happens to information processing during tonic and phasic changes in alertness is a crucial step in efforts to go beyond a general view of the reticular system as somehow involved in the process. Behavioral and physiological data from humans suggest that a number of functionally distinct mechanisms may underlie changes in alertness. An analysis of model situations may provide some ideas about these mechanisms.

In particular, we will be interested in two aspects of the question of alertness: first, tonic aspects of alertness which occur slowly and often involuntarily; second, phasic aspects of alertness which are easily under the control of the subject and occur very rapidly. The clearest situation for the study of tonic alertness is to examine diurnal variations in performance. The problem of phasic change in the level of alertness is examined during the period between a warning signal and a signal to which a response has to be made. Here, within several hundred milliseconds alertness changes dramatically.

Diurnal Rhythm

There is a marked change in many autonomic indicants over the course of the day (Colqohoun, 1971). A simple measure of the time course of these bodily changes is the oral temperature. Although there is some

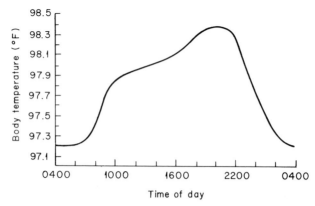

Figure 1. Body temperature as a function of time of day. [After Blake, 1971.]*

dispute about the details, the data of Figure 1 are typical in showing the steady rise of temperature throughout most of the day from 4:00 A.M. until late evening and then a steady fall to reach a low in the early morning of the next day.

A number of investigators have shown concomitant changes in performance in a variety of simple tasks. Most indices of the efficiency in detecting and responding to external signals show marked change which mirrors the rise in body temperature during the course of the day. A typical result is shown in Figure 2. Although there is general similarity in the temperature and performance curves, the post-lunch dip in performance serves to illustrate the multiple determination of performance changes.

The detailed character of the changes in performance over the day is important. Psychologists (Broadbent, 1971) have tried to discriminate two general kinds of changes in efficiency. One involves increases in signal detection or response speed which are accompanied by decreases in false detections or other errors. This kind of shift suggests a genuine increase in the basic ability to discriminate signals (d'). The second kind of change involves an increase of detections or speed which is accompanied by an increase in false detection or other errors. This type of change suggests that at some level of the nervous system there is a shift in the criterion (beta) for making a response. This shift is sometimes a conscious change in strategy by the subject but may also occur without the person's awareness (Kahneman, 1973). Another way of talking about these parameters is to suppose that a d' shift corresponds to a change in specific pathways related to the buildup of information about a signal, while the beta shift indicates the responsivity of later

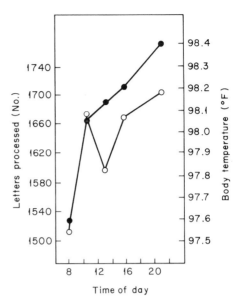

Figure 2. Relationship between time of day and both body temperature (●——●) and performance (○——○) in a letter cancellation task. The score represents the number of letters marked in a fixed time. [After Blake, 1971.]*

levels of the central nervous system (CNS) related to producing response to any signal.* The evidence so far available is not conclusive but suggests that the effects of the diurnal cycle are in d' (Blake, 1971). This is supported by signal detection studies, which show an increase in detections with no increase in false alarms as the day goes on, and also by reaction-time results which show increased speed and decreased errors. People seem to become more efficient as the day wears on.

It is important to emphasize that improvement in performance over the day is restricted to tasks which emphasize a direct response to external signals. Figure 3 shows data on the span of memory as a function of time of day. This task does not show improvement over the course of the day but instead declines. Whatever mechanism is postulated to improve the reception of external signals is apparently not a general potentiator of all performance. This point provides additional justification for distinguishing between alertness to external signals and other forms of arousal.

One physiological mechanism which might relate to these diurnal changes in performance is blood plasma levels of adrenal cortical steroids. Steroid level is maximum at 4:00 A.M. and, generally, is reduced

---

* The identification of beta and d' with different levels of processing in the nervous system is only a very approximate one. The presence of converging evidence from other operations is of importance in determining the level at which a given mechanism operates.

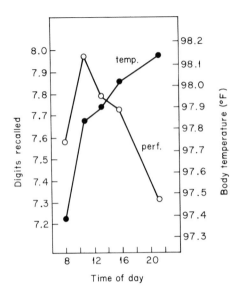

Figure 3. Relation between time of day and body temperature (●——●) and memory span performance (○——○). [After Blake, 1971.]

over the course of the day until late evening (Pertkoff, Eikens, Nugent, Fred, Nimer, Rush, & Tyler, 1959). Figure 4 shows a definite diurnal rhythmicity much like that shown in body temperature and performance.

Furthermore, there is reason to believe that the level of steroids is related to the sensory detection threshold (Henkin, 1970). Patients with an adrenal cortical insufficiency show greatly decreased sensory thresholds (increased sensitivity to stimulation). They are more effective in detecting sensory signals in several modalities than are normals (Henkin, 1970). It has also been shown that changes in adrenal cortical steroids have effects upon both peripheral and central neural conduction speed, although it is not clear that these changes would predict the improvement in reaction time (RT) which is observed. Data obtained following replacement therapy suggest that reduced levels of carbohydrate steroids are the major factor in these changes (Henkin, 1970). Of course, the changes in patients with adrenal-cortical insufficiency are much greater than changes due to time of day. However, since normal subjects undergo a drop in adrenal cortical levels from early morning until late at night, it is not unreasonable to expect them to show decreased sensory thresholds over the course of the day.

There has been some effort to test these notions by measuring evoked cortical potentials. Patients suffering from adrenal insufficiency show a decrease in latency and amplitude of evoked potential components following replacement therapy (Henkin, 1970) (see Figure 5). Some of the changes occur in components of the evoked potential which are

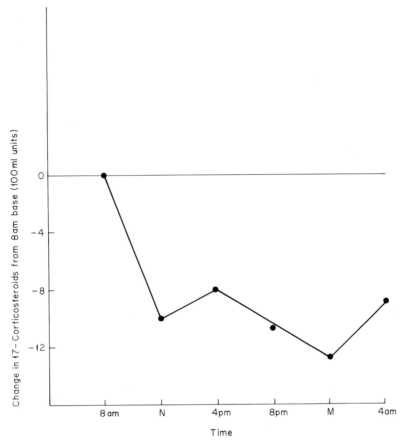

Figure 4. Relationship between time of day and change in level of plasma corticosteroids from 8:00 A.M. baseline. [After Pertkoff, Eikens, Nugent, Fred, Nimer, Rush, & Tyler, 1959.]*

likely to be due to primary cortex activity (Regan, 1972). Latencies and amplitudes of some evoked potential components have been shown to change in the normal subject over the course of the day as serum steroid levels decline (Heninger, McDonald, Guff, & Sollberger, 1969). These changes were small and variable in direction. Efforts to abolish this diurnal variation with drugs that held serum steroid levels constant were not successful (Heninger *et al.*, 1969).

Neither the behavioral nor the physiological data seem to warrant any strong conclusions about the mechanisms that underlie the diurnal changes in performance. It is clear that steroid levels can modulate sensory threshold in patients and that steroid levels change over the course of the day in normal people. More research needs to be directed

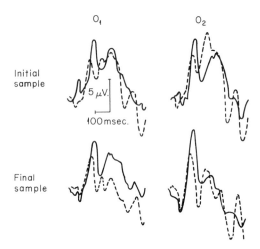

Figure 5. Average evoked response for 500-msec periods after each flash in one patient. The untreated condition is indicated by the solid line; after treatment, by the dotted line. O_1 and O_2 indicate left and right occipital electrode placements, respectively (positive reflection is up). Initial sample represents the average of the first 45 flashes, the final sample, the average of the last 45 flashes (after 20 minutes of continuous rhythmic flashes). [After Henkin, 1970.]

at determining whether or not performance changes are due to differences in sensory receptivity rather than attentional shifts and whether or not performance can be varied by manipulations in steroid levels.

Long Continued Performance

Studies using the diurnal cycle as a means of varying alertness contrast most clearly with phasic changes introduced by a warning signal. Before turning to such studies, however, it is useful to mention a category of tasks that lie somewhere in between. These are studies of long continued performance (½ to 8 hours) often called "vigilance" tasks (Broadbent, 1971; Mackworth, 1969).

The studies of vigilance can be divided into two general classes. First are those which use a high rate of sensory events, and second are those which involve only a few infrequent signals. Both of these types show clear losses in performance efficiency over time, but the mechanism(s) may be quite different. Studies with rapidly occurring signals show similar declines in performance with time on task whether the subject is actively responding to each signal or whether most signals are nontargets to which he does not respond (Mackworth, 1969). This fact makes it unlikely that the loss in performance can be described as due to a general reduction in alertness.

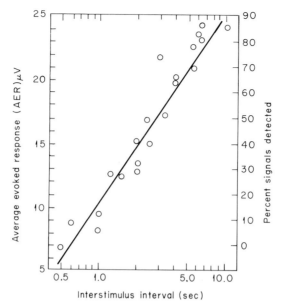

Figure 6. Effects of interstimulus interval on the average evoked response and on the percentage of signals detected during the final hour of an 80-minute vigil. In the vigilance task, the targets were presented at an average of every 4 minutes to different groups of observers; conditions differed only in the interstimulus intervals of the regularly repeated nontarget stimuli. [After Jerison, 1970.]*

If pairs of signals in the same modality arrive very close together in time there is evidence that the effective brain response to the second signal is reduced even at the level of the primary cortex (Angel, 1969). All components of the vertex evoked potential are markedly reduced when auditory signals occur at intervals of less than 10 seconds (Davis, Osterhammel, Wier, & Gjerdingen, 1972). The extent of reduction in the evoked potential is related to the similarity of the paired signals but does occur even with signal pairs in different modalities.

There is a rough general correspondence between the time course of recovery of the evoked potential (Davis *et al.*, 1972) and the decrement in performance when vigilance signals occur close together in time. This idea was dramatized by Jerison (1970) who plotted the time course of recovery of the vertex evoked potential to auditory signals along with the function relating interstimulus interval (ISI) to correct detections in a visual vigilance task. This plot is shown as Figure 6. The comparison of auditory physiological data and visual behavioral data emphasizes the looseness of this correspondence. Visual evoked potential recovery times are frequently much faster than those for auditory signals. In addition, the probability of correct detections varies with a number of other parameters beside ISI, making any one function only a rough estimate. For example, the vigilance decrement appears to occur only after a considerable period of time on the task, while the evoked potential decrement may occur to the second signal of a closely spaced pair. Nonetheless, rapid presentation of information appears to be one situation where both very clear evoked potential and behavioral decrements occur across a wide variety of tasks. And it, therefore, merits further examination.

The results obtained in high event tasks contrast markedly with those in low event vigilance tasks. In this situation, Broadbent (1971) has shown that changes in performance tend to be of the beta type, with subjects showing increased reluctance to respond as the task continues. They appear to be more sluggish or conservative in their performance as the task wears on. This performance change is accompanied by EEG and behavioral signs which are closely related to drowsiness. The EEG shows frequent signs of high amplitude sleep spindles, and the evoked potentials show an enlarged second negative component (Wilkinson, 1967) similar to what occurs during sleep.

None of these results provides a very clear analysis of the mechanisms underlying the decline in performance in vigilance tasks. But they suggest that at least two different mechanisms may be involved. One possibility is that high event tasks involve a pathway effect which specifically reduces the efficiency of processing for signals resembling the events being repeated, while low event rate tasks mainly involve a reduction in general alertness (see also p. 443). This view could be tested by interpolating infrequent signals in a different sensory modality to which the subject must respond. A pathway interpretation would suggest that time on the primary task would not affect this new performance, while a general alertness theory would predict a marked effect. Despite the dozens of studies in this area only one (Pushkin, 1972) has attempted to use this simple method, and methodological problems prevent a clear interpretation of the results. The suggestion that vigilance tasks may involve two different mechanisms should caution against any strict separation of alertness into tonic and phasic mechanisms. Indeed, the high event rate tasks resemble the tonic effect of the diurnal rhythm studies, while the low event rate tasks show effects more like the phasic studies to which we now turn.

Phasic Alertness

General State

Suppose a human subject is asked to prepare to process an incoming stimulus. There is widespread agreement from many studies about what

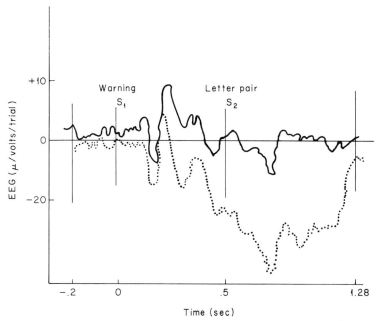

Figure 7. Solid lines represent average EEG following a warning signal (S₁) when the subject is instructed to watch the stimuli. The dotted line represents trials when the subject is instructed to respond to S₂. The negative shift is called the contingent negative variation.

happens. The EEG shows a temporary blocking of rhythmic alpha activity which is replaced by fast desynchronized activity. There is a slow negative drift in the EEG (contingent negative variation or CNV) which begins as rapidly as 100 to 200 msec following the warning signal and which approaches its minimum at a rate which is a function of the warning interval (see Figure 7). This change in the EEG is accompanied by a constellation of changes in autonomic activity, many of which are related to the general state of sympathetic dominance which accompanies nearly any difficult mental activity (Kahneman, 1973). However, it also includes an inhibitory component which is related specifically to states of high alertness for external signals and involves both cardiac deceleration and motor inhibition. This state has a marked reduction in irrelevant movement and a steady, unblinking eye (Webb & Obrist, 1970). The state of preparation for external signals is also marked by specific inhibition of spinal reflexes associated with the muscle which is to respond (Requin, 1969). Thus, the constellation of bodily changes in a high state of phasic alertness has some components of general excitability and also changes which involve inhibition or suspension of activity.

How are these changes related to the ability of subjects to respond to the stimulus which follows the warning? In one sense, the answer is clear. The response time to a signal which follows a warning is greatly reduced over a signal for which there has been no warning (Bertelson, 1967; Lansing, Schwartz, & Lindsley, 1959). The improvement in reaction time is very rapid for the first 150–200 msec following the warning. For some tasks, reaction time is minimum after about 200 msec, while for others it does not reach its minimum value until about 500 msec (Posner, Klein, Summers, & Buggie, 1973).

Although the state of high alertness is one which is observable in a number of bodily changes, no one of these changes appears, by itself, to be a good predictor of the ability of the subject to process a stimulus. Lansing et al. (1959) showed that although a warning signal improved reaction time and blocked alpha, presenting the critical stimulus when there was a spontaneous blocking in alpha did not show improved reaction time over presentation of the stimulus during alpha. Similarly, Rebert and Tecce (1973) have shown that while there is a clear negative shift in EEG (CNV) following a warning signal, correlations between reaction time and depth of CNV at any given foreperiod are often small. It appears that any single measure of physiological state present prior to a signal is a relatively poor indicant of performance in responding to that signal. This probably should not be taken to mean that these are not indicants of the alert state, nor that an alert state is unimportant for responding. What may be happening is simply that the alert state is indicated by a large variety of bodily changes, all of which are poorly correlated with performance. Thus, no single dimension is useful as a predictor of reaction time.

Information Processing Theory of Phasic Alertness

The results of studies of phasic alertness upon performance can best be understood with respect to a hypothetical buildup of information within the sensory-memory system of the subject. In cases where the signal is left on until the subject responds, there is a buildup of information over time which eventually asymptotes (upper curve, Figure 8). Phasic alertness appears to affect mainly the time when subjects can respond, rather than the rate at which information is building up in the sensory-memory system. If the task is one to which the subject can respond rapidly, a warning signal improves reaction time, but also increases errors (Figure 8). If the task is one which produces slower responding, reaction time may improve while errors remain flat (Figure 8) (Posner et al., 1973). These results are consistent with a constant

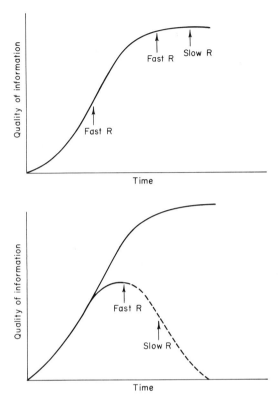

Figure 8. Hypothetical buildup of information about a signal as a function of time. The upper panel is when the signal remains present until a response. Faster RT is associated with equal or poorer quality. The lower panel also includes the condition (dotted line) when the stimulus is presented briefly and then removed. Faster RTs may be associated with increased or reduced quality of information.

buildup of information regardless of the level of alertness but a shift in the criterion (beta) for responding.

The same theory predicts rather different results in cases when the signal is present only briefly (Figure 8, lower portion). If alertness affects the time when the subject can respond and the signal is removed it is clearly possible for a faster response to be based upon a higher quality of sensory information. This constellation of results (faster reaction time and fewer errors) was reported by Fuster (1958). He presented monkeys with brief visual signals and varied alertness by electrical stimulation of the reticular formation prior to the signal. This result has been replicated with human subjects when alertness is varied by a warning signal followed by a brief visual flash (Posner *et al.*, 1973).

The information processing theory of phasic alertness outlined above proposes two systems. A sensory-memory system which is relatively unaffected by phasic alertness and a later system which is greatly affected. This view applies both to tasks which involve overt motor responses and to those involving perceptual sensitivity. Thus, it does not seem likely that the later system suggested by this analysis has much to do with muscular changes in the periphery. Rather, it must be a central system, important for both overt responding and conscious attention. Undoubtedly, this system has close ties with both association and motor cortex. Later in the chapter some possible properties of this system will be outlined.

These behavioral results suggest the view that phasic alertness as introduced by a warning signal does not involve changes in the basic receptivity of the primary cortex to sensory stimulation. Rather, according to this view, the change occurs in a later system required to integrate information and produce responses to it.

Studies of human evoked potentials following warning signals have produced contradictory results (Donald & Goff, 1971; Näätänen, 1970). However, there seems to be no clear evidence that the warning enhances either the latency or the amplitude of primary evoked potentials. There is a problem in interpreting such data in any case since there are impressive changes in background EEG with a warning signal which makes it difficult to interpret changes in evoked potential amplitude. On the other hand, there is strong evidence from animal studies (Fuster, 1958; Nakai & Domino, 1968) that reticular stimulation varies the response of primary cortical units. The time course of such changes can resemble behavioral alerting functions (Nakai & Domino, 1968). A major task will be to determine the relationship of direct reticular stimulation to the different behavioral functions of alertness outlined in this section.

Alertness and Selective Attention

According to the theory of phasic alertness outlined above, a warning improves reaction time but does not affect the buildup of information in the sensory-memory system. One way to separate the buildup of information about a letter from the time to develop a response to the letter, is to present a single letter followed after a short variable interval by a second letter. The task is to respond "same" if the letters are identical and otherwise, "different." The encoding of the first letter provides no information about the response to make when the second letter occurs, yet *both* RT and errors in making the match decline regularly as the time by which the first letter leads the second is increased from

0 to 300 msec (Posner & Boies, 1971). This improvement in RT and error contrasts markedly with changes due to shifts of phasic alertness. The improvement in the handling of the second letter provides an indirect means of observing the buildup of information (encoding) of the first letter. This experiment was run both when the subjects were alerted prior to presentation of the first letter and when they were not alerted. Subjects responded more rapidly when they had been alerted, but the extent of improvement in RT (due to buildup of information about the first letter) was completely unaffected by the level of alertness at the time the first letter was presented.

This result is confirmation of the behavioral view that buildup of information is unaffected by the level of alertness. It also suggests that the mechanism which varies the level of alertness is independent of the mechanism which improves performance due to specific information concerning the first letter. The latter mechanism is selective in the sense that it is specific to a particular pathway. This specificity is shown by the finding that the extent of the improvement is greater when the second letter matches the first than when it does not (Posner & Boies, 1971). Here we have behavioral evidence that alertness (state effects) and set (pathway effects) make independent (additive) contributions to the speed of processing.*

Of course there are also physiological reasons for viewing alertness and pathway effects as resting upon independent mechanisms (Groves & Thompson, 1970). For example, Gazzaniga and Hillyard (1973) have shown that the contingent negative variation induced by the presentation of a warning signal to one hemisphere of a split brain preparation spreads to the other hemisphere. This contrasts with information concerning the identification of a stimulus (such as a letter) which is generally not available to the opposite hemisphere. The split brain work provides additional support for the hypothesis that phasic alertness induced by a warning signal involves a subcortical mechanism, while selection based upon activation of units related to identification of a stimulus involves cortical processes.

When the physiological evidence for independent mechanisms is combined with the information processing account, the sense of independence between selection and alertness is enlarged. The physiological evidence suggests a separation of mechanism in the brain (e.g., subcortical in the case of phasic alertness and a cortical pathway in the case of selection). The psychological evidence adds to that an independence of function. Phasic alertness operates not in reducing the sensitivity

---

* For a more detailed discussion of the importance of additive contributions in assessing the independence of stages (see Sternberg, 1969).

of sensory-memory units but upon a later system, while selection involves activation of a specific cortical pathway which improves the buildup of information.

SELECTIVE ATTENTION

The study of selective attention has been a difficult one. Behavioral studies have proven somewhat more analytic than studies using physiological measures. It is possible to distinguish two different bases upon which selection could be made (Broadbent, 1971). If subjects are told to attend to a particular position in space or modality it is theoretically possible for subjects to set up a fairly peripheral gate which might allow in stimuli from one source and not from the other. Broadbent has called this situation stimulus selection. A second type of task defines selection on the basis of identification of the stimulus through contact with stored information (e.g., select the digits and ignore the letters). This can be called response- or memory-dependent selection. Of course, cues may be used which depend on less than full identification (e.g., color or tonal frequency). In accordance with the model task approach of this paper, our discussion will be confined to the purest cases of stimulus- and memory-dependent selection.

Stimulus Selection

Shadowing

If a subject's attention is kept rigidly focused on one source of signals the probability that he will respond to, be aware of, or remember signals arising from another source is greatly reduced. The strongest situation for demonstrating this phenomenon has been to have subjects shadow information presented to one ear, by repeating it back as quickly as possible after it is presented (Cherry, 1953; Treisman, 1964), while additional information is presented to the other ear. Data from these experiments show that subjects can report little about the semantic content of the unattended ear. If they are instructed to respond by pressing a key to a word they are rarely in error when the word occurs on the attended message, but have many omissions when the word occurs in the unattended message (Treisman & Geffen, 1967). Under these circumstances virtually complete blocking of the information from memory, consciousness, or behavior seems to have been achieved. It should be noted that rapid presentation of highly similar information appears to be necessary to achieve such blocking.

These early results made it reasonable that peripheral physiological mechanisms *might* be blocking or attentuating the irrelevant input and resulted in a favorable reception to evidence favoring peripheral blocking of input (e.g., Hernández-Peón, 1960). However, as the behavioral experiments progressed, it became clear that peripheral blocking of information could not be correct. For example, Lewis (1970) showed that even under conditions where the subject was neither aware of nor able to report the unattended message, the occurrence of a synonym on the unattended ear increased the reaction time to the attended word. This shows that the unattended message is classified by semantic content; since its meaning has differential effects on the attended message. Lackner and Garret (1973) and McKay (1973) have shown that a sentence presented to the attended ear (e.g., the boys are throwing stones at the bank) can be disambiguated by information presented to the unattended ear (e.g., savings), even in situations where the subjects are not able to report the content of the information on the unattended ear. A similar result (Corteen & Wood, 1972) has shown that words presented to the unattended ear which have previously been paired with shock, produce a galvanic skin response (emotional response) even under conditions when they would normally be unattended. These results cause grave doubt that any peripheral blocking mechanism could account for the selective attention results in human subjects.

Evoked Potentials

In looking at the evoked potential data in this area, it is important to make two distinctions. First, components of the evoked potential occurring in the first 150 msec after input are generally more dependent upon the sensory quality of the signal than components occurring after the first 200 msec (Regan, 1972). Alterations in intensity, modality, signal-to-noise ratio, etc., vary the amplitude of components within the first 150 msec, while instructions to attend or process the stimulus generally affect later components. Even more important than latency of the components is the location of the electrode (Regan, 1972; Walter, 1964). Electrodes located above the primary cortex are more likely to reflect sensory specific aspects of the signal than those located at the vertex. Thus, one can distinguish most clearly between affects upon the early components at electrodes located over the primary cortex of the input and affects upon the late components at electrodes located at the vertex. Even these extremes may leave somewhat confounded primary and association cortex activity. Second, it is also important to consider the role of the level of alertness in affecting the evoked potential results (Näätänen, 1970). When a subject is in a relaxed state, the background

EEG will tend toward synchrony; while in a state of high alertness the background EEG will tend to be desynchronized and will show the negative shift (CNV) if the subject expects an important signal. These differences in background could themselves account for evoked potential (EP) differences to critical stimuli if the subject can predict when the relevant stimulus will occur. Thus, it does not seem worthwhile considering studies of evoked potentials which do not control the background activity by randomizing the presentation of selected and nonselected signals.

With such studies eliminated, it appeared until quite recently that there was no important change in EP components, as a result of instruction to attend or select a signal, which might reasonably be said to reflect primary cortex activity. Careful studies by Donchin and Cohen (1967) and Hartley (1970) showed no difference between early components (first 150 msec) to selected versus nonselected signals even at vertex sites.

Quite recently, however, two reports have indicated such changes in early components (Hillyard, Hink, Schwent, & Picton, 1973; Wilkinson & Lee, 1972). Both of these studies used rapid presentation of similar signals to the two ears. It should be noted that these are the conditions which produce strong stimulus set in information-processing studies. The Hillyard paper is particularly important in evaluating the notion that components of the EP may reflect the act of stimulus selection under conditions which might produce blocking of the unattended signal from conscious attention. They found that the N_1 component of the vertex EP (about 90 msec after input) was substantially larger to tones presented to an "attended ear" than to tones presented to an "unattended" ear. This difference occurred whether the signal on the "attended" ear was a target signal which the subject was to report or an irrelevant background signal. It was also found, in agreement with studies outlined later in this chapter, that a target stimulus on the attended ear produced in addition a late positive wave, which did not occur to nontargets.

This report of an EP component related to selection of an attended ear is quite important. However, it should be kept in mind that this type of selection may only be possible with very rapid and similar signals. These are also conditions which give the most evidence for blocking of input information prior to conscious perception in behavioral studies. Rapid presentation of stimuli reduces the overall EP size (see page 451) which may itself be important in obtaining the relative increase in amplitude for the selected modalities. While the change in the N_1 component is earlier than previous reports in which selection is manipulated by the instruction to attend, it may still not index primary cortex activity since only vertex changes have so far been reported. However, the ability to separate stimulus set (selected versus unselected ear) from response set (target versus nontarget frequency) in the EP components represents an important step forward in relating EEG and behavioral analysis.

There is very strong evidence of changes in a late positive potential (P-300) under conditions of selective attention. Studies using dichotic presentation of auditory items (Smith, Donchin, Cohen, & Starr, 1970) and intramodality presentation of visual items (Donchin & Cohen, 1967) have shown an enhanced amplitude of the positive wave which in their studies occurs about 300 msec following input. Apparently a broad range of stimulus conditions can be used to obtain this effect of instructions to attend. It is clear that the late positive wave is sufficiently long after input that a stimulus could have made contact with representation in memory. Thus, the effects of attention on this component can be studied better in experiments which control more completely the processing which the subject is supposed to do with the attended stimulus. These are discussed in the next section.

Response Set

Processing Studies

A different way of looking at the process of selective attention is to provide the subject with a precise model of the stimulus that he is to select by activating the pathway related to that event. One such experiment has already been described (page 456). If the subject receives a single letter followed after a varying interval by a second letter which might or might not match it there is a marked improvement in the speed and accuracy of responses as the time between the first and second letter is increased from 0 to 300 msec. The improvement in speed is more marked for matches than for mismatches.

Two important questions may be asked about the encoding of the first letter. One question refers to the level at which the first letter is having its effect on the processing of the second letter. If this is a primarily sensory effect, one would expect the facilitation to occur only when the first letter matches the second physically (e.g., AA). This is not the case. Even if the two letters match only in name and are not identical physically (e.g., Aa), there will still be a facilitation of the processing of the second letter by the presentation of the first.

The second question is the role of intention in the facilitation. In the matching studies the subject is instructed to use the first letter to decide upon his response to the second. Encoding of the first letter serves to improve his performance. However, it is possible to show that activation of a pathway in memory will facilitate the processing of a new item which uses the same pathway even in cases where it actually harms performance. Warren (1972) presented subjects with a series of words orally (e.g., elm, oak, and maple). They were then shown visually a single word which may have been in the previous list (eg., elm) or not (e.g., man). The visual word was printed in colored ink and the subject's task was to name the ink color of the word as quickly as possible. There was a very clear increase in RT to name the color for words which had been presented orally. Apparently, the passage of the visual word through the nervous system was facilitated when the word name had been activated. The strong tendency to name the word, produced interference with the ability to name the ink color. Even when subjects were aware of this difficulty they did not appear to be able to control the interference. Thus, the facilitatory effect of one word upon a second appears to occur even in a situation where the net result is to impede the subject's processing. For this reason the effect seems to be a part of the process of selection which operates to some extent independently from specific instructions.

If a pathway is repeatedly stimulated one might expect that the facilitation discussed previously would be replaced by an inhibitory effect. There are many reports in the literature that repetition of a single word, particularly aurally, leads to a loss of meaning and a distortion of the recognition process (Esposito & Pelton, 1971). A frequent result in memory research is that massed (closely spaced) presentation of items leads to poorer long-term memory than the same number of more widely spaced presentations (Hintzman, 1974). This memory effect, however, occurs even with only two successive presentations, a condition which leads to facilitation of the pathway as shown by studies of RT and thresholds. A number of efforts to produce evidence for inhibition of a repeated pathway by techniques similar to those described for facilitation have not been successful. It is clear that many rapid repetitions of an item make responses to it less efficient at some level, but processing studies have not yet provided any analysis of the level involved.

Orienting Response

The study of the orienting response (Sokolov, 1963) provides a different literature which is also concerned with possible antecedent conditions that produce a selective advantage or disadvantage for a particular input pathway. Sokolov (1963) found that presentation of a novel stimulus led to a constellation of autonomic and EEG changes which he called the orienting response. When the stimulus was repeated, this response habituated, but would reoccur when any change in the properties of the stimulus, including its temporal position, was introduced. Sokolov's theory suggested that a neural model of the stimulus input was constructed at a relatively early level of processing (e.g., primary cortex) and signals which matched it were shunted aside. Unfortunately, the use of relatively slow autonomic responses has not made it possible to tell the level at which habituation is taking place. Moreover, there is a lack of evidence from other sources supporting habituation at the level of primary cortex (Thompson & Spencer, 1966). In order to pursue this question it would be useful to have physiological indicants of orientation earlier than those used by Sokolov. Components of the vertex evoked potential have been used to study this question (Ritter & Vaughn, 1969).

Signal Detection

Work with evoked potentials to near threshold signals has provided some evidence of a brain sign correlated with conscious detection of a signal.* When the subject detects a signal, there is an increase in the amplitude of a late positive component of the evoked potential which occurs from .2 to .5 seconds after input (P-300).

Hillyard, Squires, Bauer, & Lindsay (1971) showed a very close relationship between the probability of signal detection and the amplitude of this EEG component. Hillyard concludes that the late positive wave indexes the observer's degree of confidence or certainty in his decision that the signal has occurred. As can be seen in Figure 9, P-300 amplitude is closely related to probability of detection and increases to a maximum in the intermediate range of detection.

If one views the P-300 as an index of orienting to a signal, there is a difference between Hillyard's results and Sokolov's. In Sokolov's work, orientation occurs to a novel stimulus which is a violation of what the subject expects. In Hillyard's work, P-300 is produced when the subject succeeds in detecting the signal which he expects.

Reaction Time

Studies of evoked potentials during reaction time tasks also show the amplitude of the late positive wave is closely related to factors which might cause a subject to focus his attention on the signal. Sutton, Braren, & Zubin (1965) showed that the amplitude of P-300 was greater for low probability than for high probability signals (see Figure 10). They also found that the evoked potential was reduced when the sub-

---

* There is abundant evidence that components of the evoked potential can be decorrelated from performance by use of masking signals or drugs. Since we regard the evoked potential which occurs in these tasks as a sign or indicant of underlying internal processes, these demonstrations do not seem to vitiate this discussion.

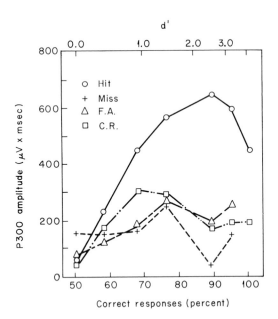

Figure 9. Association cortex potential size (P-300) for hits, misses, and control trials as a function of proportion of correct responses in a signal detection task. [After Hillyard *et al.*, 1971. Copyright 1971 by the American Association for the Advancement of Science.]

ject knew which signal was to occur. This is similar to a finding using signal detection techniques that P-300 was greater when the feedback disconfirmed the subject's expectancy (Squires, Hillyard, & Lindsay, 1973).

The P-300 is often closely related in time to the motor act of responding, but it does not seem to depend upon an actual response. Indeed this EP component is sometimes reduced when the subject must respond overtly in comparison to making no overt response. (Karlin, Martz, Brauth, & Mordkoff, 1971). Donchin, Kubovy, Kutas, Johnson, & Herning (1973) found that when subjects were forced to make a motor response to each signal, P-300 was equally large regardless of the predictability of the signals. However, when no motor response was required, P-300 was greater for unpredictable signals. A trial-by-trial analysis of overt key responses and P-300 (Ritter, Simson, and Vaughan, 1972) found that P-300 and the motor response appeared to be elaborated in parallel, with some elements of the P-300 complex preceding the motor output.

The results of studies using P-300 as a sign of orientation reveal that a very large number of antecedent conditions will produce it. Moreover,

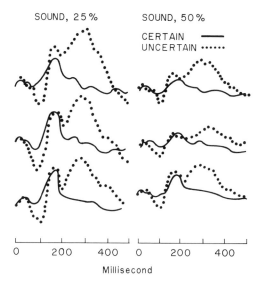

SOUND, 25%   SOUND, 50%

CERTAIN ——
UNCERTAIN ······

0   200   400   0   200   400

Millisecond

Figure 10. Average evoked response for certain (solid lines) and uncertain (dotted lines) clicks for one subject on different occasions. On the left, clicks had a 25% probability of occurrence; on the right, clicks had a 50% probability of occurrence. Electrodes are vertex to earlobe. Positive deflection is plotted up. [After Sutton et al., 1965.]

these appear to vary depending upon the task set for the subject. In detection studies the complex occurs when the subject detects an expected signal, while in RT studies it occurs more often to an unexpected or novel event. The complex may even occur without a signal, provided the subject has a strong expectancy that a signal should be time-locked to some prior event (Weinberg, Walter, & Crow, 1970). It seems fair to say that P-300 occurs whenever a signal or time interval could reasonably be said to demand close attention in a particular task and when that attention is carefully time-locked. This suggests the P-300 is the result of the selection of a signal rather than a sign of some particular antecedent condition which invariably leads to selection. The sign appears to add little to our knowledge about the causes of selection. This view might best be interpreted as supporting Goff's (1969) position that P-300 "should be thought of as being the objective brain sign related to our conscious awareness of the stimulus."* Or as Donchin et al. (1973)

---

* This should not be taken to mean that the lack of a P-300 necessarily indicates that the subject is unaware of the stimulus. Rather, it may mean only that the degree of conscious effort invested in processing the signal is related to the probability or amplitude of the component. A lack of the component means only that the degree of conscious movement was too slight or too poorly time-locked to produce this indicant.

suggested "it seems that P-300 reflects the activity of a generalized cortical computer." These statements are rather vacuous taken by themselves.
They have more meaning if it is possible to specify some objective
properties of the brain system which appear to be signified by the P-300.
In other words, it might be a useful strategy to leave aside the effort
to specify *the stimulus conditions* which will produce P-300. For if the
P-300 indexes anything like conscious effort there would have to be
a great deal of flexibility in the conditions which will produce it. Rather,
one might try to develop response–response relations; that is, to specify
the consequences for information processing when the system indexed by
P-300 is occupied by a signal. It is that effort which is discussed in the
next section.

CONSCIOUS PROCESSING

When a subject begins to attend to a stimulus in a way that would
be called effortful or conscious processing, there are clear behavioral
and physiological changes (Kahneman, 1973). Both information-processing and physiological studies have traced the consequences of conscious
processing of a signal.

Costs and Benefits of Attention

Shallice (1972) has proposed that the concepts of consciousness may
be identified with an information-processing stage in which different
"action systems" compete for dominance. If one action system is to operate at maximum efficiency, it must become dominant and hence suppress
the activity of other action systems. This is a recent statement of a
very old idea (Loeb, 1898) in which a consequence of processing of
one signal is thought to be widespread inhibition of the processing of
other signals.

Consider the following experiment (Posner & Boies, 1971). The subject
is required to process a visual stimulus. The processing of that visual
stimulus is divided into stages by the following technique: (1) the subject receives a warning signal; (2) .5 sec later a single letter is presented;
(3) 1 sec later a second letter appears which may or may not match
the first; (4) the subject's response indicates whether the two letters
are the same or not. The warning signal allows a generalized change
in alertness such as we discussed in the first section. The first letter
provides the subject with selective information (response set) about
what he is to look for to obtain a match. The second letter allows the
subject to select a response.

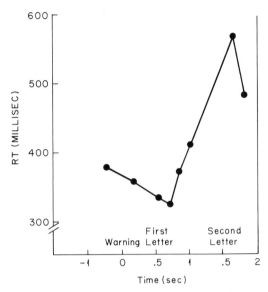

Figure 11. Reaction time to white noise probe as a function of its position within the letter-matching sequence (abscissa indicates visual events in the letter-matching task). [After Posner & Boies, 1971. Copyright (1971) by the American Psychological Association. Reprinted by permission.]

On some trials an auditory probe stimulus is presented at varying points in the sequence. In the experiment illustrated in Figure 11 the probe stimulus is a white noise burst to which the subject must respond by tapping a key. The figure displays RT to the noise burst as a function of its position in the primary task. During the warning phase, response time to the probe improves. This agrees with the nonselective view of alertness outlined in the first section. The reaction time to the irrelevant probe stimulus continues to improve for some period of time after the presentation of this letter. In this particular experiment, that turns out to be between .3 and .5 seconds following the first letter. At that time the reaction time to the probe stimulus begins to increase rather dramatically. This increase in reaction time to the probe occurs well before the second stimulus is presented. It is not due to the actual presentation of the second letter because even when responses which occur after presentation of the second stimulus are eliminated, the interference in reaction time is still obtained. After the response selection phase, the reaction time to the irrelevant probe improves until it comes down to approximately the same baseline as would be obtained if it were presented prior to the warning signal.

How does the time course of interference obtained to probes corre-

spond to the improvement in performance that processing of the first letter has upon the match? In the previous section it was shown that the
first letter produces a "response set" which improves reaction time and
reduces errors to the match. If on some trials the second letter is added
to the first after a varying interval from zero through 1 second, a picture
emerges of the buildup of information (encoding) of the first letter.
It is clear that the improvement in reaction time produced by the presentation of the first letter occurs mostly during the first 150–300 msec
following the first letter. This improvement ends before the inhibitory
effect on the processing of the probe stimulus occurs (Posner & Klein,
1973).

It should not be thought that the time course of inhibition is rigidly
time-locked to the presentation of the first letter. Rather, it depends
heavily upon the interval between the first letter and the second. If
the first letter is presented for only a brief interval, the inhibition may
occur as quickly as 50 to 150 msec following the first letter, while if
the first letter is presented for an interval of 2 sec the inhibition does
not occur until just shortly before the second letter. It is as though
whatever produces the inhibition to the irrelevant probe stimulus is
quite flexibly related to when the subject thinks the information of the
first letter is going to be used.

These results are not dependent upon a motor response being required
to the irrelevant probe stimulus. A similar experiment (Kahneman, 1970)
has shown a decrease in the detection probability of a visual probe
stimulus presented during an auditory processing task. Thus, it does
not appear that the inhibition which is caused by the primary task
depends upon motor output being required. This finding should not be
taken as evidence that the probe stimulus is blocked by a peripheral
gate. As we have seen in the previous section, there is very little reason
to suppose that the secondary stimulus is totally eliminated as a factor
in the subject's information processing. Rather, the probability of noticing
it, storing it, or responding to it is reduced.

It is possible to show in a single experiment both the facilitatory
and inhibitory characteristics of presenting a signal upon the processing
of new information. In the section on selective attention it was emphasized that presenting a letter facilitates the pathway in the memory
system and leads to more efficient processing of an identical letter. This
section stresses the inhibitory consequences which admission to the conscious processing system has upon other information.

A more complete account involves a detailed statement of the time
course of both facilitation and inhibition (Posner & Snyder, 1974). The
basic argument is as follows. Within the sensory-memory system it is

possible to facilitate processing by activating a pathway. As long as the information has not been imported into the limited capacity mechanisms such facilitation has no general inhibitory consequences for other signals.* When the subject begins to attend to a signal by use of the limited capacity conscious mechanism there is still facilitation of that pathway, but also inhibitory consequences for the processing of other signals. This cost-benefit analysis is exactly a combination of the facilitatory consequences of response set which were discussed in the previous section, and the inhibitory consequences of the limited capacity mechanism. It implies that conscious processing can be objectively defined by its widespread inhibitory consequences.

Autonomic Responses

Both the subjective feeling of effortfulness and widespread autonomic responses accompany conscious processing (Kahneman, 1973). This involves a broad constellation of changes in the autonomic nervous system including changes in heart rate, galvanic skin response, vascular dilation, and pupil size. An experimental study by Kahneman, Tursky, Shapiro, and Crider (1969) illustrates these changes (see Figure 12). Subjects were presented with a series of four digits (seconds 10–13). They were to transform these digits by adding 0, 1, or 3 and then report them back (seconds 15–18). Figure 12 shows the results of this task on pupil diameter, heart rate, and skin resistance. There is remarkable similarity in the time course of various indicators during the listening and calculating phase. A relationship to task difficulty is also apparent.

Similar results on pupil size can be found in processing sentences (see Figure 13). Here, subjects are asked to take in and comprehend the meaning of a sentence. The pupil size changes clearly reflect moment-to-moment difficulties that the subjects have in the assimilation of information from the sentence.

Kahneman (1973) points out that these striking autonomic effects are reduced and perhaps even eliminated when processing of the input signal requires no rigid time-locking. Thus the physiological indicants of conscious attention depend heavily upon time pressures being exerted. These pressures may come from requiring a hurried response, using a dim signal subject to rapid decay, or a task with a heavy memory load. The importance of time-locking should also be considered in deter-

---

* The idea is that only in accessing the conscious mechanism is there evidence for interference between any pair of stimuli regardless of modality. This does not rule out more local inhibitory processes within the memory system in cases such as repeated activation of the same pathway (see page 462).

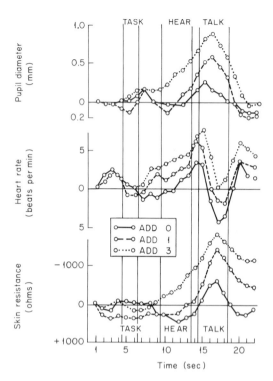

Figure 12. Second-by-second autonomic changes during digit transformation task. The digits are presented on seconds 10–13 and the responses on seconds 15–18. [After Kahneman et al., 1969. Copyright (1969) by the American Psychological Association. Reprinted by permission.]

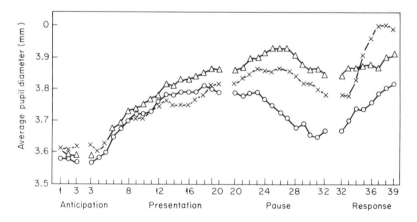

Figure 13. Average pupil dilations for the three recall conditions during and after listening to an aural sentence for three task conditions. △——△, repeat; ×——×, question after; ○——○, question before. [After Wright & Kahneman, 1971.]

470

mining when P-300 or inhibition of other signals will be a useful sign of conscious processing.

The striking effects of conscious processing upon the autonomic nervous system are accompanied by decrement in the detectability of new input (Kahneman, 1970). It may be recalled that in the section on alertness it was indicated that a somewhat similar constellation of autonomic responses accompanied a warning signal and produced an increase in the efficiency of processing a new external signal. In the case of alertness, the sympathetic-like pattern, found here during the listening and calculating phase, is accompanied by heart rate slowing and an inhibition of motor activity. We need to know much more about the similarities and differences in the constellation of bodily processes in the two conditions.

If the theory of phasic alertness which was analyzed in the first section is correct, it seems unlikely that conscious attention has any effect upon processing at the level of primary cortex. This may be one of the reasons why it is behaviorally rather difficult to maintain concentration on an internal source of signals. If it were the case that concentration on an internal source raised sensory cortex thresholds for external stimulation one would expect it to be easy to maintain concentration. It would be interesting to have more evidence about the relationship between the processing of an internal signal and the ability of new signals to activate pathways within the sensory-memory systems.

Locating the Conscious Brain Mechanism

If P-300 is thought to be related to the widespread changes in processing and bodily state related to conscious processing, it should be possible to use these signs to locate the brain mechanisms associated with the state. The term "location" can be used in two senses—first, temporal location within the ongoing stream of information processing and second, spatial location within the brain.

Temporal Location

To understand the temporal relationship between input and conscious processing we can return to the letter-matching task described previously. In this case, the first letter is followed after 1 second by a letter which either matches it or not. The response to the pair is delayed until a signal is presented half a second later, after which the subject presses one key for matches and one for mismatches. The major interest is in the vertex evoked potential to the second letter as a function of whether the pair matches or not. The behavioral analysis of response

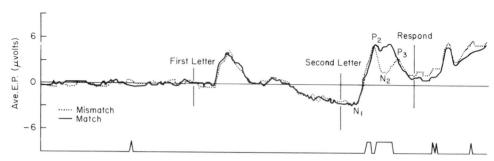

Figure 14. Average evoked potential for match and mismatch for 3072 msec from start of trial through response. Excursions on the x-axis represent significant differences between conditions. The interval between letters is 1000 msec. [After Posner *et al.*, 1973.]

set suggests that the processing of the second letter will be speeded when it matches the first. Data from vertex electrodes (Posner *et al.*, 1973) indicate that the matching pair differs from the mismatched pair starting about 160 msec after the second letter (see Figure 14). Since the matches and mismatches are randomized there is no possibility that this difference can be due to any prestimulus alertness state such as the CNV.

The difference between matching and mismatching pairs might well be summarized by suggesting that the late positive wave P-300 occurs faster for the matching pairs so that it is superimposed upon the earlier positive wave. If this were the case and if conscious attention affects the amplitude of the late positive wave as we have suggested, it should follow that when two stimuli match, the effect of the instruction to attend will influence the size of the evoked potential earlier than when the two stimuli do not match. The results of such an experiment are illustrated in Figure 15. When the subject is instructed to count matching stimuli the evoked potential size is increased starting at 200 msec after the second letter while if he is instructed to count mismatching stimuli the increase in evoked potential does not occur until about 300 msec.

The absolute time course of the P-300 found in these evoked potential studies should not be taken too seriously. The behavioral studies suggest that the time when the processing of a signal starts to produce inhibition with a probe is quite flexible. It depends upon how soon the subject is required to use the information. One should expect similar discrepancies in the time when an attentional manipulation will begin to affect evoked potentials.

Sutton *et al.* (1965) found that although the late positive wave was the main one to be affected by stimulus certainty (see Figure 10) a number of earlier components were also affected. In fact, components as early

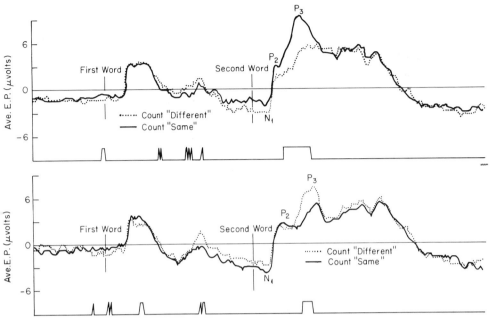

Figure 15. Upper curve shows average evoked potential for matches when instructed to count "sames" or "differents." Excursions on the x-axis represent significant differences between conditions. The interval between words is 1000 msec. Lower curve shows average evoked potential for mismatches when instructed to count "sames" or "differents." Excursions on the x-axis represent significant differences between conditions. The interval between words is 1000 msec. [After Posner et al., 1973.]

as 100 msec following the stimulus seemed to show some effect. However, these effects might be due to difference in preparation prior to the stimulus. The effect of attention is also apparent in experiments by Chapman (1973). In these experiments, Chapman required subjects to determine which of two letters was earlier in the alphabet or which of two digits was higher. The subject never knew whether the pair of items would be letters or digits thus eliminating any possibility of differential preparation. In some conditions the subjects were required to work on the letters and in other conditions to work on the digits. Chapman found that there was an enhancement of the vertex evoked potential as early as 105 msec after input when the material was relevant.* Since there was no way for the subject to determine whether the information

---

* The Chapman report of changes in EP as early as 105 msec under conditions which must clearly involve response set raises questions about whether the Hillyard et al. finding of stimulus set changes at 90 msec provides evidence of different mechanisms for stimulus and response set.

was relevant until the stimulus came on, this is evidence that response set may involve changes in EP well before 300 msec.

Several new techniques can be brought to bear upon the problem of determining the temporal characteristics of the processing mechanisms associated with P-300 and with other EP components. For example, Fox and Rudell (1970) have shown that cats can be conditioned to modify aspects of the EP. The first studies involved modification of the EP 180 msec after input, but later they showed changes in EP components as early as 50 msec after input (Rudell & Fox, 1972). Unfortunately, one cannot be certain whether the cat is making adjustments which affect the background EEG prior to input. Therefore, it is not certain that these adjustments are made as a result of the input signal. If further work can show differential modification depending upon the identification of the input signal, as in the human studies, it would be possible to use this technique to provide information on the minimum time for the input to reach mechanisms influenced by the differential reward.

Goldberg and Wurtz (1972) have shown modification of cellular activity in the superior colliculus when a monkey is to attend to the input by making an eye movement to it. Parts of these data suggest that this modification takes place within 50 msec after input, long before the start of the eye movement. Unfortunately, only on the first trial of a new block can one be sure that the monkey has not made an adjustment prior to the input signal. However, these data suggest that the system which affects eye movements may be operating on an earlier part of the input signal than the system which affects P-300. A similar finding (Evarts, 1973) is that motor cortex units show a modification of response within 50 msec when sensory information is presented to the responding member. Once again the time course suggests that such responses are occurring prior to the system which produces P-300. Of course, it is plausible to suppose that both eye movements and highly automated limb movements may be made without involving systems which mediate awareness. Hopefully, future research will be able to use these techniques to understand the relation between automated and conscious processing.

Spatial Location

The brain system which gives rise to the signs associated with conscious processing appears to be closely related to the nonspecific portions of the cortex central to the primary sensory systems. Part of the evidence for this comes from the relationship between conscious processing and the amplitude of the EEG at the vertex. Of course, the vertex responses of the EEG come from a wide area and are influenced by the sensory projection systems, but the fact that the late vertex potentials of the EEG are similar irrespective of input modality and the finding that they are recorded most strongly from association area cortex (Goff, 1969; Vaughan & Ritter, 1970) suggest that they are heavily dependent upon association cortex.

There is also evidence from the work of Thompson and Bettinger (1970) that association cortex evoked potential and unit hash activity show close relationships to some aspects of attention. They found that in recording evoked potentials from cats, the introduction of a novel stimulus systematically reduced the size of the evoked potential to a probe stimulus which followed the novel stimulus. There are many problems in the interpretation of the Thompson and Bettinger data because of the difficulty of controlling various components of attention in these animals. However, they do show systematic relationships between the size of the probe response and the degree of attention which the subject paid to a novel stimulus.

In the conclusion of their study, Thompson and Bettinger present a hypothesis which is similar to the one we have been outlining in this section. They say:

> . . . a tentative hypothesis may be proposed relating cellular activity in association areas of the cortex to attention. The multiple unit measure suggests that the degree of ongoing cell activity in association cortex is markedly increased during novel stimulation. If the not unreasonable assumption is made that amplitude of the gross association response evoked by a probe stimulus is proportional to the number of cortical cells activated, then the following hypothesis will account for the data represented above: The number of cells of the non-specific association system available to be activated by a peripheral probe stimulus is decreased in proportion to the increase in ongoing cellular activity in the system resulting from novel stimulus presentation.

Data from intracellular electrodes and non-human organisms may make it possible to develop more detailed information about the anatomical and physiological connections of the system which we are discussing as being related to conscious awareness in man.

SUMMARY

Attention is not a single concept but the name of a complex field of study. A psychobiology of attention requires an analysis of the different psychological functions of attention so that the brain systems which subserve these functions can be separated and their relationships studied.

This chapter examines alertness, selection, and conscious processing as three component functions of attention. For each of these components effort is made to suggest possible mechanisms for further study.

Alertness varies over the course of the day. These tonic variations result in specific alterations of performance, which suggest the possibility of a shift in the threshold of primary cortical units due to changed levels of carbohydrate steroids. A warning signal produces a phasic shift in alertness which also alters performance. Behavioral data suggest that these phasic changes are of a different character than those which accompany the diurnal rhythm and might be mediated by a subcortical alerting system which acts upon association or motor units.

A warning signal may alert a subject and also provide him with selective information (set) about what is to occur. Data suggest that these two functions are mediated by independent brain systems. Selection may involve the channel of entry (stimulus set) or depend upon prior identification of the signal (response set). With rapid presentation a strong stimulus set results in a complete lack of awareness of nonselected items. Even under these conditions the unselected information is undergoing complex classification by the brain. The earliest point at which the evoked potential reveals evidence for differential handling of selected and unselected information is about 90 msec after input. This early effect has been reported only for stimulus set with rapid presentation of similar items.

Selection of input by either stimulus or response set leads to the admission of the selected item into a brain system of limited capacity. The limited capacity nature of this system is evidenced by widespread interference between signals demanding its use. The system appears to be intimately related to awareness and retention of the signal in memory. The system is often indexed by an EEG sign which seems to arise in association cortex, from 150 to 500 msec following input and, somewhat later, by very widespread autonomic changes. These indicants may provide a basis for further study of the anatomical and temporal locus of this brain system.

No doubt the study of attention is still in relatively primitive condition as a field of psychobiology. The current review suggests that, like sensory psychology, the field will benefit from close ties between human performance studies and electrophysiological explorations.

ACKNOWLEDGMENTS

The research reported in this paper was supported in part by the National Science Foundation Grant GB 21020. The author is grateful to Steven W. Keele and Raymond Klein for their assistance in the research described herein.

Preliminary versions of this paper were read by Professors S. Hillyard, W. Ritter, and E. Donchin. I am most grateful for their kind assistance which greatly improved the final version.

REFERENCES

Angel, A. The central control of sensory transmission and its possible relation to reaction time. *Acta Psychologica*, 1969, **30**, 339–357.

Bartelson, P. The time course of preparation. *Quarterly Journal of Experimental Psychology*, 1967, **19**, 272–279.

Blake, M. J. F. Temperature and time of day. In W. P. Colquhoun (Ed.), *Biological rhythms and human performance*. New York: Academic Press, 1971. Pp. 109–148.

Broadbent, D. E. *Decision and stress*. London: Academic Press, 1971.

Chapman, R. M. Evoked potentials of the brain related to thinking. In F. J. McGuigan and R. Schoonover (Eds.), *Psychophysiology of thinking*. New York: Academic Press, 1973. Pp. 69–108.

Cherry, E. C. Some experiments on the recognition of speech, with one and with two ears. *Journal of the Acoustical Society of America*, 1953, **25**, 975–979.

Colquhoun, W. P. (Ed.) *Biological rhythms and human performance*. New York: Academic Press, 1971.

Corteen, R. S., & Wood, B. Autonomic responses to shock-associated words in an unattended channel. *Journal of Experimental Psychology*, 1972, **94**, 308–313.

Davis, H., Osterhammel, P. A., Wier, C. C., & Gjerdingen, D. Slow vertex potentials: Interactions among auditory, tactile, electric and visual stimuli. *EEG and Clinical Neurophysiology*, 1972, **33**, 537–545.

Donald, M. W., & Goff, W. R. Attention-related increases in cortical responsivity dissociated from the contingent negative variation. *Science*, 1971, **172**, 1163–1166.

Donchin, E., & Cohen, L. Averaged evoked potentials and intramodality selective attention. *EEG and Clinical Neurophysiology*, 1967, **22**, 537–546.

Donchin, E., Kubovy, M., Kutas, M., Johnson, R., Jr., & Herning, R. Graded changes in evoked response (P300) amplitude as a function of cognitive activity. *Perception and Psychophysics*, 1973, **14**, 319–324.

Evans, C. R., & Mulholland, T. B. *Attention in neurophysiology*. New York: Appleton, 1969.

Evarts, E. V. Motor cortex reflexes associated with learned movement. *Science*, 1973, **179**, 501–503.

Esposito, N. J., & Pelton, L. H. Review of the measurement of semantic satiation. *Psychological Bulletin*, 1971, **75**, 330–346.

Fox, S. S., & Rudell, A. P. Operant controlled neural events: Function independence in behavioral coding by early and late components of visual cortical evoked response in cats. *Journal of Neurophysiology*, 1970, **33**, 548–561.

Fuster, J. M. Effects of stimulation of brain stem on tachistoscopic perception. *Science*, 1958, **127**, 150.

Gazzaniga, M. S., & Hillyard, S. A. Attention mechanisms following brain bisection. In S. Kornblum (Ed.), *Attention and performance IV*. New York: Academic Press, 1973.

Goldberg, M. E., & Wurtz, R. H. Activity of superior colliculus in behaving monkeys. II. Effect of attention on neuronal responses. *Journal of Neurophysiology*, 1972, **35**, 560–574.

Goff, W. R. Evoked potential correlates of perceptual organization in man. C. R. Evans and T. B. Mulholland. (Eds.), *Attention in neurophysiology*. New York: Appleton, 1969.

Groves, P. M., & Thompson, R. F. Habituation: A dual process theory. *Psychological Review*, 1970, **77**, 419–450.

Hartley, L. R. The effect of stimulus relevance on cortical evoked potentials. *Quarterly Journal of Experimental Psychology*, 1970, **22**, 531–546.

Heninger, G. R., McDonald, R. K., Goff, W. R., & Sollberger, A. Diurnal variations in the cerebral evoked response and EEG. *Archives of Neurology*, 1969, **21**, 330–337.

Henkin, R. I. The neuroendocrine control of perception. In *Perception and its disorders*. Research Publication ARNMD, Vol. 48, 1970, 54–107.

Hernández-Peón, R. Neurophysiological correlates of habituation and other manifestations of plastic inhibition. *EEG and Clinical Neurophysiology*, Supplement 13, 1960, 101–114.

Hillyard, S. A., Hink, R. F., Schwent, V. K., & Picton, T. W. Electrical signs of selective attention. *Science*, 1973, **182**, 177–179.

Hillyard, S. A., Squires, K., Bauer, J., & Lindsay, P. Evoked potential correlates of auditory signal detection. *Science*, 1971, **172**, 1357–1360.

Hintzman, D. L. Theoretical implications of the spacing effect. In R. Solso (Ed.), *Theories of cognitive psychology: The Loyola Symposium*. Washington, D.C.: Winston, 1974.

Jerison, H. J. Vigilance: A paradigm and physiological speculation. In A. Sanders (Ed.), *Attention and performance III*. Amsterdam: North Holland Publishing, 1970.

Kahneman, D. *Attention and effort*. New York: Prentice Hall, 1973.

Kahneman, D. Remarks on attention control. *Acta Psychologica*, 1970, **33**, 118–131.

Kahneman, D., Tursky, B., Shapiro, D., & Crider, A. Pupillary, heart rate and skin resistance changes during a mental task. *Journal of Experimental Psychology*, 1969, **79**, 164–167.

Karlin, L., Martz, M. J., Brauth, S., & Mordkoff, A. M. Auditory evoked potentials, motor potentials and reaction time. *EEG and Clinical Neurophysiology*, 1971, **31**, 129–136.

Keele, S. W. *Attention and human performance*. Pacific Palisades, Calif.: Goodyear, 1973.

Lackner, J. R., & Garrett, M. F. Resolving ambiguity effects of biasing context in the unattended ear. *Cognition*, 1973, **1**, 359–372.

Lansing, R. W., Schwartz, E., & Lindsley, D. Reaction time and EEG activation under alerted and non-alerted conditions. *Journal of Experimental Psychology*, 1959, **58**, 1–7.

Lewis, J. Semantic processing of unattended messages using dichotic listening. *Journal of Experimental Psychology*, 1970, **85**, 225–228.

Loeb, J. As cited in Welch, J. C. On measurement of mental activity through muscular activity and the determination of a constant of attention. *American Journal of Physiology*, 1898, **1**, 283–306.

Mackay, D. G. Aspects of the theory of comprehension, memory and attention. *Quarterly Journal of Experimental Psychology*, 1973, **25**, 22–40.

Mackworth, J. F. *Vigilance and habituation*. Harmondsworth: Penguin, 1969.

McGaugh, J. L. *Psychobiology*. New York: Academic Press, 1971.

Milner, P. M. *Physiological psychology*. New York: Holt, 1970.

Moray, N. *Attention: Selective processes in vision and hearing.* London: Hutchinson Educational, 1969.

Mostofsky, D. I. (Ed.). *Attention: Contemporary theory and analysis.* New York: Appleton, 1970.

Moruzzi, G., & Magoun, H. W. Brain stem reticular formation and activation of the EEG. *EEG and Clinical Neurophysiology,* 1949, **1**, 455–473.

Näätänen, R. Evoked potential, EEG and slow potential correlates of selective attention. *Acta Psychologica,* 1970, **33**, 178–192.

Nakai, Y., & Domino, E. F. Reticular facilitation of the V.S.R. by optic tract stimulation. *Experimental Neurology,* 1968, **22**, 532–544.

Pertkoff, G. T., Eik-nes, K., Nugent, C. A., Fred, H. L., Nimer, R. A., Rush, L., Samuels, L. T., and Tyler, F. M. Studies of the diurnal variation of plasma 17-hydroxycorticosteroids in man. *Journal of Clinical Endocrinology,* 1959, **19**, 432–443.

Posner, M. I. Abstraction and the process of recognition. In G. H. Bower and J. T. Spence (Eds.), *Psychology of learning and motivation III.* New York: Academic Press, 1969.

Posner, M. I., & Boies, S. J. Components of attention. *Psychological Review,* 1971, **78**, 391–408.

Posner, M. I., & Klein, R. M. On the functions of consciousness. In S. Kornblum (Ed.), *Attention and Performance IV.* New York: Academic Press, 1973. Pp. 21–34.

Posner, M. I., Klein, R., Summers, J., & Buggie, S. On the selection of signals. *Memory and Cognition,* 1973, **1**, 2–12.

Posner, M. I., & Snyder, C. R. R. Facilitation and inhibition in the processing of signals. In P. M. A. Rabbitt (Ed.), *Attention and performance V.* New York: Academic Press, in press.

Pushkin, V. N. Vigilance as a function of the strength of the nervous system. In V. Nebylitsyn and J. Gray (Eds.), *Biological bases of individual behavior.* New York: Academic Press, 1972. Pp. 310–324.

Rebert, C. S. & Tecce, J. J. A summary of CNV and reaction time. In W. C. Callum and J. R. Knott (Eds.), *Event related slow potentials of the brain and their relation to behavior.* Amsterdam: Elsevier, 1973.

Regan, D. *Evoked potentials.* New York: Wiley, 1972.

Requin, J. Some data on neurophysiological processes involved in the preparatory motor activity to reaction performance. *Acta Psychologica,* 1969, **30**, 358–367.

Ritter, W., Simson, R., & Vaughan, H. G., Jr. Association cortex potentials and reaction time in auditory discrimination. *EEG and Clinical Neurophysiology,* 1972, **33**, 547–555.

Ritter, W., Vaughan, H. G., Jr. Averaged evoked responses in viligance and discrimination: A reassessment. *Science,* 1969, **164**, 326–328.

Routtenberg, A. Stimulus processing and response execution: A neurobehavioral theory. *Physiology and Behavior,* 1971, **6**, 589–596.

Rudell, A. P., & Fox, S. S. Operant controlled neural event: Functional bioelectric coding in primary components of cortical evoked potential in cat brain. *Journal of Neurophysiology,* 1972, **35**, 892–902.

Shallice, T. Dual functions of consciousness. *Psychological Review,* 1972, **79**, 383–393.

Sharpless, S., & Jasper, H. Habituation of the arousal reaction. *Brain,* 1956, **79**, 655–680.

Smith, D. B. D., Donchin, E., Cohen, L., & Starr, A. Auditory averaged evoked potentials in man during selective binaural listening. *EEG and Clinical Neurophysiology*, 1970, **28**, 146–152.

Sokolov, E. N. *Perception and the conditioned reflex*. New York: Macmillan, 1963.

Squires, K. C., Hillyard, S. A., & Lindsay, P. H. Cortical potentials evoked by confirming and disconfirming feedback following an auditory discrimination. *Perception and Psychophysics*, 1973, **13**, 25–31.

Sternberg, S. The discovery of processing stages. *Acta Psychologica*, 1969, **30**, 276–315.

Sutton, S., Braren, M., & Zubin, J. Evoked potential correlates of stimulus uncertainty. *Science*, 1965, **150**, 1187–1188.

Thompson, R. F. *Foundations of physiological psychology*. New York: Harper, 1967.

Thompson, R. F., & Bettinger, L. A. Neural substrates of attention. In D. L. Mostofsky (Ed.), *Attention: Contemporary theory and analysis*. New York: Appleton, 1970.

Thompson, R. F., & Spencer, W. A. Habituation: A model phenomenon for the study of neuronal substrates of behavior. *Psychological Review*, 1966, **73**, 16–43.

Treisman, A. Monitoring and storage of irrelevant messages in selective attention. *Journal of Verbal Learning and Verbal Behavior*, 1964, **3**, 449–459.

Treisman, A., & Geffen, G. Selective attention: Perception or response? *Quarterly Journal of Experimental Psychology*, 1967, **19**, 1–17.

Vaughan, H. G., & Ritter, W. The sources of auditory evoked responses recorded from the human scalp. *EEG and Clinical Neurophysiology*, 1970, **28**, 360–367.

Walter, W. The convergence and interaction of visual, auditory and tactile responses in human non-specific cortex. *Annals, New York Academy of Science*, 1964, **112**, 320–361.

Warren, R. Stimulus encoding and memory. *Journal of Experimental Psychology*, 1972, **94**, 90–100.

Webb, R. A., & Obrist, P. S. The physiological concomitants of reaction time performance as a function of preparatory interval and preparatory interval series. *Psychophysiology*, 1970, **6**, 389–403.

Weinberg, H., Walter, W. G., & Crow, H. J. Intracerebral events in humans related to real and imaginary stimuli. *EEG and Clinical Neurophysiology*, 1970, **29**, 1–9.

Wilkinson, R. T. Evoked response and reaction time. *Acta Psychologica*, 1967, **27**, 235–245.

Wilkinson, R. T., & Lee, M. V. Auditory evoked potentials and selective attention. *EEG and Clinical Neurophysiology*, 1972, 33, 411–418.

Worden, F. G. Attention and auditory electrophysiology. In E. Steller and J. Sprague (Eds.), *Progress in physiological psychology*. Vol. 1. 1966. Pp. 45–116.

Wright, P., & Kahneman, D. Evidence for alternative strategies for sentence retention. *Quarterly Journal of Experimental Psychology*, 1971, **23**, 197–213.

Chapter 16

Some Trends in the Neurological Study of Learning

G. BERLUCCHI

Istituto di Fisiologia dell'Università di Pisa and Laboratorio di Neurofisiologia del CNR, Pisa, Italy

H. A. BUCHTEL

The National Hospital for Nervous Diseases, London, England

INTRODUCTION

Generally speaking, learning may be considered to be a process by which an organism modifies its behavior or acquires entirely new behavioral patterns as a result of interactions with the environment. A more specific definition must be based, by and large, on exclusion: growth; maturation; aging; fatigue; receptor adaptation; changes in arousal, attention, motivation; and, obviously, disease, are all factors which can modify behavior but which are usually distinguished from learning. Yet these factors are often as difficult to define as those thought to indicate learning, and one can question the legitimacy and usefulness of a sharp separation between learning and all other instances of behavioral modifiability. Hebb (1953), for example, has stressed the fact that many forms of behavior which cannot be shown to be a direct product of learning are nevertheless crucially dependent for their existence on previous learning. An example is the fear of strangers which develops during infancy and which clearly depends upon the child's previous learning experience and recognition of faces which are "familiar."

On these grounds, Hebb has argued forcefully that maturation of behavior requires not just time and adequate nutrition but also early learning, which depends upon exposure to the normal environment of the species. Thus, being crucially related to previous experience, maturation cannot be easily differentiated from learning (and, in order to avoid circularity, learning cannot be defined by comparison with maturation, which is itself not due to learning).

Other more restrictive definitions of learning have limited the use of the term to particular systematic relationships between stimuli and behavioral responses. Such definitions originally served as paradigms for the laboratory study of behavior but they have subsequently been applied to analysis of behavior in general. According to these definitions, there are essentially three types of learning: (1) habituation, i.e., the waning of a response upon repeated and monotonous presentation of the stimulus evoking it, in the absence of any receptor adaptation or effector fatigue; (2) classical conditioning, i.e., the transfer of a response from one stimulus to the other when the two stimuli occur together repeatedly in a particular order; and (3) operant conditioning, i.e., the changing probability of emitting responses which are reliably followed by significant events in the environment ("reinforcers": food, shock, etc., or the special case of brain stimulation).

These definitions of learning have obvious advantages, but it is doubtful that they are exhaustive. In analyzing the role of learning in language, for instance, Hebb, Lambert, and Tucker (1971) underlined the importance of forms of learning which are not based on practice and which occur without discernible primary or secondary reinforcement. Such forms of learning, prominent in man but also demonstrable in other animals, cannot be satisfactorily interpreted within the framework of the present conditioning paradigms.

Conversely, there are changes in behavior which could, at least in principle, be attributed to processes of conditioning, and therefore learning, but are usually not. Good examples are the reorganization of behavior following central or peripheral nervous system damage, and sensory-motor rearrangements in the case of systematic distortion of sensory input. As an instance of the former, it has been shown that motor control of the eyes and posture is severely disturbed by unilateral labyrinthectomy, but eventually returns to normal; removal of the remaining laby-rinth at this time upsets the regained balance anew, but in the opposite direction (von Bechterew, 1909). Similarly, wearing inverting prisms in front of the eyes initially disrupts visually guided motor performance, but an almost complete readjustment occurs through practice; removing the prisms after readjustment results in a motor disruption opposite in sign to the initial one (for references, see Teuber, 1960).

Other phenomena of behavioral plasticity whose relation to learning is debated are those which result from use and disuse, such as denerva-tion hypersensitivity (see Sharpless, 1964) or the reorganization of mono-synaptic connections following experimental transpositions of nerve fibers (Eccles, Eccles, Shealy, & Willis, 1962).

However defined, the term "learning" must cover a large class of heterogeneous phenomena. Nevertheless, the changes in behavior from which one infers learning presumably depend upon modifications of the central nervous system, at least in vertebrates. The finding of a common neural mechanism responsible for all of them would un-doubtedly illuminate and simplify the analysis of learning in the same way as the discovery of the electrical properties of cell membranes clari-fied and facilitated the analysis of excitability in nervous and muscle tissue. In fact, our understanding of learning cannot be complete until we are able to describe the changes in nervous organization and activity which underlie the behavioral phenomena identified as learning. This is far from being possible at the moment and one must adopt more indirect approaches to the problem.

APPROACHES TO THE STUDY OF THE NEURAL BASIS OF LEARNING

A useful indirect approach to this question has been to look for the plasticity of neural functioning within the nervous system itself. The purpose of such an analysis is to show, by direct observation, active and relatively enduring changes in neural function in response either to specific experiences of the whole organism or to direct manipulations of nervous tissue. Relevant to the problem of learning are those ex-periments in which neural activities such as electroencephalographic phenomena (EEG), cortical or subcortical evoked potentials, or the dis-charge of single neurons, are directly submitted to habituation or condi-tioning procedures. In some of these experiments, unanesthetized animals and natural stimuli are used, and the neural activities may not be associ-ated with a behavioral response (see, e.g., Black, 1972). In other experiments, mostly employing a simplified preparation or animals with a primitive nervous system, stimuli are applied directly to the nervous tissue and no behavior at all may be involved, except of course, the "behavior" of the nervous system itself (Kandel & Spencer, 1968; Doty, 1969; Horn & Hinde, 1970; von Baumgarten, 1970). Ample evidence for habituation and conditioning of neural activities has been obtained in both kinds of experiments, but the correlation between such signs of neural plasticity and the behavioral phenomena of learning is still lacking or, at best, very indirect. Further, once neural plasticity has been demonstrated, there remains the problem of accounting for it in terms of known properties of cellular and subcellular constituents of the nervous tissue, and eventually in physicochemical terms.

The neuron theory has provided an adequate framework for the so-called "synaptic" hypotheses of neural, and therefore behavioral, plasticity. It is assumed by these hypotheses that plastic changes in the nervous system can all be accounted for by changes in the facility with which neurons communicate with one another by synaptic transmission. So far as we know from present evidence, synaptic transmission in most, if not all, synapses in the mammalian nervous system is a graded process. This involves the release by the presynaptic neuron of variable amounts of a chemical mediator which acts on some special sites on the membrane of the postsynaptic neuron, thereby changing its electrical state. Hypothetical processes which might account for the modifications in activity of neuronal nets possibly underlying learning include: changes in the proximity of the presynaptic and postsynaptic neurons at the synapse, due to outgrowth of axonal or dendritic processes; changes in the area of specific contacts between presynaptic and postsynaptic neurons due to expansion of the presynaptic axonal terminals or of the receptive sites on the postsynaptic membrane; changes in the synthesis and release of the mediator by the presynaptic neuron; changes in the mediator–binding action of the postsynaptic neuron; and the like. That synaptic transmission can be altered for a considerable amount of time by tetanic stimulation, habituating stimulation, disuse, and, in limited instances in simple invertebrate neural systems, by conditioning procedures, is beyond question (see Kandel & Spencer, 1968; Horn & Hinde, 1970; von Baumgarten, 1970). However, it is still not known how such synaptic changes are brought about. In any case, it would seem useful to postulate that conditioning procedures give rise to synaptic changes which are different from those underlying, for example, habituation and the effects of disuse or abnormally intense use (posttetanic potentiation). Repetitious low-frequency stimulation of multisynaptic pathways to motoneurons in the cat spinal cord leads to a decrease in motoneuronal output (a phenomenon closely resembling behavioral habituation), most probably due to synaptic depression or buildup of inhibitory synaptic activity (Wall, 1970). Disuse and tetanic stimulation of monosynaptic pathways to motoneurons in the cat spinal cord produce long-lasting increases in motoneuronal response to subsequent stimulation of such pathways, most probably because of increased efficiency of the synapse (Spencer & April, 1970).

All these plastic synaptic changes depend, in one way or another, on antecedent activity along the same neuronal pathways as those which are ultimately affected by changes in synaptic transmission. In conditioning, the situation is different. In classical conditioning, a stimulus (conditioned stimulus) becomes capable of producing a response if it is followed repeatedly by another stimulus (unconditioned stimulus) which consistently evokes the response. This can be explained by assuming that a potential neuronal mechanism for mediating the response to the conditioned stimulus is made functional when its subliminal response-producing activity is *followed by*, as opposed to being preceded by, the activity elicited by the unconditioned stimulus. Similarly, in operant conditioning, facilitation or hindrance, mediated by synaptic modifications, of neural activity underlying a particular emitted behavior, depends on *subsequent* neural events associated with reinforcement. The putative synaptic mechanisms underlying this "temporal paradox" of conditioning (see Doty, 1969) must therefore be different, at least in part, from those of habituation, posttetanic potentiation, and disuse. For the same temporal reason, they must also be different from, or involve more than, the simple mechanisms of spatial and temporal summation at the synapse.

Figure 1 provides a list of possible synaptic mechanisms which may be responsible for conditioning. These models have been designed to fit present data on behavioral and neural conditioning but, at the moment, they are only useful speculations since there is no direct evidence that such synaptic mechanisms actually exist. It should also be kept in mind that no single synaptic mechanism is likely to explain all forms of neural plasticity or to provide a substrate for all forms of learning. In fact, it is quite possible that some other mechanism or mechanisms will have to be implicated, for example, modifications of internal neuronal pacemakers (Kandel & Spencer, 1968; von Baumgarten, 1970), or other forms of nervous organization which are not envisaged by the neuron theory (see the field or systems theories of, e.g., John, 1967; Pribram, 1971).

Another approach to the study of the neural basis of learning is the search for changes in activity of the nervous system during the learning process. These "correlates" of learning (for a review see John, 1967;

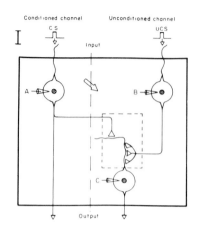

I

Conditioned channel   Unconditioned channel

C S   UCS

Input

A   B

C

Output

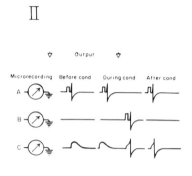

II

Output

Microrecording	Before cond	During cond	After cond
A			
B			
C			

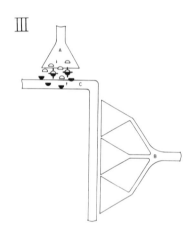

III

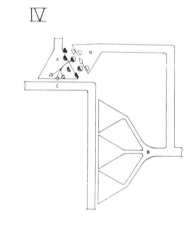

IV

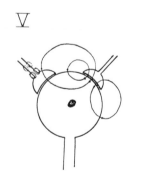

V

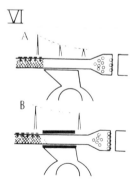

VI

VII

486

Figure 1. I and II: Classical conditioning of a single neuron. Experiments of this kind have been performed on identified neurons in the abdominal ganglion of the mollusk *Aplysia* (see Kandel & Spencer, 1968; von Baumgarten, 1970). The conditioning stimulus is an electrical shock applied to neuron A. The action potential produced by this stimulus (see IIa) causes a postsynaptic excitatory potential in neuron C which is below threshold for the emission of an action potential (see IIc). The unconditioned stimulus is an electrical shock applied to neuron B, and the action potential which is produced (see IIb) is sufficient to produce an action potential in C (see IIc). If A and then B are repeatedly stimulated in that order, neuron C becomes "conditioned" to respond to stimulation of A with the response that it originally gave only to stimulation of B, that is, an action potential (see IIc). This change in responsiveness in neuron C may last for several minutes or even hours, and is extinguished by stimulating A several times without the matching stimulation of B.

III and IV: Possible synaptic changes of an essentially chemical nature underlying the neuronal conditioning illustrated in I and II. These parts of the figure are expanded views of the area outlined by a broken line in I. It is hypothesized that when an action potential arrives at the axon terminal, the synaptic mediator released by A (open half-circles) reacts with a substance released by the postsynaptic neuron C when it is discharged by B (hatched half-circles). The combination of the presynaptic and postsynaptic substance would induce changes in the presynaptic and/or postsynaptic membranes, thereby increasing synaptic efficiency. The reaction is possible only when A and B fire almost simultaneously, with the action potential in A preceding that in C. In IV an identical interaction is supposed to occur between A and B, resulting in an increased synaptic effect of A and C by means of heterosynaptic facilitation. The schemes in I, II, III, and IV have been taken from von Baumgarten (1970); the synaptic models of conditioning illustrated in III and IV are essentially the same as suggested by Doty (1969).

V: Possible synaptic changes, essentially electrical in nature, suggested by Milner (1960) to underlie neuronal conditioning. Milner hypothesized that the synaptic terminal on the right fires the neuron (unconditioned stimulus) and that the current flowing through the membrane invades the synaptic terminal on the left. If it is assumed that an action potential running along the fiber of this terminal (conditioning stimulus) dies out before reaching the terminal itself (because, for example, of the increase in membrane surface area at the terminal), the current from the depolarized postsynaptic neuron may help the action potential to get past the critical point. This hypothesis, however, does not account for the temporal paradox of conditioning. In order to do so, it should also postulate some relatively permanent change in the membrane of the initially ineffective synaptic terminal as a result of the combination of the two currents.

VI: Glial hypothesis of synaptic changes in conditioning according to Roitbak (1970). It is assumed that the pathway of the unconditioned stimulus and that of the conditioned stimulus converge on common neurons. Action potentials traveling to the common neurons via the pathway of the unconditioned stimulus are a signal for mobilization of oligodendrocytes in the area of the neurons in common. The fiber terminals of the conditioned stimulus pathway are thought to be bare of glia in the region of their contact with the common neurons (as in A). Due to the shunting of the depolarizing current prior to conditioning, the action potential dies out before reaching the terminals, and the small amount of current which does arrive releases only a small amount of mediator (open circles in A). It is further assumed that if the bare terminals have just been active, the oligodendrocytes mobilized by the unconditioned stimulus grow around them. This provides a better membrane insulation and causes a large amount of mediator-releasing current to reach the terminals (B).

VII: Glial hypothesis of synaptic changes in learning according to Pribram (Karl H. Pribram, *Languages of the brain*, © 1971. By permission of Prentice-Hall, Inc., Englewood Cliffs, N. J.). The original hypothesis simply assumes that RNA released by the excited neuronal terminal induces cell division in the glial cell, thus making space for the growth of the axon toward the neuron beyond. The hypothesis can be applied to conditioning if it is assumed that the glial changes are possible only when the axon terminal and the neuron beyond the glial cell become active together in a fixed temporal sequence.

Thompson, Patterson, & Teyler, 1972; Leiman & Christian, 1973) are interesting in their own right. Yet usually it cannot be determined whether they are neural causes or just a consequence of learning. Or, they may simply accompany learning because they are associated with, for example, motivational or attentional variables implicit in the learning situation. Yet another approach is the investigation of the effects of neural damage on learning. We shall consider only one aspect of this approach: the search for neural mechanisms sufficient for learning in isolated parts of the neuraxis.

LEARNING AFTER REMOVAL OF NERVOUS TISSUE

We shall describe here some experiments in which learning and retention have been studied in mammals following extirpation or destruction of a large part of the neuraxis. The fundamental question in this kind of experiment is the following: Does learning, or a particular type of learning, require the integrity of a specific section of the neuraxis? The answers which experimental data provide are not as straightforward as they may at first seem. If learning is abolished by destruction or functional elimination of one part of the nervous system, it does not necessarily follow that the remaining nervous tissue is *normally* incapable of mediating such learning, for its functioning may have been disrupted or significantly modified by the lack of activity of the destroyed area. For the same reason the presence of learning after central nervous system lesion is not necessarily an indication of the normal activities of the intact tissue. In the latter case, of course, one is allowed to conclude that such tissues must be endowed with mechanisms sufficient for mediating learning, whether or not they usually perform that function. With these reservations in mind, we may start considering some of the recent data.

The problem of whether the cortex is necessary for learning has a long history. Pavlov originally claimed that decorticated dogs could not be conditioned by his method, but subsequent experiments (see Russell, 1971) proved beyond doubt that complete decortication does not abolish either acquisition or retention of classical conditioning. However, until recently there has been uncertainty as to whether or not decorticated animals can learn by means of operant conditioning. Russell (1971) has suggested that such learning is impossible without a functioning cortex but Oakley (1971) was able to condition two rabbits which had an almost complete decortication (less than 5% cortical tissue remaining) to press a treadle for food. When the subjects had to make several responses in order to receive a single reward, both decorticated and normal animals increased their pressing rates but the decorticated subjects stopped responding much sooner than the control subjects as the response–reinforcement ratio became gradually larger.

Huston and Borbély (1973) have recently reported successful operant conditioning when using hypothalamic stimulation as a reward in rats which had undergone ablation of cortex, hippocampus, septum, striatum, and part of the amygdalae. The frequency of motor activities such as moving the head, raising the tail, and sitting up could be increased by applying rewarding stimulation following their appearance. The conditioned responses, however, showed a marked resistance to extinction, possibly because of the hippocampal lesion (Kimble, 1970). The unequivocal conclusion is, therefore, that the cerebral cortex, in rats and rabbits at least, is not necessary for conditioning with the operant technique.

Decerebrated animals have been a favorite preparation of neurophysiologists since the time of Sherrington, but Kandel and Spencer (1968) have rightly pointed out how little they have been used for studies of learning. There are, in fact, data which indicate that such animals are capable of learning by the classical conditioning technique. Markel and Adam (1965) have conditioned cats, decerebrated at the mesencephalic level, to transfer a respiratory response from stimulation of the vagus nerve to stimulation of the pelvic nerve when the latter stimulus was repeatedly applied just before the former. Extinction of the acquired response was clearly possible, and adequate controls demonstrated that the conditioning effects were not due to sensitization or to repeated stimulation of the pelvic nerve alone. Intact animals which had been conditioned by the same procedure displayed the conditioned respiratory response on the first pelvic nerve stimulation after decerebration, indicating that retention did not require the telencephalon or the diencephalon. Similarly, decerebrated rats have been conditioned to display a respiratory response to an acoustic stimulus (metronome or 200-Hz tone) by applying such a stimulus just before shocking the hind limbs. Again, responses conditioned in normal animals were still present after decerebration (Markel & Adam, 1969). Even more interesting is the finding that decerebrated rats can be differentially conditioned to discriminate between tones of 200 and 2000 Hz (Markel, Anda, Juhasz, & Adam, 1969). Thus it is clear that various aspects of classical conditioning are possible even without the cerebrum.

The conditioning of responses mediated by the spinal cord in animals with a surgical separation of the cord and the encephalon has also been the object of several investigations. Habituation (for references, see Griffin, 1970) and classical conditioning (for references, see Horn, 1970) have been repeatedly reported to occur in such preparations and recently it has even been shown that responses mediated solely by the spinal cord can also be conditioned using the operant technique. This was a quite unexpected finding since operant conditioning is usually thought to be a "higher" form of learning and because the activity of the "motivation area" in the diencephalon and brain stem is generally regarded as essential for the reinforcement effects of rewards and punishments. Yet the evidence for operant conditioning mediated by the cord is unambiguous: Buerger and Fennessy (1971) have cut the rat spinal cord at the lumbar level and applied shocks to one of the hind limbs whenever it was in an extended position. Each time a shock was given, a yoked control animal, also spinalized, received an identical shock regardless of its hind-limb position. Animals in the experimental group learned to keep their hind limbs flexed, thus avoiding shock, while control group animals did not show any tendency toward hind-limb flexion. The behavior of the experimental group animals was similar, in principle, to that of neurally intact animals learning to avoid a painful stimulus, and the conclusion that operant conditioning was involved seems inescapable. Similar results have been obtained in the spinal frog (Buerger & Farel, 1972).

It comes as no surprise that learning occurs as well in cerebral tissue anterior to such sections, although there are of course technical problems in presenting stimuli and registering responses. Even so, habituation and both classical and operant forms of learning have been demonstrated in animals with a complete section through the midpontine pretrigeminal level of the brain stem. In such animals, the sensory inputs are limited to olfaction and vision, and the only observable motor responses are vertical eye movements and changes in pupil diameter. Habituation to visual (Affanni, Marchiafava, & Zernicki, 1962) and olfactory (Zernicki, Dreher, & Krzywosinski, 1967) stimuli is clearly present in such animals, and differential classical conditioning (Zernicki & Osetowska, 1963) as well as operant conditioning (Shlaer & Myers, 1972) have been successful. In the latter case, the difficulty of presenting rewards was overcome by stimulating the lateral hypothalamus which, in normal cats, can be used as a positive reinforcer.

It is clear from the experiments described that all sections of the mammalian neuraxis have the capacity to sustain plastic changes in behavior which fulfill the most rigorous criteria for learning. Of course, this is not to say that there are no differences between the encephalon and the spinal cord in terms of learning efficiency. Comparative studies of this kind are lacking and are quite difficult to design because of wide differences in both sensory input and the responses observed.

ENVIRONMENT, LEARNING, AND THE BRAIN

Many studies of central nervous system plasticity have dealt with the changes of nervous structure and function due to sensory deprivation or distortion of sensory input during the immediate postnatal period, that is, at a time when the nervous system is thought to be highly susceptible to the effects of modifying the external environment. Experiments on changes in central receptive fields due to manipulating visual input in neonatal animals are described in this book by Hirsch and Jacobson. In the following, we shall describe some studies of nervous changes which have been produced in rats as a result of different environmental conditions during rearing, and in cats as a result of experimental strabismus. The reasons for describing these apparently unrelated experiments together will become clear in the following discussion.

Rosenzweig and collaborators (see Rosenzweig, 1966, 1971) have studied the brain and behavior of rats raised for varying amounts of time after weaning in one of three different environmental conditions: an "enriched" condition, an "impoverished" condition, and a "standard" colony condition. In the first condition, animals lived together with several cage mates in a complex and changing environment, and were regularly subjected to formal behavioral testing. Animals raised in the impoverished condition lived in individual, empty cages with solid walls in a quiet, dimly illuminated room. They were fed in the same way as the rats in the enriched condition, but received no behavioral training and as little sensory stimulation as possible. Standard colony conditions involved living in groups of three to a cage without any special training or handling.

The brains of the rats raised in the enriched condition had a heavier and thicker cerebral cortex, particularly in the occipital region, and a higher ratio of cortical weight to that of the rest of the brain. In addition, rats raised in the enriched condition showed a higher activity of the enzymes acetylcholinesterase, cholinesterase, and hexokinase; fewer and larger neurons; and more glial cells in the occipital cortex. The standard colony rats exhibited characteristics intermediate between those of the other two groups, but their values tended to resemble those of the impoverished condition group. These brain differences are almost certainly not due to differences in activity per se (Zolman & Morimoto, 1965).

Behaviorally, the rats raised in the enriched condition were more successful than the rats in the other two groups in solving problems, such as learning the route through a maze or learning to reverse the response to stimuli in a two-choice discrimination. However, there is no simple relation between the brain data and the behavioral data since the structural and physiological differences were still found when the duration of treatment was shortened and also when the treatment was begun at a later time, while under such conditions some of the behavioral tests no longer discriminated between the groups or else gave results in a direction opposite to that expected. Such a dissociation between the effects on brain and behavior is probably a function of the particular behavioral tests which are chosen.

The data previously described on the effects of early environment suggest that complex stimulation during development results in an elaboration of nervous system connections which later in life may account for superior learning ability. However, on the basis of ultrastructural evidence to be described in detail later, Rosenzweig and colleagues (Rosenzweig, Møllgaard, Diamond, & Bennett, 1972) have postulated that negative as well as positive synaptic changes may result from environmental stimulation.

At the electrophysiological level, it has been seen that kittens raised with one eye horizontally deviated develop a visual cortex characterized by monocularly innervated cells, although binocular organization had presumably been present at birth (Hubel & Wiesel, 1965). This change from binocular to monocular organization of the cortex is, of course, an adaptive response to the potentially disruptive effects of double vision due to convergence on cortical units of conflicting views of the world. The "learning" in this case clearly involves the dropping out of synaptic connections, perhaps by some kind of active decrease in synaptic efficiency.

While learning in such contexts should be considered as a part of the maturational process in the Hebbian sense (see p. 482), similar brain changes can also be induced in the adult. The effects of an enriched environment have been observed in rats as old as 1 year at the beginning of the treatment (Riege, 1971) and are shown not only as a differential positive influence on the weight of cortical tissues but also as a facilitative influence on behavioral abilities. The opposite effect, that is, a time-dependent decrease in cortex values, was seen in the animals living in the impoverished environment. These changes are apparently largely restricted to the cortex since the weights of subcortical areas remained constant. It does, however, seem unlikely that subcortical centers are as rigidly organized as this result would suggest and subtle effects may have been overlooked.

A decrease in synaptic efficiency as a basis for learning has been clearly seen in the adult as well. About half of a group of cats which had undergone immobilization of one of the eyes by denervating the extraocular muscles developed a predominantly monocular innervation of neurons in the visual cortex (Buchtel, Berlucchi, & Mascetti, 1972), just as did the kittens in the Hubel and Wiesel study, previously cited. Other adult cats, again with paralytic strabismus acquired in adulthood, appeared in behavioral tests to have chosen one of the eyes as the "good" eye and were rather poor in performing tasks when using the other eye alone (Buchtel *et al.*, 1972). The visual cortex of such cats was predominantly binocular but in a modified way which seems to have involved both the dropping out of certain synapses and the strengthening or establishing (by sprouting, perhaps) of others. In these latter cats, the visual receptive fields of one or both of the eyes had apparently shifted horizontally, resulting in a total displacement of about 5 degrees of visual angle, which under certain conditions might allow a certain amount of binocular fusion (Berlucchi & Buchtel, 1973). In man there exists the visual anomaly of eccentric fixation which may well be based on a cortical reorganization of the kind seen in these cats. The dynamics of this reorganization are still being studied but they may be similar to those underlying the invasion of denervated thalamus by input from intact sensory relays in the rat (Wall & Egger, 1971).

If one believes that these morphological changes are in some way related to learning, and also that learning implies modifications in the synaptic relations between neurons, the next logical step is to deepen the level of structural analysis and see whether brains of animals raised under different conditions differ in terms of synaptic characteristics. At the light microscopic level, early postnatal stress (Schapiro & Vukovich, 1970) and an enriched environment after weaning (Greenough & Wolkman, 1973) increase, respectively, the number of dendritic spines and the dendritic branching of cells in the rat cortex. It is not yet clear whether these effects represent an acceleration of maturation or the actual establishment of a lasting improvement which could account for superior learning later in life.

At the ultrastructural level, Møllgaard, Diamond, Bennett, Rosenzweig, & Lindner (1971) have examined with the electron microscope the neuropil of Layer III of the occipital cortex in rats raised for 30 days after weaning under enriched or impoverished conditions. Measurements of synaptic relations were limited to the asymmetrical axodendritic synapse within Layer III. The enriched condition rats had 67% fewer synapses than the impoverished condition rats, but the average length of the synaptic contact was increased by 52%. The authors hypothesize that the presence of many small synapses is characteristic of an immature brain, and that fewer, larger synapses result from learning and experience which potentiate "useful" pathways and their connections while suppressing others.

SLEEP AS A PATTERN OF BRAIN ORGANIZATION NECESSARY FOR FIXATION OF LEARNING

In recent years it has become popular to regard sleep as important for the "consolidation" of learning, that is, for the transformation of recently learned material into a durable form. Paradoxical sleep, characterized by a waking EEG and occasional rapid eye movements and body jerks, has been singled out as being especially important in this regard (see chapter by Jouvet). It has been shown that the percentage of sleep time taken up by the paradoxical form may decrease owing to environmental deprivation (McGinty, 1969) or may increase after a period of intense learning (Lucero, 1970); and also that deprivation of paradoxical sleep may lead to poor acquisition (Stern, 1969) or retention (Empson & Clarke, 1970) of several kinds of learning. Changes in the percentage of paradoxical sleep are almost certainly not due to amounts of exercise or activity per se (Hobson, 1968; Lucero, 1970). Unfortunately, this subject remains somewhat controversial since other researchers have been unable to detect the effects of stimulation on paradoxical sleep (Webb & Friedman, 1971; Tagney, 1973) or of deprivation of paradoxical sleep on subsequent behavior (e.g., Albert, Cicala, & Siegel, 1970). That lasting neural modifications may depend upon, or at any rate be highly correlated with, sleep mechanisms is reinforced by the observation (Roffwarg, Muzio, & Dement, 1966) that the amount of paradoxical sleep is quite high in the infant (50%) and becomes successively less (about 25% in the adult) as maturation progresses. This change is presumed to reflect the diminishing amount of new information which must be processed as one becomes older. Further indirect support for the hypothesis comes from the finding of a relatively low percentage of paradoxical sleep in mentally retarded children and adults (see Feinberg, Braun, & Shulman, 1968).

Such findings naturally raise the question of what kinds of neural activities are actually necessary for learning and long-term retention. Again, the answer to this question is likely to be that no simple formulation is possible. It is clear that certain states favor or discourage certain kinds of learning: There is almost certainly an optimal amount of arousal for facilitating learning, and coma and sleep have a demonstrable retarding influence (learning probably does occur during sleep but it is mostly restricted to material presented during light sleep; Evans & Orchard, 1969). On the other hand, since learning can be mediated solely by the spinal cord which shows no signs of a sleep–waking cycle after its separation from the encephalon, it follows that active brain processes, such as those present during paradoxical sleep, are not necessary for learning or retention in vertebrates, provided that receptor and effector mechanisms are intact.

CONCLUSION

This review of some current trends in the study of the neural bases of learning has been, by necessity, idiosyncratic. The reader who wants a more comprehensive review is referred to books by Horn and Hinde (1970), Milner (1970), and Deutsch (1973), and the review article by Thompson *et al.*, (1972). We have not mentioned, for example, the important research on the biochemical basis of memory (see Ungar, 1970; and the chapter by Entingh *et al.* in this book) or on the effects of restricted lesions, a research topic which has expanded enormously in the past few years. Nevertheless, from this abbreviated review one should be able to appreciate that there have been significant advances in recent years, of which one of the most influential has perhaps been the widespread acceptance by physiologists and biochemists that the real function of the nervous system is to control behavior and mediate learning, and not simply to produce electrical signals or to secrete chemically interesting compounds.

ACKNOWLEDGMENT

The preparation of this article was aided by Grant No. 7001687/18 from the Consiglio Nazionale delle Ricerche, Rome.

REFERENCES

Affanni, J., Marchiafava, P. L., & Zernicki, B. Orientation reflexes in the midpontine pretrigeminal cat. *Archives italiennes de Biologie*, 1962, **100**, 297–304.

Albert, I., Cicala, G. A., & Siegel, J. The behavioral effects of REM deprivation in rats. *Psychophysiology*, 1970, **6**, 550–560.

Berlucchi, G., & Buchtel, H. A. Anomalous retinal correspondence in cats. *Brain Research*, 1973, **49**, 505–506.

Black, A. H. The operant conditioning of central nervous system electrical activity. In G. H. Bower (Ed.), *The psychology of learning and motivation.* New York: Academic Press, 1972.

Buchtel, H. A., Berlucchi, G., & Mascetti, G. G. Modification in visual perception and learning following immobilization of one eye in cats. *Brain Research*, 1972, 37, 355–356.

Buerger, A. A., & Fennessy, A. Long term alteration of leg position due to shock avoidance by spinal rats. *Experimental Neurology*, 1971, 30, 195–211.

Buerger, A. A., & Farel, P. B. Instrumental conditioning of leg position in chronic spinal frog: Before and after sciatic section. *Brain Research*, 1972, 47, 345–351.

Deutsch, J. A. (Ed.) *The physiological basis of memory*. New York: Academic Press, 1973.

Doty, R. W. Electrical stimulation of the brain in behavioral context. *Annual Review of Psychology*, 1969, 20, 289–320.

Eccles, J. C., Eccles, R. M., Shealy, C. N., & Willis, W. D. Experiments utilizing monosynaptic excitatory action on motorneurons for testing hypotheses relating to specificity of neuronal connections. *Journal of Neurophysiology*, 1962, 25, 559–580.

Empson, J. A. C., & Clarke, P. R. F. Rapid eye movements and remembering. *Nature*, 1970, 227, 287–288.

Evans, F. J., & Orchard, W. Sleep learning: The successful waking recall of material presented during sleep. *Psychophysiology*, 1969, 6, 269.

Feinberg, I., Braun, M., & Shulman, E. Electrophysiological sleep patterns in mongolism and phenylpyruvic oligophrenia (PKU). *Psychophysiology*, 1968, 4, 395.

Greenough, W. T., & Wolkman, F. R. Pattern of dendritic branching in occipital cortex of rats reared in complex environments. *Experimental Neurology*, 1973, 40, 491–504.

Griffin, J. P. Neurophysiological studies into habituation. In G. Horn and R. A. Hinde (Eds.), *Short-term changes in neural activity and behaviour*. New York: Cambridge Univ. Press, 1970.

Hebb, D. O. Heredity and environment in mammalian behavior. *British Journal of Animal Behaviour*, 1953, 1, 43–47.

Hebb, D. O., Lambert, W. E., & Tucker, G. R. Language, thought and experience. *Modern Language Journal*, 1971, 55, 212–222.

Hobson, J. A. Sleep after exercise. *Science*, 1968, 162, 1503–1505.

Horn, G. Changes in neuronal activity and their relationships to behaviour. In G. Horn and R. A. Hinde (Eds.), *Short-term changes in neural activity and behaviour*. New York: Cambridge Univ. Press, 1970.

Horn, G., & Hinde, R. A. (Eds.) *Short-term changes in neural activity and behaviour*. New York: Cambridge Univ. Press, 1970.

Hubel, D. H., & Wiesel, T. N. Binocular interaction in striate cortex of kittens reared with artificial squint. *Journal of Neurophysiology*, 1965, 28, 1041–1059.

Huston, J. P., & Borbély, A. A. Operant conditioning in forebrain ablated rats by use of rewarding hypothalamic stimulation. *Brain Research*, 1973, 50, 467–472.

John, E. R. *Mechanisms of memory*, New York: Academic Press, 1967.

Kandel, E. R., & Spencer, W. A. Cellular neurophysiological approaches in the study of learning. *Physiological Reviews*, 1968, 48, 65–134.

Kimble, D. P. Possible inhibitory functions of the hippocampus. *Neuropsychologia*, 1970, 8, 21–36.

Leiman, A. L., & Christian, C. N. Electrophysiological analyses of learning and memory. In J. A. Deutsch (Ed.), *The physiological basis of memory*. New York: Academic Press, 1973.

Lucero, M. A. Lengthening of REM sleep duration consecutive to learning in the rat. *Brain Research*, 1970, 20, 319–322.

Markel, E., & Adam, G. Elementary temporary connection in the mesencephalic cat. *Acta physiologica Academiae scientiarum hungaricae,* 1965, **26**, 81–87.

Markel, E., & Adam, G. Learning phenomena in mesencephalic rats. *Acta physiologica Academiae scientiarum hungaricae,* 1969, 36, 265–270.

Markel, E., Anda, E., Juhasz, G., & Adam, G. Discriminative learning in mesencephalic rats. *Acta physiologica Academiae scientiarum hungaricae,* 1969, **36**, 271–276.

McGinty, D. J. Effects of prolonged isolation and subsequent enrichment on sleep patterns in kittens. *EEG & Clinical Neurophysiology,* 1969, **26**, 335.

Milner, P. M. Learning in neural systems. In M. C. Yovits and S. Cameron (Eds.), *Self-organizing systems.* New York: Pergamon Press, 1960.

Milner, P. M. *Physiological psychology.* New York: Holt, 1970.

Møllgaard, K., Diamond, M. C., Bennett, E. L., Rosenzweig, M. R., & Lindner, B. Quantitative synaptic changes with differential experience in rat brain. *International Journal of Neuroscience,* 1971, **2**, 113–128.

Oakley, D. A. Instrumental learning in neodecorticate rabbits. *Nature; New Biology,* 1971, **233**, 185–187.

Pribram, K. H. *Languages of the brain.* Englewood Cliffs, New Jersey: Prentice-Hall, 1971.

Riege, W. H. Environmental influences on brain and behavior of year-old rats. *Developmental Psychobiology,* 1971, **4**, 157–167.

Roffwarg, H. P., Muzio, J. N., & Dement, W. C. Ontogenetic development of the human sleep-dream cycle. *Science,* 1966, **152**, 604–619.

Roitbak, A. I. A new hypothesis concerning the mechanisms of formation of the conditioned reflex. *Acta Neurobiologiae Experimentalis,* 1970, **30**, 81–94.

Rosenzweig, M. R. Environmental complexity, cerebral change, and behavior. *American Psychologist,* 1966, **21**, 321–332.

Rosenzweig, M. R. Effects of environment on development of brain and behavior. In E., Tabach, L. R. Aronson, and E. Shaw (Eds.), *The biopsychology of development.* New York: Academic Press, 1971.

Rosenzweig, M. R., Møllgaard, K., Diamond, M. C., & Bennett, E. L. Negative as well as positive synaptic changes may store memory. *Psychological Review,* 1972, **79**, 93–96.

Russell, I. S. Neurological basis of complex learning. *British Medical Bulletin,* 1971, **27**, 278–285.

Schapiro, S., & Vukovich, K. R. Early experience effects on cortical dendrites: A proposed model for development. *Science,* 1970, **167**, 292–294.

Sharpless, S. K. Reorganization of function in the nervous system: Use and disuse. *Annual Review of Physiology,* 1964, **26**, 357–388.

Shlaer, R., & Myers, M. L. Operant conditioning of the pretrigeminal cat. *Brain Research,* 1972, **38**, 222–225.

Spencer, W. A., and April, R. S. Plastic properties of monosynaptic pathways in mammals. In G. Horn & R. A. Hinde, *Short-term changes in neural activity and behavior.* New York: Cambridge Univ. Press, 1970.

Stern, W. C. Effects of REM sleep deprivation upon the acquisition of learned behavior in the rat. *Psychophysiology,* 1969, **6**, 224.

Tagney, J. Sleep patterns related to rearing rats in enriched and impoverished environments. *Brain Research,* 1973, **53**, 353–361.

Teuber, H. L. Perception. In J. Field, H. W. Magoun, & V. E. Hall (Eds.), *Handbook of Physiology. Section 1: Neurophysiology.* Vol. III. Washington, D. C.: American Physiological Society, 1960.

Thompson, R. F., Patterson, M. M., & Teyler, T. J. The neurophysiology of learning. *Annual Review of Psychology,* 1972, **23**, 73–104.

Ungar, G. *Molecular mechanisms of memory and learning.* New York: Plenum Press, 1970.

Von Baumgarten, R. J. Plasticity in the nervous system at the unitary level. In F. O. Schmitt (Ed.), *The neuro-sciences, second study program.* New York: Rockefeller Univ. Press, 1970.

Von Bechterew, W. *Die Funktionen der Nervencentra.* Vol. 2. Jena: Fisher, 1909.

Wall, P. D. Habituation and post-tetanic potentiation in the spinal cord. In G. Horn & R. A. Hinde (Eds.), *Short-term changes in neural activity and behaviour.* New York: Cambridge Univ. Press, 1970.

Wall, P. D., & Egger, M. D. Formation of new connexions in adult rat brains after partial deafferentation. *Nature,* 1971, **232**, 542–545.

Webb, W. B., & Friedman, J. Attempts to modify the sleep patterns of the rat. *Physiology & Behavior,* 1971, **6**, 459–460.

Zernicki, B., Dreher, B., Krzywosinski, L., & Sychowa, B. Some properties of the acute midpontine pretrigeminal cat. *Acta Biologiae Experimentalis,* 1967, **27**, 123–139.

Zernicki, B., & Osetowska, E. Conditioning and differentiation in the chronic midpontine pretrigeminal cat. *Acta Biologiae Experimentalis,* 1963, **23**, 25–32.

Zolman, J. F., & Moritomo, H. Cerebral changes to duration of environment complexity and locomotor activity. *Journal of Comparative and Physiological Psychology,* 1965, **60**, 382–387.

Chapter 17

The Function of Dreaming: A Neurophysiologist's Point of View

MICHEL JOUVET

Claude Bernard University, Lyon, France

Studies of dreaming* began to be included in the field of neurophysiological research some years ago (Dement & Kleitman, 1957; Dement, 1958; Jouvet, Michel, & Courjon, 1959). Yet, despite an enormous amount of work in both humans and animals, with psychoanalytical, psychophysiological, physiological, and biochemical methods, the problem of the function of dreaming remains one of the most important riddles of neurobiology. Very seldom in the history of neurobiology has such a gap existed between the accurate polygraphic description of a phenomenon, the knowledge of its structural basis or some of its main mechanisms, even its natural history during both phylogeny and ontogeny, and our total ignorance concerning its role. At the present time we can selectively either suppress paradoxical sleep (PS) in the cat or increase its quantity by a factor of 5 or 6, and yet we do

---

* Although there are still some discussions regarding the relationship between dreaming and paradoxical sleep, we shall use these terms interchangeably. The title of this chapter could also be "the function of paradoxical sleep" for those who don't believe that paradoxical sleep is the objective aspect of dreaming.

not know in what direction we should look for any specific disturbances. Should we admit that dreaming has no function at all? It would
seem rather peculiar for evolution to have played such a dangerous
game, since, as we shall see later, dreaming is a state during which
an organism cannot react adaptively to threatening stimuli. Yet this
dangerous game has been apparently won by a dreaming brain like
that of humans. So we have to admit that dreaming probably fulfills
some important function, but we should look more carefully to some
specific aspect of behavior in order to discover it.

   In this Chapter I shall try to summarize a theory according to which
the function of dreaming is to organize or to program genetically constituted or instinctive behavior. I do not yet have any proof of this theory
but its heuristic value may suggest several experimental verifications.

   This theory is based upon the following neurophysiological approaches:

1. The classical reductionist and synchronistic approach; i.e., (a) What
   do we know about the conditions that prime dreaming? and (b)
   What do we know about the intimate neurophysiological mechanism
   of PS in animals?
2. An historical, diachronistic approach. What do we know about the
   natural history of PS, during both phylogeny and ontogeny?
3. Finally why have we failed to observe any specific, reliable disturbances after selective dream suppression?

A NEUROPHYSIOLOGICAL DISSECTION OF
DREAMING AND ITS PRIMING MECHANISM

Sleep: The Priming of Dreaming

   In normal adult mammals including man, sleep is a necessary step
for the appearance of dreaming. The familiar succession of waking
into stage one (spindles) sleep, then into stage two (slow wave) sleep,
and finally into the so-called slow sleep with phasic pontogeniculo-occipi-
tal (PGO) activity which heralds paradoxical sleep in the cat is well-
known (Figure 1). There is also a general agreement that PS is preceded by a state of deep slow wave sleep as shown by the progressive
increase of the threshold of arousal (either by direct stimulation of
the reticular system, or by auditory stimuli). Even if the neurobiological
mechanisms underlying sleep are not yet totally understood, several converging lines of evidence strongly suggest that the 5-HT, that is, 5-hy-

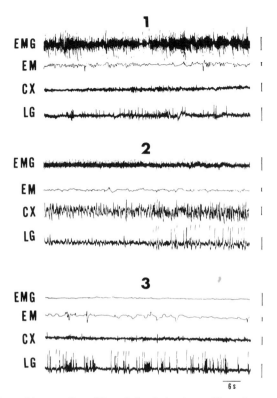

Figure 1. Polygraphic recording of the adult cat showing waking, sleep, and PS.

(1) Waking: muscular activity of the neck (EMG), eye movements (EM), fast low-voltage activity of the occipital cortex (CX), and eye-movement potentials in the lateral geniculate nucleus (LG).

(2) Slow wave sleep: decreased muscular activity of the neck, absence of eye movements, high-voltage slow waves and spindles in the cortex, high-voltage ponto-geniculate activity in the LG which directly precedes paradoxical sleep by 1 or 2 minutes.

(3) Paradoxical sleep: total disappearance of EMG activity together with rapid eye movements, fast low-voltage cortical activity similar to waking, and clusters of PGO geniculate waves.

Each line represents 1 minute of recording: vertical calibration, 50 microvolts.

droxytryptamine or serotonin, containing neurons of the rostral raphe system play a paramount role in the induction of sleep (see references in Jouvet, 1972):

1. The inhibition of synthesis of 5-HT at the level of tryptophan hydroxylase with p-chlorophenylalanine induces a state of insomnia, the intensity of which is proportional to the decrease in 5-HT. This insomnia is immediately reversible into both sleep and PS with a secondary injection of a small dose (2–5 mg/kg) of 5-hydroxytryptophan (5-HTP),

the immediate precursor of 5-HT which is readily decarboxylated in 5-HT terminals (Figure 2).

2. Any local inactivation of the system of 5-HT neurons originating from the rostral raphe leads also to insomnia. This has been shown after surgical destruction of 5-HT perikarya (Figure 3) or more selectively after the selective poisoning of 5-HT terminals which follows the intraventricular injection of 5-6-HT (Froment, Petitjean, Bertrand, Cointy, & Jouvet, 1974) a powerful and selective poison of 5-HT neurons (Baumgarten & Lachenmayer, 1972) (Figure 4).

3. The activation of 5-HT neurons may induce a dramatic increase

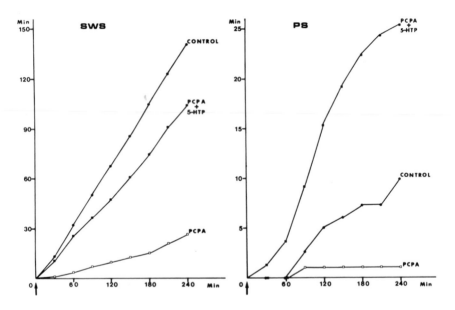

Figure 2. Pharmacological evidence for the intervention of serotoninergic mechanisms in sleep and PS. (Ordinate: Cumulative amount of SWS or PS every 30 minutes. Abscissa: Time in minutes.)

(Left) Slow wave sleep (SWS): in control cats (control), the duration of SWS during the 240 minutes of recording is 150 minutes (i.e., 60%); in cats having received two injections of 350 mg/kg of p-chlorophenylalanine (PCPA) 48 hours and 24 hours before, SWS amounts to only 30 minutes (i.e., 12%); in cats having received the same doses of PCPA, a small dose of 5 mg/kg of 5-hydroxytryptophan (5-HTP) is injected (arrow) at the beginning of the recording. The amount of SWS reaches 100 minutes (40%).

(Right) Duration of paradoxical sleep (PS) during the same period and under the same conditions. The suppression of PS is almost total in PCPA-pretreated cat (less than 1%) whereas the quantity of PS is more than 10% if 5-HTP is injected in the PCPA-pretreated cat. These experiments, which are easily reproducible, demonstrate that the inhibition of 5-HT synthesis at the tryptophan hydroxylase level by PCPA induces a state of almost total insomnia which can be reversed to result in normal amounts of both SWS and PS if the immediate precursor of serotonin is injected. (From Bobillier et al., 1973.)

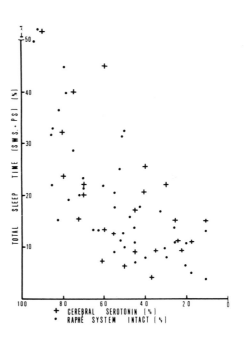

Figure 3. Neurophysiological evidence for the intervention of the raphe system in sleep and PS. Three-way correlation of the extent of destruction of the raphe system (in dots), the level of serotonin in the brain rostral to the lesion (crosses), and the amount of total sleep (SWS and PS) during the 13 days of survival after the lesion (in ordinates). The decrease of sleep is strongly correlated with the amount of destruction of the raphe system and the decrease of cerebral 5-HT. The amount of total sleep in 10 control sham-operated cats is 54%

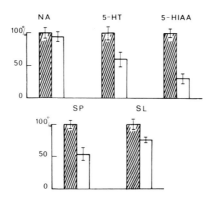

Figure 4. Decrease of sleep and PS after the selective destruction of 5-HT terminals with 5-6-hydroxytryptamine.

(Top) Amount of noradrenaline (NA), serotonin (5-HT) and 5-hydroxyindol acetic acid (5-HIAA) in the telediencephalon in cats having received 1 mg of 5-6-HT in the cerebral ventricules (white column) and in control cats having received only the Ringer solution solvent of the 5-6-HT (oblique hatching). The results are expressed in percentage of control untreated cats.

(Bottom) Percentage of PS (SP) and SWS (SL) during 15 days after the injection as compared with the control obtained during 1 week before the intraventricular injection. These results demonstrate that 5-6-HT has a selective destructive effect upon 5-HT terminals as shown by the decrease in both 5-HT and 5-HIAA, and that the inactivation of 5-HT neurons leads to a significant decrease in both SWS and PS. (Modified from Froment et al., 1973.)

in both sleep and PS. This is the case after the destruction of the
dorsal noradrenaline pathway at the level of the isthmus, which is followed by an increase in PS up to 60% of the nycthemeron (as compared
with 12% in the control). This phenomenon is accompanied by an important increase in the biosynthesis and the turnover of 5-HT in the brain
(Petitjean & Jouvet, 1970; Pujol, Stein, Blondaux, Petitjean, Froment, &
Jouvet, 1973) (Figure 5). The exact mechanisms which are responsible
for this phenomenon have not yet been discovered but the most likely
hypothesis is that there exists a direct or indirect control of some ascending noradrenaline (NA) neurons upon the rostral raphe system. The
destruction of this control would activate the 5-HT system which in turn

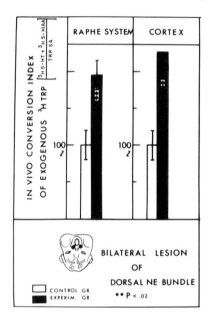

Figure 5. Increase in activity of 5-HT neurons after lesion of the dorsal noradrenaline
bundle. The lesion is schematized in a frontal section of the isthmus at the plane frontal
zero (see Figure 6). Such a lesion induces an increase in SWS and in PS which may increase
from 12% in the preoperative control to 60% during the 24–48 hours which follow the lesion.
The biochemical results are obtained 48 hours after the bilateral coagulation of the dorsal
NA bundle. The rate of synthesis of 5-HT is estimated by following the initial accumulation
of ^{3}H 5-HT and ^{3}H 5-HIAA endogenously synthetized from 1.5 mci of ^{3}H tryptophan intravenously injected 20 minutes before the sacrifice. The conversion index of tryptophan is
given by the ratio

$$\frac{^3\text{H 5-HT} + \,^3\text{H 5-HIAA}}{\text{TRP SA}}.$$

Results for the experimental group (black bars) are expressed as a percentage of the mean
value for the control group (lesion of the inferior colliculi) $\pm$ mean standard error; 5 determinations; p values are calculated by a "t" test. (Modified from Pujol et al., 1973.)

would prime the executive mechanism for initiating PS, which is located in the region of the subcoeruleus area (Figure 6).

These three experimental observations lead to the following conclusion: The activation of 5-HT neurons which leads to deep slow wave sleep is necessary for the appearance of PS. But this condition is not enough for the triggering of PS, which seems to depend also upon some "fail safe" mechanism since two other keys are necessary in order that PS can appear.

The first one concerns the waking system and possibly its noradrenergic components originating from the rostral part of the locus coeruleus

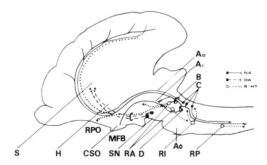

Figure 6. Monoaminergic systems regulating sleep and PS. On a sagittal map of the cat brain are highly schematized the topography of the main monoaminergic systems involved in the mechanisms of sleep and PS. Open circles and dots: 5-HT system. The anterior raphe system (RA) located in nucleus raphe dorsalis and centralis superior is responsible for the induction of SWS since its destruction is followed by an almost total insomnia. Sleep may be the results of the liberation of 5-HT either in the mesencephalic reticular formation in the preoptic region (RPO) or in the cortex. The intermediary group of the raphe system (RI), N. raphe pontis, is involved mostly in the priming of PS since its destruction is followed by an important and selective reduction of PS. Nucleus raphe pontis controls also the PGO activity probably through some terminals acting upon the locus subcoeruleus (D) (see arrows). The caudal raphe nuclei, N. raphe pallidus and obscurus (RP), which send axons to the spinal cord, do not participate in the regulation of sleep.

The complex of catecholamine (mostly NA and possibly DA)- containing neurons of the pontine tegmentum are represented by black circles and solid lines. They constitute the locus coeruleus (C) and subcoeruleus (D). The caudal part of N. locus coeruleus (see Figure 12) is responsible for the inhibition of muscular tone during PS, probably through axons descending to the inhibitory reticular formation. The locus subcoeruleus (D) send axons in the intermediary catecholamine (CA) pathway (AL) which ascends in the mesencephalic tegmentum and crosses the midline in the supraoptic decussation (CSO). These neurons are probably responsible for the PGO activity (see Figure 9). From the rostral part of the locus coeruleus ascends the dorsal noradrenaline bundle (AD) which is concerned with the regulation of waking. A lesion (B) made at the level of the isthmus (see Figure 5) induces an increase of SWS and PS possibly by destroying NA fibers which control the anterior raphe. The dopaminergic nigro-striatal system is also represented by black squares and interrupted lines. It originates from the substantia nigra (SN), ascends in the median forebrain bundle (MFB), and terminates in the striatum (S). Its destruction does not alter significantly the sleep–waking cycle but it induces a state of akinesia (Jones *et al.*, 1973). H: hypothalamus; A0: frontal plane Horsley Clarke zero. (Modified from Maeda *et al.*, 1973.)

and ascending in the dorsal noradrenaline bundle. Any external or pharmacological conditions which activate the noradrenergic component of waking (Bobillier, Pin, & Jouvet, 1973) will immediately "lighten" sleep and PS will not appear. Any disturbing noise delivered during the deep slow wave sleep, which immediately heralds PS, suppresses its appearance together with the PGO activity, even if the cat does not become aroused behaviorally or polygraphically (Figure 7). The secondary suppressor effect of amphetamine upon PS is probably mediated through the same mechanism, even if light sleep is not impaired. The difficulty of recording PS in wild animals kept in laboratory conditions and the "first night" effect, during which PS is decreased in humans, is probably due to the same mechanism.

The second key necessary for the occurrence of PS is a cholinergic mechanism, since it is well known that anticholinergic drugs can block the appearance of PS. The mechanism of the interaction of the 5-HT mechanism with the cholinergic mechanism is still not clear but it is probable that this interaction is mediated locally in the pons, since direct injection of acetylcholine or carbachol in the triggering pontine structures of PS may precipitate PS directly following waking (George, Haslett, & Jenden, 1964; Baxter, 1969).

Thus it is evident that the appearance of PS is protected by a delicate succession of biochemical mechanisms. We should now ask the following question. Why is there such a "fail safe" mechanism controlling the appearance of PS? Is it teleologically explicable? The answer is given

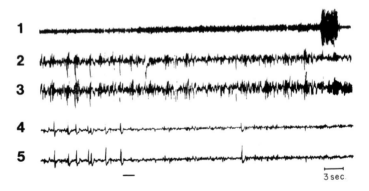

3 sec.

Figure 7. The "fail safe" mechanism of PS. Normal cat. During the period of slow sleep with PGO activity, which would normally herald PS, a noise (black line) immediately suppresses the PGO activity whereas slow sleep continues for some seconds until polygraphic arousal. It is likely that this suppression of PGO activity is mediated by noradrenergic mechanisms located in the rostral part of the locus coeruleus. 1. EMG of the neck muscles. 2 and 3. Fronto-occipital cortices. 4 and 5. Right and left lateral geniculate. Calibration: 3 seconds, 50 microvolts.

by the results of the neurophysiological analysis of the mechanisms of PS.

The Neurophysiological Mechanisms of Paradoxical Sleep

Schematically, the intimate mechanisms of dreaming may be dissociated into three principal components.

1. The first component is the most important since it explains the others. It is characterized by an intense endogenous phasic and synchronous activity of most cerebral neurons, including the pyramidal and extrapyramidal motoneurons. Such a generalized rhythmical activity which is endogenously generated implies the necessity of a cerebral "pacemaker" and also the existence of some powerful mechanisms which inhibit most of the motor output: One may dream that he is running or flying but the body does not move.

Numerous microelectrode studies have shown that PS is accompanied by increased cerebral activity (equal to or greater than that during the most intense wakefulness). Moreover it appears that a large proportion of the cerebral neurons fire or are inhibited phasically under the control of a "pacemaker" which is responsible for the so-called ponto-geniculo-occipital (PGO) activity. The bilateral "pacemaker" responsible for this activity has been located in the lateral part of the pontine tegmentum in the region of the nuclei pontis oralis and caudalis (Laurent, Cespuglio, & Jouvet, 1973). In this region are concentrated some catecholamine-containing cell bodies belonging to the nucleus locus subcoeruleus (Maeda, Pin, Salvert, Ligier, & Jouvet, 1973). Although definite proof that these neurons are *solely* responsible for PGO activity has not yet been provided, the intervention of monoaminergic mechanisms in the determination of PGO activity is strongly suggested by a large number of neuropharmacological experiments (see Jouvet, 1972). Thus, PGO activity can be dissociated from PS and may occur continuously during waking or sleeping after any pharmacological (using reserpine or p-chlorophenylalanine) or surgical lesion of the raphe nuclei alterations of monoamines which cause a decrease of 5-HT at the terminals of 5-HT neurons. A tonic inhibitory control of PGO activity by some noradrenaline-containing neurons is also possible, since the destruction of the locus coeruleus which contains norepinephrine-rich perikarya is followed by a permanent discharge of PGO activity (Roussel, 1967). This inhibitory control may explain why the activation of noradrenaline neurons during waking immediately suppresses PGO activity. Although we still do not know the basic mechanisms which are responsible for the compli-

Figure 8. Corticofugal control of PGO activity.

A) Recording of PGO activity in the VI nerve nucleus during PS in a normal cat. There is a complicated pattern of isolated PGO and bursts of PGO.

B) After total removal of the brain rostral to the pons, the same animal still presents PS episodes. The pattern of activity is strikingly altered and there are mostly bursts of PGO. The same pattern may also be observed after bilateral frontal cortical ablations. (From Gadea-Ciria, 1972.)

cated patterns of PGO occurring during PS, we know that this pattern is the result of the interactions between monoaminergic mechanisms and subcortical and cortical mechanisms, since the pattern of PGO can be altered either by vestibular lesion (Morrison & Pompeiano, 1966), or by frontal cortical ablations (Gadea Ciria, 1972) (Figure 8). If the pattern of the PGO activity conveys some information which is transmitted to most of the cerebral neurons, this information reflects both local pontine mechanisms (which are still observed during PS in a pontine cat) and the excitatory level of some cortical areas.

The ipsilateral and contralateral pathways which transmit PGO activity from the pons to the lateral geniculate bodies and occipital cortex have been recently mapped out (Laurent *et al.*, 1974) (Figures 9 and 10). They follow the same route (or are identical with) the so-called "intermediary catecholaminergic pathway" originating from the locus subcoeruleus area which ascends in the mesencephalic tegmentum and crosses the midline in the region of the supraoptic commissure (Maeda *et al.*, 1973). It has also been shown that PGO activity is related intimately to the rapid eye movements which occur during PS, i.e., the motoneurons of the oculomotor nerves are either excited or inhibited by the same pacemaker that is responsible for the extraretinal input to the lateral geniculate bodies. It is now possible to reconstruct the central PGO activity in both lateral geniculate and occipital cortex from the patterns of the rapid eye movements (as recorded with standard EOG techniques or with electromyographic recording from the extraocular muscles (Cespuglio, 1973) (Figure 11). The main problem that remains to be solved concerning the central phasic electrical activity of PS is the following: Is the PGO activity recorded with macroelectrodes a specific activity occurring *only* in the visual system or is it a regional exaggeration of

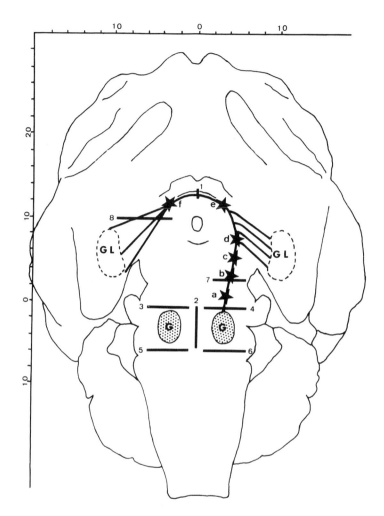

Figure 9. Organization of the PGO system. Ventral view of the cat brain showing a sketch of the direct and crossed pontogeniculate pathways at the level of the supra-optic decussation. The solid straight lines show the different sections performed (1–8) allowing delineation of the topography of these pathways; the asterisks (a–f) represent the different coagulations performed and delineate more precisely these same pathways. G: pontine "generator" of PGO activity; GL: lateral geniculate nucleus.

Sections 3 and 4, 3 and 7 suppress the PGO activity in both lateral geniculate nuclei, while 5 and 6 allow the persistence of bilateral PGO activity; 4 and 1, 4 and 8, 7 and 1 suppress the PGO activity only in the lateral geniculate situated on the right side of this figure.

Sections 3 and 8: unilateral suppression of PGO activity in the lateral geniculate situated on the left side. Sections 1 and 2: suppression of the synchronous character of PGO activity.

Section 3 associated with different coagulations gives the following results: 3 and a, b, or c result in bilateral suppression of PGO activity in both lateral geniculates; 3 and e–f result in unilateral suppression of PGO activity at the level of the lateral geniculate situated on the left side of the figure. (From Laurent et al., 1973.)

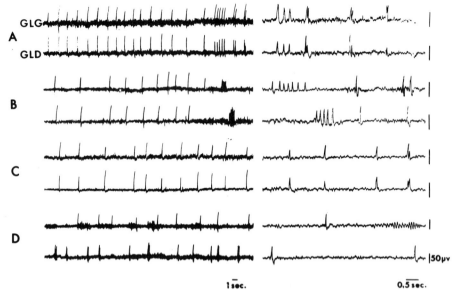

Figure 10. Recording of PGO activity at the level of left (GLG) and right (GLD) lateral geniculate nuclei after different surgical sections.

(A) Intact cat. The PGO spikes present two components separated by about 80 msec.

(B) Section at the level of the supraoptic decussation (section 1, Figure 9): The spikes are shifted in one lateral geniculate with respect to the other. Note also that the bursts of PGO are shifted as if some "command neurons" in the pons were alternately excited by the pontine generator.

(C) Midsagittal section at the level of the pons (section 2, Figure 9). Most of the spikes show only one component.

(D) Section of the supraoptic decussation and midsagittal section of the pons (sections 1 and 2, Figure 9). The PGO spikes have only one component and are desynchronized bilaterally. This proves the existence of two pontine generators.

These experiments were performed in acute experiments after injection of reserpine (0.5 mg/kg) but the same data have also been obtained during physiological PS in the encéphale isolé cat. (From Laurent et al., 1973.)

a generalized "programmed activity" which invades most of the brain? Most of the indirect data available at the present time favor the second hypothesis, i.e., in numerous areas, outside the oculomotor and visual systems, there are numerous neurons which fire in bursts synchronously with PGO activity (see references in Balzano & Jeannerod, 1970). This is the case for most of the motoneurons belonging to the pyramidal or extrapyramidal system: cortical pyramidal tract neurons (Evarts, 1964), pyramidal tracts (Marchiafava & Pompeiano, 1966), red nucleus (Gassel, Marchiafava, & Pompeiano, 1965), vestibular nuclei (Bizzi, Pompeiano, & Somogyi, 1964). Thus, it is evident that a powerful mechanism of inhibition of the motor output (sparing the oculomotor system)

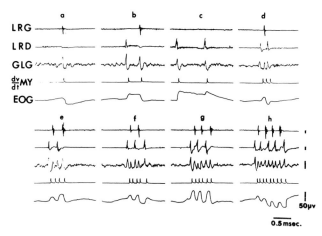

Figure 11. PGO activity and rapid eye movements. Recordings made in the encéphale isolé cat after injection of reserpine (0.5 mg/kg) from the left and right lateral recti (LRG–LRD), the left lateral geniculate (GLG), and the electroculogram (EOG). An electronic device permits one to integrate the speed of the eye movement (dv/dt).

Note the crossed innervation between both lateral recti during eye movement. It is possible to reconstruct the pattern of PGO activity from the integration of the EOG both from isolated eye movement and from the bursts. Similar results have been obtained during PS in chronic cats. (From Cespuglio, 1973.)

has developed during evolution in order to stop any gross body movements which would awaken an animal during PS. This tonic mechanism which depends upon the caudal part of the locus coeruleus acts together with the vestibular nuclei at the spinal cord level according to some complex processes of presynaptic and postsynaptic inhibition (see Pompeiano, 1967).

Fortunately the neurons that command the tonic inhibition of muscle tone are separated from the "pacemaker" neurons responsible for the phasic PGO activity and from the group of neurons responsible for the cortical activation during PS. Thus it is possible to study objectively the motor behavior of a dreaming cat.

2. The oneiric behavior of cats. The stereotaxic bilateral destruction of the caudal part of the nucleus locus coeruleus, which contains mostly noradrenergic neurons (Dahlstrom & Fuxe, 1964; Maeda *et al.*, 1973) and possibly some cholinergic neurons (Shute & Lewis, 1966), selectively suppresses the powerful motor inhibition occurring during PS. (Figure 12). After such a lesion the cats, which have normal behavior either when awake or during slow wave sleep, enter into a most dramatic stereotyped "pseudohallucinatory behavior" at the time when they would normally enter into PS (Jouvet, 1965; Henley & Morrison, 1969). They suddenly stand up, leap, and either display some aggressive behavior or play with

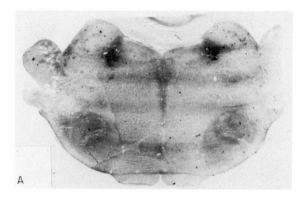

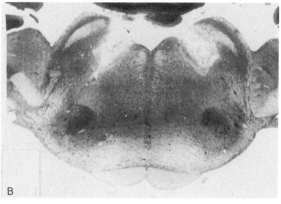

Figure 12. Control of the motor inhibition during PS.

(A) Frontal section of the pons in the cat. Glenner method. The monoamine oxidase (MAO) of the caudal locus coeruleus is colored in black.

(B) Bilateral lesion of the caudal part of the locus coeruleus which suppresses the motor inhibition during PS but does not suppress the ascending components of PS, i.e., PGO activity and cortical activation. After such lesions the cats exhibit the stereotyped behavior depicted in Figure 13.

their front paws as they might play with a mouse, or they show a rage behavior or a defense reaction against a larger predator. We have also observed, more rarely, some drinking behavior. During these episodes, even during the most intense rage behavior the nictitating membranes are relaxed and almost completely cover the pupils which are myotic. Thus, there exists a striking contrast between the motor behavior of rage and emotion and the vegetative ocular aspects of deep sleep. During these episodes the animals do not react to visual stimuli and they often collide with the walls of the observation cage. Normally these episodes last for 4 to 5 minutes (which is the average time of a PS episode). At the end, the animals either suddenly awaken (as shown by a subite mydriasis and contraction of the nictitating membranes) or return to a posture of quiet sleep (Figure 13).

These episodes are accompanied by the same central activity that distinguishes paradoxical sleep and are suppressed by the same drugs which suppress PS (MAO inhibitors) (Figure 14).

If it is admitted that this stereotyped behavior is the result of the excitation of the pyramidal and extrapyramidal systems through the direct or indirect control of the pontine pacemaker, then we should admit that this behavior is organized. It is not like an epileptic discharge. We should also admit that such behavior is programmed and that it may express some teleological project and possibly some instinct. It is also tempting to speculate that the information responsible for the stereotyped behavior is somewhat directly or indirectly related to the pattern of occurrence (the "grammar") of PGO activity. If this is the case, PGO activity could be approached by using some sophisticated linguistic method, provided of course, that we could decipher the various intermediate steps between PGO activity and the motor output (i.e., the behavior of a dreaming cat). Such a study is not totally impossible in the cat. But it remains hypothetical at the present time in man. Even if we can reconstruct the grammar of PGO activity occurring in man

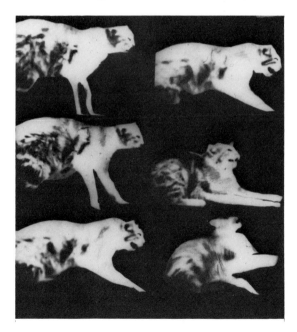

Figure 13. Stereotyped aggressive behavior during PS. See details in the text. From an 8 mm movie shown at the 23rd International Congress in Tokyo (Jouvet, 1965).

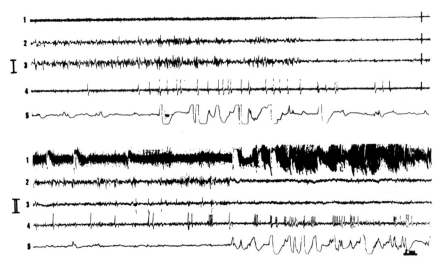

Figure 14. Polygraphic recordings of stereotyped behavior.

(I) Control recordings before the lesion. Normal succession of slow sleep, slow sleep with PGO activity, and PS with total disappearance of EMG activity.

(II) Three weeks after the lesion of the caudal part of the locus coeruleus (see Figure 12). Slow sleep is suddenly followed by stereotyped aggressive behavior (see Figure 13) during which there is an increase of muscle activity associated with the central components of PS.

Key to numbers on figures: 1, EMG of the neck. 2 and 3, Frontal and occipital cortex. 4, Lateral geniculate nucleus. 5, Electro-oculogram. Calibration 3 seconds, 50 microvolts. (From Jouvet, 1965.)

from the pattern of the rapid eye movements during dreaming we shall always be unaware of the behavior of a dreaming man but for his recall at the very end of a dream.

3. Why sleep is necessary for dreaming. It is teleologically evident that the powerful inhibition of the muscle tone which normally prevents any adaptive gross body movement during PS is a dangerous mechanism. Moreover other mechanisms contribute also to alter the responsiveness of the brain to incoming stimuli during PS. This is the case for the powerful presynaptic inhibitions which act upon most of the sensory relays: cuneate nuclei, cochlear nuclei, lateral geniculate (see references in Pompeiano, 1970). This phenomenon is probably responsible, at least in part, for the significant increase of the threshold of arousal (as compared with the immediately preceding slow wave sleep). Schematically, under the influence of the pontine system of neurons responsible for PS, the brain acts in order to stop any input and output (except the rapid eye movements and some peripheral twitches). Thus most of the neurons are prepared to receive selectively the endogenous information coming from the pons during dreaming.

It is evident that such a process can occur only during deep sleep. The priming period of deep sleep is the biochemical consequence of the absence of any signal of immediate danger since any threatening stimuli coming through the telereceptors (e.g., the barking or the odor of a dog in the case of the rabbit), would immediately act upon the waking system and activate some noradrenergic mechanism which would immediately either arouse the animal, or at least suppress the triggering of PS. It is only when an animal is in a familiar and safe environment that deep sleep may occur.

Thus sleep is the guardian of dreaming and, contrary to Freud's writings, dreaming is not the guardian of sleep. It is inconceivable that dreaming could occur during waking: no groups of animals could have survived if the periodic paralysis of PS, along with an increased threshold for arousal, had taken place at the very moment of flight or of attack. Any narcoleptic mutation, narcolepsy being the irruption of PS into waking (Rechtschaffen, Wolpert, Dement, Mitchell, & Fisher, 1963), is a potential lethal mutation.

Thus, in an adult animal, sleep is a necessary condition for the arrival of dreaming. The serotoninergic neurons responsible for sleep may play a double role, the first one being possibly a hypothetical restorative process at the level of the synapses or of the perikarya (see Moruzzi, 1972) according to some unknown intracellular second messenger effect under the control of serotonin. It is only when these processes have taken place, and only if the waking (noradrenergic) mechanisms are turned off, that the second step occurs and that most of the nervous system is ready to receive the "programmed endogenous stimulation" ascending from the pons.

It is also possible that the periods of sleep which follow dreaming may play a role in the integration of the information which has just been delivered, since there is some correlation between the duration of PS and the period of sleep which follows (Ursin, 1970).

Thus sleep and dreaming are very closely interrelated as we shall verify in the next section.

THE EVOLUTION OF DREAMING

Phylogeny

Paradoxical sleep appeared relatively late during phylogeny. There are no unequivocable polygraphic signs of PS in poikilothermic animals such as fishes, amphibians, and reptiles (Peyrethon & Dusan, 1967; Hobson, 1967), while inactivity or sleep can be recognized relatively easily on the basis of behavioral criteria. On the other hand, in almost every homoiothermic animal, from birds to mammals, PS has been recognized readily on both behavioral and polygraphic criteria (see references in Jouvet & Jouvet, 1964; Allison & Van Twyver, 1970). There are very large differences between species but in most cases these differences can be explained on the ground of ecoethological considerations. The short dreamers (10–15 minutes of PS per diem) are also short sleepers and light sleepers. They need to stay awake a great deal of the time to absorb a low caloric diet (such as grass and cellulose for herbivores). Thus there is little time left for sleep, especially for deep sleep, in those groups of animals which are mostly hunted. On the other hand, the carnivores, the hunters, usually may solve their problem of feeding relatively rapidly and they may reach deep sleep relatively easily: no animal will attack a sleeping lion. Fortunately for the neurophysiologist, the domestic cat adjusts easily to the laboratory situation and appears to be the world champion in dreaming with about 180–200 minutes per diem.

Such a survey of the phylogenetic evolution of dreaming is certainly too short and incomplete and we need much more data concerning sleep and dreaming habits in a natural environment. There seems to be, however, a general trend: increasing complexity of the nervous system is generally accompanied by an increase in the duration of PS. However, the ecoethological factors are so important that it is impossible to obtain any correlation between the amount of dreaming and any aspect of higher nervous activity. The cat and the opossum have longer periods of PS than the chimpanzee and man but their learning capability is certainly much more restricted.

Ontogeny

The evolution of PS during the ontogeny of mammals (Figure 15) provides some very important information, which should serve as a cornerstone of any theory concerning dreaming. In mammals, and probably also in birds, the amount of PS seems to be very closely related to the degree of development of the central nervous system (Jouvet-Mounier, Astic, & Lacote, 1970). Thus a newborn rat with a brain almost totally immature at birth has a very high proportion of PS during the first week after birth and PS reaches adult levels (i.e., 10% of the nychthemeron) only when the maturational process is almost completed. On the other hand, a newborn guinea pig, whose central nervous system is almost totally

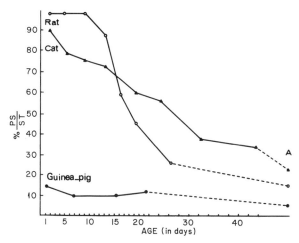

Figure 15. Ontogeny of paradoxical sleep. (Ordinate: Percent of PS. Abscissa: Age in days.) Percentages of total sleep time (ST) of time spent in PS expressed as a function of age in the rat, cat, and guinea pig. The dotted lines connect with the percent value in the adult (A). The value for a guinea pig fetus, 20 days before birth, would be analogous to the rat at birth (see Figure 16). (From Jouvet-Mounier *et al.*, 1970.)

mature at birth shows a very low proportion of PS, similar to that of the adult (6%). However, recordings taken in utero (Astic & Jouvet-Mounier, 1969) have indicated a very significant level of PS: the amount of PS of a guinea pig 20 days before birth is the same as that of a newborn rat (for an identical level of development of the brain) (Figure 16).

Thus, the amount of PS seems to be a relatively safe index of the degree of development of the brain. The brain is almost constantly dreaming when the most complex nervous structures are under development in utero or ex utero. This high level of PS in utero forces us to admit that PS probably plays some important function at the time *when genetic factors are more important than epigenetic events.*

THE FUNCTION OF DREAMING

We shall now try to organize the preceding data in an attempt to find a common denominator underlying the different neurophysiological and evolutionary aspects of dreaming. Such an attempt is certainly difficult since there are numerous theories about dreaming; but few, if any, have been successfully supported by experimental data. According to these theories (see references in Hennevin & Leconte, 1971) dreaming

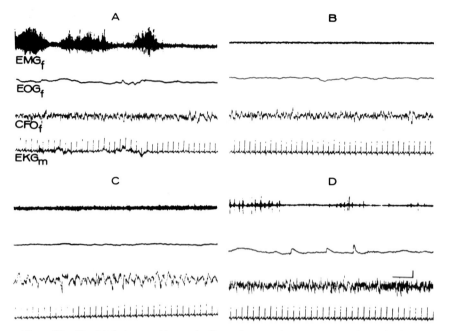

Figure 16. The fetal sleep–waking cycle. Recordings made from a fetal guinea pig (51 days, 10 days before birth). (A) Waking. (B) Light sleep. (C) Deep slow sleep. (D) Paradoxical sleep.

Key to labels: EMG_f: EMG of the neck muscle of the fetus. EOG_f: Electro-oculogram of the fetus. CFO_f: Frontal cortex of the fetus. EKG_m: EKG of the mother.

Note that all the components of PS are present in the fetus. PS in one fetus is independent of the state of any other fetus or of the mother. Calibration 1 second, 50 microvolts. (From Astic and Jouvet-Mounier, 1969.)

or PS may play a specific function in binocular vision, in learning, or in memory, or may have a nonspecific role in cortical homeostasis or a role of endogenous stimulation during ontogenesis: the excitation of some neurons during PS would afford intense stimulation to the central nervous system (CNS). Such stimulation would be particularly crucial during the periods in utero and shortly after birth, before appreciable exogenous stimulation is available to the CNS. Paradoxical sleep may assist in structural maturation and differentiation of key sensory and motor areas (Roffwarg, Muzio, & Dement, 1966). This theory has certainly the merit of emphasizing the importance of PS during development. But it does not explain the following experimental results. Despite the intense stimulation of the oculomotor system at least from the time of birth and of the visual system after the second week (Adrien, 1973) through PGO activity, PS does not seem to be sufficient for the development of the visual system of the kitten after birth if external stimuli are lacking. When newborn animals are reared in darkness, such treatment affects the morphology of neurons in the visual system. There is a reduction in the number of spines of apical dendrites (Valverde, 1967). On the other hand, if the lids of one eye of a kitten are sutured shortly after birth, the occluded eye becomes functionally disconnected from the cortex (Wiesel & Hubel, 1965). Thus the visual system needs the effects of experience, or epigenetic stimuli, in order to develop normally (see Chapter 4 by Hirsch and Jacobson). PS, which is present in dark-reared animals or even in totally deafferented animals (Vital-Durand & Michel, 1971), does not play a sufficient role for the functional development of visual neurons. The perinatal theory also does not explain why PS is still observed in adult life when epigenetic stimuli are present.

Dreaming as a Genetic Programming of Instinctive Behavior

It is evident that the striking increase of PS in utero, or immediately after birth in immature animals, is an indication that PS is linked with maturation—i.e., during the time when neurological development is regulated by a dialectical interplay between the genetic readout and the biochemical milieu. Nevertheless there is evidence that some mechanisms resembling embryogenesis, such as cellular growth, differentiation, and genetic switching, continue to operate during some periods of behavior modification (Altman, 1966). Moreover some macromolecular biosynthesis, which seems to occur during stable alterations of behavior (Glassman, 1969), suggests that the orderly switching along gene-directed biosynthesis pathways, which characterizes maturation, also participates in the postnatal structuring of the brain. However there seem to be two kinds of processes during postnatal maturation. The first one seems to possess a critical period and depends upon some epigenetic stimuli from the external milieu. This is the case with the maturation of the visual system.

In the second kind of process, a programmed genetic readout may ensue as a function of prior epigenetic events. Dreaming or PS would serve to program or organize the integration of all the complicated sequences of motor behavior that are necessary for the genetically constituted behaviors (or instincts) which may appear during each degree of historical or hormonal development—postnatal behavior such as the pecking behavior of chicks, the repertoire of song in the deafened chaffinch (Nottebohm, 1970), and the selection of homesite, the defense of the territory, hunting, homing, and mating in mammals (see references in Eibl-Eibesfeldt, 1967). The function of dreaming would be to stimulate,

possibly under the influence of PGO activity, the coding of which may
be genetically programmed, some strategic neurons (e.g., some hypothetical command neurons which would command the net of interneurons and
motoneurons responsible for innate behavior; see Chapter 1 by Hoyle),
according to the epigenetic development of the CNS. In summary,
*dreaming would represent, at the level of integration of motor organiza-
tion of innate behavior, what the epigenetic stimuli represent for the
maturation of sensory systems.*

It is most likely that such a programming of command neurons would
activate, select, instruct, or release important motor activity. This is
the case when one considers the "oneiric behavior" of a cat after destruction of the caudal part of the locus coeruleus. Thus it is evident that
PS constitutes the ideal frame for such a genetic programming, either
in utero or even in the adult, according to the important plastic modifications which have been effected through learning or through hormonal
changes. It appears logically inevitable that such a genetic organization
which affects the motoneurons can be achieved only in a closed circuit,
any afferent input or efferent output (except rapid eye movements)
being inhibited. It seems difficult to imagine that such a programming
or reorganization could occur during waking at a time when most of
the command neurons and interneurons are being stimulated by epigenetic stimuli from the external milieu.

If we accept the hypothesis that dreaming is related to the genetic
organization of innate behavior, then we must explain why PS probably
does not exist in poikilothermic animals, which have a very large repertoire of "innate" behavior. The answer to this problem is difficult. It
may be that PS is present but that we do not recognize its polygraphic
aspects. Rapid eye movements seem to occur during behavioral sleep
in some fishes (Tauber & Weitzman, 1969) and reptiles (Tauber,
Roffwarg, & Weitzman, 1968) and this may represent some PS episodes
of short duration. Another explanation is possible. In poikilothermic
animals there is extraordinary regeneration in the central nervous system
as shown by the experiments of Sperry (1944–1951) and Jacobson and
Gaze (1965) (see Chapter 4 by Hirsch and Jacobson). The restoration
of vision is independent of function or experience, for, if the optic nerve
is cut and the eye is inverted, a frog recovers with inverted visuomotor
reflexes, which are never corrected by experience. Thus both the development *and* the regeneration of retinotectal connections depend directly
on the rigid specification of the genome. Such a possibility does not
seem to exist in homoiothermic animals. In poikilothermic animals the
programming of behavior will be entirely built in the genome of the
neurons, while in homoiothermic animals it would be restricted to a

special class of neurons, the monoaminergic neurons, which, curiously enough, seem to exhibit the same regenerative properties as the neurons of poikilothermic animals.

Dreaming and Plasticity

We summarized in the first section the experimental evidence indicating that 5-HT-containing neurons are responsible both for sleep and for the priming of dreaming, and catecholamine (CA)-containing neurons are responsible for the *executive* mechanisms of PS. These monoaminergic systems appear to be quite appropriate for any genetic programming of behavior. On the one hand, both 5-HT and CA neurons appear immediately before PS can be recognized on polygraphic criteria in the fetal guinea pig (Maeda & Astic, 1972). On the other hand, monoaminergic neurons seem to have, even in the adult mammal (rat), a considerable capacity to regenerate. Thus the section of ascending monoaminergic bundles is followed by a secondary sprouting which innervates again the structures that were temporarily denervated (Katzman, Bjorklund, Owman, Stenevi, & West, 1971). In some cases, the biochemical signals which "attract" monoaminergic neurons are generated by some alteration of the nervous system. Thus the unilateral denervation of the septal nuclei, after unilateral section of the fornix, is followed by the appearance of new NA-containing terminals which seem to occupy the place of degenerated (nonaminergic) terminals (Moore, Bjorklund, & Stenevi, 1971).

It may be speculated that this phenomenon is only an exaggeration (demonstrated by histochemistry) of some physiological processes which would occur normally during the interactions between genetic programming of behavior and epigenetic alterations of the CNS. In such a case the alteration of some synapses (through disuse or learning) would attract the monoaminergic terminals which in turn would subject these synapses to genetic programming during PS. This might explain the increase in PS which occurs during some learning tasks (Lucero, 1970).

The Suppression of Dreaming

If PS fulfills an important function in adult life, then we should be able to observe some specific disturbances after its selective suppression. In fact, it is somewhat paradoxical and frustrating that no specific alteration of behavior has yet been recognized. Indeed it is possible to suppress PS selectively either instrumentally with the so-called "swimming pool" techniques (Vimont, Jouvet-Mounier, & Delorme, 1966) or with potent drugs, such as MAO inhibitors or α-methyl DOPA. In adult animals or men, it has been repetitively shown that the instrumental suppression of PS is followed by a secondary rebound (Dement, 1960; Vimont *et al.*, 1966). Thence the concept arises that PS fulfills a need, since the debt of PS (and even the debt of PGO) is almost entirely repaid during the subsequent rebound. However such a rebound is not obligatory, since it is either nonexistent or very short-lasting after PS suppression with pharmacological techniques. The behavioral results of PS deprivation have been the subject of many reviews (see references in Hennevin and Leconte, 1971). It is safe to assume, at the present time, that we don't know of any specific disturbances which are caused by PS deprivation *per se* (and not by the stress of the deprivation). The results concerning disturbances of memory and learning are often contradictory (Stern, 1970; Greenberg, 1970). A detailed analysis concerning the effect on learning of selective suppression of PS with α-methyl DOPA in some genetic strains of mice (CBA and 57 BR) has recently been completed. Again the results were quite equivocal since this procedure slightly impairs learning in one species while it accelerates retention significantly in the other (Kitahama, 1973). Finally there is increasing evidence obtained from adult men under treatment for narcolepsy with either chlorimipramine or MAO inhibitors that no significant specific memory or intellectual disturbances may be observed even after some months of total or subtotal PS suppression verified almost every night with polygraphic recordings (Passouant, Cadilhac, & Ribstein, 1972; Fisher, Kahn, Edwards, & Davis, 1972).

In fact, if dreaming is the time during which there occur some interactions between genetic programming of innate behavior and epigenetic events, it is quite conceivable that we will not observe any disturbances in *adult* men or in laboratory animals. In an adult man, who is submitted to some intense treatment which suppresses dreaming, it is often assumed that the subtle disturbances of personality (alterations of aggression or of sexual behavior) which are observed are provoked by the drug itself. This is certainly possible but such alterations may also be provoked by the absence of reprogramming of some instinctive behavior.

This is probably the same for animals kept in the neurophysiological laboratory in which ethological studies are not often used: it is possible that the contradictory results concerning the effects of PS deprivation upon learning may be due to the fact that some learning situations may involve instinctive behavior more "deeply" than others, or that the instinctive behavior might need more reorganization in some cases than in others. What do we know about the process by which a mouse chooses to go toward the dark side of a maze? Is such a choice innate and programmed during PS? If this is the case one may understand that any kind of learning based upon this behavior might be altered through PS deprivation, while the learning mechanisms, per se, would not.

Any theory of dreaming should have some heuristic value, and in designing experiments for testing the hypothesis that PS serves to program genetically constituted behavior the following problems need to be considered:

1. If the aggressive stereotyped oneiric behavior which occurs in the cat after destruction of the caudal part of the locus coeruleus is genetically programmed, then it should occur also in adult animals which have been kept in isolation from the time of birth and have never had any opportunity when awake to play or to fight or to kill a mouse.

2. Does the suppression of PS, in utero, or immediately after birth, suppress or alter considerably the appearance of imprinting in newborn chicks? Does it alter the repertoire of the song in deafened chaffinches? And if so, is there a critical period corresponding to this programming?

3. The pecking behavior of the newborn chick is apparently innate— but it may be suppressed if the chick is prevented from pecking for 2 weeks (Padilla, 1935). Does the suppression of PS considerably shorten this period?

4. Finally, it is possible that the register of "instinctive" behavior might be triggered or released by hormonal factors. Does PS suppression at the time of puberty alter aggressive behavior or the defense of territory? Does PS suppression in a pregnant cat or rat suppress or alter its subsequent maternal behavior?

These are questions which have never been really considered in most of the PS-deprivation experiments. Finding the answers to these questions may be possible but certainly difficult, since we have learned that the instrumental deprivation of PS is not feasible in newborn animals (rats or kittens) for more than a few hours. On the other hand the pharmacological suppression of PS, which is easier, may not be "selective" enough, and if positive results are obtained, it would be difficult to distinguish between a purely neuropharmacological effect and the result of PS suppression. This is why the selective local poisoning in utero of the monoaminergic neurons responsible for PS appears to be the best method, but the technical problems are still very hard to solve.

Whatever may be learned from these experiments, one thing is obvious: We shall not be able to explain satisfactorily the functioning of the brain as long as we do not understand why we dream 100 minutes every night.

ACKNOWLEDGMENTS

The research in this department has been sponsored by DRME (72108), INSERM (U 52), and CNRS (LA 162).

REFERENCES

Adrien, J. Ontogénèse des activités phasiques du sommeil paradoxal. *Thèse de Neurophysiologie* Lyon, 1973.

Allison, T., & Van Twyver, H. The evolution of sleep. *Natural History*, 1970, **79**, 57–65.

Altman, J. Autoradiographie and histological studies of postnatal neurogenesis. II. A longitudinal investigation of the kinetics, migration and transformation of cells incorporating tritiated thymidine in infant rats. *Journal of Comparative Neurology*, 1966, **128**, 431–475.

Astic, L., & Jouvet-Mounier, D. Mise en évidence du sommeil paradoxal in utero chez le cobaye. *Comptes Revues de l' Académie des Sciences* (Paris), 1969, **264**, 2578–2581.

Balzano, E., & Jeannerod, M. Activité multi-unitaire de structures sous-corticales pendant le cycle veille-sommeil chez le chat. *Electroencephalography and Clinical Neurophysiology*, 1970, **28**, 136–145.

Baumgarten, H. G., & Lachenmayer, L. Chemically induced degeneration of indoleamine-containing nerve terminals in rat brain. *Brain Research*, 1972, **38**, 228–232.

Baxter, B. L. Induction of both emotional behavior and a novel form of REM sleep by chemical stimulations applied to cat mesencephalon. *Experimental Neurology*, 1969, **23**, 220–230.

Bizzi, E., Pompeiano, O., & Somogyi, I. Spontaneous activity of single vestibular neurons of unrestrained cats during sleep and wakefulness. *Archives Italiennes de Biologie*, 1964, **102**, 308–330.

Bobillier, P., Froment, J. L., Seguin, S., & Jouvet, M. Effets de la P: Chlorophenylalanine et du 5-hydroxytryptophane sur le sommeil et le métabolisme central des monoamines et des protéines chez le chat. *Biochemical Pharmacology*, 1973, **22**, 3077–3090.

Cespuglio, R. Organisation de l'activité ponto-géniculo-occipitale au niveau du système oculo-moteur. *Thèse de Neurophysiologie*, Lyon, 1973. 141 pp.

Dahlstrom, A., & Fuxe, K. Evidence for the existence of monoamine neurons in the central nervous system. I. Demonstration of monoamines in the cell bodies of brain stem neurons. *Acta Physiologica Scandinavica*, 1964, **62**, suppl. 232.

Dement, W. The occurrence of low voltage, fast, electroencephalogram patterns during behavioral sleep in the cat. *Electroencephalography and Clinical Neurophysiology*, 1958, **10**, 291–296.

Dement, W. The effect of dream deprivation. *Science*, 1960, **131**, 1705–1707.

Dement, W., & Kleitman, N. Cyclic variations in EEG during sleep and their relation to eye movements, body motility and dreaming. *Electroencephalography and Clinical Neurophysiology*, 1957, **9**, 673–690.

Eibl–Eibesfeldt, I. *Grundiss der Vergleichenden Verhaltens-forschung.* Munich: Piper Verlag, 1967.

Evarts, E. V. Temporal patterns of discharge of pyramidal tract neurons during sleep and waking in the monkey. *Journal of Neurophysiology*, 1964, **27**, 152–172.

Fisher, C., Kahn, E., Edwards, A., & Davis, D. Total suppression of REM sleep with nardil in a patient with intractable narcolepsy. In M. H. Chase, W. C. Stern, & P. L. Walter (Eds.), *Sleep research.* Los Angeles: Brain Inf. Service, 1972.

Froment, J. L., Petitjean, F., Bertrand, N., Cointy, C., & Jouvet, M. Effets de l'injection intracérébrale de 5-6-hydroxytryptamine sur les monoamines cérébrales et les états de sommeil du chat. *Brain Research,* 1974, **67**, 405–417.

Gadea–Ciria, M. Etude sequentielle des pointes ponto-géniculo-occipitale au cours du sommeil paradoxal chez le chat normal et après lésions corticales et sous-corticales. *Thèse de Neurophysiologie,* Marseille, 1972. 219 pp.

Gassel, M. M., Marchiafava, P. L., & Pompeiano, O. Activity of the red nucleus during deep desynchronized sleep in unrestrained cats. *Archives Italiennes de Biologie,* 1965, **103**, 369–397.

George, R., Haslett, W. L., & Jenden, D. J. A cholinergic mechanism in the brainstem reticular formation: Induction of paradoxical sleep. *International Journal of Pharmacology,* 1964, **3**, 451–552.

Glassman, E. The biochemistry of learning an evaluation of the role of RNA and protein. *Annual Review of Biochemistry,* 1969, **38**, 605–646.

Greenberg, R. Dreaming and memory. *International Psychiatry Clinics,* 1970, **7**, 258–267.

Henley, K., & Morrison, A. Release of organized behavior during desynchronized sleep in cats with pontine lesion. *Psychophysiology,* 1969, **6**, 245.

Hennevin, E., & Leconte, P. La fonction du sommeil paradoxal. Faits et hypothèses. *Année Psychologique,* 1971, **2**, 489–519.

Hobson, J. A. Electrographic correlates of behavior in the frog with special reference to sleep. *Electroencephalography and Clinical Neurophysiology,* 1967, **22**, 113–121.

Jacobson, M., & Gaze, R. M. Selection of appropriate tectal connections by regenerating optic nerve fibers in adult goldfish. *Experimental Neurology,* 1965, **13**, 418–430.

Jones, B. E., Bobillier, P., Pin, C., & Jouvet, M. The effects of lesion of catecholamine containing neurons upon monoamine content of the brain and EEG and behavioural waking in the cat. *Brain Research,* 1973, **58**, 157–177.

Jouvet, M. Behavioral and EEG effects of paradoxical sleep deprivation in the cat. In *Proceedings of the 23rd International Congress of Physiological Sciences.* Vol. 4. Excerpta Medica. Internat. Congress Series, No. 87. Tokyo, 1965. Pp. 344–355.

Jouvet, M. The role of monoamines and acetylcholine containing neurons in the regulation of the sleep waking cycle. *Ergebnisse der Physiologie,* 1972, **64**, 166–307.

Jouvet, M., & Delorme, J. F. Locus coeruleus et sommeil paradoxal. *Comptes Revues de la Societé de Biologie* (Paris), 1965, **159**, 895–899.

Jouvet, M. & Jouvet, D. Le sommeil et les rêves chez les animaux. In H. Ey (Ed.), *Psychiatrie animale.* Paris: Desclée de Brouwers, 1964.

Jouvet, M., Michel, F., & Courjon, J. Sur un stade d'activité électrique cérébrale rapide au cours du sommeil physiologique. *Comptes Revues de la Societé de Biologie* (Paris), 1959, **153**, 1024–1028.

Jouvet–Mounier, D., Astic, L., & Lacote, D. Ontogenesis of the states of sleep in rat, cat, and guinea pig during the first postnatal month. *Developmental Psychobiology,* 1970, **2**, 216–239.

Katzman, R., Bjorklund, A., Owman, CH., Stenevi, U., & West, K. A. Evidence for

regenerative axon sprouting of central catecholamine neurons in the rat mesencephalon following electrolytic lesion. *Brain Research,* 1971, **25,** 579–596.

Kitahama, K. Contribution à l'étude de la relation sommeil-apprentissage. Effets de
la privation de sommeil paradoxal par alpha methyl DOPA. *Thèse Sciences,*
Lyon, 1973. 98 pp.

Laurent, J. P., Cespuglio, R., & Jouvet, M. Délimitation des voies ascendantes de
l'activité ponto-géniculo-occipitale chez le chat. *Brain Research,* 1974, **65,** 29–52.

Lucero, M. A. Lengthening of REM sleep duration consecutive to learning in the
rat. *Brain Research,* 1970, **20,** 319–322.

Maeda, K., & Astic, L. Développement des neurones monoaminergiques centraux
chez le foetus de cobaye. *Comptes Revues de la Societé de Biologie* (Paris),
1972, **166,** 1014–1017.

Maeda, T., Pin, C., Salvert, D., Ligier, M., & Jouvet, M. Les neurones contenant
des catécholamines du tegmentum pontique et leurs voies de projection chez
le chat. *Brain Research,* 1973, **57,** 119–152.

Marchiafava, P. L., & Pompeiano, O. Pyramidal influences on spinal cord during
desynchronized sleep. *Archives Italiennes de Biologie,* 1964, **102,** 500–509.

Moore, R. Y., Bjorklund, A., & Stenevi, U. Plastic changes in the adrenergic innervation of the rat septal area in response to denervation. *Brain Research,* 1971,
33, 13–36.

Morrison, A. R., & Pompeiano, O. Vestibular influences during sleep. IV. Functional
relations between vestibular nuclei and lateral geniculate nucleus during desynchronized sleep. *Archives Italiennes de Biologie,* 1966, **104,** 425–458.

Moruzzi, G. The sleep–waking cycle. *Ergebnisse der Physiologie,* 1972, **64,** 1–164.

Nottebohm, F. Ontogeny of birds song. *Science,* 1970, **167,** 950.

Padilla, S. G. Further studies on the delayed pecking of chicks. *Journal of Comparative Psychology,* 1935, **20,** 413–443.

Passouant, P., Cadilhac, J., & Ribstein. Les privations de sommeil avec mouvements
oculaires par les antidépresseurs. In *Les mediateurs chimiques.* XXIX Reunion
Neurologique Internationale, Paris, Masson, 1972. Pp. 173–192.

Petitjean, F., & Jouvet, M. Hypersomnie et augmentation de l'acide 5-hydroxyin-
dolacétique cérébral par lésion isthmique chez le chat. *Comptes Revues de la
Societé de Biologie* (Paris), 1970, **164,** 2288–2293.

Peyrethon, J., & Dusan-Peyrethon, D. Etude polygraphique du cycle veille-sommeil
d'un teleosteen (tinca-tinca). *Comptes Revues de la Societé de Biologie* (Paris),
1967, **161,** 2533–2537.

Pompeiano, O. The neurophysiological mechanisms of the postural and motor events
during desynchronized sleep. *Research Publications, Association for Nervous and
Mental Disease,* 1967, **45,** 351–423.

Pompeiano, O. Mechanism of sensori motor integration during sleep. *Progress in Physiological Psychology,* 1970, 3, 1–179.

Pujol, J. F., Stein, D., Blondaux, Ch., Petitjean, F., Froment, J. L., & Jouvet, M.
Biochemical evidences for interaction phenomena between noradrenergic and
serotoninergic systems in the cat brain. In E. Usdin (Ed.), *Frontiers in catecholamines research.* New York: Pergamon Press, 1973.

Rechtschaffen, A., Wolpert, E. A., Dement, W. C., Mitchell, S. A., & Fisher, C.
Nocturnal sleep of narcoleptics. *Electroencephgraphy and Clinical Neurophysiology,* 1963, **15,** 599–609.

Roffwarg, H. P., Muzio, J. N., & Dement, W. C. Ontogenetic development of the
human sleep–dream cycle. *Science,* 1966, **52,** 604–619.

Roussel, B. Monoamines et sommeils: Suppression du sommeil paradoxal et diminution de la noradrénaline cérébrale par lésion des noyaux locus coeruleus. *Thèse Médecine*, Lyon, Tixier ed., 1967, 141 pp.

Shute, C. C. D., & Lewis, P. R. Cholinergic and monoaminergic systems of the brain. *Nature* (Lond.), 1966, **212**, 710–711.

Sperry, R. W. Optic nerve regeneration with return of vision in anurans. *Journal of Neurophysiology*, 1944, **7**, 57–69.

Stern, W. C. The D-state, dreaming and memory. *International Psychiatry Clinics*, 1970, **7**, 240–257.

Tauber, E. S., Roffwarg, H. P., & Weitzman, E. D. Eye movements and electroencephalogram activity during sleep in diurnal lizards. *Nature*, 1966, **212**, 1612–1613.

Tauber, E. S., & Weitzman, E. D. Eye movements during behavioral inactivity in certain bermuda reef fish. *Communications in Behavioral Biology*, 1969, **3**, 131–135.

Ursin, R. Sleep stage relations within the sleep cycles of the cat. *Brain Research*, 1970, **19**, 91–99.

Valverde, F. Appical dendritic spines of the visual cortex and light deprivation in the mouse. *Experimental Brain Research*, 1967, **3**, 337–352.

Vimont, P., Jouvet–Mounier, D., & Delorme, F. Effets EEG et comportementaux des privations de sommeil paradoxal chez le chat. *Electroencephalography and Clinical Neurophysiology*, 1966, **20**, 439–449.

Vital–Durand, F., & Michel, F. Effets de la désafférentation périphérique sur le cycle veille-sommeil chez le chat. *Archives Italiennes de Biologie*, 1971, **109**, 166–186.

Wiesel, T. N., & Hubel, D. H. Single-cell responses in striate cortex of kittens deprived of vision in one eye. *Journal of Neurophysiology*, 1963, **26**, 1003–1017.

Chapter 18

Cognition and Peripheralist–Centralist Controversies in Motivation and Emotion

STANLEY SCHACHTER

Columbia University

INTRODUCTION

Many years ago, piqued by the disorderly cataloguing of symptoms that characterized the then classic works on emotion, William James offered what was probably the first simple, integrating, theoretical statement on the nature of emotion. This well-known formulation stated simply that "the bodily changes follow directly the perception of the exciting fact, and that our feeling of the same changes as they occur *is* the emotion" (James, 1890, p. 449.) Since James's proposition equates bodily changes and visceral feelings with emotion, it must follow first, that the different emotions will be accompanied by recognizably different bodily states, and second, that the manipulation of bodily state, by drugs or surgery, will also manipulate emotional state. These implications have, directly or indirectly, guided much of the research on emotion since James's day. The results of this research, on the whole, provided little support for a purely visceral formulation of emotion, and led Cannon (1927, 1929) to his brilliant and devastating critique of the James–Lange theory—a critique based on these points:

1. The total separation of the viscera from the central nervous system does not alter emotional behavior.
2. The same visceral changes occur in very different emotional states and in nonemotional states.
3. The viscera are relatively insensitive structures.
4. Visceral changes are too slow to be a source of emotional feeling.
5. The artificial induction of visceral changes that are typical of strong emotions does not produce the emotions.

Although new data have weakened the cogency of some of these points, on the whole Cannon's logic and findings make it inescapably clear that a completely peripheral or visceral formulation of emotion, such as the James–Lange theory, is simply inadequate to cope with the facts. In an effort to deal with the obvious inadequacies of a purely visceral or peripheral formulation of emotion, Ruckmick (1936), Hunt, Cole, and Reis (1958), Schachter (1959), and others have suggested that cognitive factors may be major determinants of emotional states. In this chapter, I shall attempt to spell out the implications of a cognitive-physiological formulation of emotion and to describe a series of experiments designed to test these implications.

To begin, let us grant, on the basis of much evidence (see Woodworth and Schlosberg, 1954, for example), that a general pattern of sympathetic discharge is characteristic of emotional states. Given such a state of arousal, it is suggested that one labels, interprets, and identifies this state in terms of the characteristics of the precipitating situation and of one's apperceptive mass. This suggests, then, that an emotional state may be considered a function of a state of physiological arousal[*] and a cognition appropriate to this state of arousal. The cognition, in a sense, exerts a steering function. Cognitions arising from the immediate situation as interpreted by past experience provide the framework within which one understands and labels one's feelings. It is the cognition that determines whether the state of physiological arousal will be labeled "anger," "joy," or whatever.

---

[*] Though the experiments to be described are concerned largely with the physiological changes produced by the injection of adrenaline, which appear to be primarily the result of sympathetic excitation, the term physiological arousal is used in preference to the more specific "excitement of the sympathetic nervous system" because there are indications, to be discussed later, that this formulation is applicable to a variety of bodily states.

In order to examine the implications of this formulation let us consider how these two elements—a state of physiological arousal and cognitive factors—would interact in a variety of situations. In most emotion-inducing situations, of course, the two factors are completely interrelated. Imagine a man walking alone down a dark alley when a figure with a gun suddenly appears. The perception-cognition "figure with a gun" in some fashion initiates a state of physiological arousal, this state of arousal is interpreted in terms of knowledge about dark alleys and guns, and the state of arousal is labeled "fear." Similarly, a student who unexpectedly learns that he has made Phi Beta Kappa may experience a state of arousal which he will label "joy."

Let us now consider circumstances in which these two elements, the physiological and the cognitive, are, to some extent, independent. First, is the state of physiological arousal alone sufficient to induce an emotion? Best evidence indicates that it is not. Marañon (1924), in a fascinating study [which was replicated by Cantril and Hunt (1932) and Landis and Hunt (1932)], injected 210 of his patients with the sympathomimetic agent adrenaline and then asked them to introspect. Of his subjects, 71% simply reported physical symptoms with no emotional overtone; 29% of the subjects responded in an apparently emotional fashion. Of these, the great majority described their feelings in a way that Marañon labeled "cold" or "as if" emotions; that is, they made statements such as "I feel *as if* I were afraid" or "*as if* I were awaiting a great happiness." This is a sort of emotional "déjà vu" experience; these subjects are neither happy nor afraid, but only feel "as if" they were. Finally, a very few cases apparently reported a genuine emotional experience. However, in order to produce this reaction in most of these few cases, Marañon points out, "one must suggest a memory with a strong affective force but not so strong as to produce an emotion in the normal state. For example, before the injection, in several cases, we spoke to our patients about their sick children or dead parents, and they responded calmly to this topic. The same topic presented later, during the adrenal commotion, was sufficient to trigger emotion. This adrenal commotion places the subject in a situation of 'affective imminence.'" Apparently, then, to produce a genuinely emotional reaction to adrenaline, Marañon was forced to provide such subjects with an appropriate cognition.

Though Marañon does not explicitly describe his procedure, it is clear that his subjects knew that they were receiving an injection, and in all likelihood they knew that they were receiving adrenaline and probably had some familiarity with its effects. In short, though they underwent the pattern of sympathetic discharge common to strong emotional states, at the same time they had a completely appropriate cognition or explanation of why they felt this way. This, I would suggest, is the reason so few of Marañon's subjects reported any emotional experience.

Consider next a person in a state of physiological arousal for which no immediately explanatory or appropriate cognitions are available. Such a state could result were one to inject a subject with adrenaline covertly, or, to feed the subject, without his knowing it, a sympathomimetic drug such as ephedrine. Under such conditions a subject would be aware of palpitations, tremor, face flushing, and most of the symptoms associated with a discharge of the sympathetic nervous system. In contrast to Marañon's subjects, he would be utterly unaware of why he felt this way. What would be the consequence of such a state?

Schachter (1959) has suggested that just such a state would lead to the arousal of "evaluative needs" (Festinger, 1954); that is, an individual in this state would feel pressures to understand and label his bodily feelings. His bodily state grossly resembles the condition in which it has been at times of emotional excitement. How would he label his present feelings? It is suggested, of course, that he will label his feelings in terms of his knowledge of the immediate situation.* Should he at the time be with a beautiful woman he might decide that he was wildly in love or sexually excited. Should he be at a gay party, he might, by comparing himself to others, decide that he was extremely happy and euphoric. Should he be arguing with his wife, he might explode in fury and hatred. Or, should the situation be completely inappropriate, he might decide that he was excited about something that had recently happened to him, or, simply, that he was sick. In any case, it is my basic assumption that emotional states are a function of the interaction of such cognitive factors with a state of physiological arousal.

This line of thought, then, leads to the following propositions:

1. Given a state of physiological arousal for which an individual has no immediate explanation, he will "label" this state and describe his feelings in terms of the cognitions available to him. To the extent that cognitive factors are potent determiners of emotional states, one might anticipate that precisely the same state of physiological arousal could be labeled "joy" or "fury" or any of a great number of emotional labels, depending on the cognitive aspects of the situation.

2. Given a state of physiological arousal for which an individual has a completely appropriate explanation (e.g., "I feel this way because

---

* This suggestion is not new. Several psychologists have suggested that situational factors should be considered the chief differentiators of the emotions. Hunt, Cole, & Reis (1958) probably make this point most explicitly in their study distinguishing among fear, anger, and sorrow in terms of situational characteristics.

I have just received an injection of adrenaline"), no evaluative needs will arise and the individual is unlikely to label his feelings in terms of the alternative cognitions available.

Finally, consider a condition in which emotion-inducing cognitions are present but there is no state of physiological arousal. For example, an individual might be completely aware that he is in great danger but for some reason (drug or surgical) might remain in a state of physiological quiescence. Does he experience the emotion "fear"? This formulation of emotion as a joint function of a state of physiological arousal and an appropriate cognition, would, of course, suggest that he does not, which leads to my final proposition.

3. Given the same cognitive circumstances, the individual will react emotionally or describe his feelings as emotions only to the extent that he experiences a state of physiological arousal.*

THE EXPERIMENTS

The experimental test of these propositions requires: (1) the experimental manipulation of a state of physiological arousal or sympathetic activation; (2) the manipulation of the extent to which the subject has an appropriate or proper explanation of his bodily state; and (3) the creation of situations from which explanatory cognitions may be derived.

In order to satisfy these experimental requirements, Schachter and Singer (1962) designed an experiment that was cast in the framework of a study of the effects of vitamin supplements on vision. As soon as a subject arrived, he was taken to a private room and told by the experimenter:

> In this experiment we would like to make various tests of your vision. We are particularly interested in how certain vitamin compounds and vitamin supplements affect the visual skills. In particular, we want to find out how the vitamin compound called "Suproxin" affects your vision.
>
> What we would like to do, then, if we can get your permission, is to give you a small injection of Suproxin. The injection itself is mild and harm-

---

* In his critique of the James-Lange theory of emotion, Cannon (1929) makes the point that sympathectomized animals and patients do seem to manifest emotional behavior. This criticism is, of course, as applicable to the above proposition as it was to the James-Lange formulation. The issues involved will be discussed later in this chapter.

less; however, since some people do object to being injected we don't want
to talk you into anything. Would you mind receiving a Suproxin injection?

If the subject agreed to the injection (and all but one of 185 subjects
did), the experimenter continued with instructions we shall describe,
and then left the room. In a few minutes a doctor (a genuine M.D.)
entered the room, briefly repeated the experimenter's instructions, took
the subject's pulse, and then injected him with Suproxin.

Depending upon the experimental condition, the subject received one
of two forms of Suproxin—epinephrine or a placebo.

Epinephrine or adrenaline is a sympathomimetic drug, the effects of
which, with minor exceptions, are almost a perfect mimicry of a discharge
of the sympathetic nervous system. Shortly after injection, systolic blood
pressure increases markedly, heart rate increases somewhat, cutaneous
blood flow decreases while muscular and cerebral blood flow increase,
blood sugar and lactic acid concentrations increase, and respiration rate
increases slightly. The major subjective symptoms noted by the subject
are palpitation, tremor, and sometimes a feeling of flushing and acceler-
ated breathing. With a subcutaneous injection (in the dosage administered to our subjects) these effects usually begin within 3 to 5 minutes
of injection and last anywhere from 10 minutes to an hour. For most
subjects the effects are dissipated within 15 to 20 minutes after injection.

Subjects receiving epinephrine received a subcutaneous injection of
$\frac{1}{2}$ cc of a 1:1000 solution of Winthrop Laboratories' Suprarenin, a saline
solution of epinephrine bitartrate. Subjects in the placebo condition received a subcutaneous injection of $\frac{1}{2}$ cc of saline solution.

Manipulating an Appropriate Explanation

By an "appropriate" explanation, I refer to the extent to which the
subject has an authoritative, unequivocal explanation of his bodily condition. Thus, a subject who had been informed by the physician that
as a direct consequence of the injection he would feel palpitations,
tremor, and so on would be considered to have a completely appropriate
explanation. A subject who had been informed only that the injection
would have no side effects would have no appropriate explanation of
his state. This dimension of appropriateness was manipulated in three
experimental conditions, which shall be called:

1. Epinephrine Informed (Epi Inf)
2. Epinephrine Ignorant (Epi Ign)
3. Epinephrine Misinformed (Epi Mis)

Immediately after the subject had agreed to the injection and before the physician entered the room, the experimenter gave one of the following speeches, depending on the condition:

1. *Epinephrine Informed.*

> I should also tell you that some of our subjects have experienced side effects from the Suproxin. These side effects are transitory; that is, they will last only for about 15 or 20 minutes. What will probably happen is that your hand will start to shake, your heart will start to pound, and your face may get warm and flushed. Again, these are side effects lasting 15 or 20 minutes.

While the physician was giving the injection, she told the subject that the injection was mild and harmless, and repeated the description of the symptoms that the subject could expect as a consequence of the injection. In this condition, then, subjects have a completely appropriate explanation of their bodily state. They know precisely what they will feel and why.

2. *Epinephrine Ignorant.* In this condition, when the subject agreed to the injection, the experimenter said nothing more about side effects and simply left the room. While the physician was giving the injection, she told the subject that the injection was mild and harmless and would have no side effects. In this condition, then, the subject had no experimentally provided explanation for his bodily state.

3. *Epinephrine Misinformed.*

> I should also tell you that some of our subjects have experienced side effects from the Suproxin. These side effects are transitory, that is, they will last only for about 15 or 20 minutes. What will probably happen is that your feet will feel numb, you will have an itching sensation over parts of your body, and you may get a slight headache. Again, these are side effects lasting 15 or 20 minutes.

And again, the physician repeated these symptoms while injecting the subject.

None of these symptoms, of course, are consequences of an injection of epinephrine and, in effect, these instructions provide the subject with a completely inappropriate explanation of his bodily feelings. This condition was introduced as a control condition of sorts. It seemed possible that the description of side effects in the Epi Inf condition might make the subject introspective and possibly slightly troubled. Differences in the dependent variable between the Epi Inf and Epi Ign conditions might then be due to such factors rather than to differences in appropriateness. The false symptoms in the Epi Mis conditions should similarly make the subject introspective, but the instructions in this condition do not provide an appropriate explanation of the subject's state.

Subjects in all of the above conditions were injected with epinephrine. Finally, there was a placebo condition, in which subjects were injected with saline solution and were then given precisely the same treatment as subjects in the Epi Ign condition.

Producing an Emotion-Inducing Cognition

My initial hypothesis suggested that, given a state of physiological arousal for which the individual has no adequate explanation, cognitive factors can lead the individual to describe his feelings with any of a number of emotional labels. In order to test this hypothesis, it was decided to manipulate two emotional states that can be considered quite different—euphoria and anger.

There are, of course, many ways to induce such states. In my own program of research, we have concentrated on social determinants of emotional states. We have demonstrated in other studies that people evaluate their own feelings by comparing themselves with others around them (Wrightsman, 1960; Schachter, 1959). In the experiment being described, an attempt was again made to manipulate emotional state by social means. In one set of conditions, the subject was placed with a stooge who had been trained to act euphorically. In a second set of conditions the subject was placed with a stooge trained to act in an angry fashion.

Euphoria

Immediately* after the subject had been injected, the physician left the room and the experimenter returned with a stooge whom he introduced as another subject. The experimenter then said:

> Both of you had the Suproxin shot and you'll both be taking the same tests of vision. What I ask you to do now is just wait for 20 minutes. The reason for this is simply that we have to allow 20 minutes for the Suproxin to get from the injection site into the bloodstream. At the end of 20 minutes, when we are certain that most of the Suproxin has been absorbed into the bloodstream, we'll begin the tests of vision.

---

* It was, of course, imperative that the sequence with the stooge begin before the subject felt his first symptoms, since otherwise the subject would be virtually forced to interpret his feelings in terms of events preceding the stooge's entrance. Pretests had indicated that for most subjects, epinephrine-induced symptoms began within 3 to 5 minutes after injection. A deliberate attempt was made then to bring in the stooge within 1 minute after the subject's injection.

The room in which this was said had been deliberately put into a state of mild disarray. As he was leaving, the experimenter apologetically added, "The only other thing I should do is to apologize for the condition of the room. I just didn't have time to clean it up. So, if you need any scratch paper or rubber bands or pencils, help yourself. I'll be back in 20 minutes to begin the vision tests."

As soon as the experimenter had left, the stooge introduced himself again, made a series of standard icebreaking comments, and then launched his routine:

He reached first for a piece of paper, doodled briefly, crumpled the paper, aimed for a wastebasket, threw, and missed. This led him into a game of "basketball," in which he moved about the room crumpling paper and trying out fancy basketball shots. Finished with basketball, he said, "This is one of my good days. I feel like a kid again. I think I'll make a plane." He made a paper plane, spent a few minutes flying it around the room, and then said, "Even when I was a kid, I was never much good at this." He then tore off the tail of his plane, wadded it up, and making a slingshot of a rubber band, began to shoot the paper. While shooting, he noticed a sloppy pile of manila folders. He built a tower of these folders, then went to the opposite end of the room to shoot at the tower. He knocked down the tower, and while picking up the folders, he noticed behind a portable blackboard a pair of hula hoops. He took one of these for himself, put the other within reaching distance of the subject and began hula hooping. After a few minutes of this he replaced the hula hoop and returned to his seat, at which point the experimenter returned to the room.

All through this madness an observer, through a one-way mirror, systematically recorded the subject's behavior and noted the extent to which the subject joined in the stooge's whirl of activity.

Subjects in each of the three "appropriateness" conditions and in the placebo condition were submitted to this setup. The stooge, of course, never knew to which condition any particular subject had been assigned.

Anger

Immediately after the injection, the experimenter brought a stooge into the subject's room, introduced the two, and after explaining the necessity for a 20-minute delay for "the Suproxin to get from the injection site into the bloodstream," he continued, "We would like you to use these 20 minutes to answer these questionnaires." Then handing out the questionnaires, he concluded, "I'll be back in 20 minutes to pick up the questionnaires and begin the tests of vision."

The questionnaires, five pages long, began innocently, requesting face-sheet information, and then grew increasingly personal and insulting, asking questions such as :

> With how many men (other than your father) has your mother had extramarital relationships?
> 4 and under _____; 5–9 _____; 10 and over _____.

The stooge, sitting directly opposite the subject, paced his own answers so that at all times subject and stooge were working on the same question. At regular points in the questionnaire, the stooge made standardized comments about the questions. His comments started innocently enough, but grew increasingly querulous. Finally, in a rage, he tore up his questionnaire, slammed it to the floor, saying "I'm not wasting any more time. I'm getting my books and leaving," and stamped out of the room.

Again an observer recorded the subject's behavior.

In summary, this is a 7-condition experiment that for two different emotional states allows us: (1) to evaluate the effects of "appropriateness" on emotional inducibility, and (2) to begin to evaluate the effects of sympathetic activation on emotional inducibility. In schematic form the conditions are the following:

Euphoria	Anger
Epi Inf	Epi Inf
Epi Ign	Epi Ign
Epi Mis	Placebo
Placebo	

The Epi Mis condition was not run in the anger sequence. This was originally conceived as a control condition and it was felt that its inclusion in the euphoria conditions alone would suffice as a means of evaluating the possible artifactual effect of the Epi Inf instructions.

The subjects were all male college students taking classes in introductory psychology at the University of Minnesota. The records of all potential subjects were reviewed by the Student Health Service in order to ensure that no harmful effects would result from the injections.

Measurement

Two types of measurements of emotional state were obtained. Standardized observation through a one-way mirror was used to assess the subject's behavior. To what extent did he join in with the stooge's pattern of behavior and act euphoric or angry? The second type was a self-report questionnaire in which, on a variety of scales, the subject indicated his mood of the moment.

These measurements were obtained immediately after the stooge had finished his routine, at which point the experimenter returned, saying:

> Before we proceed with the vision tests, there is one other kind of information we must have. We have found that there are many things besides Suproxin that affect how well you see in our tests. How hungry you are, how tired you are, and even the mood you're in at the moment—whether you feel happy or irritated at the time of testing—will affect how well you see. To understand the data we collect on you, then, we must be able to figure out which effects are due to causes such as these and which are caused by Suproxin.

He then handed out questionnaires containing a number of questions about bodily and emotional state. To measure mood, the following two were the crucial questions:

1. How irritated, angry, or annoyed would you say you feel at present?

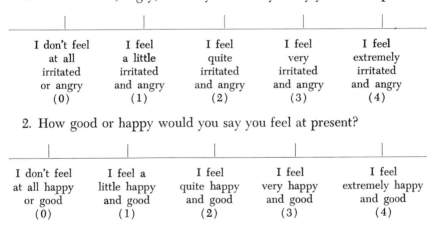

| I don't feel at all irritated or angry (0) | I feel a little irritated and angry (1) | I feel quite irritated and angry (2) | I feel very irritated and angry (3) | I feel extremely irritated and angry (4) |

2. How good or happy would you say you feel at present?

| I don't feel at all happy or good (0) | I feel a little happy and good (1) | I feel quite happy and good (2) | I feel very happy and good (3) | I feel extremely happy and good (4) |

THE EFFECTS OF THE MANIPULATIONS ON EMOTIONAL STATE

Euphoria

The effects of the several manipulations on emotional state in the euphoria conditions are presented in Table 1. The scores recorded in this table are derived, for each subject, by subtracting the value of the point he checks on the "irritation" scale from the value of the point he checks on the "happiness" scale. Thus, if a subject were to check the point "I feel a little irritated and angry" on the "irritation" scale,

TABLE 1 Self-Report of Emotional State in the
Euphoria Conditions[a]

Condition	N	Self-report scales
Epi Inf	25	0.98
Epi Ign	25	1.78
Epi Mis	25	1.90
Placebo	26	1.61

[a] Values of p: Epi Inf versus Epi Mis, $<.01$;
Epi Inf versus Epi Ign, .02;
Placebo versus Epi Mis, Epi
Ign, or Epi Inf, n.s.
All p values reported in this paper are two-
tailed.

and the point "I feel very happy and good" on the "happiness" scale,
his score would be $+2$. The higher the positive value, the happier and
better the subject reports himself to be feeling. Though an index is
employed for expositional simplicity, it should be noted that each of
the two components of the index yields results completely consistent
with those obtained by using this index.

Let us examine first the effects of the "appropriateness" instructions.
A comparison of the scores of the Epi Mis and Epi Inf conditions makes
it immediately clear that the experimental differences are not due to
artifacts resulting from the "informed" instructions. In both conditions
the subject was warned to expect a variety of symptoms as a consequence
of the injection. In the Epi Mis condition, where the symptoms were
inappropriate to the subject's bodily state, the self-report score is almost
twice that in the Epi Inf condition, where the symptoms were completely
appropriate to the subject's bodily state. It is reasonable, then, to attribute differences between informed subjects and those in other conditions
to differences in manipulated appropriateness rather than to artifacts
such as introspectiveness.

It is clear that, consistent with expectations, subjects were more susceptible to the stooge's mood, and consequently more euphoric, when
they had no explanation of their own bodily states than when they
had an explanation. The means of both the Epi Ign and Epi Mis conditions are considerably larger than the mean of the Epi Inf condition.

Comparing the placebo condition to the epinephrine conditions, we
note a pattern that will be repeated throughout the data. Placebo subjects
are less euphoric than either Epi Mis or Epi Ign subjects, but somewhat

more euphoric than Epi Inf subjects. These differences are not, however, statistically significant. I shall consider the epinephrine–placebo comparisons in detail later in this chapter. For the moment, it is clear from these self-reports that manipulating "appropriateness" has had a very strong effect on euphoria.

The analysis of the observational data is reported in detail elsewhere (Schachter & Singer, 1962). Here it is sufficient to note that on all behavioral indices devised—e.g., the amount of time the subject spends on stooge-initiated activity, "creative euphoria" (the extent to which the subject initiates euphoric activities of his own devising)—the same pattern of between-condition relationships holds. Subjects in the Epi Mis and Epi Ign conditions behave more euphorically than subjects in the Epi Inf condition. Placebo subjects again fall between Epi Ign and Epi Inf subjects.

Anger

In the anger conditions, we should again expect that the subject will catch the stooge's mood only in those conditions where he has been injected with epinephrine and has no appropriate explanation for the bodily state thus created. Subjects in the Epi Ign conditions should, then, be considerably angrier than those in the Epi Inf or the placebo condition. Data on behavioral indications of anger are presented in Table 2. These figures are derived from a coding of the subject's comments and behavior during the experimental session with the angry stooge. The nature of the index devised is described in detail elsewhere (Schachter & Singer, 1962). For present purposes, it is sufficient to note that a positive value of this index indicates that the subject agrees with the stooge's comments and is angry. The larger the positive value, the angrier the subject. A negative value indicates that the subject either disagrees with the stooge or ignores him.

It is evident in Table 2 that expectations are confirmed. The value for the Epi Ign condition is positive and large, indicating that the subjects have become angry, whereas in the Epi Inf condition the score is slightly negative, indicating that these subjects have completely failed to catch the stooge's mood. Placebo subjects fall between Epi Ign and Epi Inf subjects. On the self-report scales of mood, this pattern is repeated, though on this measure, placebo subjects do not differ significantly from either Epi Ign or Epi Inf subjects.

Now that the basic data of this study have been presented, let us examine closely how well they conform to theoretical expectations. If my hypotheses are correct and if this experimental design provided

TABLE 2   Behavioral Indications of Emotional
State in the Anger Conditions[a]

Condition	N	Anger index
Epi Inf	22	−0.18
Epi Ign	23	+2.28
Placebo	22	+0.79

[a] Values of p: Epi Inf versus Epi Ign, $<.01$;
Epi Ign versus placebo, $<.05$;
Placebo versus Epi Inf, n.s.

a perfect test for these hypotheses, it should be anticipated that in the euphoria conditions the degree of experimentally produced euphoria should vary in the following way:

$$\text{Epi Mis} \geqq \text{Epi Ign} > \text{Epi Inf} = \text{placebo}$$

And in the anger conditions, anger should conform to the following pattern:

$$\text{Epi Ign} > \text{Epi Inf} = \text{placebo}$$

In both the euphoria and the anger conditions, emotional level in the Epi Inf condition is considerably less than that achieved in any of the other Epi conditions. The results for the placebo condition, however, are ambiguous, since the placebo subjects consistently fall between the Epi Ign and the Epi Inf subjects. This is a particularly troubling pattern because it makes it impossible to evaluate unequivocally the effects of the state of physiological arousal, and indeed raises serious questions about the entire theoretical structure. Although the emotional level is consistently greater in the Epi Mis and Epi Ign conditions than in the placebo condition, this difference is significant at acceptable probability levels only on the behavioral indices in the anger conditions.

In order to explore the problem further, let us examine experimental factors that might have acted to restrain the emotional level in the Epi Ign and Epi Mis conditions. Clearly the ideal test of the first two hypotheses requires an experimental setup in which the subject has flatly no way of evaluating his state of physiological arousal other than by means of the experimentally provided cognitions. Had it been possible to produce physiologically a state of sympathetic activation by means other than injection, one could have approached this experimental ideal more closely than in the present setup. As it stands, however, there is always a reasonable alternative cognition available to the aroused subject—he feels the way he does because of the injection. To the extent that the subject seizes on such an explanation of his bodily state, we should expect that he will be uninfluenced by the stooge.

It is possible, fortunately, to examine the effect of this artifact. In answers to open-end questions in which subjects described their own mood and physical state, some of the Epi Ign and Epi Mis subjects clearly attributed their physical state to the injection, saying, e.g., "the shot gave me the shivers." In effect, these subjects are self-informed. Comparing these subjects with the remaining subjects in a condition, one finds in the anger Epi Ign condition that self-informed subjects are considerably less angry than the remaining subjects. Similarly in the euphoria Epi Mis and Epi Ign conditions, self-informed subjects are considerably less euphoric than their non-self-informed counterparts. If one eliminates such self-informed subjects, the differences between the placebo and Epi Ign or Epi Mis conditions become highly significant statistically in both the anger and the euphoria set of conditions. Clearly, indications are good that this self-informing artifact has attenuated the effects of epinephrine.

Let us examine next the fact that, consistently, the emotional level in placebo conditions is higher than that in the Epi Inf conditions. Theoretically, of course, it should be expected that the level in the two conditions will be equally low, for by assuming that emotional state is a joint function of a state of physiological arousal and the appropriateness of a cognition, we are, in effect, assuming a multiplicative function, so that if either component is at zero, emotional level is at zero. This expectation should hold, however, only if we can be sure that there is no sympathetic activation in the placebo conditions. This assumption, of course, is completely unrealistic, since the injection of placebo does not prevent sympathetic activation. The experimental situations were fairly dramatic, and certainly some of the placebo subjects must have experienced physiological arousal. If this general line of reasoning is correct, it should be anticipated that the emotional level of subjects who give indications of sympathetic activity will be greater than that of subjects who do not.

Since in all conditions a subject's pulse was taken before the injection and again after the session with the stooge, there is one index of sympathetic activation available—change in pulse rate. The predominant pattern in the placebo conditions was, of course, a decrease in pulse rate. It will be assumed, therefore, that in the placebo conditions, those subjects whose pulses increase or remain the same give indications of sympathetic arousal, whereas those subjects whose pulses decrease do not. Comparing, within placebo conditions, such self-aroused subjects with those who give no indication of sympathetic activation, we find in the anger condition that those subjects whose pulses increase or remain the same are considerably and significantly angrier than those subjects whose pulses decrease. Similarly, in the euphoria placebo condition, the self-aroused subjects are considerably and significantly more euphoric than the subjects who give no indication of sympathetic activation. As expected, sympathetic activation accompanies an increase in emotional level.

It should be noted, too, on the several indices, that the emotional levels of subjects who show no signs of sympathetic activity are quite close to the emotional levels of subjects in the parallel Epi Inf conditions. The similarity of these sets of scores and their uniformly low level of indicated emotionality would certainly make it appear that both factors are essential to an emotional state. When either the level of sympathetic arousal is low or a completely appropriate cognition is available, the level of emotionality is low.

Let us summarize the major findings of this experiment and examine the extent to which they support the propositions offered at the beginning of this chapter. It has been suggested, first, that given a state of physiological arousal for which an individual has no explanation, he will label this state in terms of the cognitions available to him. This implies, of course, that by manipulating the cognitions of an individual in such a state, we can manipulate his feelings in diverse directions. Experimental results support this proposition, since after the injection of epinephrine, those subjects who had no explanation for the bodily state thus produced, proved readily manipulable into the disparate feeling states of euphoria and anger.

From this first proposition, it must follow that given a state of physiological arousal for which the individual has a completely satisfactory explanation, he will not label this state in terms of the alternative cognitions available. Experimental evidence strongly supports this expectation. In those conditions in which subjects were injected with epinephrine and told precisely what they would feel and why, they proved relatively immune to any effects of the manipulated cognitions. In the anger condition, these subjects did not become at all angry; in the euphoria condition, these subjects reported themselves to be far less happy than subjects with an identical bodily state but no adequate knowledge of why they felt the way they did.

Finally, it has been suggested that, given constant cognitive circumstances, an individual will react emotionally only to the extent that he experiences a state of physiological arousal. Without taking account of experimental artifacts, the evidence in support of this proposition is consistent but tentative. When the effects of "self-informing" tendencies in epinephrine subjects and of "self-arousing" tendencies in placebo subjects are partialed out, the evidence strongly supports the proposition.

The pattern of data, then, falls neatly into line with theoretical expectations. However, the fact that we were forced to rely to some extent on internal analyses in order to partial out the effects of experimental artifacts inevitably makes these conclusions somewhat tentative. In order further to test these propositions on the interaction of cognitive and physiological determinants of emotional state, a series of additional experiments was designed to rule out or overcome the operation of these artifacts.

The first of these experiments was designed by Schachter and Wheeler (1962) to test, by extending the range of manipulated sympathetic activation, the proposition that emotionality is positively related to physiological arousal. It seemed clear from the results of the study just described that the self-arousing tendency of placebo subjects tended to obscure the differences between placebo and epinephrine conditions. A test of the proposition at stake, then, would require a comparison of subjects who have received injections of epinephrine with subjects who are rendered incapable, to some extent, of self-activation of the sympathetic nervous system. Thanks to a class of drugs known generally as autonomic blocking agents, such blocking is, to some degree, possible. If it is correct that a state of sympathetic discharge is a necessary component of an emotional experience, it should be anticipated that whatever emotional state is experimentally manipulated should be experienced most strongly by subjects who have received epinephrine, next by placebo subjects, and least of all by subjects who have received injections of an autonomic blocking agent.

In order to conceal the purposes of the study and the nature of the injection, the experiment was again cast in the framework of a study of the effects of vitamins on vision. As soon as a subject (again, subjects were male college students) arrived, he was taken to a private room and told by the experimenter:

> I've asked you to come today to take part in an experiment concerning the effects of vitamins on the visual processes. Our experiment is concerned with the effects of Suproxin on vision. Suproxin is a high-concentrate vitamin C derivative. If you agree to take part in the experiment, we will give you an injection of Suproxin and then subject your retina to about fifteen minutes of continuous black and white stimulation. This is simpler than it sounds: we'll just have you watch a black and white movie. After the movie, we'll give you a series of visual tests.
>
> The injection itself is harmless and will be administered by our staff doctor. It may sting a little at first, as most injections do, but after this you will feel nothing and will have no side effects. We know that some

people dislike getting injections, and if you take part in the experiment, we want it to be your own decision. Would you like to?

All subjects agreed to take part. There were three forms of Suproxin administered—epinephrine, placebo, and chlorpromazine.

1. Epinephrine: Subjects in this condition received a subcutaneous injection of ½ cc of a 1:1000 solution of Winthrop Laboratories' Suprarenin.
2. Placebo: Subjects in this condition received a subcutaneous injection of ½ cc of saline solution.
3. Chlorpromazine: Subjects in this condition received an intramuscular injection of a solution consisting of 1 cc (25 mg) of Smith, Kline & French Thorazine and 1 cc of saline solution.

The choice of chlorpromazine as a blocking agent was dictated by considerations of safety, ease of administration, and known duration of effect. Ideally, one would have wished for a blocking agent whose mechanism and effect were precisely and solely the reverse of those of epinephrine—a peripherally acting agent that would prevent the excitation of sympathetically innervated structures. Though it is certainly possible to approach this ideal more closely with agents other than chlorpromazine, such drugs tend to be dangerous, or difficult to administer, or of short duration.

Chlorpromazine is known to act as a sympathetic depressant. It has a moderate hypotensive effect, with a slight compensatory increase in heart rate. It has mild adrenergic blocking activity, since it reverses the pressor effects of small doses of epinephrine and depresses responses of the nictitating membrane to preganglionic stimulation. Killam (1959) summarizes what is known and supposed about the mechanism of action of chlorpromazine as follows: "Autonomic effects in general may be attributed to a mild peripheral adrenergic blocking activity and probably to central depression of sympathetic centers, possibly in the hypothalamus." Popularly, of course, the compound is known as a "tranquilizer."

It is known that chlorpromazine has effects other than the sympatholytic effect of interest to us. For purposes of experimental purity, this is unfortunate but inevitable in this sort of research. It is clear, however, that the three conditions do differ in the degree of manipulated sympathetic activation.

Rather than the more complicated devices employed in the previous experiment, an emotion-inducing film was used as a means of manipulating the cognitive component of emotional states. In deciding on the type of film, two extremes seemed possible—a horror-, fright-, or anxiety-provoking film, or a comic, amusement-provoking film. Since it is a common stereotype that adrenaline makes one nervous and that the tranquilizer, chlorpromazine, makes one tranquil and mildly euphoric, the predicted pattern of results with a horror film would be subject to alternative interpretation. It was deliberately decided, then, to use a comedy. If my hypothesis is correct, it should be anticipated that epinephrine subjects would find the film somewhat funnier than placebo subjects who, in turn, would be more amused than chlorpromazine subjects.

The film chosen was a 14-minute excerpt from a Jack Carson movie called "The Good Humor Man." This excerpt is a self-contained, comprehensible episode involving a slapstick chase scene.

Three subjects, one from each of the drug conditions, always watched the film simultaneously. The projection room was deliberately arranged so that the subjects could neither see nor hear one another. Facing the screen were three theater-type seats separated from one another by large, heavy partitions. In a further attempt to maintain the independence of the subjects, the sound volume of the projector was turned up to mask any sounds made by the subjects.

The subjects' reactions while watching the film were used as the chief index of amusement. During the showing of the movie an observer, who had been introduced as an assistant who would help administer the visual tests, systematically scanned the subjects and recorded their reactions to the film. He observed each subject once every 10 seconds, so that over the course of the film 88 units of each subject's behavior were categorized. The observer simply recorded each subject's reaction to the film according to the following scheme:

1. Neutral: straight-faced watching of film with no indication of amusement.
2. Smile.
3. Grin: a smile with teeth showing.
4. Laugh: a smile or grin on face accompanied by bodily movements usually associated with laughter, e.g., shaking shoulders, moving head.
5. Big laugh: belly laugh; a laugh accompanied by violent body movement such as doubling up, throwing up hands.

In a minute-by-minute comparison, two independent observers agreed in their categorization of 90% of the 528 units recorded in six different reliability trials.

The observer, of course, never knew which subject had received which injection.

The observation record provides a continuous record of each subject's reaction to the film. As an overall index of amusement, the number of units in which a subject's behavior was recorded in the categories "smile," "grin," "laugh," and "big laugh" are summed together. The means of this amusement index are presented in Table 3. The larger the figure, the more amusement was manifest. Differences are in the anticipated direction. Epinephrine subjects gave indications of greater amusement than placebo subjects who, in turn, were more amused than chlorpromazine subjects.

Though the trend is clearly in the predicted direction, epinephrine and placebo subjects do not differ significantly in this overall index. The difference between these two groups, however, becomes apparent when we examine strong ("laugh" and "big laugh") reactions to the film; we find an average of 4.84 such units among the epinephrine subjects and of only 1.83 such units among placebo subjects. This difference is significant at better than the .05 level of significance. Epinephrine subjects tend to be openly amused at the film, placebo subjects to be quietly amused. Some 16% of epinephrine subjects reacted at some point with belly laughs, whereas not a single placebo subject did so. It should be noted that this is much the state of affairs one would expect from the disguise injection of epinephrine—a manipulation which, as has been suggested, creates a bodily state "in search of" an appropriate cognition. Certainly laughter can be considered a more appropriate accompaniment to the state of sympathetic arousal than can quietly smiling.

It would appear, then, that the degree of overt amusement is directly related to the degree of manipulated sympathetic activation.

A further test of the relationship of emotionality to sympathetic activity was made by Singer (1963), who in a deliberate attempt to rule out the operation of the self-informing artifact, conducted his study on

TABLE 3   The Effects of Epinephrine, Placebo, and
Chlorpromazine on Amusement[a]

Condition	N	Mean amusement index
Epinephrine	38	17.79
Placebo	42	14.31
Chlorpromazine	46	10.41

[a] Values of p: Epi versus placebo, n.s.;
Epi versus chlorpromazine, $<.01$;
Placebo vs. chlorpromazine, $<.05$.

rats—a species unlikely to attribute an aroused physiological state to an injection. Among other things, Singer examined the effects of injections of epinephrine (an intraperitoneal injection of epinephrine suspended in peanut oil in a concentration of .10 mg per kilogram of body weight), chlorpromazine (chlorpromazine hydrochloride in saline in a concentration of 2.0 mg per kilogram of body weight), and placebo on the reactions of rats to standard frightening situations. His technique was simple. In fright conditions, he placed his animals in a box containing a doorbell, a door buzzer, and a flashing 150-watt bulb. After a brief interval a switch was tripped, setting off all three devices simultaneously for a 1.5-minute interval. In nonfright conditions, of course, the switch was never tripped.

Singer's results are presented in Table 4. The figures presented in this table represent an index whose components are generally accepted indicators of fright, such as defecation, urination, and the like. The larger the figure the more frightened the animal. Clearly there is a substantial drug-related difference in the fright condition, and no difference at all in the nonfright condition. In the fear conditions, adrenaline-injected rats are more frightened than placebo rats $(p < .05)$, who in turn are more frightened than chlorpromazine rats $(p < .05)$. The

TABLE 4 Sympathetic Arousal and Fright

| | Mean fright index in | | | |
| | Fear condition | | Nonfear condition | |
Drug treatment	N	Mn	N	Mn
Epinephrine	18	13.20	18	7.33
Placebo	12	11.38	12	7.16
Chlorpromazine	6	9.72	6	7.94

| | Analysis of variance | | | |
Source	df	MS	F	p
Drug	2	12.41	3.32	$< .05$
Fear	1	411.30	109.97	$< .001$
D × F	2	19.61	5.24	$< .01$
Error	66	3.74		

drug–stress interaction is significant at better than the .01 level of confidence. It would certainly appear that under these experimental circumstances the state of fear is related to sympathetic activity. Further evidence for this relationship is found in a study, conducted by Latané and Schachter (1962), which demonstrated that rats injected with epinephrine were notably more capable of avoidance learning than rats injected with a placebo. Using a modified Miller–Mowrer shuttle-box, these investigators found that during an experimental period involving 200 massed trials, 15 rats injected with epinephrine avoided shock an average of 101.2 trials, whereas 15 placebo-injected rats averaged only 37.3 avoidances.

Let us summarize the facts of the studies that have been presented. We know that :

1. Given a state of epinephrine-induced sympathetic arousal, subjects may be manipulated into states of euphoria and anger if they have not been provided with an appropriate explanation of their bodily state.

2. Given the same state of arousal, subjects are virtually nonmanipulable into such emotion or mood states if they have a proper explanation of their bodily feelings (e.g., "My heart is pounding because of the injection").

3. Making allowance for experimental artifacts, subjects injected with placebo are less manipulable into euphoric and angry states than are subjects injected with epinephrine and given no explanation of their feelings.

4. Amusement at a movie is directly related to the degree of arousal as manipulated by injections of epinephrine, placebo, and chlorpromazine.

5. In fear-inducing situations, the intensity of fear and fear-related behavior in rats is, within nondebilitating dose limits, directly related to manipulated sympathetic arousal. In non-fear-inducing situations, manifestations of fear are few and are unrelated to sympathetic arousal.

Given this assortment of facts, the evidence in support of the propositions that generated these studies seems strong and convincing. It does appear experimentally useful to conceive of emotional states as a function of both cognitive or situational factors, and of physiological arousal.

PERIPHERAL AND CENTRAL THEORIES OF EMOTION AND MOTIVATION

Other than their relevance to my own line of thought these facts do, it seems to me, have critical implications for some of the central controversies that over the years have dominated research and debate in the fields of emotion and motivation. Since William James's day the study of emotion has tended to polarize around two sets of facts:

1. Emotional states are accompanied by various marked peripheral or visceral physiological changes such as modification of blood pressure, heart rate, endocrine levels, and the like.

2. Directly manipulating various lower brain structures, by lesion or electric stimulation techniques, may directly manipulate emotional states such as rage and fear.

This pair of complementary facts has been the focus of the famous James–Cannon debate over the peripheral versus the central nature of emotion. James, Lange, and their followers, even those of today, have maintained that the feeling state accompanying the various peripheral bodily "changes as they occur *is* the emotion." Cannon, Bard, and probably most contemporary brain physiologists maintain that peripheral activity is irrelevant, and that emotional states are controlled by the activation of particular structures in the central nervous system.

It is of interest to note that this peripheral–central dimension of controversy has also been the focus of much of the scientific activity in the study of bodily states such as hunger or thirst, which, like emotion, are accompanied by "feelable," measurable, peripheral activity. The gist of the vast body of research on the physiological correlates of hunger, for example, can be summarized in a fashion that perfectly parallels my summary of the study of the physiological correlates of emotion:

1. Food deprivation leads to peripheral physiological changes such as modification of various blood constituents, increase in gastric motility, changes in body temperature, and the like.

2. Directly manipulating hypothalamic structures, by lesion or electric stimulation techniques, can directly manipulate the amount eaten.

And, again, this pair of facts has provided the basis for a long-term, active controversy on peripheral versus central mechanisms of hunger regulation. (See Rosenzweig, 1962, for an absorbing scientific history of this controversy.)

Whether for emotion, or hunger, or thirst, this peripheralist–centralist controversy has been marked, then, by the opposition between those who choose to identify a particular state with some visceral process, processes, or structure, and those who choose to identify the same state with some brain process or structure. Note, in both cases, the state is identified with a particular physiological process, or structure, or change. Though no one has bothered to make the assumption explicit, both peripheralists and centralists accept what, in other context, I have called the assumption of "identity" (Schachter, 1970); that is, the assumption that there is a one-to-one relationship between a set or pattern of physiological processes or biochemical changes and a specific behavior or psychological state. It is this assumption of identity which is, I believe, most seriously called into question by my series of experiments on emotion and adrenaline which demonstrate that precisely the same physiological state, a state of epinephrine-induced arousal, can, depending on cognitive circumstances, be interpreted as euphoria, anger, amusement at a movie, fear, or, as in the Epi Inf conditions, no emotion or mood state at all. It is this assumption of identity which is at the heart of the peripheralist–centralist difficulties and which, as I will try to demonstrate, is responsible for the fact that neither a purely central nor a purely peripheral point of view can possibly cope with the existing facts.

Let us first review the recognized inadequacies of a purely visceral formulation of emotion and examine the extent to which the addition of cognitive factors allows us to cope with these shortcomings. Since Cannon's critique (1927, 1929) has been the most lucid and influential attack on a visceral view of emotion, I shall focus discussion around Cannon's five criticisms of the James–Lange theory. Each of these critical points, it will be noted, is essentially an attack on the "identity" implications of James's view of matters.

A REEXAMINATION OF CANNON'S CRITIQUE OF A VISCERAL FORMULATION OF EMOTION

Criticisms Overcome by Cognitive Considerations

1. Cannon's criticism that "artificial induction of the visceral changes typical of strong emotions does not produce them" is based on the results of Marañon's (1924) study and its several replications. The fact that the injection of adrenaline produces apparently genuine emotional states in only a tiny minority of subjects is, of course, completely damning for a theory that equates visceral activity with affect. This is, on the other hand, precisely the fact that inspired the series of studies described earlier in this chapter. Rather than being a criticism, the fact that the injection of adrenaline, in and of itself, does not lead to an emotional state is one of the strong points of the cognitive–physiological formulation, since, with the addition of cognitive propositions, we are able to

specify and manipulate the conditions under which such an injection will or will not lead to an emotional state.

2. Cannon's point that "the same visceral changes occur in very different emotional states" is again damning for a purely visceral viewpoint. Since we are aware of a great variety of feeling and emotional states, it must follow from a purely visceral formulation that the variety of emotions will be accompanied by an equal variety of differentiable bodily states. Though the evidence as of today is by no means as one-sided as it appeared in Cannon's day, it does seem that the gist of Cannon's criticism is still correct. Following James's pronouncement, a formidable number of studies were undertaken in search of the physiological differentiators of the emotions. The results, in those early days, were usually failure to find any discriminable patterns. All of the emotional states experimentally manipulated were characterized by a general pattern of activation of the sympathetic nervous system, but there appeared to be no clear-cut physiological discriminators of the various emotions.

More recent work has given some indication that there may be differentiators. Ax (1953) and Schachter (1957) studied fear and anger. On a large number of indices both of these states were characterized by a similar level of sympathetic activation, but on several indices they did differ in the degree of activation. Wolf and Wolff (1943) studied a subject with a gastric fistula and were able to distinguish two patterns in the physiological responses of the stomach wall. It should be noted, though, that for many months they studied their subject during and following a great variety of moods and emotions but were able to distinguish only two patterns.

Whether there are physiological distinctions among the various emotional states must still be considered an open question. Recent work might be taken to indicate that such differences are at best rather subtle, and that the variety of emotion, mood, and feeling states is by no means matched by an equal variety of visceral patterns—a state of affairs hardly compatible with the Jamesian formulation. On the other hand, the question of the physiological differentiability of the various emotions is essentially irrelevant to the present formulation, which maintains simply that cognitive and situational factors determine the labels applied to any of a variety of states of physiological arousal.

The experimental search for the physiological differentiators of emotional states has involved such a substantial, long-term effort that I would like to comment further on the problem. Taken together, these experiments have yielded inconclusive results. Most, though not all, of these studies have indicated no differences among the various emotional states. Since, as human beings, rather than as scientists, we have no difficulty identifying, labeling, and distinguishing among our feelings, the results of these studies have long seemed rather puzzling and paradoxical. Perhaps, because of this, there has been a persistent tendency to discount such results as being due to ignorance or to methodological inadequacy, and to pay far more attention to the very few studies that demonstrate some sort of physiological differences among emotional states than to the very many studies that indicate no differences at all. It is conceivable, however, that these results should be taken at face value and that emotional states may, indeed, be generally characterized by a high level of sympathetic activation with few, if any, physiological distinguishers among the many emotional states. If this is so, the cognitive–physiological formulation I have outlined and the findings of the studies I have described may help to resolve the problem. Obviously these studies do not rule out the possibility of differences among the emotional states. However, given precisely the same state of epinephrine-induced sympathetic activation, we have, by means of cognitive manipulations, been able to produce in our subjects the very disparate states of euphoria, anger, and amusement at a movie. It may, indeed, be the case that cognitive factors are major determiners of the emotional "labels" we apply to a common state of sympathetic arousal.

A novelist's view of this position is Ambler's (1958) description of a fugitive who introspects:

> Rather to his surprise, he found that being wanted for murder produced in him an effect almost identical to that of a dentist's waiting-room—a sense of discomfort in the intestinal region, a certain constriction in the chest. He supposed that the same glands discharged the same secretions into the blood stream in both cases. Nature could be absurdly parsimonious.

If these speculations are correct, nature may indeed be far more parsimonious than Ambler suggests.

3. Cannon's point that "the viscera are relatively insensitive structures" is again damaging to a formulation which virtually requires a richness of visceral sensation in order to be able to match the presumed richness of emotional experience. For the present formulation, of course, the criticism is irrelevant. Just so long as there is some visceral or cardiovascular sensation, the cognitive–physiological hypotheses are applicable.

The introduction of cognitive factors does allow us, then, to cope with three of Cannon's criticisms of a purely visceral formulation. Let us turn next to Cannon's remaining two points, which are quite as troublesome for a cognitive–physiological view of emotion as for the Jamesian view.

Visceral Separation and Emotion

Cannon's remaining criticisms are these: "Visceral changes are too slow to be a source of emotional feeling" (i.e., the latency period of arousal of many visceral structures is longer than the latency of onset of emotional feelings reported in introspective studies), and "total separation of the viscera from the central nervous system does not alter emotional behavior." Both criticisms make essentially the same point, since they identify conditions in which there are apparently emotions unaccompanied by visceral activity. The data with which Cannon buttresses his latter criticism are based on his studies (Cannon, Lewis, & Britton, 1927) of sympathectomized cats, and Sherrington's (1900) study of sympathectomized dogs. For both sets of experimental animals "the absence of reverberation from the viscera did not alter in any respect the appropriate emotional display; its only abbreviation was surgical" (Cannon, 1929, p. 349). In the presence of a barking dog, for example, the sympathectomized cats manifested almost all of the signs of feline rage. Finally, Cannon notes the report of Dana (1921) that a patient with a spinal-cord lesion and almost totally without visceral sensation still manifested normal emotionality.*

For either the Jamesian or the present formulation, such data are crucial, since both views demand visceral arousal as a necessary condition for emotional arousal. When faced with this evidence, James's defenders (e.g., Wenger, 1950; Mandler, 1962) have consistently made the point that the apparently emotional behavior manifested by sympathectomized animals and men is well-learned behavior, acquired long before sympathectomy. There is a dual implication in this position: first, that sympathetic arousal facilitates the acquisition of emotional behavior, and, second, that sympathectomized subjects act but do not feel emotional. There is a small but growing body of evidence supporting these contentions. Wynne and Solomon (1955) have demonstrated that sympathectomized dogs acquire an avoidance response considerably more slowly than control dogs. Furthermore, on extinction trials most of their 13 sympathectomized animals extinguished quickly, whereas not a single one of 30

---

* More recent work supporting Cannon's position is that of Moyer and Bunnell (Moyer, 1958; Moyer & Bunnell, 1958, 1959, 1960a, 1960b), who have in an extensive series of studies of bilaterally adrenalectomized rats, consistently failed to find any indication of differences between experimental and control animals on a variety of emotionally linked behaviors such as avoidance learning. The effects of adrenalectomy are by no means clear-cut, however, for other investigators (Levine & Soliday, 1962) have found distinct differences between operated and control animals.

control dogs gave any indications of extinction over 200 trials. Of particular interest are 2 dogs who were sympathectomized after they had acquired the avoidance response. On extinction trials these 2 animals behaved precisely like the control dogs—giving no indication of extinction. Thus, when deprived of visceral innervation, animals are quite slow in acquiring emotionally linked avoidance responses and, in general, quick to extinguish such responses. When deprived of visceral innervation only after acquisition, the animals behave exactly like normal dogs—they fail to extinguish. A true Jamesian would undoubtedly note that these latter animals have learned to act as if they were emotional, but would ask if they feel emotional.

This apparently unanswerable question seems on its way to being answered in a thoroughly fascinating study of the emotional life of paraplegics and quadriplegics conducted by Hohmann (1962, 1966). Hohmann studied a sample of 25 patients of the Spinal Cord Injury Service of the Veterans Administration Hospital at Long Beach, California. The subjects were divided into five groups according to the height of the clinically complete lesions as follows:

Group I, with lesions between the second and seventh cervical segmental level, have only the cranial branch of the parasympathetic nervous system remaining intact.

Group II, with lesions between the first and fourth thoracic segmental level, have, in addition to the above, at least partial innervation of the sympathetically innervated cardiac plexus remaining intact.

Group III, with lesions between the seventh and twelfth thoracic segmental level, have, additionally, at least partial innervation of the splanchnic outflow of the sympathetics remaining intact.

Group IV, with lesions between the first and fifth lumbar segmental level, have, in addition, at least partial sympathetic innervation of the mesenteric ganglia.

Group V, with lesions between the first and third sacral segments, have, in addition, at least partial innervation of the sacral branch of the parasympathetic nervous system.

These groups, then, fall along a continuum of visceral innervation and sensation. The higher the lesion, the less the visceral sensation. If the present conception of emotion is correct, one should expect to find decreasing manifestation of emotion as the height of the lesion increases.

With each of his subjects Hohmann conducted an extensive, structured interview, which was "directed toward his feeling rather than toward the concomitant ideation." Hohmann asked each subject to recall an emotion-arousing incident prior to the injury and a comparable incident following the injury. The subjects were then asked to compare the intensity of their emotional experiences before and after injury. Changes in reported affect comprise the body of data. I have adapted Hohmann's data for presentation in Figure 1. Following Hohmann's coding schema, a report of no change is scored as 0; a report of mild change (e.g., "I feel it less, I guess") is scored −1 for a decrease and +1 for an increase; a report of strong change (e.g., "I feel it a helluva lot less") is scored as −2 or +2.

Hohmann's data for the states of fear and anger are plotted in Figure 1. It can be immediately seen that the higher the lesion and the less the visceral sensation, the greater the decrease in emotionality. Precisely the same relationship holds for the states of sexual excitement and grief. The sole exception to this consistent trend is "sentimentality," which,

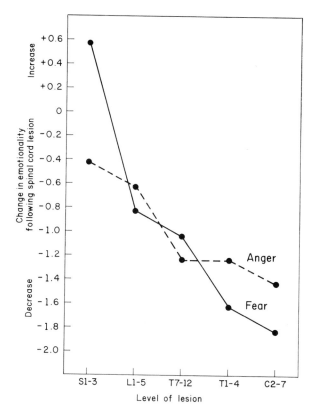

Figure 1. Changes in emotionality as related to height of spinal-cord lesion. (Adapted from Hohmann, 1962.)

I suspect, should be considered a cognitive rather than a "feeling" state. It is clear that for these cases, deprivation of visceral sensation has resulted in a marked decrease in emotionality.

If, in an attempt to assess the absolute level of emotionality of these cases, one examines their verbalized introspections, one notes again and again that subjects with cervical lesions describe themselves as acting emotionally but not feeling emotional. A few typical quotes follow:

> It's sort of cold anger. Sometimes I act angry when I see some injustice. I yell and cuss and raise hell, because if you don't do it sometimes, I've learned people will take advantage of you, but it just doesn't have the heat to it that it used to. It's a mental kind of anger.
>
> Seems like I get thinking mad, not shaking mad, and that's a lot different.
>
> I say I am afraid, like when I'm going into a real stiff exam at school, but I don't really feel afraid, not all tense and shaky, with that hollow feeling in my stomach, like I used to.

In effect, these subjects seem to be saying that when the situation demands it, they make the proper emotional-appearing responses but they do not feel emotional. Parenthetically, it should be noted that these quotations bear an almost contrapuntal resemblance to the introspections of Marañon's subjects who, after receiving an injection of adrenaline, described their feelings in a way that led Marañon to label them "cold" or "as if" emotions. Many of these subjects described their physical symptoms and added statements such as "I feel as if I were very frightened; however, I am calm."

The two sets of introspections are like opposite sides of the same coin. Marañon's subjects report the visceral correlates of emotion, but in the absence of veridical cognitions do not describe themselves as feeling emotion. Hohmann's subjects describe the appropriate reaction to an emotion-inducing situation, but in the absence of visceral arousal do not seem to describe themselves as emotional. It is as if they were labeling a situation, not describing a feeling. Obviously, this contrasting set of introspections is precisely what should be anticipated from a formulation of emotion as a joint function of cognitive and physiological factors.

The line of thought stimulated by the Wynne and Solomon (1955) and the Hohmann (1962) studies may indeed be the answer to Cannon's observation that there can be emotional behavior without visceral activity. From the evidence of these studies, it would appear, first, that autonomic arousal greatly facilitates the acquisition of emotional behavior but is not necessary for its maintenance if the behavior is acquired prior to sympathectomy and, second, that in the absence of autonomic arousal, behavior that appears emotional will not be experienced as emotional.

On the whole, it does appear that the explicit recognition of cognitive variables does allow us to cope with the generally recognized shortcomings of a purely peripheral or visceral formulation of emotion. More importantly, though, this cognitive–physiological, interactionist formulation does permit us to predict and cope with phenomena (e.g., the results of the euphoria–anger experiment) which are well beyond the scope of any purely visceral or purely central theory.

Though I realize that this presentation smacks of neo-Jamesianism, I rather hope not to become embroiled in any revival of the peripheralist–centralist battles, for I believe neither view is, of itself, adequate to cope with present day experimental data and that cognitive or situational factors are of equal importance for either point of view. In the past (Schachter, 1964) I have, in fact, been inclined to interpret the results of my experiments within a modified peripheralist framework. I have, however, become increasingly convinced that this interpretation was almost an accident of the experimental techniques my students and I employed. The injection of adrenaline is a means of directly activating peripheral structures—an experimental fact which makes likely a peripheralist interpretation of these particular results.

Central Theories and the Assumption of Identity

Let us turn next to the central theories—that variety of theory that from Cannon's day on has involved one or another of the lower brainstem structures as crucial to emotion or motivation. In essence, such theories have rested upon the numerous demonstrations that brain lesions or brain stimulation can induce intense emotional and drive states. From the present point of view, the crucial question, of course, is this—does such brain manipulation inevitably lead to modification of specified emotion or drive states, or are the consequences of such manipulation contingent upon stimulus, environmental, or cognitive circumstances? The effects of peripheral manipulation of bodily state are demonstrably dependent on external circumstances, and the implications of this fact constitute the substance of the chief criticisms leveled at purely visceral formulations. Are the central theories any less vulnerable to precisely the same order of criticisms?

Though I claim flatly no expertise, my talks with researchers in this area and my reading of the publications to which they have directed me have convinced me that a purely central theory of emotion or motivation is as inadequate at coping with all of the facts as a purely peripheral theory. The external circumstances surrounding the experimental animal appear to play an extraordinary role in determining whether or not brain manipulation has motivational or emotional effects. In the field of hunger, for example, we know that there is a feeding control center in the ventromedial area of the hypothalamus. Experimentally produced lesions in this area lead to hyperphagia and immensely obese animals. This is one of a large number of related findings which has led many scholars (e.g., Rosenzweig, 1962) to the conclusion that feeding behavior is entirely under central control. However, as Miller, Bailey, and Stevenson (1950) and Teitelbaum (1955) have demonstrated, such lesions lead to overeating and obesity only when the available food is palatable. When the food is unpleasant (coarse in texture or adulterated with quinine) the experimental animals eat considerably less than control animals presented with the same diet. It would appear that this feeding control center operates in intimate interaction with environmental stimuli.

In the area of emotion, experiments explicitly dealing with this point seem rare, but what I have found so far certainly supports the suspicion that external circumstances play a major part in determining whether or not electrical brain stimulation leads to emotional display. Von Holst's and von Saint Paul's (1962) studies of aggression in the rooster neatly illustrate the point. When presented with a stuffed weasel (a natural enemy) an unstimulated rooster ignores the stuffed animal. When electrically stimulated, the rooster attacks it. If the stuffed animal is absent, the rooster will, after sustained stimulation, attack its keeper's face. However, in von Holst's words (p. 61):

> If all substitutes for an enemy are lacking—when there is, so to speak, no hook on which to hang an illusion—the rooster exhibits only motor restlessness. Moreover, the same motor restlessness is observed if one stimulates brain areas associated with hunger, thirst, courtship, or fighting* under conditions in which the environment does not permit the unreeling of the entire behavior sequence. For this reason it is often necessary to vary the external conditions to be sure which particular behavior sequence—which complex drive—has in fact been activated.

---

* The one exception noted by von Holst is escape behavior which "can be evoked in the absence of the appropriate external object (or its substitute) if the brain stimulation is sufficiently intense." Von Holst seems so convinced of the importance of external circumstances that he suggests for this exception: "It is probable that the absent object of fear is being hallucinated."

Even more convincing are the studies of Hutchinson and Renfrew (1966) which demonstrate that stimulation of a single area in the lateral hypothalamus can lead to either predatory attack or to eating behavior, depending entirely on the nature of the situational cues. If a cat is stimulated in the presence of a rat, it will attack; if stimulated in the presence of food, it will eat; if stimulated in the presence of both food and a rat, it will either eat or attack, depending entirely on which of the cues is physically closest.

It would appear, then, that the central theorists of emotion are faced with many of the same problems as are the peripheralists. Direct stimulation of brainstem structures does produce emotional behavior, but only in the presence of appropriate external stimuli. The experimental production of the peripheral correlates of emotion also produces emotional behavior but, again, only in the presence of appropriate external stimuli. Neither a purely central nor a purely peripheral view of emotion is adequate for coping with the facts. Nor is any sort of compromise formulation—one suggesting that both peripheral and central processes are important—likely to be any more successful. Any physiologically based formulation of emotion must specify the fashion in which physiological processes interact with stimulus, cognitive, or situational factors.

If we are eventually to make sense of these areas, I believe we will be forced to adopt a set of concepts with which most physiologically inclined scientists feel somewhat uncomfortable and ill at ease, for they are concepts which are difficult to reify, and about which it is, at present, difficult to physiologize. We will be forced to deal with concepts about perception, about cognition, about learning, and about the social situation. We will be forced to examine a subject's perception of his bodily state and his interpretation of it in terms of his immediate situation and his past experience.

In order to avoid any misunderstanding, let me make completely explicit that I am most certainly not suggesting that such notions as perception and cognition do not have physiological correlates. I am suggesting that at present we know little about these physiological correlates, but that we can and must use such nonphysiologically anchored concepts if we are to make headway in understanding the relations of complex behavioral patterns to physiological and biochemical processes. If we don't, my guess is that we will be just about as successful at deriving predictions about emotion or any other complex behavior from a knowledge of biochemical and physiological conditions as we would be at predicting the destination of a moving automobile from an exquisite knowledge of the workings of the internal combustion engine and of petroleum chemistry.

ACKNOWLEDGMENTS

The material presented in this chapter is condensed from earlier publications (Schachter, 1971). My thanks to Academic Press for permission to reprint.

The research reported was supported by Grant MH 05203 from the National Institute of Mental Health, United States Public Health Service, and by Grants G 23758 and GB 29292 from the National Science Foundation.

REFERENCES

Ambler, E. *Background to danger.* New York: Dell, 1958.

Ax, A. F. The physiological differentiation between fear and anger in humans. *Psychosomatic Medicine*, 1953, **15**, 433–442.

Cannon, W. B. The James–Lange theory of emotions: A critical examination and an alternative theory. *American Journal of Psychology*, 1927, **39**, 106–124.

Cannon, W. B. *Bodily changes in pain, hunger, fear and rage.* (2nd ed.) New York: Appleton, 1929.

Cannon, W. B., Lewis, J. T., & Britton, S. W. The dispensability of the sympathetic division of the autonomic nervous system. *Boston Medical and Surgical Journal*, 1927, **197**, 514–515.

Cantril, H., & Hunt, W. A. Emotional effects produced by the injection of adrenalin. *American Journal of Psychology*, 1932, **44**, 300–307.

Dana, C. L. The anatomic seat of the emotions: A discussion of the James–Lange theory. *American Medical Association Archives of Neurological Psychiatrics*, 1921, **6**, 634–639.

Festinger, L. A theory of social comparison processes. *Human Relations*, 1954, **7**, 114–140.

Hohmann, G. W. The effect of dysfunctions of the autonomic nervous system on experienced feelings and emotions. Paper presented at the Conference on Emotions and Feelings at the New School for Social Research, New York, October, 1962.

Hohmann, G. W. Some effects of spinal cord lesions on experienced emotional feelings. *Psychophysiology*, 1966, **3**, 143–156.

Hunt, J. McV., Cole, M. W., & Reis, E. S. Situational cues distinguishing anger, fear, and sorrow. *American Journal of Psychology*, 1958, **71**, 136–151.

Hutchinson, R. R., & Renfrew, J. W. Stalking attack and eating behavior elicited from the same sites in the hypothalamus. *Journal of Comparative and Physiological Psychology*, 1966, **61**, 360–367.

James, W. *The principles of psychology.* New York: Holt, 1890.

Killam, E. K. The pharmacological aspects of certain drugs useful in psychiatry. In J. O. Cole & R. W. Gerard (Eds.), *Psychopharmacology: Problems in evaluation.* National Academy of Sciences, National Research Council Publication 583, 1959, pp. 20–45.

Landis, C., & Hunt, W. A. Adrenalin and emotion. *Psychological Review*, 1932, **39**, 467–485.

Latané, B., & Schachter, S. Adrenalin and avoidance learning. *Journal of Comparative and Physiological Psychology*, 1962, **65**, 369–372.

Levine, S., & Soliday, S. An effect of adrenal demedullation on the acquisition of a conditioned avoidance response. *Journal of Comparative and Physiological Psychology*, 1962, **55**, 214–216.

Mandler, G. Emotion. In R. Brown, E. Galanter, E. Hess, & G. Mandler. *New directions in psychology*. New York: Holt, 1962. Pp. 267–343.

Marañon, G. Contribution à l'étude de l'action émotive de l'adrénaline. *Revue française d'Endocrinologie*, 1924, **2**, 301–325.

Miller, N. W., Bailey, C. J., & Stevenson, J. A. F. Decreased "hunger" but increased food intake resulting from hypothalamic lesions. *Science*, 1950, **112**, 256–259.

Moyer, K. E. Effect of adrenalectomy on anxiety-motivated behavior. *Journal of Genetic Psychology*, 1958, **92**, 11–16.

Moyer, K. E., & Bunnell, B. N. Effect of injected adrenalin on an avoidance response in the rat. *Journal of Genetic Psychology*, 1958, **92**, 247–251.

Moyer, K. E., & Bunnell, B. N. Effect of adrenal demedullation on an avoidance response in the rat. *Journal of Comparative and Physiological Psychology*, 1959, **52**, 215–216.

Moyer, K. E., & Bunnell, B. N. Effects of adrenal demedullation on the startle response in the rat. *Journal of Genetic Psychology*, 1960, **97**, 341–344.(a)

Moyer, K. E., & Bunnell, B. N. Effect of adrenal demedullation, operative stress, and noise stress on emotional elimination. *Journal of Genetic Psychology*, 1960, **96**, 375–382.(b)

Rosenzweig, M. R. The mechanisms of hunger and thirst. In L. Postman (Ed.), *Psychology in the making*. New York: Alfred A. Knopf, 1962. Pp. 73–143.

Ruckmick, C. A. *The psychology of feeling and emotion*. New York: McGraw-Hill, 1936.

Schachter, J. Pain, fear, and anger in hypertensives and normotensives: A psychophysiologic study. *Psychosomatic Medicine*, 1957, **19**, 17–29.

Schachter, S. *The psychology of affiliation*. Stanford, California: Stanford Univ. Press, 1959.

Schachter, S. The interaction of cognitive and physiological determinants of emotional state. In L. Berkowitz (Ed.), *Advances in experimental social psychology*. Vol. 1. New York: Academic Press, 1964.

Schachter, S. The assumption of identity and peripheralist–centralist controversies in motivation and emotion. In M. Arnold (Ed.), *Feelings and emotions*. New York: Academic Press, 1970.

Schachter, S. *Emotion, obesity and crime*. New York: Academic Press, 1971.

Schachter, S., & Singer, J. E. Cognitive, social, and physiological determinants of emotional state. *Psychological Review*, 1962, **69**, 379–399.

Schachter, S., & Wheeler, L. Epinephrine, chlorpromazine, and amusement. *Journal of Abnormal and Social Psychology*, 1962, **65**, 121–128.

Sherrington, C. S. Experiments on the value of vascular and visceral factors for the genesis of emotion. *Proceedings of the Royal Society of London*, 1900, **66**, 390–403.

Singer, J. E. Sympathetic activation, drugs and fright. *Journal of Comparative and Physiological Psychology*, 1963, **56**, 612–615.

Teitelbaum, P. Sensory control of hypothalamic hyperphagia. *Journal of Comparative and Physiological Psychology*, 1955, **43**, 156–163.

von Holst, E., von Saint Paul, U. Electrically controlled behavior. *Scientific American*, March, 1962, 50–59.

Wenger, M. A. Emotion as visceral action: An extension of Lange's theory. In M. L.

Reymert (Ed.), *Feelings and emotions: The Moosehart Symposium.* New York: McGraw–Hill, 1950. Pp. 3–10.

Wolf, S., & Wolff, H. G. *Human gastric function.* New York: Oxford Univ. Press, 1943.

Woodworth, R. S., & Schlosberg, H. *Experimental psychology.* New York: Holt, 1954.

Wrightsman, L. S. Effects of waiting with others on changes in level of felt anxiety. *Journal of Abnormal and Social Psychology,* 1960, **61,** 216–220.

Wynne, L. C., & Solomon, R. L. Traumatic avoidance learning: Acquisition and extinction in dogs deprived of normal peripheral autonomic function. *Genetic Psychology Monographs,* 1955, **52,** 241–284.

Chapter 19

Brain Mechanisms and Behavior

MICHAEL S. GAZZANIGA

State University of New York at Stony Brook

During the past 20 years there have been some remarkable and intriguing developments in the study of the brain and behavior. These advances have been made at every level in psychobiology and most of them are described in this book. In the following, what I would like to set forth is some of the data and brain theory that has emerged from the neuropsychological examinations of both animals and man. It is unavoidably a selective and personal view of the field. In particular, a brief review of the studies on commissurotomy will serve as a point of departure for considering the problems of the cerebral distribution of mental functions, the extent of cortical reorganization possible following brain damage, and the problem of relating cerebral mass to cognitive capability.

BRAIN BISECTION: A GENERAL REVIEW

The interhemispheric commissures have fascinated neurologists and psychologists for years. As recently as 1950, however, Lashley wrote the structures off as not serving any important neurological or psychological function in the mammalian brain. He cited Akelaitis' (1941, 1943,

1944) work as revealing that section in whole or in part produced no striking changes in behavior.

It was in this climate that Myers and Sperry (1953) performed their classic experiment on the cat which showed that visual discriminations trained to one hemisphere of a callosum-chiasm section cat could not be performed by the other (see Sperry, 1966). This work was extended and developed in the 1950s to include monkeys and even chimpanzees. Myers' and Sperry's series of studies were truly dramatic and stood in marked contrast to the Akelaitis series in the 1940s carried out in man.

In the early 1960s, Dr. Joseph Bogen, encouraged by a critical review of the Akelaitis reports, began a study of a new limited series of patients along with Dr. P. J. Vogel (1962). The opportunity to study these patients from a neuropsychological view originally fell to R. W. Sperry and this author at Cal Tech. It was an exciting time and over a 5-year period the patients were intensively examined on sensory, motor, and general

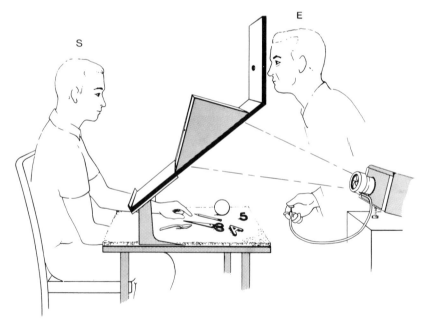

Figure 1. General testing apparatus originally used in testing of split-brain patients. I would monitor the subject's fixation and, when the eyes were on target, would flash a picture no longer than 100 msec into one or both visual fields. The subject was then free to choose with either hand an object that matched the stimulus, be it a word or a picture. Because stereognostic information from the left hand projects exclusively to the contralateral right hemisphere, a complete transaction can take place in a visual–tactile matching test without the left half-brain knowing what had occurred. (Adapted from Gazzaniga, 1970.)

cognitive functions (Gazzaniga, Bogen, & Sperry, 1962, 1963, 1965, 1967; Gazzaniga & Sperry, 1967; Bogen & Gazzaniga, 1965).

These studies have continued on the Bogen series of patients both at Cal Tech and elsewhere and have yielded a variety of insights into brain function (Milner, Taylor, & Sperry, 1968; Levy–Agresti & Sperry, 1968; Nebes & Sperry, 1972; Trevarthen & Sperry, 1969; Gazzaniga, 1970; Gazzaniga & Hillyard, 1971; Gazzaniga & Hillyard, 1973; Gazzaniga, 1967). In brief, the thrust of this work demonstrates there is a sharp breakdown in communication between the hemispheres of visual, somato-sensory, motor, and cognitive information. Information presented to the left hemisphere was normally named and described while information presented to the right went nameless and the left hemisphere was unable to say what the right hemisphere was seeing or doing (Figure 1). Devel-opment of nonverbal response tests, however, made possible the demon-stration that the right perceived the nature of those tasks and could execute an appropriate response. The studies went on to show that there were suggestions of marked differences in the way the hemispheres processed information with the left handling the verbal information and the right the visual-spatial tasks (Gazzaniga *et al.*, 1965; Bogen & Gaz-zaniga, 1965). More recent work suggests the intriguing possibility that problems that can be solved by either mode are handled by quite differ-ent cognitive strategies as a function of which hemisphere works on the task (Levy, Trevarthen, & Sperry, 1972; Nebes, 1971).

PARTIAL COMMISSUROTOMY AND CEREBRAL LOCALIZATION OF FUNCTION

Animal Studies

Of the variety of questions raised by split-brain studies, none is more important than determining the effects of partial commissurotomy on interhemispheric integration tasks. Animal work has suggested there is a discrete specificity of function within the forebrain commissures. The posterior or splenial areas subserve visual transfer as does the anterior commissure (Downer, 1962; Black & Myers, 1964; Gazzaniga, 1966; Sul-livan & Hamilton, 1974). The area slightly anterior to the splenium is responsible for somatosensory transfer (Myers, 1962). The functional significance of the rest of the callosum remains largely unknown although there is a strong suggestion that it helps, among other things, to maintain a motivational equilibrium between the two hemispheres (Gibson, 1972).

Perhaps the most improbable and intriguing results from the partial section studies come from a series of reports by Hamilton and his colleagues (Hamilton & Brody, 1973). They have shown that split-brain monkeys, with small segments of the callosum anterior to the splenium left intact, are able to perform an interhemispheric match-to-sample on a pattern discrimination, while at the same time being unable to transfer pattern discriminations interhemispherically. Thus, a "+" versus "0" discrimination will not transfer through the callosum to the untrained half-brain while the animal can match a "+" or a "0" flashed to one hemisphere with a "+" and "0" sample presented to the other. This remarkable dissociation shows there is not, contrary to Lashley's (1950) opinion, a total equivalence within a sensory system.

One interpretation of these data is that, as visual information proceeds into the brain, it is constantly being recoded into new forms such that "X" becomes "X'." On the match experiments, the recoded "X'" transfers and as such is recognized by the opposite hemisphere which has performed the same operation on its "X." Thus, the match is possible. In a straight transfer experiment, however, "X" is transferred but as such is of no assistance to the learning and memory mechanisms of the opposite brain which has had no meaningful experience with "X'." In some sense, the untrained hemisphere has the answer in the form of "X'" continually available to it—but doesn't know what the question is. In a neurological vein, these data suggest that the learning and memory processes are initiated early in the primary visual system itself.

Human Studies

In humans, there are some classic accounts of neuropsychological deficits with partial callosal section as well as later reports (Geschwind & Kaplan, 1962; Gazzaniga & Freedman, 1973; Gordon, Bogen, & Sperry, 1971; Gazzaniga et al., 1974a). In the late 1930s, Maspes (1948) and Trescher and Ford (1937) described cases where a partial posterior callosal lesion resulted in a breakdown in the interhemispheric transfer of visual information. Indeed, Maspes reported that while his patients were unable to read in the left visual field, they were able to describe various objects presented in the left visual field.

A striking breakdown in the interhemispheric transfer of visual information following splenial section was recently studied by Gazzaniga and Freedman (1973). The patient who was left-handed, with speech in the right hemisphere, was easily able to report verbally left-field visual information of all kinds while being unable to report identical stimuli presented to the right visual field. In this patient, there was no breakdown in interhemispheric tactile communication. This contrasts markedly with Geschwind and Kaplan's (1962) patient who had a dramatic breakdown in tactile function but no interhemispheric visual deficits. Their patient had tumor involvement in the anterior portions of the corpus callosum. Side-by-side comparisons of clinical data such as these allow the determination of where in the corpus callosum various sensory functions are transferred.

In further testing of our patient, we ran a modification of the Hamilton experiment on interhemispheric matching. We found some evidence that the subject could perform a match as to whether two geometric shapes were the same or different, no matter whether they were placed in the same or different visual fields. In this test, geometric shapes such as a square, triangle, and circle were flashed in pairs with both appearing in the same visual field or with one presented to the left of fixation and one to the right. The subject was able to judge correctly whether the two were the same or different, no matter where they were flashed. Yet, he tended to make errors when verbally describing the stimuli in the right visual field on the "different" trials. In other words, the right hemisphere speech system could not recognize the transferred information in the form it came over but could judge it was different from what was being processed in the right. This state of affairs would allow for a correct response of a "same or different" judgment but would not afford the patient's right hemisphere information as to "what" was different about the stimuli.

At quite another level, this same kind of finding was seen in other studies. Engrams laid down in the right hemisphere in the absence of speech and language as the result of the left unilateral injection of amytal cannot be accessed by the language system upon its return to normal functioning (Gazzaniga, 1972). Again, it would seem information stored in one form or neural mechanism is not easily accessible to another form or mechanism such as the language system.

MAPPING THE FLOW OF COGNITIVE
INFORMATION WITH COMMISSUROTOMY

One of the powerful uses of partial commissurotomy as outlined in the preceding discussion is its application in the examination of where in the corpus callosum information transfers. Since the area of the projection of the callosal fibers is usually known, this gives good hints as to what part of the brain is involved in a particular task (Figure 2). It also allows examination of the question of how much backtracking there is in the cerebral flow of information. For example, if the visual

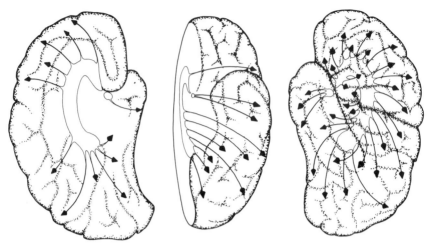

Figure 2. Indicates area of callosal projections in man as drawn from anatomical studies in animals. Surgical disconnection of the posterior or splenial region produces a visual split-brain syndrome but leaves other sensory functions, such as touch, intact.

areas are disconnected through a splenial section, can a visual–tactile match occur when the stereognostic information is presented exclusively to the hemisphere opposite the one receiving the visual stimulus? Or, conversely, can partial commissurotomy be used to explore the questions of whether lateralized mental duties carried out by specific structures, such as, for example, the right frontal or temporal lobes, can be carried out by the left hemisphere following frontal or temporal callosal disconnections? Likewise, can olfactory discriminations presented to the right hemisphere be known to the left following anterior commissure section?

These questions and more have been asked of a series of patients, most of whom have been operated on by Dr. Donald Wilson of Dartmouth Medical School (Wilson, personal communication, 1973). Dr. Wilson, using a new surgical approach he developed, has operated on several patients with the aim of limiting the interhemispheric spread of epileptic seizures. Encouraged by the medical results of the Bogen series as well as the Lussenhop *et al.* series (1970), Wilson has carried out both complete and partial sections of the callosal and anterior commissure fibers. This new series of patients is currently under investigation, and some of these results will be discussed later in this chapter.

Visual–Tactile Matches and Vice Versa

In the splenial section case previously described (Gazzaniga & Freedman, 1973), the patient was unable to cross-communicate visual information, but visual–tactile matches were easily performed when tactile information was placed in one hand and visual information was presented in either visual field. Since the match was made with split sensory input and since the visual information per se did not transfer in this subject, we must conclude the tactile information transferred over to the hemisphere viewing the stimulus. It is not possible, of course, to say in which domain the actual judgment is made, that is whether the tactile information is made available to the visual system or vice versa. A less simplistic interpretation, of course, would be that the information is funneled into another brain system within the hemisphere for final processing.

In a patient from the Wilson series where we have concluded that a portion of the splenial area remained intact while the rest of the forebrain commissures have been sectioned, tactile information could not transfer interhemispherically whereas visual information could. In this case, visual–tactile matches which required interhemispheric integration were also easily carried out. Here, it must be assumed, the visual information transferred for processing to the hemisphere exclusively holding stereognostic information.

This kind of data, of course, underlines how important it is to keep in mind the multiple channels that are available to the interexchange of information in the brain. Block one, and yet another remains that can do the job. Also, one cannot help but observe that it should not be too long before this kind of analysis could allow for even greater specificity in determining the nature of the interhemispheric codes. Namely, could a lesion that allows the interhemispheric transfer of recoded visual information, as in Hamilton's experiments, be able to perform a visual–tactile match? Answers to questions such as these ought to shed more light on the extent to which neurally recoded information is functionally available to other brain systems.

Lateralized Lobular Functions

Milner (1963) has elegantly shown striking dysfunctions following left and right frontal lobe lesions. Specifically, lesions of the right and left frontal lobes produce perseveration revealed by the inability easily to shift conceptual hypotheses on the Wisconsin Card Sorting Test. Moreover, left frontals, as opposed to right frontals, have marked deficits in word fluency and spontaneous speech.

To one patient with anterior callosal and anterior commissure disconnection we administered the Card Sorting Test in two forms—the normal approach and one where the stimuli were lateralized and quickly flashed onto the left or right visual field. The patient showed absolutely no hesitancy in responding to either test and performed well within the normal range. Another patient, however, where it is believed some splenial fibers remained, showed a deficit on the Card Sorting Task. However, since defects were also observed on the word fluency test, the presence of an overall general frontal decrement is suggested. Still another patient, completely split, but who had suffered from cerebral palsy since the age of five, showed no deficit.

These preliminary data suggest that the proper synchronous function of the frontal lobes is not disrupted following the interhemispheric disconnection of the frontal callosal projections. This evidence contrasts markedly, of course, with the observations on the basic sensory systems of vision, touch, and possibly also olfaction, where the appropriate callosal section does block interhemispheric transfer. The model that can be entertained here, of course, is that the superordinate regulating mechanisms in behavior, such as those involving the frontal lobes, diffusely influence many brain systems that extend way beyond the boundaries of frontal callosal projections. It would seem that most of the cerebral mantle has access to and makes use of the processed output of the frontal lobes and, as such, can communicate it to the opposite half brain in a meaningful way, through a variety of callosal channels.

A similar point could be made on another observation of ours where we again followed up a study carried out by Dr. Milner. We recently administered to the Wilson series the tactile–spatial match-to-sample test that Milner and Taylor (1972) gave to the Bogen (Bogen, Fisher, & Vogel, 1965) patients. Milner and Taylor found a dramatic right hemisphere superiority on this task with the left being essentially unable to perform the task. In our test on the partially sectioned patients, the left proved easily able to do the task—even in the case where only splenial fibers remained intact but with the interhemispheric somatosensory communication system disconnected. Here again, the data suggest that a superordinate function specialized to one hemisphere can be contributed to the other over almost any callosal channel.

Summary

Revealing basic brain mechanism through studying the partially commissure-sectioned case appears to be a most promising enterprise. The animal work of Hamilton and others that appears to be teasing out psychological-brain process heretofore not imagined just hints at what the potential seems to be. Study of the partially disconnected patient seems equally revealing and productive in showing how many high-level cognitive activities are managed in the cerebral flow of information.

With respect to the issue of localization of function, it would seem clear that those cerebral areas clearly involved in the immediate processing of raw sensory information can be selectively and specifically isolated and disconnected. In other words, the informational products of the long axonal Type I cells of Golgi, which Jacobson claims are the brain cells under strict genetic control, can be isolated, whereas the products of more complex and integrative mental activities which are managed by the more mutable Golgi Type II cells seem not to be so specifically disposed. Thus, these data suggest the lateralized specialities of the various left and right brain areas can make their contribution to the cerebral activities of the opposite hemisphere through almost any callosal area no matter what its size and location. Indeed, this interpretation suggests to me that the long-standing issue of the extent of localization could be better understood by considering the dichotomy in genetic specification as offered by Hirsch and Jacobson (see Chapter 4 in this book). Those cerebral processes that develop because of the mutability potential of Golgi Type II cells are hard to localize because of their ubiquitous nature whereas those tied to specific genetically determined systems are not.

CEREBRAL DYNAMICS AND MENTAL FUNCTIONS

Inherent in the discussion of the cerebral basis of mental functions is the foregoing problem of the nature and extent of cerebral localization. Most clinical data suggest that the brain is not a homogeneous system with every part equally involved with every function. Figure 3, for example, summarizes a variety of classic observations on the dominant hemisphere which argue for the view that cerebral lesions in particular areas give rise to more or less specific dysfunctions. To some extent, however, these descriptions, made by neuropsychologists and neurologists through the years, of behavioral disorders resulting from specific brain lesions were a direct function of the psychological views on language and cognitive behavior current at the time of testing. This phenomenon has been dramatically recounted by Bogen (1969) who listed various dichotomies of behavior and how, through the years, they have been assigned to either the left or right cerebral hemispheres. In the long run, this may say more about man's propensity to correlate any two dichotomies than how neurological systems subserve behavior.

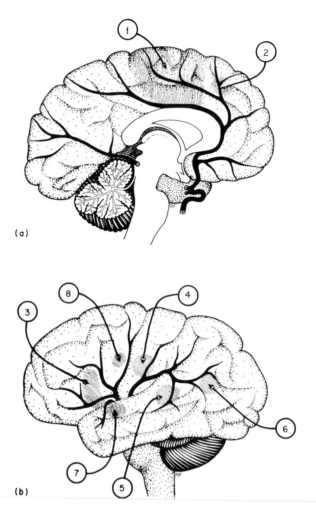

(a)

(b)

Figure 3. Lesions affecting language and speech processes.

(1) Area supplied by the anterior cerebral artery—involves fronto-parietal region and the corpus callosum with only transient loss of speech as a result of inclusion of the supplementary motor area.

(2) Supplementary motor area of either hemisphere—stimulation of which produces arrest of ongoing speech and initiation of repetitive, nonvoluntary vocalizations (similar to epileptogenic lesions in the left hemisphere of that area). Lesions result in abnormalities in initiation, continuation, and inhibition of speech.

(3) Broca's area—lower part of the pre-motor zone—typically produces agrammartism, poor articulation, and abnormal writing of events required both for pronunciation of words and fluent speech.

(4) Retrocentral area—lesions results in apraxia of the lips and tongue, leads to disintegration of speech as a whole.

(5) Temporal region—lesions affect the ability to generalize and differentiate phonetic sounds, cause disintegration of phonetic hearing. It always produces abnormal speech

In recent years, the psychological processes involved in language and speech have been analyzed in more sophisticated and complex ways than ever before. Realizing that language behavior is not the simple output of a discrete system in the brain, the neuropsychologist finds himself looking for a set of different correlations regarding brain damage and verbal behavior. Instead of the old dichotomy of receptive and expressive language, of verbal and nonverbal processes, and the like, the neuropsychologist is finding that left hemisphere damage in the classic language and speech areas leaves the patient with enormous syntactic and semantic capacity that can be realized through other response systems (Gazzaniga, Velletri, & Premack, 1971; Glass, Gazzaniga, & Premack, 1972). Imagery and language have received wide attention (Paivio, 1969; Bower, 1972) and some studies suggest that this assisting cognitive system to language behavior is managed in quite different cerebral sites than the classic language area (Seamon & Gazzaniga, 1973). Word-match capability as well as other mental duties are also functions not performed in the left dominant hemisphere (Gibson, Dimond, & Gazzaniga, 1972; for review, see Gazzaniga, 1973). What emerges from these data is a picture of brain function which is quite different from that generally held in the past. This new view is that various cerebral functions managed in discrete areas throughout the brain are thought to be in a dynamic relation with other cerebral sites. While the idea that there are specific cerebral centers specialized for particular cognitive functions is not directly challenged here, it is maintained that a variety of spheres of influence in the brain, some with functions that overlap those of others, are all interacting in a dynamic way to produce a final integral behavioral response.

RECOVERY OF FUNCTION

There seems to be, in part, a swing back to older views of basic central nervous system maintenance and growth principles. Prior to Sperry's long series of studies emphasizing the presence of a high degree

---

production and poor reading and writing.

(6) Temporo-parieto-occipital region—lesions do not change the external articulated speech, but prevent mental integration of separate elements (disturb simultaneous synthesis). Leads to disintegration of rational speech and to disturbance of the understanding of logical, grammatical constructions (semantic aphasia).

(7) Bilateral lesion of the temporal lobe, posterior $2/3$ of the first and second temporal convolutions, plus posterior half of the Isle of Reil, extending to the interior parietal lobe—shows word deafness, severe motor aphasia, more muteness than paraphasia.

(8) Near Broca's in the motor area of the lateral side of the left hemisphere—electrical stimulation results in perseveration of speech.

of neurospecificity, the wildest claims had been made about the extent of functional recovery possible, following either central or peripheral nerve damage (Sperry, 1966). For years, it was these studies, as well as Harlow's (1971) at a psychological level, that put a halt to the runaway interpretation that function precedes form. However, many other studies—for example, the work on visual deprivation (Hirsch & Spinelli, 1971; Blakemore & Cooper, 1970)—have focused attention on the question of the degree of plasticity of the central nervous system (CNS) (see Chapter 4 by Hirsch and Jacobson). These studies have shown us how discrete visual experience during rearing modifies the basic organization of the primary visual system. In addition to this work are the studies of Schneider (1973), Raisman (1969), Moore, Bijorland, and Stenevi (1971), and others which detail the extent to which new neural growth is possible following central neural lesions. As a result, the idea in everyone's mind is that we may now have a grasp of the physical basis of recovery in the CNS—not to mention the insights such work affords us on the broader question of the physical basis of learning and memory.

Indeed, Hirsch and Jacobson, as previously pointed out, argue that adaptive behavior, in general, is the product of changes in the microneurons or Type II cells of Golgi. These cells, it is believed, remain adaptive while the long axon cells responsible for the major transmission of information into and out of the CNS are under early, and exacting, genetic control and specification. How long this state of flexibility obtains for the microneurons is not known. It supposedly extends into the teens in humans, a supposition that supports the speculation that it is the mechanism responsible for the kind of speech and language recovery seen in the rehemispherization of these processes following early brain damage. It could also explain why the relocalization of speech and language rarely occurs after 12 years of age. In this connection, it is worth noting the studies of bird song which Nottebohm (1970) conducted using canaries. He has shown that up to the age of 1 year, the song can be taught to birds that have not heard the song before. Birds deprived for longer periods cannnot be so trained. Yet, if the birds are castrated when young, thus altering the testosterone level, they are able to learn the song well into the second year. Here, we see an exciting model for the experimental manipulation of how and why the CNS at some time "wires out" adaptive changes in communicative behavior.

In my view, however, all of the fascinating basic work in neural development does not directly bear on the question of recovery of function in the CNS as the term is normally used in a clinical sense. Before proceeding, however, let us look at the status of clinical functional recovery.

Clinical Recovery

There are various claims concerning the mechanisms and extent of nervous system recovery. Luria *et al.* (1969) feel that temporarily depressed areas can be "deinhibited" both by training and with the aid of pharmacological agents such as atropine and neostigmine. Yet most neurologists are skeptical of applying these methods to patients and, in general, believe the extent of long-term recovery from a lesion is a function of the individual's own capacity to recover and has little to do with external therapy. In studies, for example, examining the value of rehabilitation on motility and other sensory-motor functions on stroke patients, it has been concluded that no more improvement resulted than if the patient had been left alone (Stern, McDowell, Miller, & Robinson, 1971). The same case can be made for speech and language rehabilitation following stroke (Sarno *et al.*, 1971). Indeed, in the clinical setting it is hard to improve upon von Monakow's concept of diaschisis, where recovery is viewed as the reestablishment of temporarily impaired neural systems—not the vast reorganization of neural systems through substitution or retraining, or as the result of new growth. There have been, in recent years, both physiological and metabolic studies that support von Monakow's ideas. Recordings from cortical areas distant from cerebral lesions, for example, find the areas transiently depressed followed by return to normal levels of firing (Kempinsky, 1958). In stroke, it is observed that there is a marked transient decrement in metabolic rate in the brain areas opposite the lesions (Hoedt–Rasmussen & Skinhoj, 1964).

Yet, even if the basic ideas of von Monakow prove correct, some clinical instances of recovery involving the higher cognitive processes following massive brain damage probably come about through other mechanisms and involve neither deinhibition nor actual structural changes. In what follows, instances of recovery of function will be reported which can be brought about quickly after a brain lesion, by prelesion prophylactic measures; and that can be obtained long after diaschistic processes are thought to be active by the use of proper behavioral training routines.

In general, all of the data to be reported lead one to the view that recovery in the adult, arising from nonphysiological improvement, is the result of preexisting behavioral mechanisms not necessarily previously routinely involved in a particular act now covering for the mental activity under question. For instance, it will be maintained that the implicit functional syntactic mechanism present and active in decoding a meaningful pictorial array is probably able to come to the assistance of the organism when the syntactic mechanism for language has been destroyed through stroke or lesion. But before getting into the clinical work, let me lay a broader base for this view with some recent work in animals.

Animal Research

We believe that the behavioral dysfunction supposedly resulting from discrete lesions in the brain can frequently be quickly circumvented by changing the environmental or behavioral contingencies (Gazzaniga, 1973b). For example, we recently pursued this idea in one of the most exhaustively studied systems in physiological psychology, the lateral hypothalamic syndrome. Bilateral lesions here, of course, produce an adipsic animal who will neither drink nor eat, and, if left alone postoperatively, would die (Figure 4). Such animals are nursed for an extended period of time and, with enough coaxing, some will eventually be able to sustain life postoperatively. We, however, observe that within a few days after the lesions, most rats will have a higher probability of running than they have of drinking. Thus the adipsic rat will show essentially no probability of taking a lick from a water spout but within a half-hour period will run between 100 and 150 seconds. We then made these

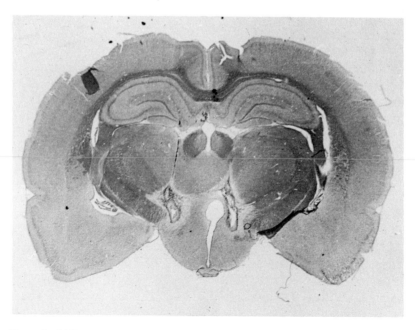

Figure 4. A bilateral lesion in one of the rats used in the study described. While the animal would not drink spontaneously, it would drink in order to have the opportunity to run.

two behavioral events contingent such that if the animal wanted the opportunity to run, he had to drink, which in turn released a brake on a wheel that allowed the animal to run. Dramatically, the adipsic rats immediately began to drink in order to have the opportunity to run.

Or, consider the inferotemporal lobe syndrome. Clark and Gazzaniga (1974) have recently extended the kind of insight afforded in the preceding experiment into discrimination problems trained to animals undergoing inferotemporal lesions. In the beginning, we assumed caged monkeys were like rats and would relish the opportunity to run. Instead, we seemed to have discovered the phylogenetic origin of "don't rock the boat." Here, when the animal has the opportunity to run, a preferred response turned out to be to adopt a vertical spread-eagle position so as to minimize movement in the wheel! This required a change in contingencies such that if the monkey made a correct choice, the wheel would be locked so no movement was possible.

Specifically, three monkeys were trained on a pattern discrimination for food reward on a discrimination panel that was placed inside of a large activity wheel. When the discrimination was learned, an added contingency was introduced. The wheel, driven by a motor, would automatically start to turn at the onset of the stimulus. As described, if the correct choice was made, the wheel locked during the intertrial interval. The animals, under these conditions, decreased their latency in making their responses and immediately made a perfect score even after all food reward was withdrawn.

All animals then underwent bilateral inferotemporal ablation. To our great surprise, the animals were instantly able to perform perfectly the discrimination to a food reward alone, as well as, of course, to the not-to-run contingency. Our expectation was that we would see a dissociation of performance between the food condition and the not-to-run contingency.

It would appear from this that the training of a visual task with two explicitly different kinds of rewards insulates the organism from showing the classic impairment following bitemporal lesions. It is as if the preoperative dual training encouraged the organism to use a number of conceptual strategies to solve the problem, creating a cerebral redundancy such that impairment to one part of the brain could in no way do exclusive damage to all of the paths used in problem solution. Indeed, the well-known beneficial effects of preoperative overtraining on postoperative scores may be the result of a similar mechanism. During the long overtraining period, the animals may well decide to solve the problem through a different kind of strategy than the one originally used. This, of course, could never be delineated by the present experimental design. At the same time, this kind of interpretation is commonplace in complex discrimination training in humans where it is shown, using other testing methods, that both children and adults are constantly changing their hypotheses along the way as they learn a particular visual discrimination (Levine, 1966). In short, the old analysis of learning phenomena which urged simple behavioristic interpretations with the corresponding simplistic neurological models won't do any more, for the data are giving way to the view that distinctly separate mental processes are active during even the simplest kind of discrimination training.

Cognition following Stroke

The problem of determining the amount and kind of cognitive function remaining after severe brain damage to the left dominant hemisphere is difficult and, in the past, little credit has been given to what the right hemisphere is capable of in this regard. Encouraged by our earlier studies on the cognitive capacity of the right hemisphere, as described in the foregoing (Bogen & Gazzaniga, 1965; Gazzaniga et al., 1967; Gazzaniga & Sperry, 1967; Gazzaniga, 1970), we undertook a series of studies on the severely left-brain-damaged patient in an effort to determine what, in fact, the cognitive limits were. We predicted that, with the right behavioral testing technique, much more extensive behavioral capacity would be evident than is usually claimed, and this has indeed been our experience. Using Premack's (1971) language training system which he developed for the chimp, we ran a series of tests on global aphasic patients and quickly discovered these patients could learn to perform many language-like operations (Glass et al., 1972).

Before beginning language training, a viable social relationship must be established between the patient and the trainer. The importance of this phase cannot be overemphasized for if the motivational setting is inappropriate, no learning will occur. In psychological parlance, if a patient is emotionally flat and shows no preference, then it is impossible to arrange a contingency where manipulating and learning X will produce desired reward Y. Indeed, it would seem fair to say that all too frequently neuropsychological assessment procedures ignore this factor. Tests are designed, norms are established on a normal population, and the relation all this has to testing a brain-damaged patient who surely is in a complex ever-changing motivational state is frequently remote.

Using paper cut-out symbols, errorless training procedures were administered in the initial training. For example, in teaching "same versus different," two similar objects, say two erasers, were placed on a table in front of the patient. Placed in between was another symbol, a question marker, which comes to mean "missing element." The subjects learned to slide the question marker out from between the two test objects and insert in its place the symbol meaning "same." At first, this is the only response allowed. Subsequently, an eraser and a screwdriver are placed in front of the patient and the patient must remove the question marker and insert the symbol meaning "different." Following this training, the two symbols are both available on each trial and the subject must now make the correct response to the two varying, "same" or "different," stimuli. When the stimuli used in training are then changed, it is observed the subjects can use the symbols correctly no matter what test objects are used by the examiner.

These procedures then enable one to teach any of a number of language operations to the global aphasic patient. The negative, yes, no, the question, and simple sentences were all successfully mastered by the patient (Figure 5). Before teaching the sentences, the patient's lexicon was increased by teaching him a few nouns, verbs, and personal names.

Figure 5. After the examiner stirs some water in a glass, as opposed to pouring it or stirring Tang or pouring it, the subject must describe the act by arranging three of the six symbols correctly, which in effect says "Mike stirs water."

Each of these words was taught by associating a symbol with an object, action, or agent in the context of a simple social transaction. An object was placed before the patient along with the symbol for the object and the patient was required to place the symbol on the writing surface, after which he was given the object.

It is of interest to note that training in the use of symbols referent to actions (verbs) was consistently much more difficult than training in the use of symbols referent to nouns. Noun symbols were learned in a few trials whereas verbs sometimes took weeks to learn. To some extent, of course, this is not too surprising. To know a verb is to know the whole context, subject and object, whereas to know a noun is simply to know a single object. The difficulty we experienced in training the patient to use symbols referent to actions is also reminiscent of the finding that the right hemisphere of the split-brain patient was unable to process natural language verbs.

These data clearly suggest that the severely left-brain-damaged patient can perform a wide variety of conceptual tasks. Because of the large extent of left damage, it would seem likely that the intact right hemisphere is surely involved in many of these tasks. We know from other studies that the right hemisphere has enormous cognitive power (Gazzaniga & Sperry, 1967; Bogen & Gazzaniga, 1965; Levy et al., 1972; Milner & Taylor, 1972) and, indeed, the imagery mechanism associated with language behavior appears to be a right-hemisphere process (Seamon & Gazzaniga, 1973).

In other studies, use was made of Sternberg's (1966) serial processing model of short-term memory processes. In brief, he found that as a memory set increased in size—for example, from one item to three items—a probe examining whether a word or letter was part of the set took longer to yield an answer the larger the memory set size. Seamon reasoned that if the instructions to a subject were varied, different response patterns would be evident. Instead of asking the subject to keep repeating the instructions verbally, as is usually the case in the Sternberg design, he told them to create, with the memory set words, an interactive image where all the words in the set "touched" one another in the image (Seamon, 1972). Thus "tree" and "bird" should find the bird in the tree, not flying by it. Changing the instructions in this way resulted in equivalent response times no matter how large the memory set.

This remarkable observation encouraged us, of course, to examine the possibility that there may be a left–right difference here. For years, we had felt that it was the right hemisphere that was specialized for handling the visual abilities of mental life and, in this context, we investigated whether different response times would function as a feature of both our instructions for encoding the original material and the visual-field hemisphere first receiving the probe.

Results of the study clearly showed it is the right hemisphere that is specialized for the image process and the left for verbal directions (Figure 6).

For present purposes, these studies indicate how a cognitive system working in parallel with the language system could be helpful following left-brain damage. In addition, and perhaps more importantly, we see how, by manipulating the encoding instructions, wholly different brain systems are called upon to process information. In a sense, then, one can "shunt" around a brain lesion by setting up the environmental contingencies differently and, thereby, requiring a different part of the brain to be used in the solution of a problem.

Summary

For the present, we are faced with the problem of how to account for clinical improvement in terms of recovery of function. Does it reflect a process where the central and dominant language processing systems have recovered to the extent of allowing the observed behavior? Or, are these cognitive talents the product of other existing behavioral strate-

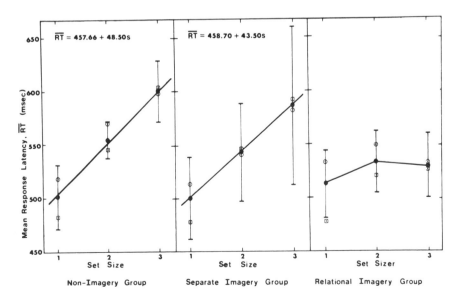

Figure 6. The right hemisphere proves superior in the normal in processing imagery information while the left excels at instructions emphasizing a verbal rehearsal strategy. (After Seamon and Gazzaniga, 1973.)

gies that are capable of handling the job but have previously been involved in other more supportive roles? With the latter view, the recovery period becomes more the time needed to allow for the realignment of these cognitive processes than the time needed for physical repair.

CORTICAL MASS AND COGNITIVE CAPABILITY

One of the most intriguing problems facing the neuropsychologist is to understand how the amount of brain tissue correlates with cognitive capability. The old approach of correlating brain weight with intelligence quotient (IQ) and the like has given way in recent years to a more approachable question—namely, is one-half of a split brain as competent as a whole brain? Put differently, what is the cost to the organism, if any, for having the brain split?

A number of studies on cats (Sechzer, 1970; Meikle, 1964; Voneida & Robinson, 1970) and dogs (Mosidze, Rizhinashvili, Totibadze, Kevanishvili, & Akbarolia, 1971) report that split-brain mammals show cognitive deficits compared to normals. Standing in contrast, however, are three studies on discrimination learning in monkeys documenting learning times, which reveal no differences between splits and normals (Hamilton & Gazzaniga, 1964; Myers, 1965; Gazzaniga, 1966).

In humans with commissurotomy, early studies suggested there were no major cognitive deficits resulting from hemisphere disconnection (Gazzaniga et al., 1965, 1967). On the other hand, short-term memory (STM) studies suggested each hemisphere was below normal in its information-handling capacity (Gazzaniga, 1968). Yet, when both hemispheres were working together, the scores fell into the normal range. The study by Milner and Taylor described earlier, which reported that the split-brain patient is better in the right hemisphere than in the left hemisphere in carrying out a spatial–tactile STM task, also showed that on the whole a half brain performs inferiorly to normals or brain-damaged cases.

Using their test, we examined the three patients in the Wilson series described previously. The results clearly showed that when the callosum is not completely sectioned, the general deficits seen in the fully commissure-sectioned patient are not present. Moreover, either hemisphere can perform the problem equally well. Even in the case where only splenial fibers remain, there was no sign of a decrement in performing this task by the left hemisphere.

Still, since previous work has not clearly resolved these questions, as well as the more general question of the callosal contribution to intellectual function, we have begun a series of experiments on monkeys to clarify the effects of commissurotomy on learning and memory. The following study examines the effects of commissurotomy on STM using a multiple delayed matching paradigm (Nakamura & Gazzaniga, 1974).

Short-Term Memory and Multiple Delayed Matching to Sample

In brief, the subjects were six rhesus monkeys. One had undergone section of the anterior commissure, corpus callosum, hippocampal commissure, and optic chiasm prior to testing. The other five began as normals, two of which were split in the course of the study. At the beginning of training, all monkeys were conditioned to perform a single color-matching task with no delay. A red or green light would appear at a top screen until it was turned off by the monkey's pressing on the screen. Immediately, the red and green choice colors would appear at two screens placed just below. A press on the screen of the color matching the earlier presentation would be rewarded with water. Subsequently, a delay period was introduced between the initial response of the monkey and the projection of the choice colors. As the animals met criterion performances, the delay was increased up to 18 seconds. On the initial condition, both eyes were open and the monkeys were tested on the delayed matching task until average weekly performance peaked.

Figure 7 shows the averaged data comparisons for splits versus nor-

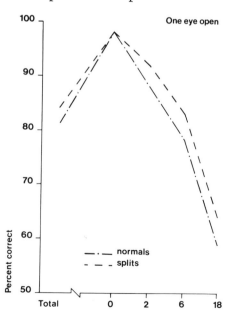

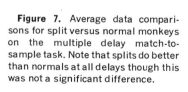

Figure 7. Average data comparisons for split versus normal monkeys on the multiple delay match-to-sample task. Note that splits do better than normals at all delays though this was not a significant difference.

mals. Each of the points to the left showing total percent correct is the result of at least 800 trials per animal and the points of the various delays denote, in turn, at least 200 trials per animal. The splits do better than normals at all delays, though this difference is not significant. There was also no significant interaction between operative condition and delay. Thus, split-brain animals are not worse than normals at any delay.

We have, then, a clear indication that the performance level following complete forebrain commissurotomy does not result in inferior behavior. Indeed, there was a suggestion that the performance had improved. Further work will, hopefully, make clear how acquisition of information might be affected by disconnection of the two half brains.

Cognitive Specialization and Motivation

When lateral cognitive specialization is apparent in the split brain, the question remains how much of the difference can be attributed to motivation? This strikes us as important because we have successfully manipulated the motivational state of animals unilaterally through brain lesions (Gibson & Gazzaniga, 1972; Gibson, 1972). With measurable differences here, in behavior such as eating, differences would clearly be forthcoming in the effectiveness of learning and performing cognitive tasks (Nakamura & Gazzaniga, 1974). These animal studies raise the question as to how much left–right differences could reflect left–right differences in motivation rather than in cognitive capability.

Summary

The overriding issue which cuts through so much of the foregoing is the question of the nature and extent of localization of function of the nervous system. While tying structure to function is one of the long-standing activities of neuropsychology, it still seems to be an elusive enterprise. The reasons are many and, in part, have to do with the problem of determining what it is that is supposed to be localized in a particular brain area. It would seem clear that even the simplest behavioral act is the result of a multitude of overlapping mental activities, such that destruction or impairment in one domain can often find the organism performing as if nothing had happened. What does seem to emerge from the studies reviewed here is that those informational systems in the brain which are involved in conducting primary input (and output) messages are susceptible to dramatic breakdowns in function and these can be easily localized and detected. When, however, the behavior in question is a superordinate process which involves interactions with cognition and motivational states, such as concept formation, or visual–spatial analysis, or logical capability, and the like, all processes which surely involve the small cortical cells, then it is extremely difficult to isolate through which channels such information can be mediated.

It seems to me that neuropsychology, instead of needing less complicated models to help us in the analysis of brain processes, needs a far more sophisticated set of concepts and ideas if it is to penetrate the sheath of mystery surrounding the problem of brain and behavior.

ACKNOWLEDGMENTS

This study has been aided by United States Public Health Service Grant No. MH 17883-04.

REFERENCES

Akelaitis, A. J. Studies on corpus callosum: Higher visual functions in each homonymous field following complete section of corpus callosum. *Archives of Neurology and Psychiatry*, 1941, **45**, 788.

Akelaitis, A. J. Studies on corpus callosum: Study of language functions (tactile and visual lexia and graphia) unilaterally following section of corpus callosum. *Journal of Neuropathology and Experimental Neurology*, 1943, **2**, 26.

Akelaitis, A. J. Study of gnosis, praxis and language following section of corpus callosum and anterior commissure. *Journal of Neurosurgery*, 1944, **1**, 94.

Benton, A. L. Problems of test construction in the field of aphasia. *Cortex*, 1967, **3**, 32–58.

Black, P., & Myers, R. E. Visual function of the forebrain commissures in the chimpanzee. *Science*, 1964, **146**, 799.

Blakemore, C., & Cooper, G. F. Development of the brain depends on the visual environment. *Nature*, 1970, **288**, 477–478.

Bogen, J. E. The other side of the brain. II. An oppositional mind. *Bulletin of the Los Angeles Neurological Society*, 1969, **34**, 135–161.

Bogen, J. E., Fisher, E. D., & Vogel, P. J. Cerebral commissurotomy: A second case report. *Journal of the American Medical Association*, 1965, **194**, 1328–1329.

Bogen, J. E., & Gazzaniga, M. S. Cerebral commissurotomy in man: Minor hemisphere dominance for certain visual–spatial functions. *Journal of Neurosurgery*, 1965, **23**, 394–399.

Bogen, J. E., & Vogel, P. J. Cerebral commissurotomy in man: Preliminary case report. *Bulletin of the Los Angeles Neurological Society*, 1962, **27**, 169.

Bower, G. H. Mental imagery and associative learning. In L. Gregg (Ed.), *Cognition in learning and memory*. New York: Wiley, 1972.

Clark, E., & Gazzaniga, M. S. Preventing visual discrimination defects in monkeys with inferotemporal lesions. 1974, in preparation.

Critchley, M. *The parietal lobes*. London: Arnold, 1953.

Downer, J. L. de C. Interhemispheric integration in the visual system. In V. B. Mountcastle (Ed.), *Interhemispheric relations and cerebral dominance*. Baltimore: Johns Hopkins Press, 1962.

Gazzaniga, M. S. Interhemispheric communication of visual learning. *Neuropsychologia*, 1966, **4**, 261–262.

Gazzaniga, M. S. The split brain in man. *Scientific American*, 1967, **217**, 24–29.

Gazzaniga, M. S. Short term memory and brain bisected man. *Psychonomic Science*, 1968, **12**, 161–162.

Gazzaniga, M. S. *The bisected brain*. New York: Appleton, 1970.

Gazzaniga, M. S. One brain—two minds? *American Scientist*, 1972, **60**, 311–317.

Gazzaniga, M. S. Brain theory and minimal brain dysfunction. *Annals of the New York Academy of Science*, 1973, **205**, 89–92. (a)

Gazzaniga, M. S. Recovery of function in the nervous system. Clark University, *NSF Symposium*, September, 1973. (b)

Gazzaniga, M. S., Bogen, J. E., & Sperry, R. W. Some functional effects of sectioning the cerebral commissures in man. *Proceedings of the National Academy of Science*, 1962, **48**, 1765–1769.

Gazzaniga, M. S., Bogen, J. E., & Sperry, R. W. Laterality effects in somesthesis following cerebral commissurotomy in man. *Neuropsychologia*, 1963, **1**, 290–315.

Gazzaniga, M. S., Bogen, J. E., & Sperry, R. W. Observations of visual perception after disconnection of the cerebral hemispheres in man. *Brain*, 1965, **88**, 221.

Gazzaniga, M. S., Bogen, J. E., & Sperry, R. W. Dyspraxia following division of the cerebral commissures in man. *Archives of Neurology*, 1967, **16**, 606–612.

Gazzaniga, M. S., & Freedman, N. Observations of visual processes following posterior callosal section. *Neurology*, 1973, **23**, 1126–1130.

Gazzaniga, M. S., & Hillyard, S. A. Language and speech capacity of the right hemisphere. *Neuropsychologia*, 1971, **9**, 273–280.

Gazzaniga, M. S., & Hillyard, S. A. Attention mechanisms following brain bisection. In S. Kornblum (Ed.), *Attention and performance IV*. New York: Academic Press, 1973.

Gazzaniga, M. S., Risse, G. L., Springer, S. P., Clark, E., & Wilson, D. H. Psychological and neurological consequences of partial and complete commissurotomy. *Neurology* (in press).

Gazzaniga, M. S., & Sperry, R. W. Simultaneous double discrimination response following brain bisection. *Psychonomic Science*, 1966, **4**, 261–262.

Gazzaniga, M. S., & Sperry, R. W. Language after section of the cerebral commisures. *Brain*, 1967, **90**, 131–148.

Gazzaniga, M. S., Szer, I., & Crane, A. Modifying drinking behavior in the adipsic rat. *Experimental Neurology*, 1974, **42**, 484–489.

Gazzaniga, M. S., Velletri, A. S., & Premack, D. Language training in brain-damaged humans. *Federation Proceedings Abstracts*, 1971, **30**(2), 265.

Geschwind, N., & Kaplan, E. A human cerebral deconnection syndrome. *Neurology*, 1962, **12**, 675.

Gibson, A. R. Independence of cortico–hypothalamic feeding mechanisms in brain bisected monkeys. Doctoral thesis, New York Univ., 1972.

Gibson, A. R., Dimond, S. J., & Gazzaniga, M. S. Left field superiority in word matching. *Neuropsychologia*, 1972, **10**, 379–381.

Gibson, A. R., & Gazzaniga, M. S. Hemispheric differences in eating behavior in split-brain monkeys. *Physiologist*, 1972, **14**, 150.

Glass, A. S. Cognition following stroke. Doctoral thesis, New York Univ., 1973.

Glass, A. S., Gazzaniga, M. S., & Premack, D. Artificial language training in global aphasics. *Neuropsychologia*, 1972, **11**, 95–103.

Gorden, H. W., Bogen, J. E., & Sperry, R. W. Absence of disconnexion syndrome in two patients with partial section of the neocommissures. *Brain*, 1971, **94**, 327–336.

Hamilton, C. R., & Brody, B. A. Separation of visual functions within the corpus callosum of monkeys. *Brain Research*, 1973, **49**, 185–189.

Hamilton, C. R., & Gazzaniga, M. S. Lateralization of learning of colour and brightness discriminations following brain bisection. *Nature*, 1964, **201**, 220.

Harlow, H. *Learning to love*. San Francisco: Albion, 1971.

Hirsch, H. V. B., & Spinelli, D. N. Modification of the distribution of receptive field orientation in cats by selective visual exposure during development. *Experimental Brain Research*, 1971, **12**, 504–527.

Hoedt–Rasmussen, R., & Skinhoj, E. Transneuronal depression of the cerebral hemispheric metabolism in man. *Acta Neurologica Scandinavica*, 1964, **40**, 41–46.

Kempinsky, W. H. Experimental study of distant effects of acute focal brain injury. *Archives of Neurology and Psychiatry*, 1958, **79**, 376–389.

Lashley, K. S. In search of the engram. In *The neuropsychology of Lashley*. New York: McGraw–Hill, 1950.

Levine, M. Hypothesis in humans during discrimination learning. *Journal of Experimental Psychology*, 1966, **71**, 331–338.

Levy, J., Trevarthen, C., & Sperry, R. W. Perception of bilateral chimeric figures following hemispheric deconnexion. *Brain*, 1972, **45**, 61–78.

Levy-Agresti, J., & Sperry, R. W. Differential perceptual capacities in major and minor hemisphere. *Proceedings of the National Academy of Science*, 1968, **61**, 1151.

Luria, A. R., Nayden, V. L., Tsvetkova, L. S., & Vinarskaya, E. N. Restoration of higher cortical function following local brain damage. In P. J. Winken & G. W. Bruyn (Eds.), *Handbook of clinical neurology*. Amsterdam: North Holland Pub., 1969.

Lussenhop, A. J., de la Cruz, T., & Fenichel, G. M. Surgical disconnection of the cerebral hemispheres for intractable seizure. *Journal of the American Medical Association*, 1970, **213**, 1630–1636.

Maspes, P. E. Le syndrome experimental chez l'homme de la section du splenium du corps calleux alexia visuelle pure hemispherique. *Revue Neurologique*, 1948, **80**, 100–113.

Meikle, T. H., Jr. Failures of interocular transfer of brightness discrimination. *Nature*, 1964, **202**, 1243–1244.

Milner, B. Psychological deficits produced by temporal lobe excision. *Proceedings of the Association Research in Nervous and Mental Disease*. Baltimore: Williams & Wilkins, 1958.

Milner, B. Effects of different brain lesions on card sorting. *Archives of Neurology and Psychology*. 1963, **9**, 90–100.

Milner, B., & Taylor, L. Right-hemisphere superiority in tactile pattern-recognition after cerebral commissurotomy: Evidence for nonverbal memory. *Neuropsychologia*, 1972, **10**, 1–16.

Milner, B., Taylor, L., & Sperry, R. W. Lateralized suppression of dichotically presented digits after commissural section in man. *Science*, 1968, **161**, 184–185.

Moore, R. G., Bijorland, A., & Stenevi, U. Plastic changes in adrenergic innervation of the rat septal area in response to denervation. *Brain Research*, 1971, **33**, 13–35.

Mosidze, V. M., Rizhinashvili, N. K., Totibadze, N. K., Kevanishvili, Z., & Akbardia, K. K. Some results of studies on split brain. *Physiology and Behavior*, 1971, **7**, 763–772.

Myers, R. E. Transmission of visual information within and between the hemispheres:

A behavioral study. In V. B. Mountcastle (Ed.), *Interhemispheric relations and cerebral dominance*. Baltimore: Johns Hopkins Press, 1962.

Myers, R. E. The neocortical commissures and the interhemispheric transmission of information. In E. G. Ettlinger (Ed.), *Functions of the corpus callosum*. London: Churchill, 1965.

Myers, R. E., & Sperry, R. W. Interocular transfer of a visual form discrimination habit in cats after section of the optic chiasm and corpus callosum. *Anatomical Record*, 1953, **175**, 351–52.

Nakamura, R. K., & Gazzaniga, M. S. Comparative aspects of short term memory mechanisms. In J. A. Deutsch (Ed.), *Short term memory*. New York: Academic Press, 1974.

Nebes, R. D. Superiority of the minor hemisphere in commissurotomized man for the perception of part–whole relations. *Cortex*, 1971, **7**, 333–349.

Nebes, R. D., & Sperry, R. W. Superiority of the minor hemisphere in commissurotomized man for the perception of part–whole relations. *Cortex*, 1972, **7**, 333–349.

Nottebohm, R. Ontogeny of bird song. *Science*, 1970, **167** (3920), 950–956.

Paivio, A. Mental imagery in associative learning and thought. *Psychological Review*, 1969, **76**, 241–263.

Premack, D. Reinforcement theory. In M. R. Jones (Ed.), *Nebraska Symposium on Motivation*. Lincoln: Univ. of Nebraska Press, 1966.

Premack, D. Language in chimpanzee? *Science*, 1971, **172**, 808–822.

Raisman, G. Neuronal plasticity in the septal nuclei in the rat. *Brain Research.*, 1969, **14**, 25–48.

Robindon, M. Factors influencing stroke rehabilitation. *Stroke*, 1971, **2**, 213–218.

Sarno, J. E., Sarno, M. E., & Levita, E. Evaluating language improvement after completed stroke. *Archives of Physical Medicine and Rehabilitation*, 1971, **52**, 73–78.

Sarno, M. T. Speech therapy and language recovery in severe aphasia. *Journal of Speech and Hearing Disorders*, 1970, **13**, 607–625.

Schneider, G. Early lesions of the superior colliculus: Factors affecting the formation of abnormal retinal projections. *Brain, Behavior and Evolution*, 1973, (in press).

Seamon, J. G. Imagery codes and human information retrieval. *Journal of Experimental Psychology*, 1972, **96**, 468–470.

Seamon, J. G., & Gazzaniga, M. S. Coding strategies and cerebral laterality effects. *Cognitive Psychology*, 1973, **5**, 249–256.

Sechzer, J. A. Prolonged memory and split-brain cats. *Science*, 1970, **169**, 889–892.

Sperry, R. W. Embryogenesis of behavioral nerve nets. In R. L. Dehaan & Ursprung-itienrich (Eds.), *Organogenesis*, 1966.

Stern, P., McDowell, F., Miller, J. M., & Robinson, M. Factors influencing stroke rehabilitation. *Stroke*, 1971, **2**, 213–218.

Sternberg, S. High-speed scanning in human memory. *Science*, 1966, **153**, 652–654.

Sullivan, M. V., & Hamilton, C. R. Memory establishment via the anterior commissure of monkeys. *Physiology and Behavior*, 1974 (in press).

Trescher, J. F., & Ford, F. R. Colloid cyst of the third ventricle: Report of a case. *Archives of Neurology and Psychology*, 1937, **37**, 959.

Trevarthen, C. B., & Sperry, R. W. Perceptual unity of the ambilant visual field in human commissurotomy patients. Unpublished paper, 1969.

Voneida, T. J., & Robinson, J. S. Effect of brain bisection on capacity for cross comparison of patterned visual input. *Experimental Neurology*, 1970, **26**, 60–71.

Zangwill, O. Intelligence in aphasia. In A. V. S. de Rauk & M. O'Connor (Eds.), *Foundation Symposium on Disorder of Language*. London: Churchill, 1964.

Chapter 20

On the Origins of Language*

DAVID PREMACK

University of California, Santa Barbara

INTRODUCTION

Consider two main ways in which you could benefit from my knowledge of the conditions next door. I could return and tell you, "the apples next door are ripe." Alternatively, I could come back from next door, chipper and smiling. On still another occasion I could return and tell you, "a tiger is next door." Alternatively, I could return mute with fright, disclosing an ashen face and quaking limbs. The same dichotomy could be arranged on numerous occasions. I could say, "the peaches next door are ripe," or say nothing and manifest an intermediate amount of positive affect since I am only moderately fond of peaches. Likewise, I might report, "a snake is next door," or show an intermediate amount of negative affect since I am less shaken by snakes than by tigers.

For simplicity, consider that everything of interest is next door, making locational information irrelevant. The question to be answered therefore is always, What is it? never, Where is it? Also, for simplicity, assume

* Paper read at Conference on Behavioral Basis of Mental Health, Galway, Ireland, 1972. Reprinted from "Concordant preferences as a precondition for affective but not for symbolic communication (or How to do experimental anthropology)," *Cognition*, 1973, 1, 251–264. Reprinted by permission of Mouton Publishers.

that everything I tell you is true, even as every affective state I display is genuine and not simulated. These qualifications are not necessary for the simple argument to be made here but they smooth the way.

Information of the first kind consists of explicit properties of the world next door; information of the second kind of affective states that I will assume can be positive or negative, and can vary in degree. Since changes in the affective states are caused by changes in the conditions next door, the two kinds of information are obviously related. In the simplest case, we could arrange that exactly the condition referred to in the symbolic communication be the cause of the affective state. If "cause" is too simple, then consider that the tiger, apple, or snake, etc. is the dominant factor in the affective display; much of what I will say here could be given more sophisticated formulation but without contributing materially to the basic argument.

The use you could make of my statements needs no comment; but the use you could make of my affective states is almost equally obvious. You could go next door when my state was positive, not go when it was negative. The speed and certainty with which you went could be proportional to the intensity of my positive state. The certainty with which you did not go, even perhaps the distance you went to the opposite side of the house or the number of doors you locked behind you, could be proportional to the intensity of my negative state. Or you could go on all occasions but carry a bucket on some occasions, a spear on others. All of this is foretold from a simple fact stated earlier: the two kinds of information are correlated; they are caused by the same factors.

Conveniently locating everything of interest next door is a hypothetical arrangement, of course, but we can see the same affective system operate to good advantage in an actual experimental situation. Menzel (1971) has recently reported some ingenious experiments concerning the social behavior of a group of young chimpanzees in a 1-acre compound. After hiding such objects as food or snakes in the compound, the experimenter took one animal into the compound and showed it the hidden object. He then returned the informed animal to the rest of the group, which was held in a restraining cage on the edge of the compound, and released all animals together (regrettable from the point of view of communication studies but compatible with Menzel's objectives). The animals succeeded in finding the hidden object, snake no less than food, significantly more often than when they were released into the compound without the benefit of an informed animal.

In other experiments, Menzel hid two caches of food, one larger than the other, showed one to animal A the other to animal B, and again released all animals at the same time. Typically they not only found both objects, but found the larger one first. In addition, Menzel noted that the quality and even perhaps quantity of the object hidden could be predicted by an uninformed observer from the exit behavior of the animals (personal communication). From the moment the animals were released there were detectable differences in posture, gait, and vocalizations.

Why did the knowing animal not simply steal off and enjoy the food to itself? Immature chimps, and perhaps even caged adult ones, are apparently afraid to venture too far into the compound alone. Menzel's work shows that they tend not to exceed a certain distance from one another, actually not a distance so much as a time needed to overcome a distance. The informed animal's fear combined nicely with the group's ignorance of the location of the hidden objects, giving rise to a high mutual need for communication. The informed animal was afraid to go alone, while the uniformed animals were unlikely to find the hidden object if they went alone. Each party needed the other.

Man has both affective and symbolic communication. Indeed, conflict between the two—e.g., a father saying to his son through gritted teeth, "I do agree with you, more than you realize"—has been proposed as a source of mental illness (Mehrabian, 1970). All other species, except when tutored by man (Gardner & Gardner, 1969; Premack, 1970), have only the affective form. Even affective information alone could be of great value in a world where changes in location can be costly in terms of energy, risk, or both. As a rule, the individual cannot simply venture forth each time a forager returns in a positive state. Instead he must weigh possible gains against possible losses, taking into account the valence and intensity of the speaker's state on the one hand and his knowledge of predators or distance to be traveled on the other. If predators were abundant, a leader's positive state would have to be exceptionally high to induce a positive decision. Conversely, if predators were few, a listener could indulge his curiosity, going forth unreservedly to learn what had occasioned a leader's mildly negative state.

In principle, an affective system would permit an animal's choice behavior to accurately reflect the objective probabilities of his world. This is so despite the fact that a speaker's affective state probably does not distinguish quality from quantity but is proportional to their resultant. For instance, a large amount of a $+4$ item would probably occasion the same affective state as a lesser amount of a $+7$ item. Yet this need not be seen as a weakness of the affective system. If my preferences are the same as yours, I would be as indifferent as you in a choice between large $+4$ and small $+7$, and thus would be willing to pay the same price for both commodities in terms of risk and energy.

The affective system could not rely entirely on unconditional factors but for maximum efficiency would seem to require some amount of learning. For example, since all members of a group are not likely to be of the same temperament, listeners should not respond merely to a magnitude of affect but rather to magnitude plus source. Contextual sensitivity of this kind would enable the listener to react in the same way to .4 intensity from sluggish Henry as to .8 intensity from excitable George. On the other hand, in the field this problem may be reduced if not averted by the fact that a listener would not be guided by the affective state of any one speaker. Instead, the speaker's state would elicit affective states in all his listeners, the intensities of which would vary with their respective temperaments. If each listener more or less integrated over the several affective states, responding, say, to the average intensity of the group, the source of the individual states could be ignored. Only in the laboratory where a listener was restricted to a single speaker would it be necessary to take the source of the affect into account. Yet if one belonged to a group that varied, either through occasionally leaving one group to join another or through changes in membership brought on by the movement of others, the desirability of responding contextually would arise again. Only this time the contextual factor to which a listener should be sensitive would be the group rather than the individual.

Even when helped by learning, the affective system is capable only of answering "what"-questions and not "where"-questions. But this does not limit the applicability of the affective system as sharply as it might seem. There are at least three ways in which to circumvent the need for locational information while retaining the value of the what-information. First, everything of interest can be next door, in a department store on 6th Street, or across from the fire station, i.e., in some agreed upon location. Second, an informed leader through fear of venturing too far alone or for other reasons can lead his uninformed peers to the hidden objects. Third, the successful forager need not return to the group but can reveal his location by calls, manifesting his affective state on an auditory basis. The latter is undoubtedly the case most often found in the field; nonetheless all three cases share this property: the value of what-information is preserved despite the fact that the affective system does not code for directional information. These examples may help to underscore the uniqueness of the bee's putative communication system (von Frisch, 1967). The proportionality between the rate at which the bee waggles and the quantity, quality, or distance of the food is merely another instance of the motivational system now turned to a communication purpose quite like the affective system discussed here. But the coding for direction which the bee's system is said to contain is unique and to my knowledge could not be derived from the classic properties of motivational systems.

We come now to the main point of the paper. Affective communication depends upon a simple precondition: All members of the group must have concordant preference orders for the items about which they communicate. When members of a group are agreed about what is positive and negative, and the order of their magnitudes, then, in effect, any member of the group can use the affective state of any other member to predict his own affective state. But if you and I do not order items comparably then neither of us can use the valence or intensity of the other one's excitement to predict his own.

Suppose, for example, you are very fond of strawberries but I detest them. You return in a high positive state. Knowing nothing of your peculiar tastes I become equally excited in anticipation of a highly positive item. On such an occasion I am especially likely to go next door or to follow you into the compound, only to suffer the disappointment of strawberries. Your excitement has not proved to be a good basis for predicting my excitement.

Is this a problem which learning could resolve? Typically, when owing to some change in circumstances, the unlearned behavior of a species becomes maladaptive, we turn to learning as the most powerful corrective device for restoring adaptive behavior. Could the affective system be restored by learning in the case of disconcordant preferences? Only in those special cases where there was a systematic relation between the preferences of several individuals. For example, if the preferences of one party were the perfect inverse of those of another party, the two parties could learn, either swiftly through a rule induced on a few exemplars, or slowly by trial and error, to adjust to this difference. Comparable adjustments could be made in principle for any systematically related preference orders.

But if the relations were not systematic it is not clear that learning could contribute substantively to the problem. Consider a successfully communicating group in which nonsystematic preferences were introduced. Although the speaker's affective response to snake or food may be largely unlearned, a listener's response to a speaker should have both learned and unlearned components. Thus a listener who had rushed next door on the occasion of a speaker's positive excitement, only to find food that he did not care for, or worse a snake, would learn to inhibit his response to that speaker's positive excitement. But would that solve his problem? On a susequent occasion he might learn belatedly that the same speaker's positive excitement, which he chose to ignore, was the occasion for an encounter with bananas, an item which he cared for greatly. He might also discover that some of the speaker's negative states were occasioned by items of which he was quite fond. Yet it would not do simply to respond positively to all of the speaker's negative states since at least some of them would be associated with negative conditions. And the listener would have no way of distinguishing "good" occasions from "bad" ones. In the long run, when preferences differ nonsystematically, acting on a speaker's positive affective states would lead a listener to negative and positive stimuli with about equal frequency; acting on his negative affective states would have the same outcome. Similarly, the average intensity of the stimuli a listener encountered would be the same for all intensities of the speaker's states.

Could the problem of disconcordant preferences be resolved by a call system that was not restricted to affective states which were either simply positive or negative? Suppose the species had one call for food, another call for danger, etc. The functional effect of such calls is tantamount to an agreement between members of the species to call the same things food, the same things dangerous, etc. Yet calls of this kind do not differentiate one member of the food class from another member, nor one member of the dangerous class from another member. Thus the call system would protect an anticipation of, say, strawberries against the discovery of a tiger, or vice versa. But it would not protect anticipation of bananas against discovery of strawberries, or anticipation of snake against discovery of tiger. To avoid within-class as well as between-class confusion would require either concordant preferences or calls that were specific not only to classes but to members of the classes.

The disastrous effect upon the affective system of disconcordant preferences would be comparable to that of a chaotic world the content of which changed before a listener had an opportunity to act upon a speaker's message. Indeed, when first beset by the consequences of disconcordant preferences, a listener might well conclude that the world had changed, that it was no longer a reliable place. For often before he could get next door, apples would have turned into snakes and conversely; or so it might seem to the listener whose companion's preferences had been altered without his knowledge.

ICONIC AND SYMBOLIC COMMUNICATION

Both symbolic and iconic communication escape the simple precondition upon which affective communication depends. Organisms that disagree radically about values can nevertheless guide one another through the world provided they communicate either symbolically or iconically. Communication in this case does not depend upon a unanimity of values but merely upon the consistent application of names or icons to the items in the world. For instance, if you tell me the turnips next door are ready, your possible dislike of them would not detract from the information. If I want to try some I can, your finicky message notwithstanding. Symbolic communication escapes the precondition because the listener is presented not with (only) the speaker's affective response to a condition but with a statement of the condition. In the iconic case he is presented with a piece of the condition. If not a tree full of ripe apples, then an apple core or even a leaf of the tree; or if not a tiger then perhaps a product of the tiger, droppings or claw marks. But the tiger might be constipated, the heroic speaker might die in an attempt to bring home a whisker, or with a more plebeian speaker listeners might die for lack of a warning. Icons are inefficient. Worse, concepts are not equally susceptible to iconic representation; some, such as the logical connectives, could not be represented in that manner at all. Admittedly, symbolic communication is pervaded by iconicity (e.g., Durbin, 1971; Wescott, 1971), but the ultimate unacceptability of the pure iconic approach is incontestable. All this is beside the main point, however, which is simply that organisms agreed about values can guide one another through the world with affective communication alone; whereas organisms disagreed about values can still guide one another through the world provided they communicate symbolically.

EXPERIMENTAL ANTHROPOLOGY

The difference in preconditions for affective and symbolic communication can be used to do what might reasonably be called experimental anthropology. Years ago a moritorium was declared on speculating about the origin of language (Hewes, 1972). Learned societies sought to help man resist the temptation of speculating about the unknowable by prohibiting all such publication. Today I think we can go beyond idle speculation, not only about origins of language but of human milestones generally. We can test models concerning the origins of language, agriculture, religion, art, etc., and though perhaps we can never say how in fact they did originate, we can assign weights to the alternatives on an experimental basis.

A model of the origin of language or of any other human activity will have two components. The first will state the cognitive skills that are a prerequisite for the activity. The second will state the selective pressures which, if imposed upon organisms with the prerequisite skills, will lead to the development of the activity in question. The first component deals with the problem-solving ability of the species or its information-processing capacity generally. The second deals with environmental pressures, problems that are posed a species by changes in the world.

In the rest of this paper, I will take the conclusion from the first half of the paper, and show how it can be utilized as a selective pressure in experiments on the origins of language. In addition, I will make some tentative proposals for a general model of the origin of human activities.

LABORATORY–FIELD COMBINATION

Consider a joint laboratory–field approach to the origin of language. Even though we cannot yet enumerate the cognitive skills that are prerequisites for language, laboratory studies have already shown that the chimpanzee can be taught some of the principal exemplars of language (Gardner & Gardner, 1969; Premack, 1970). We know, for example, that the chimp is capable of symbolization, of using one event to represent another; of responding differentially to different word orders; of concatenating and rearranging words in ways that are necessary for the production of sentences. A successful comparison of human and animal intelligence requires that we be able to state the cognitive preconditions for these linguistic performances and ultimately all basic human activities. Our progress in achieving this objective could be measured by our ability to predict, for example, that species with certain cognitive skills could be taught language whereas those without these skills could not.

Since we know that the chimp can be taught symbolic communication, it is sensible to apply a selective pressure to this species which may lead it to develop symbolic communication on its own. In the field, using Menzel's procedures, we could induce in a small group of chimps a high need to communicate. As we have seen, the need to communicate which this procedure induces is normally handled nicely on an affective basis. Followers can accurately anticipate both the valence and the intensity of their own future excitement from the current excitement of the informed animal. But we also know that we could undermine the affective system simply by introducing disconcordances in the preferences, thus leaving an unresolved need to communicate.

In the laboratory we could deprive and satiate the animals on different foods. An animal normally keen on, but now satiated on, bananas would be disappointed to find the bananas to which it was led by a highly excited animal that was not satiated on banana. Preferences could also be manipulated by contingencies, by arranging that an animal be able to obtain a highly preferred food only by first eating a nonpreferred one. This would increase the animal's preference for a normally nonpreferred food and lead it to bring back "false" reports (false positive) concerning what was in the compound. With an appropriate combination of these procedures we could arrange that no animal's preference order be a function of the preference order of any other animal. And if this were not enough, we could also change any animal's preference order from time to time by changing the satiation and contingency procedures from time to time. In this way, in animals known to be capable of being taught symbolic communication, we could produce a high need to communicate, while at the same time eliminating the normal mechanism for doing so. Could the chimpanzees then invent iconic or symbolic communication themselves, first when the hints from the experimenter were strong, later when they were made progressively weaker?

Perhaps the first time this experiment is done, the animals should be left entirely to their own resources. If they proved incapable of developing a substitute as seems highly likely, or, as is also possible, developed one that we could not decipher, we could attempt to structure their problem solving, not only as an aid to them but also to make certain that we could follow their solution. For instance, we could allow the informed animal to return with a piece of the hidden object. This may duplicate the field situation in which the forager returns not only with its affective state, but also with vestiges of the source of its affective state, for example, with the smell of the food on its breath or body. (So an observer could use the forager's breath to tell him what was next door and the intensity of the forager's excitement to tell him how much was there.) We might sidetrack briefly to explore the chimp's overall ability to use iconic representations. Starting with icons whose relation to the hidden object was that of part–whole, we could progressively weaken the relation, ending up with cases where the icon was merely an associate of the hidden object. Ultimately we could study metaphors. In addition, by giving the informed animal not one object to return with, but a number of alternatives from which to choose, we could study the informed animal's ability to choose wisely, to pick items that its uninformed companions could use.

When an informed animal returned with icons that were informative as its affective states no longer were, we could observe the possible transition from a reliance on affective states to a reliance on icons. In the course of this transition, we might observe a general degradation of responsivity to emotional cues, since these cues would no longer possess the functional significance they once did.

WORDS

In a quite different approach we would provide the informed animal not with icon, metaphor, or a choice among different possible ones, but with words taught both it and the other members of the group in the laboratory. In this case the animal could return to the uninformed group with the word naming the hidden object that it had been shown. Although the proportionality between a speaker's affective state and the hidden object would no longer be an aid to its companions, with words the informed animal could tell the other animals exactly what was there. Possible differences in their evaluations of the hidden object should become irrelevant. Animals with a high preference for the object named by a given word on a particular trial should go forth with the informed leader; those with low preferences for the item should stay home.

INVENTED WORDS

Once the animals succeeded in using icons or words that had been taught them in the laboratory, we could raise the critical question, Can they devise their own symbols? The invention of symbols would seem to involve a change more difficult than that involved in other kinds of innovation. Changes in food preparation, tool use, and the like, well documented in primate groups (e.g., Marler, 1965), can be made by one animal and then transmitted by social modeling to other animals. But a symbol cannot be invented and transmitted in this way. At least two individuals must use a symbol in the same way in order for it to be effective. I may use a blue triangle to represent apple in my private thinking and problem solving while you use a red square for the same purpose, but we could not communicate about apples until we used the same symbol, or found a way to establish the equivalence of our different symbols. Symbols seem, therefore, more likely to be social inventions rather than individual ones. An alternative would be for one inventor to transmit his idea to other animals; an improbable alternative in light of the degree of instruction that would be demanded and the fact that didactic intervention of that kind is apparently totally unknown outside of man. Indeed, instruction is considered to play only a minor role in the child's acquisition of language.

How can we arrange for the kind of joint invention that the symbol seems likely to require? Two animals could be shown a hidden food at the same time and supplied an arbitrary object as the only possible item with which to represent the hidden food. These animals would be in a position to share the same potential symbol. In addition, animals that chanced to follow the first pair into the compound could associate the arbitrary object shown them with the food discovered in the compound. The association is more likely to develop if the interval was short, or perhaps merely if the food were new. In some species (Garcia, Ervin, & Koelling, 1966), associations develop between avoidance responses and foods despite long intervals provided the food is new and it results in gastrointestinal upset. The sickness may be unnecessary, however, and the association between stimuli and food may develop over unusual intervals merely if the food is new.

In another approach to the invention of symbols, words or icons would not be brought back from the compound by the informed animal but would be selected by that animal from alternatives stored in the restraining cage. After being returned to the restraining cage, the informed animal would look over the alternatives available to him there and select the one he considered to best represent the object he had seen in the compound. If the alternatives were consistently stored in specific locations, the removal of either the words or icon need not prevent the informed animal from communicating with his peers. He could put his hand in the appropriate place, or merely point in a given direction, and in this way perhaps devise gestures that would substitute for the previous words or icons.

AESTHETICS AND THE DISCOVERY OF
BASIC CAUSAL RELATIONS

In the other human milestones—art, religion, agriculture—comparable analyses would be made: cognitive skills on the one hand and selective pressures on the other. Since the logic of these cases is not different from that of language, I will not take them up but will turn to a slightly different problem. One precondition for certain human activities is a knowledge of basic cause–effect relations. In agriculture, for example, the most basic causal relation is that between the seed and the plant. Consider the nature of the circumstance in which relations of this kind are likely to be discovered and ultimately used.

Agriculture is considered to have replaced hunting and gathering in areas where population density made earlier forms of provisioning untenable (e.g., Binford, 1971). Population density may have led to agriculture, but is this same pressure likely to have led to the discovery of the seed–plant relation? Bushmen of today are reported to know the seed–plant relation yet they continue to hunt and gather nonetheless. Though more efficient than hunting or gathering, agriculture is actually more arduous. People turn to it, I suspect, because they have to and not because they have just discovered the seed–plant relation.

Discoveries of basic cause–effect relations such as the seed–plant relation seem more likely to occur under the aegis of aesthetic or exploratory dispositions than utilitarian ones, and thus to occur in contexts far removed from those in which the knowledge is ultimately used. If this is so and the causal knowledge is often not used directly, some functional repository would seem necessary, a system for preserving knowledge that a group carried but was not yet using. Finally, there is the terminal phase in which appropriate selective pressures operate upon existing knowledge to produce technologies representing solutions to practical problems. This suggests a three-state model in which the basic steps in the development of human technologies are discovery, retention, and use.

A principal root of the aesthetic disposition is a preoccupation with the discontinuities of space and with the possibilities of their transformation. We need not go to the human artist in whom these dispositions are institutionalized; in a minute way they can be seen even in a rat. Placed in a small box in which a lever projects from a wall, the rat rises on its hind legs, sniffing and sweeping its vibrassae across the wall. In dropping back to the floor, its front legs contact the lever, which gives slightly under the pressure, causing the rat to stiffen and its hair to bristle; there is a momentary excitement. Having discovered this break in the texture of space, the rat is likely to return to reinstate it. The event in the rat is small but it can be magnified in the monkey and still more in the chimp. Consider a monkey that has pressed the same lever hundreds of times, producing no extrinsic consequence. One day the lever sticks before returning to resting position. The visibly excited monkey presses 30 times in the space of a few minutes trying presumably to restore the change in the visual transit of the lever.

There is no end to this kind of event in the chimp; I will offer only one example. Sarah, a ten-year-old female chimp, who is the subject of a long-term language project, occasionally finds cuts on the hand of her trainer. She squeezes the cut expertly, not by opposing her thumb and forefinger in the human manner, but by placing her index fingers on opposite sides of the cut. The pressure accomplished in this manner can be very finely graded. As she squeezes, her attention is rapt; she looks up from the cut only to peer into the eyes of the trainer (who looks back puzzled and a bit frightened, not of Sarah but of the intensity of her preoccupation). The chimp leaves off pressing just as a thin red line appears along the cut, outlining it against the rest of the skin. Presumably she would go on in this manner, raptly attentive, making subtle changes in space—if we had a device that could offer her multiple cuts. But no one has been willing to inflict a series of even small cuts in his hand simply to confirm the obvious.

Rather than elaborate examples from chimp behavior, I will provide one example from human behavior, one which, as you will see, applies directly to agriculture. In discussing the present thesis with Dr. Barbara Partee (gifted UCLA linguist) she was reminded of an event from her childhood which she has kindly consented to have reported. Walking in the woods in the late fall, she found an unusual clump of small trees. Returning with a small saw, she cut down the trees in the center and stuck them in with the other trees to form "a fort" as she recalls it. In the spring she rediscovered the clump of trees, finding not only the original trees in bloom but those she had sawed off and transplanted as well. In this way, under the aegis of a disposition to operate upon and "improve" space, she discovered rooting, one of the oldest forms of horticulture.

Fossilized seeds, recently discovered on the graves of Neanderthal Man, have proved to be the seeds of flowers, suggesting that 50 thousand years ago man was already placing flowers on the graves of his dead. Who can say but that in the context of burial, seed may have fallen in fresh earth, sprouted, and led man to discover the seed–plant relation. The initial discovery of that relation is lost in prehistory; we cannot reasonably hope to recover it. My point is simply to note, first, the urgency of the aesthetic disposition in man and even chimp, and second that the disposition is of a kind to lead man into activities where he is likely to discover basic cause–effect relations.

If knowledge of great utilitarian potential is first discovered in nonutilitarian contexts, it seems reasonable to provide a repository for it, such that it may be preserved for later use. The repository may be ritual, religion, or even art; I have no clear idea. The problem has clear aspects of psychological interest however. What factors make it likely that knowledge acquired in one context will be preserved in some other context; and what factors make it likely that knowledge will be used in contexts different from those in which it was discovered, preserved, or both? The literature on problem solving suggests some of the difficulties that can arise in transporting an idea from one domain to another. Also we know the power of metaphor, a power by no means restricted to art but found also in scientific discovery. Can we systematize these matters and show, for example, how causal relations discovered in one context are more likely to be utilized than causal relations discovered or preserved in some other context?

The last assumption in the model is simply that cause–effect relations which may have been a part of group knowledge for years will come to provide the basis of technology when activated by appropriate pressures. These three assumptions provide the tentative basis of a model as to how knowledge may be discovered, carried, and ultimately used. There are psychological issues of considerable interest locked in these assumptions and it is my hope to free them with the help of chimpanzees in a combined laboratory–field approach.

CONCLUSION

An argument is shown that could provide a litmus paper-like test for determining whether or not a species communicates symbolically when appropriately constrained (as opposed to, is capable of acquiring language with human intervention). In affective communication, a listener can predict his own emotional state from that of a speaker. If a speaker, having found food in one case and predators in another, returned in a high positive state in one case and a high negative state in another, listeners could venture forth and stay home respectively, in anticipation of objects that would induce comparable affective states in them. The effect could be comparable to that of the symbolic communication "there are apples out there" and "there is a tiger out there." However, the affective system depends upon a precondition which the symbolic one does not. A listener can predict his own affective state from that of a speaker only if both parties have concordant preferences. Therefore, induce a high need to communicate while at the same time destroying the normally concordant preferences of the species; if the species still communicates, it is symbol positive. In addition, a model is presented suggesting that basic cause–effect relations, of a kind underlying human technologies, are discovered in nonutilitarian contexts under the aegis of an aesthetic disposition. We explore the experimental application of these ideas to chimpanzees in a simulated field situation.

REFERENCES

Binford, L. R. Post-Pleistocene adaptations. In S. Struever (Ed.), *Prehistoric agriculture*. New York: The Natural History Press, 1971.

Durbin, M. Some non-arbitrary aspects of language. Paper presented at the American Anthropological Association, New York, 1971.

Garcia, J., Ervin, F., & Koelling, R. Learning with prolonged delay of reinforcement. *Psychonomic Science*, 1966, **5**, 121–122.

Gardner, R. A., & Gardner, B. T. Teaching sign language to a chimpanzee. *Science*, 1969, **165**, 664–672.

Hewes, G. W. An explicit formulation of the relationship between tool-using, tool-making and the emergence of language. Unpublished manuscript. Univ. of Colorado, Boulder, 1972.

Marler, P. Communication in monkeys and apes. In I. De Vore (Ed.), *Primate behavior*. New York: Holt, 1965.

Mehrabian, A. *Tactic of social influence*. Englewood Cliffs, New Jersey: Prentice-Hall, 1970.

Menzel, E. W. Social organization of a group of young chimpanzees. Paper read at the meeting of the American Anthropological Association, New York, 1971.

Premack, D. A functional analysis of language. *Journal of Experimental Analysis of Behavior*, 1970, **14**, 104–125.

von Frisch, K. *The dance language and orientation of bees*. Cambridge, Massachusetts: Harvard Univ. Press, 1967.

Wescott, R. Linguistic iconism. *Language*, 1971, **47**, 416–428.

Chapter 21

Do We Need Cognitive Concepts?

RICHARD L. GREGORY

University of Bristol, England

INTRODUCTION

Although the study of perception is, of all the diverse problems of psychology, the most readily investigated by controlled experiment, it meshes uneasily with the body of scientific thought. Physiology is, however, a different case. It is seen to relate closely to physics: which, perhaps, is why there are many attempts to describe all perceptual phenomena in terms of concepts appropriate to the techniques of physiology. These techniques, especially single cell recording, have revealed so much, clearly of the first importance, that a strong case might be made for assuming that this is the most direct way to understanding all of perception. On the other hand there are many people thinking in terms of "cognitive concepts," which may be hardly at all concerned with the details of the underlying physiology of the sense organs or the brain.

Considerable heat is generated by the conflicting opinions of how we should describe and explain perceptual phenomena, and which should be singled out for special study. So although perception is a lively experimental subject, we remain enmeshed in philosophical issues, perhaps too seldom made explicit. The purpose of this chapter is to try to bring out some of the philosophical issues, especially concerning the uneasy relation between "cognitive" and "physiological" concepts, in the light of some basic phenomena of perception. For the present, it may be appropriate to describe the study of perception by the old name for the physical sciences: Experimental Philosophy.

Some aspects of perception do seem to be very different from the physical world: if only because the most dramatic of perception's phenomena—the great variety of illusions—are departures, or deviations, from the world of objects as described by the physical sciences. Clearly, illusions, like errors in science, cannot be part of what *is*, in any normal sense, for an illusion (like an error) represents what is *not*. But this should not place them beyond our understanding. It is certainly possible to explain, and sometimes to predict, many kinds of errors; so we should not on this account be pessimistic about bringing perceptual phenomena to heel, nor should we relegate them to metaphysics—though perhaps we will require concepts not found in the physical sciences, including even those of physiology. The issue as to whether special—cognitive— concepts are needed for perception is clearest when illusions are discussed. I shall lead up to this point gradually, as there are important background issues. My theme will be that cognitive concepts may be not only useful but necessary, and that they are (or should be) related to the *methods* of the physical sciences, as adopted in science for developing predictive hypotheses from limited data. Cognitive concepts should thus appear not alien to science—but like its methods.

When scientific methods are applied inappropriately, or with false assumptions, they can generate systematic errors, which may be the equivalent to various kinds of perceptual error, or illusion. A curious corollary of this notion is that *discoveries* of science (even perhaps those of physiology) may be less useful for understanding these aspects of perception than are the *procedures* by which science makes such effective use of data to predict and control events. This is, however, to be expected if the perceptual brain is confronted with problems similar to those of the physicist, also faced with limited data to derive knowledge of objects in the external world.

The above paragraphs contain statements and suggestions which would be hotly disputed (or dismissed with a shrug) by many experts on brain function and perception. In particular, rejection would come from those holding that perceptions are "passive selections of the world of objects;" that visual perceptions are "selections from the ambient array"; or that there are no conceptual problems to perception, beyond fully understanding the mechanisms of feature detectors tuned to specific features such as contours, and to complex objects up to the proverbial grandmother. Theories that perception is a matter of *selecting* from what is, in the external world, or from what is being currently signaled by the senses, have the advantage of holding out promise of description in terms of defined stimuli, immediately available for detection and measurement. Cognitive theories, on the other hand, hold out no such promise, and so may appear suspect from the outset. They implicitly deny that perception can be understood from stimuli alone, because they assert that perception is far richer than current stimuli or sensory data.

This claim that perception is "richer" than and "goes beyond" available sensory data, becomes important if the "additions" are more appropriate than we should expect by chance. Granted there is no "extra sensory" information available to organisms, how could perception be richer than the data available to the senses? An adequate cognitive account must answer this point, if it stresses "enrichment" as a fact of perception. It is an empirical question whether perception is indeed richer than stimuli, for which we must marshal experimental evidence. So at the outset we are faced with philosophical and with experimental questions: hence the justification for calling the study of perception Experimental Philosophy.

SOME THEORY-LADEN TERMS

In discussions of perception, as in philosophy (where also there is a confusing wealth of opinion combined with disheartening poverty of fact), we find in the technical vocabulary many near-synonyms. Sometimes indeed our language becomes suspect as we wobble between terms each loaded with decayed meaning, as likely to poison us from the past as to feed a future thought. Since to communicate and to think we must use some of these loaded terms, I shall attempt a brief account of what some convey within various paradigms of perception. It is particularly interesting that many terms exist, with subtly different meanings or implications, for *sensory inputs*. Three are in common use: (1) *stimuli;* (2) *sensory signals;* (3) *sensory data.* (The classical philosophical term "sense data" is now obsolete, or obsolescent, for perhaps no one would now follow Locke and Berkeley in thinking that *sense experience* is the input, or the raw data of perception.) The three currently used near-synonyms for "sensory inputs" are each linked to theories of perception. So these terms serve as warning flags of concepts to come. I shall wave them briefly, to show their colors.

1. *Stimuli* generally means: patterns (in space or time) of physical activity affecting organs of sense in a more or less normal manner and generally evoking a response.

2. *Sensory signals* generally means: activity (in afferent nerves) conveying information of patterns of stimulation of sense organs to other and generally "higher" regions of the nervous system. The signals may evoke responses by reflex arcs or they may be stored, perhaps for later use; and they may be modified, either immediately or later when in stored form.

3. *Sensory data* generally means that neural signals are conveying information of states of affairs of the external world, according to known or accepted properties of objects. Implicit but essential to this concept is the notion that the neural signals are acted upon according to a stored *data base,* at least roughly corresponding to common properties of objects. For example, to have "sensory data" that there is a book upon the table, implies on this paradigm a great deal of stored information (a large data base) relevant to books and tables—including *unsensed* properties to which behavior may be appropriate. It is this data base and its use which takes us into cognitive theories, and away from the stimulus–response paradigm as adequate for perception.

It is worth mentioning some related control engineering terms. The biological term "sensed" may be translated by the engineering term "monitored." Each implies that directly relevant (and usually continuous) inputs are available, for control or estimation. It is just this *direct relevance* and *availability* which are assumed for "stimuli," especially when coupled with "response." They are not however assumed for "sensory data," which may be but indirectly related to what is needed and not continuously or on-demand available. The corresponding engineering term here might be "information": but this word is of course extremely ambiguous, though it does have a rigorously defined and relevant meaning within Shannon's Information Theory (Shannon & Weaver, 1949).

Considering now the *output:* there are similarly loaded context, or paradigm-related words, though these seem less easily identified or defined. Among the many "output" words, we find: *response, performance,* and *behavior.* We also find, for more specific kinds of output: *act, action, achievement, failure, mistake,* and *solution.* Some of these imply judgments of the degree of success of the output in achieving some more or less specified goal.

If we now turn to what is supposed to be *going on inside* to link the input with the output, we find an astonishing variety of terms and concepts. We find, of course, the whole range of physiological and psychological theories of perception and behavior. In close association with "stimuli" and "response" are *reflex* and *reflex arc;* also *conditioned reflex* and *unconditioned reflex,* conditioning being invoked (following Pavlov) to form links between stimulus and response with repeated occurrence of time-related stimuli, such as the classical food and bell.

This stimulus–response linked by reflex arc notion is universally employed for describing the behavior of simple organisms. The question arises: How far up the evolutionary scale does it remain adequate? Watson and Skinner are remarkable in asserting that it remains adequate right to the top—even to the chap writing the book saying that it is adequate for him and for his readers. Given, however, that it *is* generally accepted near the bottom of the evolutionary scale, the onus is clearly on others to show that it is misleading (and just why it is wrong) to continue with the stimulus–reflex–response paradigm right up to man. The stimulus–response paradigm has at least four strong considerations working for its acceptance:

1. It is universally accepted for simple organisms, which thus serve as an initial "existence theorem" for it.

2. It has a physiological basis, which is well understood in many cases, and might be plausibly extrapolated to cases where direct evidence is not available.

3. Stimuli are physical patterns of energy, which since they affect sense organs can also affect instruments and so be readily detected and measured in the laboratory with standard techniques. (It is important to note that this is not so for "signals" or for "data": Although they have physical substrates, in that energy transfer is involved, they are *not* directly measurable with instruments.)

4. It is an essentially simple concept and familiar in many engineering devices, for imputs are usually arranged to control outputs directly.

These four reasons favoring a stimulus–response (or direct-control-of-output-by-input) theory are sufficiently strong to demand good evidence and clear arguments from those claiming its inadequacy. At this point we enter a battle of paradigms. Having summarized four shots for one side, we may now summarize (or fire) four shots for the other. They will be offered as evidence that perception (or behavior) is not stimulus or sensory-input controlled but is, rather, the result of data-based internal processes—for which stimuli play a rather minor part.

EVIDENCE FOR COGNITIVE PROCESSES

1. *Behavior may be self-initiated,* for example, signing one's name, or drawing a picture from memory or imagination. Where, here, is the initiating stimulus? If there is one, it could be little more than a trigger

to set off far more than is present in the stimulus, and so complete description in terms of stimuli could not be adequate.

2. *Behavior and perception are often appropriate to non-sensed features of objects, or situations.* Sensory inputs are almost never continuous, but, at best, sample what is going on, though behavior continues through gaps in the input. But how could perception or behavior be "stimulus determined," or a matter of "stimulus–response reflexes," if they continue with no input? Further, to perceive and "respond" to, say, a book on a table, is to act according to all manner of object characteristics that are not available to the eye, but are only available after one knows what books and tables are like. Here knowledge must be important.

3. *Skilled behavior can run with zero delay between "input" and "output."* Although there is the irreducible physiological delay of about .3 seconds, due to neural transmission rates and synaptic switching times, in such tasks as eye–hand tracking without preview, there is zero delay when the situation (in this case the track) is familiar. Therefore the output cannot be under direct control of the inputs—it must be controlled by some kind of running prediction of what the inputs (or rather the relevant external events) are likely to be in the near future.

4. *Perception makes effective use of absence of stimulus features, especially when the absence is improbable;* but the *absence* of expected input cannot be a stimulus—for stimulus is defined as physical patterns of activity at the receptors. Considering now *data* rather than *stimuli,* it is important to note that absence of signals (especially of expected ones) can provide important data. Indeed absence of signals may be as important as their presence for providing data. (This reminds us of Sherlock Holmes' comments on the dog that did *not* bark in the night, in *Silver Blaze.* The absence of barking led Holmes to a man who was trusted by the dog. The absence of barking was as good, as data or evidence, as the presence of something such as a footprint—but being absent it was not a *stimulus* for Holmes.) This is an example from behavior rather than perception, but the same holds *mutatis mutandis* for perception. A most striking (and so common that it is often missed) related perceptual phenomenon is ability to accept or assume completeness of objects partially hidden by nearer (opaque) objects. Indeed, partial absence of the patterns of further objects can be evidence for nearer objects. This evidence from absence can be crucial when only edges are available, especially in line drawings which would be incomprehensible without this Sherlock Holmes' ability of perception.

These four points we submit as evidence against theories of perception in terms only of stimuli, "selections of the world," or of feature detectors responding to what is present. It is, however, most important to note that acceptance of cognitive concepts, though stressing the value of stored data and the power to predict, does not conflict with the evidence of theories for "direct" control by stimuli for all situations. For example, eye blink to a flash of light, or a sudden sound would still be described as a stimulus–response reflex. This concept only becomes inadequate when the response is to a (recognized or assumed) *source* of the sound, or the flash. It is this identification of origins of stimuli—of *objects,* having pasts and futures and related non-sensed properties allowing prediction—which makes perception more than pattern recognition, requiring cognitive concepts for an adequate theory.

COGNITIVE THEORIES OF PERCEPTION

The weight of the argument so far is that perception is predictive (both in time, and to current non-sensed properties of objects) and that it functions by data rather than by stimuli, because absence of input can be as important as its presence. This leads us directly to an emphasis on assessed probabilities—which are not stimuli—and to a large data base for giving the "richness" of perception and appropriateness of behavior when available sensory data are intermittent or not directly relevant. These kinds of considerations were basic to the thinking of Bartlett (1932) and to the "internal models" notion put forward by Craik (1943). Craik was perhaps the first to suggest that the brain *models* aspects of the external world, and that neural models may control behavior and provide perception. We may see Hebb's (1949) notion of "phase sequences" as an attempt to give "internal models" a physiological basis. For both Craik and Hebb, they are largely inductively derived and represent the most common, average states of worldly affairs. Craik (1943) tried to avoid metaphysical overtones by emphasizing that his "internal models" are to be regarded as physical mechanisms:

> By a model we thus mean any physical or chemical system which has a similar relation-structure to that of the process it imitates. By "relation-structure" I do not mean some obscure non-physical entity which attends the model, but the fact that it is a physical working model which works in the same way as the process it parallels. . . . On our model theory neural or other mechanisms can imitate or parallel the behaviour and interaction of physical objects and so supply us with information on physical processes which are not directly observable to us [51–52].

Craik also says that he "can see no great difficulty in understanding how anything so 'different' from physical objects as concepts and reasoning can tell us more about those physical objects; for I see no reason to suppose that the processes of reasoning *are* fundamentally different from the mechanism of physical nature." Here I would not follow him. This is not a view of reasoning which would be generally, or surely need be, held. It reflects, curiously, on the earlier worry of Berkeley (1709) in a simpler situation. He wrote: "We are not to think that brutes, or children or even grown reasonable men, whenever they perceive an object to approach, or depart, from them, do it by virtue of *geometry* and *demonstration*" (*A New Theory of Vision*, XXIV). But, of course, this inhibition to computing by symbols has melted away with its demonstrations by mechanisms following rules of inference as controlled by "software."

Boden (1972) in a cogent discussion, to which I am indebted, quotes in this connection an interesting passage from William McDougall (1911):

> Any familiar object, such as my dog, may be seen in many positions and from many angles and distances, and in each of an indefinite multitude of such cases the visual impression may evoke the same reaction (e.g. the calling of his name), though in each case the sum of physical stimuli constituting the impression on the sense-organ is unique. . . . The object is always recognised as the same though the actual retinal image differs in every case. [He refers this comment to Hans Driesch; *Body and Mind*, pp. 268–269].

McDougall then goes on with a statement diametrically opposed to the position of Craik, or "unconscious inference" as mediated by computer–like logical processes, with: "It is absolutely impossible to understand this fact on the assumption of any kind of preformed material recipient in the brain, corresponding to the stimulus in question." We can now see, through computer technology, that such transformations *are* possible, though admittedly their complexity cannot be simulated fully by existing computers.

Recent experiments by Shepard and Metzler (1971) have investigated the interesting human ability to transform "mentally" three-dimensional shapes. Shepard presented an unfamiliar shaped object, beside a copy rotated through a certain angle (or sometimes the comparison object was of somewhat different form, readily confused with the first). The subjects were asked to judge whether the second object was the same as the first but rotated, or was a different object. To perform this task, the subjects "mentally" rotated the first object and compared their "internally rotated" object with the second shape. This internal rotation was a slow process, and its rate could be measured. Shepard and Metzler found that subjects could not rotate their internal representations faster than 30° per sec: the time required increasing linearly with the angle the object had been rotated. One would give a lot to know just *what*, in the perceptual system, is being rotated, "neurally" or "mentally"! Is it an analogue Craikian model, or a sequence of digital operations?

This power to perform rotational transforms is basic to motion perception, maintaining what one is tempted to call "object hypotheses" through changing viewpoints and motions of objects giving changing stimuli, but remarkably constant perception. This takes us far from a direct pattern-perception account. The present author has made (somewhat informal) observations suggesting some rules that seem to be followed for maintaining object integrity through rotation, when texture is removed and perspective controlled (with or without stereoscopic information) by shadow projection (Gregory, 1970a). These observations show that certain assumptions are held which "lock" certain object features in spite of changing input—for example, that converging lines rotating together, projected on a screen, are seen as parallel; converging by perspective in depth. Surprisingly, no equivalent assumption is held for corners. Familiar objects, however, even of complex shapes, maintain their integrity provided there is sufficient information for the object to be recognized. A remarkable case of nonrecognition and continuous loss of integrity (shown by apparent changes of apparent shape, rather than rotation) is a texture-free solid opaque cube. Its profile is not adequate information for "cube recognition" and it transforms in shape in weird ways; until just a little surface texture (or stereoscopic disparity) is added, when immediately it becomes a rigid, solid rotating cube. It is remarkable how much agreement there is between observers in these situations evoking highly complex transformations, taking perception far, though systematically away from, the (ever-changing) stimulus pattern. Even the errors are very similar between observers.

To suggest that such perceptual powers as these (and they must be seen to be believed, or appreciated) are given by unique properties of brain cells does not seem helpful (if only because unique cases are virtually closed to science), or likely. With a cognitive point of view, we may see these powers as given by *procedures* or *operations,* somehow *carried out* by brain cells for perception but to be understood in a wider context. Such procedures may be realized by computers, which may perform the crucial procedures, though by different means. This is the promise of "artificial intelligence": especially "scene analysis," and "robot vision." The approach has its origins in the thinking of the remarkable polymath of the last century, Hermann von Helmholtz (1856–1866), with his notion of perception as "unconscious inference."

COMPUTER ANALOGIES—THE ARTIFICIAL
INTELLIGENCE (AI) APPROACH

The notion put forward by Helmholtz that perception essentially involves "unconscious inference" from sensory clues was almost totally re-
sisted for nearly a century; but it has recently emerged in the dramatic
and testable form of computer programs for "scene analysis" of pictures
and "robot vision"—using video signals from TV cameras feeding computers controlling limbs handling objects, according to limited optical information, much as for sighted organisms. The first thing to state is that
these projects have turned out to be extremely difficult. So far, results
are modest, especially for robot vision with normal (shadow-casting)
lighting, and in particular for sets of objects having widely varied forms.
It is however true that much has been learned which seems of immediate
relevance to perception—especially in what, we are arguing, are its "cognitive" aspects.

As Max Clowes (1972) pointed out: "The early phases of picture and
pattern analysis by computer were largely pursued within the framework
of classification. During the middle sixties a sustained attempt was made,
utilizing analogies with formal methods of sentence analysis (Clowes,
1971)." The insight here is that lines, corners, edges, and so on may
be informational units, like phonemes, and that scenes (sets of objects
viewed from a given position) may be "parsed" like sentences. The
corollary notion is that we derive meaning from retinal images rather
as we derive meaning from sentences, and that both work by following
the rules of a "generative grammar" much as described by Chomsky
for language. But the current view is that, for both language and scene
analysis, far more stored information of object characteristics (data base)
is required than was supposed necessary by Chomsky and the first enthu-
siasts of generative grammars. The rules and forms of sentences are
not sufficient: the meanings of words as based on experience of object
situations are important also. This direction of thought can lead to regarding perceptions (and sentences), entertained and tested by perceptual processes, as like hypotheses in science. Perceptions are predictive;
they make effective use of little information (Gregory, 1970b) and when
incorrect, can have marked similarities to perceptual errors or illusions:
they also can suffer or display ambiguity and scale distortions, and can
be paradoxical. The world itself, however, can be none of these—which
is a strong philosophical-type argument for a Representative theory of
perception: especially that perceptions are, or are like, hypotheses. It
might be argued that, although perhaps philosophically it is interesting
to give perceptions (and sentences) the same logical status as hypothe-

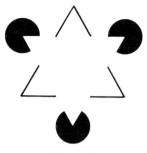

Figure 1

ses, this discussion is a waste of time if it does not suggest new experiments. Does it suggest experiments, or help to relate phenomena? What does *regarding perceptions as predictive hypotheses* suggest? We will consider two examples.

Example I: Illusory Contours

Figure 1 (after Kanizsa, 1950) presents a broken triangle and three sector disks. What is *seen*, in addition, is a large whiter-than-white illusory triangle, having sharp illusory contours joining the in-line edges of the sectors. A similar illusory figure also occurs when the sector disks are replaced with dots (Figure 2); the contours now join the dots, but only when the broken line triangle is present. If the dots are removed, no such illusory contour appears in the gaps of the line triangle—so in-line features, or dots placed at the apices of an imaginary triangle, are not sufficient to evoke these illusory effects. This observation strongly suggests that here we are not dealing with "trigger features" of any simple kind. In particular, it does not seem likely that striate line detectors (Hubel & Wiesel, 1968) would be triggered by these patterns, especially across the dots. (This should be tested electrophysiologically: but I shall assume that they will turn out not to be directly triggered, as seems at present most likely.) Does scene analysis make a contribution

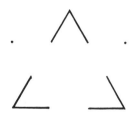

Figure 2

to understanding this phenomenon? A program by Falk (1971) accepts video data of line diagrams of three-dimensional objects in which some lines are missing. It has a "preprocessor" which makes predictions of where incomplete or missing lines should be, and fills them in by its own invention. It then proceeds to analyze its invented scene, after it has added missing lines. But, of course, if it were to add a line not in fact missing, then it would proceed according to its false postulate, and might go on to generate elaborate error. This program has the power to tackle the difficult problem of identifying partially masked objects, arranged in three-dimensional space, by working from further to nearer objects. This at first sight curious procedure is necessary for the machine because it requires (necessarily hypothetical) knowledge of the complete shapes of the partially hidden objects—which of course are always the more distant—in order to identify the masking problems to be solved. This procedure is an extension of Guzman's (1968) earlier program giving object-identification of polyhedrons from complete line drawings. The new procedures are known as JOIN, ADDCORNER, and ADDLINE. Lines and corners are added according to rules appropriate to most objects: but they cannot always be appropriate. When inappropriate to a particular situation, they *must* create illusory lines, or corners, unless there are adequate check procedures for rejection at later stages of the scene analysis. Presumably such check procedures could never cover all contingencies—and so we are bound to get some procedure-generated illusions of false invention. It is particularly interesting that these include what are usually taken as basic perceptual data—contours, corners, and so on (cf. Gregory, 1973). (This would be most upsetting to Bishop Berkeley!)

Referring again to Figure 1, we may suggest that we have an ADDLINE preprocessor, which here invents nonexistent lines because such gaps are unlikely. There is more likely to be a nearer, triangular-shaped object producing the gaps in the line figure by masking. The human ADDLINE program, similar to Falk's scene analysis program, for similar reasons accepts the absence of stimulation as data for an object, which it proceeds to invent, or postulate. It would be interesting to know whether Falk's program would also postulate the same illusory triangle when presented with this figure. Scene analysis has just got to the stage when such direct comparison tests can begin to be carried out—although much caution is required as the programs are still far too easily fooled; for example by shadows, which are often mistaken for contours of objects by the machines though not by us.

This explanation of these illusory contours and figures suggests an immediate perceptual test. What happens if we add information competing with the supposition of a nearer masking object? Harris and Gregory (1973) using a stereoscopic technique, have forced the three dots of Figure 2 back to lie behind the broken triangle, and find that the contours then disappear. When brought forward, they are as strong as in the original figure, though sticking out from it—when they could still mask the gaps. This is exactly what is to be expected on this view of the matter, for, when pushed back, masking hypotheses would no longer hold. There is, however, an interesting complication: the illusory figure may persist when the dots are forced back—when the illusory figure appears to curve forward from the dots, to remain in front of the gaps in the broken line triangle which it could then still be masking. We also find that different illusory contours can be generated separately in each eye—and that pairs of differently curved illusory contours can be "fused" to give stereoscopic depth even though there are no corresponding stimuli for fusion. It seems to follow that stereopsis can function from separate hypothetical contours, postulated by each eye system: So stereopsis does not necessarily depend on binocular stimuli inputs to "binocular cells" of the striate cortex. Perhaps it is not too surprising if preprocessing occurs separately in each eye system; for it is only through recent evolution that the eyes have moved forward, to give common shared views of the world, allowing stereopsis through disparity. In many animals, the eyes get entirely different views when preprocessing must be separate for each eye system.

Having put forward this view of stereopsis, at least in some circumstances as working not from raw retinal signals but from data derived from preprocessing, according to probable object situations, it must be added that this account may turn out to be wrong. There may be some other way of interpreting these data, to save stereopsis from "cognitive processes," that cannot be understood without knowing the stored conditional probabilities and the procedures by which stored data are used to build hypotheses of the external world. On our cognitive view no sensory data are raw: all data are "cooked." This must indeed make perceptual experiments difficult to control or interpret. But they *are* difficult to control and interpret!

Example 2: Scale Distortions

It is generally agreed that visual depth given by perspective convergence of lines and texture gradients is "cognitive," by depending upon assumptions that such features of the retinal image are usually related to distance in the external world. This is certainly so for typical scenes; but it is a simple matter to construct objects or draw pictures where these features are not related to depth or are related in curious, atypical ways. If the normally depth-related features continue to be accepted as depth data, then we should expect trouble (and this trouble will not be physiological in origin). We should not, however, expect "invented" lines, as in the example just given of illusory contours; but we might expect to find *scale distortions*. For a dramatic example we may take the weird changes of shape and size of the Ames rotating trapezoid window (Ittelson, 1952). The interesting question is: Can (perhaps all) the classical distortion illusions be regarded as due to misscaling by preprocessing strategies set up by typical depth data presented in atypical situations for which they are misleading? This would be a cognitive-type explanation of the distortion illusions (the Muller-Lyer, Ponzo, Hering, Orbison, and so on) which would be explained in terms of *procedures* for deriving and using sensory data, rather than the physiological *mechanisms* carrying out the procedures. If the source of the trouble is procedural we should expect similar trouble (scale distortions) to occur also in robot vision, where the procedures are similar though the mechanisms carrying them out are very different from those of organisms (Gregory, 1967). To explain these illusions we should need to know the procedures, while we could wait for adequate knowledge of the mechanisms carrying them out without undue impatience, if indeed the key does not lie in these particular mechanisms.

Perhaps the most plausible "inappropriate procedures" explanation of the classical distortion illusions is in terms of scale corrections known to be introduced to maintain object size roughly constant over distance, in spite of the shrinking of retinal image size with distance. On this theory (Gregory, 1963), perspective convergence of lines (and possibly texture gradients) associated with distance serve as visual data for setting the scale correction to maintain "constancy." When these distance-data are presented on the flat plane of a picture, or line diagram, they should produce expansion according to increasing distance represented. But as the figure is flat, the resulting scale corrections would be *inappropriate*, for the drawing is a flat object, lying on the page. So it should be distorted, in ways related to the three-dimensional form it represents. There should be *increase* in size with represented *increase* in distance: since the geometrical shrinking of retinal images with increasing object distance is normally compensated by perceptual expansion, to give "size constancy." Clearly this expansion would be inappropriate for pictures, because there is no shrinking of the image to compensate for flat pictures.

Alternative "physiological mechanism" theories have, however, been put forward, and these may be just as plausible. Possibly they are more generally accepted as at least along the right lines than is this cognitive theory. In particular, there is a strong case to be made for what might be called the "inappropriate lateral inhibition" theory put forward by Blakemore (Blakemore, Carpenter, & Georgeson, 1972; and see Georgeson & Blakemore, 1973, for an important explicit attack of the cognitive approach from their terms). It would not be possible to go into this issue here in sufficient detail to give a fair account: what matters for the present purpose is to point out that this is a currently live example of a "cognitive" versus "mechanism" debate—where the prize is a wealth of illusory phenomena for the winner.

WHAT IS COGNITION?

We begin to see this "cognitive aspect" of brain function as requiring, for its description, such concepts as grammar, rule-following, strategy, and procedure. Also, useful analogies may come from games, or computers, which are physically very different from the nervous system. The criteria of what are relevant for analogy and understanding lie at the logical level of the procedures by which information is stored, selected, and used.

As a possible example, consider a game of chess. In physical terms this is clearly so different from a brain as to be laughable as an analogy—but there may be similarities which could be revealing. A chess game may be described in several very different ways—and for various purposes one way could be more useful than another. We may describe it as a series of positions of identifiable pieces on the squares of a board; or we may describe it in terms of the "power," the "threat," the "potential", and so on of these pieces, according to the rules of the game, including the allowed moves of each piece, and the known or guessed general strategies of the players.

For both of these descriptions, the positions of the chessmen are limited by physical restraints, set by gravity and the strength of the board (which restrict the positions to two dimensions), and by friction, and so on. But the borders of the squares do not set up physical restraints in this sense. A chessman sitting on a line (and so occupying two, or up to four squares) would, however, at once be moved—because he would be *ambiguous,* with respect to the state of the game. Similarly, a man might move to produce an *impossible* situation, though it would not be physically impossible. (Similarly, $2 + 5 = 9$ is impossible according to the rules of normal arithmetic, but not for physical reasons. So we have to provide special means for preventing calculating machines from adopting symbolically impossible states.)

When rule following (for games, arithmetic, or logic) is carried out by machine, the restraints of the machine must match the restraints of what is allowed by the formal rules of the game, or the calculus. To carry out an optimum (or reasonbly efficient) strategy, the states must be still further restricted from what is physically possible. The *answer* may, however, be physically impossible. Restraints are set in computers by "software." The software exists as punched holes, or magnetic regions on tape, which are "read" to control the machine. This control is given by limiting the machine's restraints so that the steps it follows are *logically* appropriate for solving the problem. There must also be a set of starting states, representing the initial assumptions and data defining the problem, and what kind of solution is required. This is the "data base." These states are not models, in a toy-like sense, of the problem or the situation to be solved (except for analogue computers where in some sense this can be true). The relation between the physical state of the software (holes on tape and so on) and the state of affairs it represents, for the machine or for the world, is extremely subtle. It is a *language,* and the grammar and the syntax must be known for this relation to be appreciated. Here we differ from Craik.

One of the implications of a cognitive approach to brain function is that we should look for an equivalent "brain language." Much electrical activity may come to look like "words" arranged in a "grammar" underlying thought and perception. This takes us back to "artificial intelligence," and specifically to Clowes' remark that for scene analysis we should find a grammar for pictures.

When should we say that a mechanism is carrying out "cognitive" procedures? How can we recognize cognition? Although vitally important for setting up and interpreting experiments, this is not at all easy to answer. We might start by asking what other kinds of procedures there may be: Or are all procedures to some degree cognitive?

I take "procedure" to mean "proceeding (more or less successfully) toward some kind of identifiable goal, or to a conclusion to a problem." For example we may adopt procedures for opening a wine bottle, employing some kind of mechanism (our fingers, and a device such as a corkscrew, which may have several designs). The optimum procedure will depend upon the nature of the bottle, and the forces and so on required, in relation to the characteristics of the available mechanisms (the corkscrew and the fingers). What is accepted as an adequate, or the optimum, sequence of procedures will depend upon how the goal is defined and what is accepted as "success," as well as the cost of the operation. If we should cut our fingers, shake up the wine, or take 5 minutes rather than a few seconds to open the bottle, then the cost will have been high. When the cost is high, either the procedures adopted or the mechanism available may be blamed. It is important to note that these are extremely different kinds of blame. To improve the situation for future occasions, we might change the mechanism (use a different kind of corkscrew or possibly try to cure the arthritis in our fingers) or we might adopt a different set of procedures, while using the same mechanisms. Such a change of procedures may be called: "learning how to use a corkscrew." Further, a change of procedure may suggest, or require, a different mechanism: and a change of mechanism may require different procedures for its use. We see these interactions between mechanisms and procedures throughout technology and throughout organic evolution.

So far, we have a man working the corkscrew. The whole process might however be mechanized. Would we wish to describe a fully automatic machine—such as this bottle-opening machine—as "cognitive"? So far, there seems to be no reason for saying this. We might, however, wish to say that the machine is carrying out procedures, of more or less efficiency, for accomplishing the task. It follows, that we cannot *equate* the carrying out of procedures with being cognitive, or as requiring cognition. So we cannot say more than that *some* procedure-carrying-out is cognitive. But which procedures are these? Which kinds of procedure-carrying-out are cognitive?

The word "cognitive" means, from its origin, "knowing" or "understanding." We would not wish to say that the bottle-opening machine *knows* about bottles or corks, or that it *understands* them, or what it is doing. We would also deny that the machine estimated the cost/benefit of its procedures. It is, surely, this lack of knowledge on the part of the machine which makes us reject the notion that it is cognitive.

Suppose now that the machine receives knowledge of the bottle, the corks, and so on from sensors (such as photo-electric cells, television cameras, and contact switches) and that it acts according to this information. Then we might begin to think of it as cognitive. But if it is to be capable of accepting signals as data, for making decisions appropriate to characteristics of various corks and bottles, these signals must serve to select from alternative possibilities for appropriate action. This implies that several possible actions and outcomes are, in some way, represented in the machine. In short, for the monitoring signals to be used, the machine must have a stored data base (however crude) of information about bottles, and the merits of alternative opening procedures. Now we have, in these terms, a cognitive machine working according to strategies based on knowledge—however primitive. It might perhaps use its stored knowledge (derived by its own learning, or fed to it by tuition) to open several kinds of bottle or even to select wines appropriate for occasions or needs. We begin to see here an "artificial intelligence." Although this is still a mechanism, to describe it we should need to know about how it makes use of its knowledge.

If the machine makes mistakes, we might attribute the mistakes to *lack of appropriate knowledge*, just as much as to *failure of mechanism*. The notion of errors made from employing inappropriate knowledge is the psychological concept: "negative transfer," of training or skill. For understanding the origins of such errors we may look for differences between the past situations for which the strategies were appropriate and the present situation for which they are inappropriate. It is hopeless to look for failures of mechanism to explain errors due to misapplied knowledge or inappropriate procedures.

A cognitive theory of perceptual illusions is of this kind. Such illusions are phenomena which can (in general) be measured and described much as can the results of mechanism failure; but their explanation is of a logically different kind. To understand them we must know the inappropriateness of the operating procedures. Discrepancies between past situations and the present can provide clues. (For example, if vision employs perspective convergence for scale setting, we might expect this to produce distortions, which provides an explanation though we lack knowledge of the relevant mechanisms. Indeed, without knowing the strategies, we cannot appreciate the tasks the mechanisms are carrying out: we cannot appreciate their function.) Of course we should aim to understand both the mechanisms and the procedures of perception: the double description is necessary for acceptable physiological psychology.

CONCLUSION

How can we recognize that an organism, or a machine, is functioning cognitively? The essential is to establish that its behavior is not being controlled entirely by its current input signals, but in part by knowledge more or less appropriate to the situation it is handling. Our problem is how to recognize control by knowledge, to distinguish it from direct control by signals. For this we must devise operational criteria.

Can errors, or illusions, be keys to distinguish control by signal (for which we require only "mechanism" concepts) from control by knowledge, or knowledge-based procedures (for which we require also "cognitive" concepts)? Errors and illusions can certainly be produced by failure, or limitations, or restraints, of mechanisms—so we have to find criteria for distinguishing cognitive-type from such mechanism-type errors. This should be possible because (on a cognitive account) retinal or other signals provide *data*, which may remain invariant in spite of large changes in the signal. Indeed, extremely different signals may provide identical data—though they activate the mechanism very differently. So by changing the stimulus pattern or changing the situation, and finding *what remains invariant*, we might hope to separate mechanism-type from cognitive-type errors or illusions. By varying (*1*) the stimulus pattern, and (*2*) the supposed data-significance of given stimuli, we might discover whether it is the pattern per se or (misleading) data conveyed by the pattern, producing the error. We have strong reason to believe, for example, that afterimages can have a purely mechanistic explanation because they are very closely related to the stimulus pattern, changing as it is changed. On the other hand, hollow molds of faces, appearing as normal (the nose apparently sticking out as in normal faces though in fact it is hollow in the mold) is surely a probability-biasing effect, no faces in nature being hollow. We find that the form of the hollow mold (the stimulus pattern) can be varied over wide limits (until face recognition from the contours of the mold face is lost) before this illusion is destroyed—when it appears correctly hollow. This maintenance of the error over a wide range of stimulus patterns satisfies our criterion for a cognitive-type illusion.

We are now in a position to return, finally, to the curiously tricky problem of how to classify the distortion illusions. It is now clear why this is such a difficult case. In the first place, these illusions are remarkably unaffected by changes of the size of the display—they are even maintained where large eye movements are needed to incorporate the whole of the figures into perception—which satisfies our criterion for a *cognitive-type* illusion. On the other hand, the distortions do change systematically with change of the angles between the crucial lines in the stimulus patterns. This satisfies our criterion for a *mechanism-type* illusion. So both seem to be satisfied! Why should this be?

Suppose that these angles normally *represent* something—say, object distance, as related to perspective convergence for commonly shaped objects. If such (generalized knowledge-based) distance-data are employed to set visual scale, related distortions should be generated when the data do not apply to the (queer) objects—as when the perspective is given by lines on a flat picture plane. But indicated distance in normal conditions is precisely related to the angles—so on this cognitive account (cognitive because dependent upon knowledge-based assumptions of the world) we should expect the distortions to be closely related to the angles—as would also be expected on a mechanism account. In this perhaps unusual case, we cannot distinguish between the two kinds of illusion simply by changing the angles of the figures. We can however change the situation, making it more or less appropriate to the *data* supposed (on the cognitive account) to be signaled by the stimulus patterns. We should expect the distortions whenever perspective-like features are presented without the corresponding distances of typically shaped objects. This is just what we have for flat perspective pictures. This example suggests that to recognize cognitive-type illusions we should consider not only effects of varying the stimulus pattern but also the situation it may be representing. This is by no means a straightforward or simple matter.

We may now summarize other (and hopefully more direct) ways of recognizing cognition, other than from errors, or illusions, which, however, are in many ways of great interest. Among the most important is *evidence of prediction:* especially prediction to features or situations which do not occur in stereotyped patterns. Evidence of such prediction comes from: (*1*) behavior occurring with no time lag between input signals and output "response" (or where the delay is considerably less than the transit time of signals through the system; (*2*) when behavior continues appropriately through extensive gaps in the input signals; (*3*) when behavior is appropriate to the situation, but the available input signals are but indirectly related to situations. Perception and skilled behavior of higher organisms show all these characteristics of cognition (Cf. Gregory, 1970,b).

Finally, we have *ability to answer questions.* This can give the most clear-cut and precise information of what knowledge is stored and how it is elicited, hence the importance of examinations. But unfortunately, for perception and skill, what questions are being posed by reality, and how they are being answered, is often far from clear—hence the need for experiments.

As we begin to understand cognitive processes as carried out by neural activity, so it becomes possible to devise other mechanisms to carry out these processes. This is the biological interest of robotics, or machine or artificial intelligence. Once such a cognitive machine has been built, it may be possible to converse with it to discover its procedures and its knowledge, the most dramatic example of which may develop (not always appropriately or correctly) unknown to its maker. We are of course familiar with this in children. For a machine, the most dramatic example is provided by a program of remarkable power developed by Terry Winograd (1972). The machine learns how to recognize and stack objects, such as cubes and pyramids. It learns names and phrases, and may even make up its own, with which it communicates with and learns from human beings. Here is part of a dialogue between Winograd and his robot, recorded on a teletype controlled in turn by man and machine.

1. Pick up a big red block.
   O.K.
2. Grasp the pyramid.
   I DON'T UNDERSTAND WHICH PYRAMID YOU MEAN.
   (*There are three pyramids in the scene, so the program does nothing.*)
3. Find a block which is taller than the one you are holding and put it into the box.
   (*The system has to figure out what is meant by "one" and "it" as well as using deduction to decide which block it is holding.*)
   BY "IT," I ASSUME YOU MEAN THE BLOCK WHICH IS TALLER THAN THE ONE I AM HOLDING.
   O.K.
   (*The hand has to set down the red block to pick up the taller blue one.*)
4. What does the box contain?
   THE BLUE PYRAMID AND THE BLUE BLOCK.
12. Stack up two pyramids.
    I CAN'T.
    (*The robot tries and fails.*)
43. Is there anything to the right of the red pyramid?
    YES, FOUR THINGS: THE BOX, THE RED CUBE, THE BLUE PYRAMID, AND THE BLOCK NAMED SUPERBLOCK.
44. Thank you.
    YOU'RE WELCOME!

As man teaches machines to see, machines teach us what it is to be able to see. The first machines to show intelligence teach us that their knowledge—kept up to date by data—is as important as their mechanisms. Without knowledge, they are blind and dumb and so are we.

REFERENCES

Bartlett, F. C. *Remembering.* New York: Cambridge Univ. Press, 1932.

Berkeley, Bishop G. *A new theory of vision.* 1709. New York: Everyman's Library, 1910.

Blakemore, C., Carpenter, R. H. S., & Georgeson, M. A. Lateral inhibitions between orientation detectors in the human visual system. *Nature,* 1972, **228,** 37–39.

Boden, M. A. *Purposive explanation in behavior.* Harvard, 1972.

Clowes, M. B. Picture descriptions. In B. Meltzer & N. V. Findler (Eds.), *Artificial intelligence and heuristic programming.* Edinburgh: Edinburgh Univ. Press, 1971.

Clowes, M. B. Scene analysis and picture grammars. In F. Nake & A. Rosenfield (Eds.), *Graphic languages.* Amsterdam: North Holland, 1972.

Craik, K. J. W. *The nature of explanation.* New York: Cambridge Univ. Press, 1943.

Falk, G. Scene analysis based on imperfect edge data. *Second International Joint Conference on Artificial Intelligence, Proceedings.* London: British Computer Soc., 1971. Pp. 8–16.

Georgeson, M. A., & Blakemore, C. Apparent depth and the Muller-Lyer illusion. *Perception,* 1973, **2**(2), 225–234.

Gregory, R. L. Distortion of visual space as inappropriate constancy scaling. *Nature,* 1963, **119**, 678.

Gregory, R. L. Will seeing machines have illusions? In N. L. Collins & D. Michie (Eds.), *Machine intelligence I.* Edinburgh: Oliver and Boyd, 1967.

Gregory, R. L. *The intelligent eye.* London: Weidenfeld, 1970. (a)

Gregory R. L. On how so little information controls so much behavior. *Ergonomics,* 1970, **13**(1), 25–35. (b)

Gregory, R. L. The confounded eye. In R. L. Gregory & E. H. Gombrich (Eds.), *Illusion in nature and art.* London: Duckworth, 1973.

Gregory, R. L., & Harris, J. P. Illusory contours and stereo depth. *Perception and Psychophysics,* 1974, **15**(3), 411–416.

Guzman, A. Analysis of curved line drawing using context and global information. In B. Meltzer & D. Michie (Eds.), *Machine intelligence VI.* Edinburgh: Edinburgh Univ. Press, 1971.

Harris, J. P., & Gregory, R. L. Fusion and rivalry of illusory contours. *Perception,* 1973, **2**(2), 235–247.

Hebb, D. O. *Organisation of behavior.* London: Chapman and Hall, 1949.

Helmholtz, H. von. *Handbook of physiological optics.* J. P. C. S. Southall (Ed.), 1856–1866. London: Dover reprint, 1963.

Hubel, D. H., & Wiesel, T. N. Receptive fields and functional architecture of monkey striate cortex. *Journal of Physiology,* 1968, **195**, 215–243.

Ittelson, W. H. *The Ames demonstrations in perception.* Princeton: Princeton Univ. Press, 1952.

Kanizsa, G. Margini quasi-percettivi in campi con stimulazioni omegenea. *Rivista di Piscologia,* 1955, **49**, 7–30.

McDougall, W. *Body and mind: A history and defence of animism.* London: Methuen, 1911.

Shannon, C. E., & Weaver, W. *The mathematical theory of communication.* Urbana: Univ. of Illinois Press, 1949.

Shepard, N. R., & Metzler, J. Mental rotation of three-dimensional objects. *Science,* 1971, **171**, 701–703.

Winograd, T. *Understanding natural language.* Edinburgh: Edinburgh Univ. Press, 1972.

Subject Index